LINEAR SYSTEMS THEORY
SECOND EDITION

Ferenc Szidarovszky
A. Terry Bahill

CRC Press

Boca Raton   Boston   New York   Washington, D.C.   London

Acquiring Editor: *Bob Stern*
Project Editor: *Susan Fox*
Cover Design: *Denise Craig*
Prepress: *Carlos Esser and Gary Bennett*

Library of Congress Cataloging-in-Publication Data

Szidarovszky, Ferenc.
    Linear systems theory / Ferenc Szidarovszky, A. Terry Bahill. —
2nd ed.
       p. cm.
    Includes bibliographical references and index.
    ISBN 0-8493-1687-1 (alk paper)
    1. System theory.   2. Control theory.   3. Dynamics.   I. Bahill,
Terry. II. Title
Q295.S995   1997
003—dc21
                                    97-42062
                                       CIP

Contents

Authors

Ferenc Szidarovszky has been a Professor of Systems and Industrial Engineering at the University of Arizona in Tucson, AZ, since 1988. He received his Ph.D. in mathematics from the Eotvos University of Sciences, Budapest, Hungary, in1970. He received a second Ph.D. in economics from the Budapest Economics University, Hungary, in 1977. His research interests are in the field of dynamic economic systems, game theory, numerical analysis, and their applications in natural resources management. He is the author of five textbooks and three monographs published in Hungarian, as well as *Principles and Procedures of Numerical Analysis* (with Sidney Yakowitz), Plenum, 1978, *Introduction to Numerical Computations* (with Sidney Yakowitz), Macmillan, 1986 (second edition, 1989), *Techniques of Multiobjective Decision Making in System Management* (with Mark Gershon and Lucien Duckstein), Elsevier, 1986, *The Theory of Oligopoly with Multi-Product Firms* (with Koji Okuguchi), Springer, 1990, and *The Theory and Applications of Iteration Methods* (with Iannis K. Arguros), CRC Press, 1993.

A. Terry Bahill has been a Professor of Systems and Industrial Engineering at the University of Arizona in Tucson, AZ, since 1984. He received his Ph.D. in electrical engineering and computer science from the University of California, Berkeley, in 1975. He is a Fellow of the Institute of Electrical and Electronic Engineers (IEEE). His research interests are in the fields of modeling physiological systems, eye-hand-head coordination, validation of expert systems, concurrent engineering, and systems design theory. He is the author of *Bioengineering: Biomedical, Medical, and Clinical Engineering*, Prentice-Hall, 1981, *Keep your Eye on the Ball: The Science and Folklore of Baseball* (with Bob Watts), W.H. Freeman, 1990, *Verifying and Validating Personal Computer-Based Expert Systems*, Prentice-Hall, 1991, *Engineering Modeling and Design* (with Bill Chapman and Wayne Wymore), CRC Press, 1992, and *Metrics and Case Studies for Evaluating Engineering Designs* (with Jay Moody, Bill Chapman, and David Van Voorhees), Prentice-Hall, 1997.

Preface

How is this book different from scores of other books on systems theory? First, it is more rigorous mathematically. All developments are based on precise mathematical arguments, many being innovative and original. But no derivations are included just because they are elegant; each derived theorem or lemma is used later in the book. Second, the theory is general; often it applies to (1) linear and nonlinear systems, (2) continuous and discrete systems, and (3) time invariant and time varying systems. Third, modern, computer-oriented methods are presented, not graphical techniques. Fourth, it has examples from most major fields of engineering, economics, and social sciences, with special emphasis on electrical engineering.

This book is self-contained. It starts with a solid mathematical foundation in Chapters 1 and 2. These chapters have all the mathematical developments that will be used later in the book. However, these chapters can be skimmed if the results are already known to the reader or if the reader is uninterested in the mathematical details. This rigorous presentation of basic principles allows fast, efficient development of later material.

This text was primarily written for first-year graduate students in electrical engineering, although it is suitable for graduate students in other engineering fields as well as those in economics, social sciences, or mathematics. It would also be suitable for some advanced undergraduate students. The prerequisites are the fundamentals of calculus and matrix algebra only. The instructor can select the examples and applications that match the students' backgrounds. For example, for a course in electrical engineering the instructor may wish to present the second, third, fourth, sixth and ninth problems of the Engineering Applications sections. These five systems are introduced in Chapter 3 and are repeated in various forms in each of the subsequent chapters. For such a class the instructor might present some of the other problems just for the fun of it. On the other hand, an economics professor might ignore the engineering applications and study instead the economics examples.

This book is an outgrowth of courses taught over the last two decades to electrical, mechanical, systems, chemical and biomedical engineers at the University of Arizona and Carnegie Mellon University and to economics students at the Budapest University of Economics, Hungary.

The objective of this book is to help students develop their capabilities for modeling dynamic systems, examining their properties, and applying this knowledge to real-life situations. These objectives are served by four main features of the book. The theoretical foundation and the theory of nonlinear and linear systems are given in a comprehensive, precise way. A unified approach is presented for continuous and discrete systems. We develop a unified treatment of controllability and observability that is valid for both time-invariant and time-varying systems. In selecting the theoretical material and methodology to be covered in this book, we concentrated mainly on modern, computer-oriented techniques and omitted the old-fashioned, pre-computer age (mostly graphical) methods.

For example, in this text we present only four techniques for assessing stability of a system: Lyapunov functions, the boundedness and convergence of the state transition functions, the location in the complex plane of the eigenvalues of the coefficient matrix or the location in the complex frequency plane (s-plane) of the poles of the transfer function, and the Hurwitz criterion. Proving stability with Lyapunov functions is general: it also works for nonlinear and time-varying systems. It is good for proving stability and asymptotical stability. However, proofs based on Lyapunov functions are difficult, and failure to find a Lyapunov function that proves a system is stable does not prove that the system is unstable. The second technique we present requires checking if the state transition function is bounded and even converges to the zero matrix if $t \rightarrow \infty$. This method can be used for both time-invariant and time varying linear systems. The third technique is based on finding the location in the complex plane of the eigenvalues of the time-invariant coefficient matrix or finding the poles of the transfer function. This task is sometimes difficult because it requires finding and factoring the characteristic equation of the system. However, many computer packages are now available to do this job. Finally, we present the Hurwitz technique because it can assess the stability of a system without factoring polynomials. Routh's criterion is similar to the Hurwitz approach, so it is not presented here. These techniques were developed in the nineteenth century. In the twentieth century, many more techniques were developed to help assess the stability of a system without factoring polynomials, such as Bode plots, Nyquist's criterion, Sylvester's condition, Kalman's extension of Lyapunov's criterion, and the root locus technique. We do not present any of these because for real world problems no one would ever apply them by hand. One would use a computer program that implemented the method. And if one were to use a computer program, you might as well have it solve the characteristic equation. Many commercial software packages implement common linear systems techniques. We think the users of this book would benefit by having one of these computer programs available. Advertisements in technical publications such as the *IEEE Control Systems Magazine* and *IEEE Spectrum* and the article by Foster [13] describe many appropriate software packages.

In this book, theory is illustrated by simple numerical examples that are easy to follow and help the student understand the essence of the methodology. At the end of each chapter we present real-life applications. They are selected from most major fields of engineering with special emphasis on electrical engineering, social sciences, and economics. In addition to the illustrative examples and applications, we present homework problems at the end of each chapter. The last five or six homework problems in each chapter need a deep understanding of the material and an ability to develop mathematical proofs.

The organization of the book is as follows. Chapter 1 presents the mathematical background that will be used in later chapters. It is self-contained and presents all the material that will be needed later. Chapter 2 contains the basics of differential and difference equations as well as Laplace and Z transforms. In Chapter 3, characterizations of nonlinear and linear systems are discussed both in state space form and by using the transfer function method. The stability of dynamic systems is analyzed in Chapter 4. Conditions are derived for marginal stability, asymptotical stability, global asymptotical stability, uniform stability, uniform exponential stability and BIBO stability. We present a unified general approach that can be used for continuous and discrete and for nonlinear and linear systems. Controllability and observability are studied in Chapters 5 and 6, respectively. In Chapter 6 we also introduce the concept of duality, which has many theoretical and practical consequences. In solving and examining the properties of dynamic systems, special forms, called canonical forms, are often used. These canonical forms are introduced in Chapter 7. System realizations and minimal realizations are discussed in Chapter 8, where conditions for the realizability of weighting patterns and transfer functions are also introduced. Special system structures, such as the use of observers for constructing feedback compensators, are analyzed in Chapter 9. In Chapter 10 we introduce four advanced topics: nonnegative systems, Kalman filters, adaptive control, and neural networks.

There are several major additions to this second edition. New stability concepts are introduced and analyzed in Chapter 4. We added illustrative examples in all chapters, and we added five to six "theoretical" homework problems to each set. These theoretical problems can be very useful for stimulating graduate students to think about the main concepts and the main results of the chapters. Engineering applications

Number six and nine are new, and their different aspects are examined in each chapter. With these new case studies we put a greater emphasis on electrical engineering application.

There is enough material for a full academic year course, but for graduate students the essence of the book can be covered in a three-unit, one-semester course. The first edition of this book, with certain omissions, has been used in a one-semester, three-unit, graduate course at the University of Arizona. A chapter-dependency chart is illustrated in Figure 0.1, which should help instructors tailor individual course programs.

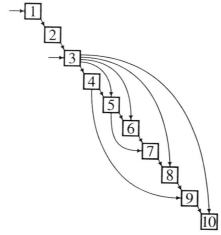

Figure 0.1   Chapter dependency chart.

In preparing the manuscript we obtained significant help from our students, especially from Jerome Yen, Dan Liu and Ling Shen who carefully read the manuscript, checked for misprints and understandability, and prepared the solutions for the homework problems. We also thank Jerome Yen for writing the neural network programs used in Chapter 10. Mo Jamshidi provided us with helpful critiques. The figures were composed by Morgan, Cain and Associates of Tucson, Arizona. Our most special thanks should be addressed to our families, who provided the needed support and personal understanding while we prepared the manuscript.

Ferenc Szidarovszky
and
A. Terry Bahill
Tucson, AZ

Introduction

The technological revolution of the past century occurred because people learned to control large systems composed of nature, machines, people, and society. When they used and improved devices to help them control such systems, they found that they could control bigger and bigger systems. To show this evolution to bigger systems let us consider transportation systems. A horse-drawn coach was handled quite well by the coachman, using simple direct controls. The early automobile was similarly controlled in a direct manner. However, modern automobiles have a multitude of devices to help the driver control the vehicle: electronic automatic transmission, power steering, and four-wheel anti-lock brakes, to mention a few. To fly an airplane the pilot needs control devices to translate his manual actions into the large forces required to move the wing control surfaces. Some of the devices he uses amplify his strength and others augment his intelligence. On many airplanes the intelligence devices are so good the pilot can put the airplane on autopilot and let the plane fly itself. As systems become more complicated, the human does less of the controlling and the machine does more. If we make a manned voyage to Mars, humans will play a small role in controlling the spacecraft.

The design of control devices is called *control engineering*. Early control devices were mechanical, and their design was mainly intuitive. However, recent control devices are algorithms embedded in computers, and their design is very mathematical. The linear systems tools presented in this book comprise the basic mathematical portion of the control engineer's toolkit. This introduction discusses these tools, but does so in nonmathematical terms. It is meant to motivate the main concepts and help the reader see the forest through the trees.

Chapters 1 and 2 of this book present the mathematical basis for understanding the concepts, methods, and derivations in later chapters. The material presented may at first be foreign to you, but we encourage you to bear with us, because after you master this material the rest of the book will be easy.

Chapter 3 shows how we model dynamic systems. The term "dynamic" means that the patterns describing the system change with time and the characteristics of the patterns at any time period are interrelated with those of earlier times. A system is a process that converts inputs to outputs. A system accepts inputs and, based on the inputs and its present state, creates outputs. A system has no direct control over its inputs. An example of a system used in everyday life is a traffic light. It accepts inputs, such as pedestrians pushing the walk button or cars driving over sensors, and based on its current state, creates outputs that are the colors of the lights in each direction. Defining the state of a system is one of the most important, and often most difficult, tasks in system design. The state of the system is the smallest entity that summarizes the past history of the system. The state of the system and the sequence of inputs allows computation of the future states of the system. The state of a system contains all the information needed to calculate future responses without reference to the history of inputs and responses. For example, the current balance of your checking account is the state of that system. There are many ways that it could have gotten to the current value, but when you are ready to write a check that history is irrelevant. The names of the states are often composed of a set of variables, called *state variables*. For systems described by difference or differential equations, these state variables are often the independent variables of these dynamic equations. If the time scale is assumed to be continuous, then the system is described with differential equations, whereas if the time scale is assumed to be discrete (as for computers), then the system is described with difference equations. For sequential logic circuits (computers) the outputs of the memory elements are usually the state variables. However, it is important to note that the choice of state variables is not unique. Most physical systems can be described with many different sets of state variables.

In Chapter 4 we analyze stability and instability of systems. Without giving a formal definition we can say that in an unstable system the state can have large variations and small inputs may produce very large outputs. A common example of an unstable system is illustrated by someone pointing the microphone of a public address (PA) system at a speaker; a loud high-pitched tone results. Often instabilities are caused by too much gain. So to quiet the PA system, decrease the gain by pointing the microphone away from the speakers. Discrete systems can also be unstable. A friend of ours once provided an example. She was sitting in a chair reading and she got cold. So she went over and turned up the thermostat on the heater. The house warmed up. She got hot, so she got up and turned down the thermostat. The house cooled off. She got cold and turned up the thermostat. This process continued until someone finally suggested that she put on a sweater (reducing the gain of her heat loss system). She did and was much more comfortable. We called this a discrete system because she seemed to sample the environment and produce outputs at discrete intervals about 15 minutes apart.

In Chapter 5 we mathematically analyze controllability. Informally, a system is controllable if we can construct a set of inputs that will drive the system to any given state. A real world example of an uncontrollable system is illustrated by a mother in a grocery store with two toddlers. The mother cannot control the states; she can never get the exact behavior she wants. But continual action by an intelligent controller can restrain the children to acceptable behavior. Notice that this is not an unstable system; the children are always within the confines of the store. Stability and controllability are not the same.

In Chapter 6 we mathematically analyze observability. Informally, observability means that by controlling the inputs and watching the outputs of a system we can determine what the states were. A person driving a car is a nonobservable system. Most aspects of the car can be observed, but we cannot put electrodes inside the driver's skull to observe the driver's states and control signals. When engineers must control nonobservable systems, they sometimes build observers. In one prosthetic system, electrical signals recorded from the upper arm of an amputee were used to control a prosthetic arm. This technique was not successful until a computer was placed in between the human's arm and the prosthetic device. The computer contained a type of observer that modeled the body's internal states.

There are different ways to say the same thing. Here are three ways to describe events in a particular baseball game. Joe hit two home runs. Joe homered twice. Two home runs were hit by Joe. These all say about the same thing. But each would be best in certain situations. Similarly, the canonical forms, presented in Chapter 7, show how the same system can be represented in many different mathematical ways. Most canonical forms can be used for most systems, but for any given situation one may be more useful.

When building a new system, management would like to know if the system the engineers design is the simplest one that satisfies the customer's requirements. Suppose a customer asked for a system that receives radio station KUAT FM, and the engineers come up with a design for an AM–FM-tape system. Is the engineer's system minimal? No, a simpler system could be built. (However, given the realities of manufacturing, it may not be less expensive.) In Chapter 8 we present mathematical tools to help build a system and determine if it is minimal.

As shown in Chapter 9, adding feedback loops can reduce the sensitivity to variations in certain parameters, increase rejection of output disturbances, and change system dynamics. For example, have a friend hold a book in her hand with her elbow at her hip and her forearm perpendicular to her body. Instruct her to close her eyes and hold her arm steady. In a few minutes, as her muscles get tired, her arm will sag. Now allow her to open her eyes and look at the book. She will be able to hold her arm steadier because the visual feedback loop has reduced her sensitivity to muscle fatigue. Once again ask her to close her eyes. This time push down on the book. The book will move quite some distance before she can reject your disturbance and return it to its original position. Now allow her to open her eyes and do the same thing. Using visual feedback she will be better at rejecting your disturbances. Finally, when feedback is used to change system dynamics, it is usually used to speed up the system. Imagine using flash cards to help a fifth grader learn the multiplication tables. First do it without feedback. Show him a card, wait for his answer, put it down, and show him another. He will learn, but very slowly. Next give him feedback.

> For example, if you show him the card with 7×3, and he says, "21."
> Respond with, "Very Good." :-)
> <That symbol is a smiling face turned on its side.>
> However, if he says, "22."
> Respond with, "No. It is 21." :-(
> <That is a frowning face turned on its side.>

He will learn faster with feedback. Notice that you can tailor the feedback to get almost any dynamic you want. If you give him ten cents for each correct answer and take away five cents for each incorrect answer, he will learn much faster.

In Chapter 10 we present four advanced tools of systems theory. If the state variables of a model are, for example, human body temperature and blood pressure, then we know that they should never become negative; if they could, then the model is wrong. In Section 10.1 we present simple mathematical tools that can be used to check if a system's state variables always remain nonnegative, and we discuss the main properties of such systems.

In Section 10.2 we present a Kalman Filter. It represents a class of adaptive systems often used in signal processing to separate signals from noise. These filters are designed to extract signals from white noise. However, white noise would have equal energy at all frequencies, therefore it is impossible to make; so, real systems use bandlimited white noise, which is called pink noise. The human auditory system uses adaptive filters to extract signals from noise. At a party most people are able to listen to one person (the signal) in spite of many background conversations (noise). However, when they walk away and start conversing with another person (with different frequencies, intonations, and accents), they must change their filters. Some of us have difficulty understanding the first few sentences with a new conversant, while our filters are still adapting. Following one voice out of many is easy in person, but difficult over a telephone, unless the speaker has increased the signal-to-noise ratio by talking directly into the mouthpiece. Some hearing aids have been fitted with Kalman Filters. They are better at separating signals from noise.

Section 10.3 presents a different type of adaptive system, a type used to control time-varying systems. Many systems change with time due to, for example, bearing wear, warming of lubricants, or fatigue of muscles. When controlling such systems the controller must also change with time. One of the biggest challenges in designing adaptive systems is proving that the resulting systems are stable. In this section we present techniques for designing stable adaptive control systems.

Artificial neural networks are computer systems composed of a very large number of adaptive units connected in parallel. They are useful for pattern recognition and have been used, for example, in banks to verify signatures on checks. Because they can recognize patterns of input variations and specify appropriate outputs, they have also been used as controllers in control systems. The basic operation of neural networks is explained in Section 10.4.

This brief overview of systems theory was provided to help motivate the mathematical analysis that follows. We hope that knowledge of these mathematical techniques will help engineers design systems for the betterment of mankind.

chapter one

Mathematical Background

1.1 Introduction

This chapter provides the foundation for understanding the mathematical details to be discussed in later chapters of this book. It is devoted to two fundamental topics of applied mathematics: metric spaces and matrices. We will see in later chapters that many problems in dynamic systems theory can be solved by evaluating the solutions of linear and nonlinear algebraic equations as well as by computing the solutions of ordinary differential and difference equations. The most commonly used techniques are iterative, which determine a sequence of real (or complex) numbers, vectors, or functions that converge to the desired solutions. The theory of metric spaces and contraction mapping establishes the basis of such methods. This theory will be outlined in the first part of this chapter. In the second part we present the basic properties of linear structures, which will be needed in analyzing linear systems. In this section, norms, transformations, and function of matrices are discussed.

1.2 Metric Spaces and Contraction Mapping Theory

Metric spaces and special mappings defined in metric spaces play very important roles in the solution methodologies of linear and nonlinear algebraic, difference, and differential equations. The solutions of these equations are real or complex vectors, scalars, or functions defined on discrete or continuous time scales. Therefore, the convergence analysis of iteration methods for solving such equations requires the concept of a certain kind of distance between vectors, scalars, and functions. A unified approach using metric spaces is given in this section. The elements of this theory are outlined below.

1.2.1   Metric Spaces

If A and B are any sets, then $A \times B$ denotes the *Cartesian product* of A and B:

$$A \times B = \{(x, y) \mid x \in A, y \in B\} . \tag{1.1}$$

If A and B are single-dimensional intervals, then $A \times B$ is the rectangle shown in Figure 1.1.

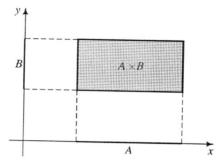

Figure 1.1   Cartesian product in one dimension.

DEFINITION 1.1   *A pair (M, ρ) is called a metric space if M is a set, and ρ is a real valued function defined on $M \times M$ with the following properties:*

(i)   *For all $x, y \in M$, $\rho(x, y) \geq 0$, and $\rho(x, y) = 0$ if and only if $x = y$.*

(ii)   *For all $x, y \in M$, $\rho(x, y) = \rho(y, x)$.*

(iii)   *For all $x, y, z \in M$, $\rho(x, z) \leq \rho(x, y) + \rho(y, z)$.*

Function ρ is called *metric*, which represents the distance between elements of set M. Property (i) requires that the distance of different elements is always positive, and the distance of any element from itself is zero. Property (ii) represents the symmetry of the distance, and Property (iii) is known as the triangle inequality, which states that the direct distance between two elements is never larger than the sum of their distances from a third element.

Example 1.1

Let M be the set $\mathbf{R}$ (or $\mathbf{C}$) of real (or complex) numbers, and define $\rho(x, y) = |x - y|$. Properties (i) and (ii) are satisfied obviously, and the triangle inequality is the obvious consequence of the well-known inequality $|a + b| \leq |a| + |b|$, when we select $a = x - y$, $b = y - z$, and $a + b = x - z$. Hence, (M, ρ) is a metric space.

Example 1.2

Let M be the set $\mathbf{R}^n$ (or $\mathbf{C}^n$) of the n-dimensional real (or complex) vectors.

(A) Select the metric function as

$$\rho_\infty(\mathbf{x}, \mathbf{y}) = \max_i |x_i - y_i| \,, \tag{1.2}$$

where x_i and y_i are the ith components of vectors $\mathbf{x}$ and $\mathbf{y}$, respectively. Properties (i) and (ii) are satisfied, and (iii) can be verified as follows. From the previous example we know that for all i,

$$|x_i - z_i| \leq |x_i - y_i| + |y_i - z_i| \,. \tag{1.3}$$

Assume that

$$\rho_\infty(\mathbf{x}, \mathbf{z}) = \max_i |x_i - z_i| = |x_{i_0} - z_{i_0}| \,.$$

Then

$$\rho_\infty(\mathbf{x}, \mathbf{z}) = |x_{i_0} - z_{i_0}| \leq |x_{i_0} - y_{i_0}| + |y_{i_0} - z_{i_0}|$$

$$\leq \max_i |x_i - y_i| + \max_i |y_i - z_i| = \rho_\infty(\mathbf{x}, \mathbf{y}) + \rho_\infty(\mathbf{y}, \mathbf{z}) \,.$$

Figure 1.2(a) illustrates this distance.

(B) Select now the function

$$\rho_1(\mathbf{x}, \mathbf{y}) = \sum_{i=1}^{n} |x_i - y_i| \,. \tag{1.4}$$

We can easily show that (M, ρ_1) is also a metric space because properties (i) and (ii) are satisfied, and (iii) can be proven by adding inequalities (1.3) for $i = 1, 2, \ldots, n$. Figure 1.2(b) shows this distance.

(C) The most commonly used metric function defined on n-dimensional vectors is given as

$$\rho_2(\mathbf{x}, \mathbf{y}) = \left\{ \sum_{i=1}^{n} |x_i - y_i|^2 \right\}^{1/2} \,. \tag{1.5}$$

Properties (i) and (ii) are obvious again, and the triangle inequality can be proven by applying the Cauchy–Schwarz inequality, which can

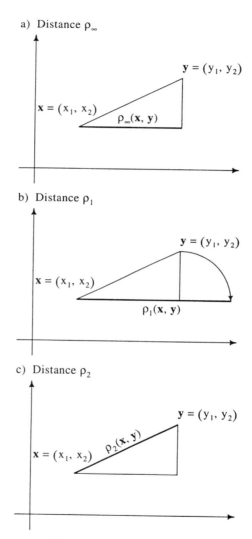

Figure 1.2   Distances in $\mathbf{R}^2$.

be stated as follows: If $\mathbf{u} = (u_i)$ and $\mathbf{v} = (v_i)$ are real (or complex) n-dimensional vectors, then

$$\left(\sum_{i=1}^{n} |u_i| \cdot |v_i| \right)^2 \leq \sum_{i=1}^{n} |u_i|^2 \cdot \sum_{i=1}^{n} |v_i|^2 .$$

Use the triangle inequality and the Cauchy–Schwarz inequality for vectors $\mathbf{x} - \mathbf{y}$ and $\mathbf{y} - \mathbf{z}$ to get

$$\rho_2(\mathbf{x}, \mathbf{z})^2 = \sum_{i=1}^{n} |x_i - z_i|^2 \leq \sum_{i=1}^{n} [|x_i - y_i| + |y_i - z_i|]^2$$

$$= \sum_{i=1}^{n} |x_i - y_i|^2 + \sum_{i=1}^{n} |y_i - z_i|^2 + 2 \sum_{i=1}^{n} [|x_i - y_i| \cdot |y_i - z_i|]$$

$$\leq \sum_{i=1}^{n} |x_i - y_i|^2 + \sum_{i=1}^{n} |y_i - z_i|^2 + 2 \left\{ \sum_{i=1}^{n} |x_i - y_i|^2 \right\}^{1/2}$$

$$\left\{ \sum_{i=1}^{n} |y_i - z_i|^2 \right\}^{1/2}$$

$$= \left[\left\{ \sum_{i=1}^{n} |x_i - y_i|^2 \right\}^{1/2} + \left\{ \sum_{i=1}^{n} |y_i - z_i|^2 \right\}^{1/2} \right]^2$$

$$= [\rho_2(\mathbf{x}, \mathbf{y}) + \rho_2(\mathbf{y}, \mathbf{z})]^2 .$$

Figure 1.2(c) illustrates this distance.

Hence, $\mathbf{R}^n$ (or $\mathbf{C}^n$) is a metric space with distances ρ_∞, ρ_1, and ρ_2. Note that the above three metric functions ρ_∞, ρ_1, and ρ_2 are all special cases of the more general Minkowski distance

$$\rho_p(\mathbf{x}, \mathbf{y}) = \left\{ \sum_{i=1}^{n} |x_i - y_i|^p \right\}^{1/p} , \tag{1.6}$$

where $p \geq 1$ is a given constant.

Example 1.3

Let M be the set $\mathbf{C}[a, b]$ of the continuous functions on the finite closed interval $[a, b]$. Define

$$\rho(f, g) = \max_{x \in [a,b]} |f(x) - g(x)|, \tag{1.7}$$

which is illustrated in Figure 1.3.

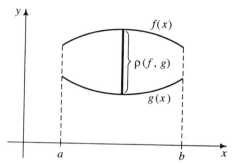

Figure 1.3   Distance of continuous functions.

We now prove that (M, ρ) is a metric space. Note first that the continuity of function $|f(x) - g(x)|$ implies that the maximum exists. Properties (i) and (ii) are obvious, and the triangle inequality can be proven as in the case of distance ρ_∞ of n-dimensional vectors.

Example 1.4

If M is any set, then we can define

$$\rho(x, y) = \begin{cases} 1, & \text{if } x \neq y \\ 0, & \text{if } x = y. \end{cases} \tag{1.8}$$

Properties (i), (ii), and (iii) can be proven easily. The resulting metric space (M, ρ) is called *discrete*.

Let (M, ρ) be a metric space, and $M_1 \subset M$. Define function ρ_1 on $M_1 \times M_1$ as $\rho_1(x, y) = \rho(x, y)$ $(x, y \in M_1)$. Then (M_1, ρ_1) is also a metric space and is called the *subspace* generated by subset M_1.

An *open ball* with center $x \in M$ and radius $r > 0$ is defined as

$$B(x, r) = \{y \mid y \in M, \rho(x, y) < r\}, \tag{1.9}$$

and the set

$$\bar{B}(x, r) = \{y \mid y \in M, \rho(x, y) \leq r\} \tag{1.10}$$

is called the *closed ball* with center x and radius r.

Let $M_1 \subseteq M$. A point $x \in M_1$ is called an *interior point* of M_1 if for some $r > 0$, $B(x, r) \subseteq M_1$. A point $x \in M$ is called a *boundary point* of M_1 if for all $r > 0$, $B(x, r)$ contains points that belong to M_1 and points that do not belong to M_1.

A set M_1 is called *open* if each point of M_1 is interior. A set M_1 is *closed* if $M - M_1$ is open. Note that the empty set is considered to be both open and closed.

DEFINITION 1.2   *A sequence $\{x_n\}$ of elements of M is said to be convergent and to have the limit point $x^* \in M$, if $\rho(x_n, x^*) \to 0$ as $n \to \infty$. This property is denoted as $x_n \to x^*$ or as $\lim_{n \to \infty} x_n = x^*$.*

First we prove that the limit point of any convergent sequence is unique. In contrast to the assertion, assume that x^* and x^{**} are both limit points of sequence $\{x_n\}$. Then

$$0 \leq \rho(x^*, x^{**}) \leq \rho(x^*, x_n) + \rho(x_n, x^{**}) = \rho(x_n, x^*) + \rho(x_n, x^{**}) \ .$$

Since both terms of the right-hand side converge to zero, $\rho(x^*, x^{**}) = 0$. Hence, $x^* = x^{**}$.

Let $M_1 \subseteq M$ be a closed set, and assume that for all $n \geq 1$, $x_n \in M_1$ and sequence x_n converges to an $x^* \in M$. We will next prove that $x^* \in M_1$. Assume in contrast to the assertion that $x^* \notin M_1$. Since M_1 is closed, $M - M_1$ is open. Therefore, there is a ball $B(x^*, r)$ with some $r > 0$ that is in $M - M_1$. Convergence $x_n \to x^*$ implies that for sufficiently large values of n, $\rho(x_n, x^*) < r$, that is, $x_n \in B(x^*, r)$ implying that $x_n \in M - M_1$ contradicting the assertion that $x_n \in M_1$. This property of closed sets can be formulated by saying that a closed set M_1 contains all limit points of sequences from M_1.

DEFINITION 1.3   *A sequence $\{x_n\}$ of elements of M is called a Cauchy sequence if $\rho(x_n, x_m) \to 0$ as $n, m \to \infty$.*

THEOREM 1.1

If $\{x_n\}$ is convergent, then it is also a Cauchy sequence.

PROOF   Let the limit point of $\{x_n\}$ be denoted by x^*. Then

$$0 \leq \rho(x_n, x_m) \leq \rho(x_n, x^*) + \rho(x^*, x_m) = \rho(x_n, x^*) + \rho(x_m, x^*) \ .$$

Because both terms of the right-hand side converge to zero, we conclude that $\rho(x_n, x_m) \to 0$ as $n, m \to \infty$. ∎

REMARK 1.1   A Cauchy sequence is not necessarily convergent, as the case of metric space (M, ρ) with $M = (0, \infty)$ and $\rho(x, y) = |x - y|$ illustrates. Consider sequence $x_n = 1/n$ $(n \geq 1)$, which has no limit point in M (the zero limit does not belong to M), but for $n, m \to \infty$, $|(1/n) - (1/m)| \to 0$. ∎

DEFINITION 1.4   *A metric space (M, ρ) is called complete if all Cauchy sequences of elements in M have limit points in M.*

It is well known from calculus that in the set of real (or complex) numbers all **Cauchy sequences** are convergent, so **R** (or **C**) is complete. The convergence of vectors in any of the discussed distances means component-wise convergence. Because each component is a real (or complex) number, $\mathbf{R}^n (or\ \mathbf{C}^n)$ is also complete. The convergence in $C[a, b]$ is the well-known uniform convergence. It is also well know from calculus that the limit function of uniformly convergent continuous functions defined on a closed interval $[a, b]$ is also continuous implying that $C[a, b]$ with distance (1.7) is also complete.

In the next theorem we prove that $\rho(x, y)$ is a continuous two-variable function.

THEOREM 1.2
If $x_n \to x^*$ and $y_n \to y^*$ for $n \to \infty$, then $\rho(x_n, y_n) \to \rho(x^*, y^*)$ for $n \to \infty$.

PROOF   By applying the triangle inequality we have

$$\rho(x_n, y_n) \leq \rho(x_n, x^*) + \rho(x^*, y_n) \leq \rho(x_n, x^*) + \rho(x^*, y^*) + \rho(y^*, y_n) ,$$

that is,

$$\rho(x_n, y_n) - \rho(x^*, y^*) \leq \rho(x_n, x^*) + \rho(y_n, y^*) .$$

By interchanging x_n with x^* and y_n with y^* we conclude that

$$\rho(x^*, y^*) - \rho(x_n, y_n) \leq \rho(x^*, x_n) + \rho(y^*, y_n) .$$

Hence

$$0 \leq |\rho(x_n, y_n) - \rho(x^*, y^*)| \leq \rho(x_n, x^*) + \rho(y_n, y^*) ,$$

where the right-hand side tends to zero as $n \to \infty$. Thus, $\rho(x_n, y_n) \to \rho(x^*, y^*)$ for $n \to \infty$.   ∎

In the next section, mappings between metric spaces will be defined and their main properties will be examined. These properties then will be applied in proving the contraction mapping theorem, which will be very useful in showing the existence of a unique state trajectory in continuous systems.

1.2.2   *Mappings in Metric Spaces*

Any iteration method consists of the repeated application of a certain mapping. In order to analyze the convergence of iteration procedures, the basic properties of such mappings have to be investigated. This section is devoted to this subject. Assume that (M, ρ) and (M', ρ') are two (not necessarily different) metric spaces. The *domain* $D(A)$ and *range* $R(A)$ of a single-valued mapping A from M to M' are defined as follows:

$$D(A) = \{x \mid x \in M \text{ and } A(x) \text{ is defined}\} \ ,$$

$$R(A) = \{x' \mid x' \in M' \text{ and there exists } x \in D(A) \text{ such that } x' = A(x)\} \ .$$

Obviously $D(A) \subseteq M$ and $R(A) \subseteq M'$. The domain and range of mappings are illustrated in Figure 1.4.

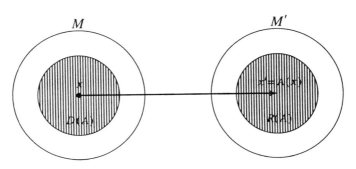

Figure 1.4   Domain and range of mappings.

DEFINITION 1.5   *Mapping A is said to be continuous at a point $x \in D(A)$, if for every sequence $\{x_n\}$ from $D(A)$ converging to x, $A(x_n) \to A(x)$ as $n \to \infty$.*

Similarly, a mapping A is said to be continuous if it is continuous at every
$x \in D(A)$.

DEFINITION 1.6   *Mapping A is called bounded if there exists a nonneg-*
ative constant K such that for all $x, y \in D(A)$,

$$\rho'(A(x), A(y)) \leq K \cdot \rho(x, y) . \qquad (1.11)$$

Bounded mappings are illustrated in Figure 1.5.

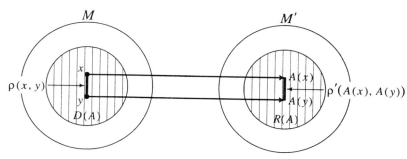

Figure 1.5   Bounded mappings.

Note first that every bounded mapping is continuous, because if $x^*, x_n \in$
$D(A)$ $(n \geq 1)$ and $x_n \to x^*$, then

$$0 \leq \rho'(A(x_n), A(x^*)) \leq K \cdot \rho(x_n, x^*) .$$

Since the right-hand side converges to zero, $A(x_n) \to A(x^*)$ as $x_n \to x^*$.
It is easy to see that a continuous mapping is not necessarily bounded.
As an example, consider $M = M' = \mathbf{R}$, $\rho(x, y) = |x - y|$, $\rho' \equiv \rho$, and
$A(x) = x^2$. In this case, for all $x \neq y$,

$$\frac{\rho'(A(x), A(y))}{\rho(x, y)} = \frac{|x^2 - y^2|}{|x - y|} = |x + y| ,$$

which can be arbitrarily large, if x and y are sufficiently large positive
numbers.

A special class of bounded mappings is defined next.

DEFINITION 1.7   *Mapping A is called a contraction if it is bounded with*
a constant $0 \leq K < 1$.

Example 1.5

Let $M = M' = \mathbf{R}$, $\rho(x,y) = |x - y|$, $\rho' = \rho$, furthermore let A be a differentiable function on some interval $D \subseteq M$. Assume that for all $x \in D$, $|A'(x)| \leq K$, where A' is the derivative of A. Then for all $x, y \in D$,

$$\rho'(A(x), A(y)) = |A(x) - A(y)| = |A'(\xi)| \cdot |x - y|$$

$$\leq K \cdot |x - y| = K \cdot \rho(x,y) , \tag{1.12}$$

where ξ is between x and y. Hence, mapping A is bounded with constant K.

Example 1.6

Let $M = M' = \mathbf{R}^n$, $\rho' \equiv \rho \equiv \rho_p$ $(p = 1, 2, \infty)$, and assume that $\mathbf{A}$ is a differentiable function on a convex set $D \subseteq M$. Assume furthermore that for all $\mathbf{x} \in D$, $|(\partial A_i / \partial x_j)(\mathbf{x})| \leq \alpha_{ij}$, where $\mathbf{A} = (A_i)$ and $\mathbf{x} = (x_j)$. The mean-value theorem of the derivatives of multivariable functions implies that for all $\mathbf{x}$ and $\mathbf{y} \in D$,

$$A_i(\mathbf{x}) - A_i(\mathbf{y}) = \sum_{j=1}^{n} \frac{\partial A_i}{\partial x_j}(\boldsymbol{\xi}_i) \cdot (x_j - y_j) \qquad (i = 1, 2, \dots, n) .$$

Select first $p = 1$, then

$$\rho_1(\mathbf{A}(\mathbf{x}), \mathbf{A}(\mathbf{y})) = \sum_{i=1}^{n} |A_i(\mathbf{x}) - A_i(\mathbf{y})| \leq \sum_{i=1}^{n} \sum_{j=1}^{n} \left| \frac{\partial A_i}{\partial x_j}(\boldsymbol{\xi}_i) \right| \cdot |x_j - y_j|$$

$$= \sum_{j=1}^{n} \left(|x_j - y_j| \cdot \sum_{i=1}^{n} \left| \frac{\partial A_i}{\partial x_j}(\boldsymbol{\xi}_i) \right| \right)$$

$$\leq K_1 \cdot \sum_{j=1}^{n} |x_j - y_j| = K_1 \rho_1(\mathbf{x}, \mathbf{y}) ,$$

where

$$K_1 = \max_j \sum_{i=1}^{n} \alpha_{ij} . \tag{1.13}$$

Select next $p = 2$, then the Cauchy–Schwarz inequality implies that

$$\rho_2(\mathbf{A}(\mathbf{x}), \mathbf{A}(\mathbf{y}))^2 = \sum_{i=1}^{n} |A_i(\mathbf{x}) - A_i(\mathbf{y})|^2 = \sum_{i=1}^{n} \left| \sum_{j=1}^{n} \frac{\partial A_i}{\partial x_j}(\boldsymbol{\xi}_i) \cdot (x_j - y_j) \right|^2$$

$$\leq \sum_{i=1}^{n} \left(\sum_{j=1}^{n} \left| \frac{\partial A_i}{\partial x_j}(\boldsymbol{\xi}_i) \right|^2 \right) \left(\sum_{j=1}^{n} |x_j - y_j|^2 \right)$$

$$\leq \sum_{i=1}^{n} \left(\sum_{j=1}^{n} \left| \frac{\partial A_i}{\partial x_j}(\boldsymbol{\xi}_i) \right|^2 \right) \cdot \rho_2(\mathbf{x}, \mathbf{y})^2 \,,$$

that is,

$$\rho_2(\mathbf{A}(\mathbf{x}), \mathbf{A}(\mathbf{y})) \leq K_2 \cdot \rho_2(\mathbf{x}, \mathbf{y})$$

with

$$K_2 = \left\{ \sum_{i=1}^{n} \sum_{j=1}^{n} \alpha_{ij}^2 \right\}^{1/2} . \tag{1.14}$$

Select finally $p = \infty$. Then

$$\rho_\infty(\mathbf{A}(\mathbf{x}), \mathbf{A}(\mathbf{y})) = \max_i |A_i(\mathbf{x}) - A_i(\mathbf{y})| \leq \max_i \sum_{j=1}^{n} \left| \frac{\partial A_i}{\partial x_j}(\boldsymbol{\xi}_i) \right| \cdot |x_j - y_j|$$

$$\leq \max_i \sum_{j=1}^{n} \left| \frac{\partial A_i}{\partial x_j}(\boldsymbol{\xi}_i) \right| \cdot \max_j |x_j - y_j| \leq K_\infty \rho_\infty(\mathbf{x}, \mathbf{y}) \,,$$

where

$$K_\infty = \max_i \sum_{j=1}^{n} \alpha_{ij} . \tag{1.15}$$

Hence, $\mathbf{A}$ is a bounded mapping with constants K_p $(p = 1, 2, \infty)$. Note that in the special case of the linear function

$$\mathbf{A}(\mathbf{x}) = \mathbf{A}\mathbf{x} + \mathbf{f} \,,$$

where $\mathbf{A} = (a_{ij})$ is an $n \times n$ real matrix and $\mathbf{f}$ is an n-element real vector, $(\partial A_i / \partial x_j)(\mathbf{x}) = a_{ij}$. Consequently, we may select $\alpha_{ij} =$

$|a_{ij}|$, hence the above bounds have the special forms:

$$K_1 = \max_j \sum_{i=1}^{n} |a_{ij}|,$$

$$K_2 = \left\{ \sum_{i=1}^{n} \sum_{j=1}^{n} |a_{ij}|^2 \right\}^{1/2} \tag{1.16}$$

and

$$K_\infty = \max_i \sum_{j=1}^{n} |a_{ij}|.$$

It is easy to verify that these bounds hold even if **A**, **f**, and **x** are complex.

Example 1.7

Let $M = M' = C[a, b]$ with the metric defined in Example 1.3. Assume that function k is continuous on $[a, b] \times [a, b]$ and f is continuous on $[a, b]$. Define mapping A on M as

$$A(y)(x) = \int_a^b k(x, s) y(s)\, ds + f(x), \tag{1.17}$$

where $y \in M$ is a continuous function, and the left-hand side is the value of mapping $A(y)$ at x. Hence, we choose $D(A) = M$ and $R(A) \subseteq M'$. It is easy to show that mapping A is bounded, since for all $y, z \in M$,

$$\rho(A(y), A(z)) = \max_{x \in [a,b]} |A(y)(x) - A(z)(x)|$$

$$\leq \max_{x \in [a,b]} \int_a^b |k(x, s)| \cdot |y(s) - z(s)|\, ds$$

$$\leq \max_{x \in [a,b]} \int_a^b |k(x, s)|\, ds \cdot \max_{x \in [a,b]} |y(x) - z(x)| = K \cdot \rho(y, z),$$

where

$$K = \max_{x \in [a,b]} \int_a^b |k(x, s)|\, ds.$$

Consequently, mapping A is bounded.

Example 1.8

Let M be the set $C^1[a, b]$ of the continuously differentiable functions on $[0, 2\pi]$ with the distance defined in Example 1.3. Define mapping $A(y) = y'$ with $D(A) = M$. We shall now verify that this mapping is not bounded. Consider functions $y(x) = \sin nx$ $(n \geq 0)$, $z(x) \equiv 0$. Then

$$A(y)(x) = n \cos nx \quad \text{and} \quad A(z)(x) = 0 ,$$

therefore,

$$\rho(A(y), A(z)) = \max_{x \in [0, 2\pi]} |n \cdot \cos nx - 0| = n .$$

Since

$$\rho(y, z) = \max_{x \in [0, 2\pi]} |\sin nx - 0| = 1 ,$$

no finite K satisfies Definition 1.6. Hence, mapping A is not bounded.

The last example shows that differentiation as a mapping is not bounded. This is the reason why in Section 2.1.1 differential equations will be be rewritten as integral equations. The resulting integral mappings will be not only bounded, but also contractions, and therefore the results of the next section can be easily applied.

1.2.3   *Contraction Mappings and Fixed Points*

The main result of this section is formulated as a theorem, which gives sufficient conditions for the existence of the unique solution of fixed-point problems of the form $x = A(x)$, where A is a mapping with $D(A)$ and $R(A)$ being the subsets of the same set M. As we will see later, the computation of equilibrium states and state trajectories of a dynamic system requires the solution of such fixed-point problems.

THEOREM 1.3

Assume that metric space (M, ρ) is complete, $M_1 \subseteq M$ is a closed set. Let mapping A be a contraction such that $D(A) = M_1$ and $R(A) \subseteq M_1$. Then the fixed-point problem $x = A(x)$ has a unique solution in M_1, furthermore it can be found as the limit of the iteration sequence

$$x_1 = A(x_0), \quad (x_0 \in M_1 \text{ arbitrary})$$

$$x_2 = A(x_1)$$

$$x_{n+1} = A(x_n)$$

$$\vdots \tag{1.18}$$

PROOF   The proof contains several steps.

(a)   First we prove that the iteration sequence is a Cauchy sequence. The repeated application of the triangle inequality implies that for $m > n$,

$$\rho(x_n, x_m) \leq \rho(x_n, x_{n+1}) + \rho(x_{n+1}, x_{n+2}) + \cdots + \rho(x_{m-1}, x_m) \,. \tag{1.19}$$

From the definition of contractive mappings we see that for all $k \geq 1$,

$$\rho(x_k, x_{k+1}) = \rho(A(x_{k-1}), A(x_k)) \leq K \cdot \rho(x_{k-1}, x_k)$$

$$= K \cdot \rho(A(x_{k-2}), A(x_{k-1}))$$

$$\leq K^2 \cdot \rho(x_{k-2}, x_{k-1}) \leq \cdots \leq K^k \rho(x_0, x_1) \,.$$

Combining this relation with (1.19) yields the inequality

$$\rho(x_n, x_m) \leq \rho(x_0, x_1) \cdot [K^n + K^{n+1} + \cdots + K^{m-1}]$$

$$\leq \rho(x_0, x_1) \cdot [K^n + K^{n+1} + \cdots] = \rho(x_0, x_1) \cdot \frac{K^n}{1 - K} \,.$$

Since $0 \leq K < 1$, the right-hand side converges to zero as $n \to \infty$, which implies that $\rho(x_n, x_m) \to 0$ as $m > n$ and $n \to \infty$.

(b)   Since metric space (M, ρ) is complete, sequence $\{x_n\}$ converges to an element $x^* \in M$. Note that for all n, $x_n \in M_1$. Since M_1 is a closed set, $x^* \in M_1$.

(c)   Next we show that x^* is a fixed point of A. Mapping A is bounded, therefore it is continuous. Letting $n \to \infty$ in the iteration equation $x_{n+1} = A(x_n)$ we have $x_{n+1} \to x^*$ and $A(x_n) \to A(x^*)$, and the uniqueness of the limit implies that $x^* = A(x^*)$.

(d)   Finally we verify that the fixed point is unique. Assume that x^* and x^{**} are fixed points of A such that $x^* \neq x^{**}$. Since A is a contraction mapping,

$$\rho(x^*, x^{**}) = \rho(A(x^*), A(x^{**})) \leq K \cdot \rho(x^*, x^{**})$$

with $K < 1$. Divide both sides of this inequality by $\rho(x^*, x^{**}) > 0$ to get the relation

$$1 \leq K ,$$

which contradicts the definition of K. Hence the proof is complete.

∎

COROLLARY 1.1

The iterations of mapping A can be defined by the following recursion:

$$A^1 \equiv A, \quad A^{k+1}(x) = A(A^k(x)) \qquad (all \ x \in M_1) . \tag{1.20}$$

It is easy to verify that the assumption of the theorem that A is a contraction can be replaced by the weaker condition that for some $k \geq 1$, A^k is a contraction.

The iteration process (1.18) is illustrated in Figure 1.6 and can be summarized as follows:

Step 1   Select an initial approximation $x_{old} \in M_1$.

Step 2   Compute $x_{new} = A(x_{old})$.

Step 3   If $\rho(x_{new}, x_{old})$ is less than an error tolerance ε, then accept x_{new} as the solution and stop. Otherwise set $x_{old} = x_{new}$ and go back to Step 2.

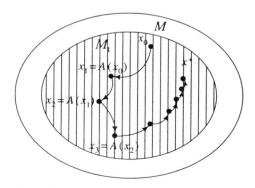

Figure 1.6   Illustration of fixed-point iteration.

At the conclusion of this section, a numerical example is presented.

Example 1.9

Consider the special case when $M = M_1$ is the closed interval $[1, 2]$ and mapping A is defined as

$$A(x) = \frac{1}{2} \left(x + \frac{2}{x} \right) .$$

Obviously M is complete and M_1 is closed. For all $x \in [1, 2]$,

$$A(x) \leq \frac{1}{2} \left(2 + \frac{2}{1} \right) = 2$$

and

$$A(x) \geq \frac{1}{2} \left(1 + \frac{2}{2} \right) = 1 ,$$

since $x \leq 2$, $2/x \leq 2/1 = 2$, $x \geq 1$, and $2/x \geq 2/2 = 1$. That is, $A(x) \in M_1$. Furthermore,

$$\left| \frac{d}{dx} A(x) \right| = \frac{1}{2} \left| 1 - \frac{2}{x^2} \right| \leq \frac{1}{2} < 1 ;$$

therefore, Example 1.5 implies that mapping A is a contraction. Thus, all conditions of Theorem 1.3 are satisfied. Select $x_0 = 2$, then the iteration sequence is the following:

$$x_1 = \frac{1}{2} \left(2 + \frac{2}{2} \right) = 1.5 ,$$

$$x_2 = \frac{1}{2} \left(1.5 + \frac{2}{1.5} \right) \approx 1.4166667 ,$$

and in a similar manner

$$x_3 \approx 1.4142157, \qquad x_4 \approx 1.4142136 ,$$

and so on. Note that the only fixed point in $[1, 2]$ of mapping A is $\sqrt{2}$, which equals x_4 to the accuracy shown.

Finally, we note that some applications of these results to the theory of iteration methods are discussed, for example, in [2, 42], and in [8].

1.3   Some Properties of Vectors and Matrices

In the theory of linear systems, the metric properties of finite-dimensional vectors and matrices have important roles. For example, the stability analysis of dynamic systems requires the investigation of the convergence of the state vector as time approaches infinity, which can be performed easily by using the concepts of the first part of this section. In the second part, special matrix transformations and decompositions are discussed. They will be useful in transforming linear systems to special forms. In the third part of this section, matrix functions are introduced and analyzed, which will be applied to solve linear difference and differential equations, governing discrete and continuous systems, in closed form.

1.3.1   Norms of Vectors and Matrices

In the previous section distances of vectors were introduced, but they do not measure explicitly the magnitude of a single vector. However, in analogy to the definition of the absolute values of real numbers being their distances from zero, the length (or norm) of a vector $\mathbf{x}$ is defined as its distance $\rho(\mathbf{x}, \mathbf{0})$ from the zero vector.

DEFINITION 1.8   *The p-norms ($p = 1, 2, \infty$) of an n-element real or complex vector $\mathbf{x} = (x_i)$ are defined as follows:*

$$\|\mathbf{x}\|_1 = \sum_{i=1}^{n} |x_i| ,$$

$$\|\mathbf{x}\|_2 = \left\{ \sum_{i=1}^{n} |x_i|^2 \right\}^{1/2} , \tag{1.21}$$

and

$$\|\mathbf{x}\|_\infty = \max_i |x_i| .$$

These vector norms are illustrated in Figure 1.7. We note that $\|\mathbf{x}\|_2$ can also be written as $\sqrt{\mathbf{x}^*\mathbf{x}}$, where $\mathbf{x}^*$ is the conjugate transpose of $\mathbf{x}$ (or the usual transpose of $\mathbf{x}$ in the real case).

The main properties of vector norms are given next.

THEOREM 1.4

The p-norms of n-element real or complex vectors satisfy the following properties:

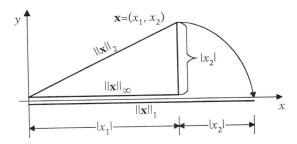

Figure 1.7   Vector norms in $\mathbf{R}^2$.

(i) $\|\mathbf{x}\| \geq 0$, *and* $\|\mathbf{x}\| = 0$ *if and only if* $\mathbf{x} = \mathbf{0}$, *where* $\mathbf{0}$ *is the zero vector with all elements equal to zero.*

(ii)   *For an arbitrary real (or complex) number* α, $\|\alpha\mathbf{x}\| = |\alpha| \cdot \|\mathbf{x}\|$.

(iii) $\|\mathbf{x} + \mathbf{y}\| \leq \|\mathbf{x}\| + \|\mathbf{y}\|$.

PROOF

(i) $\|\mathbf{x}\| = \rho(\mathbf{x}, \mathbf{0})$ is always nonnegative, and is zero if and only if $\mathbf{x} = \mathbf{0}$.

(ii)   This properly follows immediately from the definition of the vector norms.

(iii) $\|\mathbf{x}+\mathbf{y}\| = \rho(\mathbf{x}, -\mathbf{y}) \leq \rho(\mathbf{x}, \mathbf{0}) + \rho(\mathbf{0}, -\mathbf{y}) = \|\mathbf{x}\| + \|-\mathbf{y}\| = \|\mathbf{x}\| + \|\mathbf{y}\|$.

∎

REMARK 1.2   Any real-valued function defined on the set of the n-element real (or complex) vectors satisfying conditions (i), (ii), and (iii) is called a vector norm. It can be proved (see Problem 1.21) that any two vector norms defined on the set of the n-element real (or complex) vectors are equivalent to each other, that is, if $\|\cdot\|$ and $\|\cdot\|'$ are two norms, then there exist positive constants a_1 and a_2 such that for all vectors $\mathbf{x}$,

$$a_1\|\mathbf{x}\| \leq \|\mathbf{x}\|' \leq a_2\|\mathbf{x}\| .$$

∎

DEFINITION 1.9   Let $\| \cdot \|$ *be a given norm of n-dimensional real (or* complex) vectors. The matrix norm generated by this vector norm is given as

$$\|\mathbf{A}\| = \max\{\|\mathbf{A}\mathbf{x}\| \mid \|\mathbf{x}\| = 1\}\,.$$

Since vector norms are continuous and the set $S = \{\mathbf{x} \mid \|\mathbf{x}\| = 1\}$ is closed and bounded, the maximum exists.

The fundamental properties of matrix norms are summarized next.

THEOREM 1.5

Let $\|\cdot\|$ be a matrix norm generated by a vector norm. Then

(i) $\|\mathbf{A}\| \geq 0$, *and* $\|\mathbf{A}\| = 0$ *if and only if* $\mathbf{A} = \mathbf{O}$, *where* $\mathbf{O}$ *is the zero matrix with all elements equal to zero.*

(ii) *For an arbitrary real (or complex) number* α, $\|\alpha\mathbf{A}\| = |\alpha| \cdot \|\mathbf{A}\|$.

(iii) $\|\mathbf{A} + \mathbf{B}\| \leq \|\mathbf{A}\| + \|\mathbf{B}\|$.

(iv) $\|\mathbf{A}\mathbf{B}\| \leq \|\mathbf{A}\| \cdot \|\mathbf{B}\|$.

PROOF   Properties (i) and (ii) are obvious, and (iii) is the consequence of the inequality

$$\|\mathbf{A} + \mathbf{B}\| = \max_S \|(\mathbf{A} + \mathbf{B})\mathbf{x}\| = \max_S \|\mathbf{A}\mathbf{x} + \mathbf{B}\mathbf{x}\|$$

$$\leq \max_S(\|\mathbf{A}\mathbf{x}\| + \|\mathbf{B}\mathbf{x}\|) \leq \max_S \|\mathbf{A}\mathbf{x}\| + \max_S \|\mathbf{B}\mathbf{x}\|$$

$$= \|\mathbf{A}\| + \|\mathbf{B}\|\,.$$

The last property can be shown as follows:

$$\|\mathbf{A}\mathbf{B}\| = \max_S \|\mathbf{A}\mathbf{B}\mathbf{x}\| = \max_S \|\mathbf{A}(\mathbf{B}\mathbf{x})\| \leq \max_S(\|\mathbf{A}\| \cdot \|\mathbf{B}\mathbf{x}\|)$$

$$= \|\mathbf{A}\| \cdot \max_S \|\mathbf{B}\mathbf{x}\| = \|\mathbf{A}\| \cdot \|\mathbf{B}\|\,.$$

∎

REMARK 1.3   Any real-valued function defined on the set of the $n \times n$ real (or complex) matrices satisfying conditions (i)–(iv) is called a matrix norm. ∎

For $p = 1, 2, \infty$, let $\| \cdot \|_p$ denote the matrix norm generated by the vector norm $\| \cdot \|_p$.

THEOREM 1.6
If $\mathbf{A}$ is an $n \times n$ real (or complex) matrix with (i, j) element a_{ij}, then

$$\|\mathbf{A}\|_1 = \max_j \sum_{i=1}^n |a_{ij}|$$

and

$$\|\mathbf{A}\|_\infty = \max_i \sum_{j=1}^n |a_{ij}| \, .$$

Let $\mathbf{A}^* = \bar{\mathbf{A}}^T$ denote the conjugate transpose of $\mathbf{A}$, and let λ^* denote the largest eigenvalue of matrix $\mathbf{A}^*\mathbf{A}$. That is, λ^* is the largest real number such that $\mathbf{A}^*\mathbf{A}\mathbf{x} = \lambda^*\mathbf{x}$ with some nonzero vector $\mathbf{x}$. Then

$$\|\mathbf{A}\|_2 = \sqrt{\lambda^*} \, .$$

PROOF   Let $\|\mathbf{x}\|_1 = \sum_{i=1}^n |x_i| = 1$. Then

$$\|\mathbf{A}\mathbf{x}\|_1 = \sum_{i=1}^n \left| \sum_{j=1}^n a_{ij} x_j \right| \leq \sum_{i=1}^n \sum_{j=1}^n |a_{ij}| \cdot |x_j|$$

$$= \sum_{j=1}^n \sum_{i=1}^n |a_{ij}| \cdot |x_j| = \sum_{j=1}^n \left(|x_j| \cdot \sum_{i=1}^n |a_{ij}| \right)$$

$$\leq \sum_{j=1}^n |x_j| \cdot \max_j \sum_{i=1}^n |a_{ij}| = \max_j \sum_{i=1}^n |a_{ij}| \, .$$

In order to show that the last expression on the right-hand side is the maximum of $\|\mathbf{A}\mathbf{x}\|_1$ we find a vector $\mathbf{x}$ with unit norm that gives equalities everywhere in the above inequality. Assume that

$$\max_j \sum_{i=1}^n |a_{ij}| = \sum_{i=1}^n |a_{ij_0}| \, ,$$

then the selection

$$x_j = \begin{cases} 1 \text{ if } j = j_0 \\ 0 \text{ if } j \neq j_0 \end{cases}$$

is satisfying.

Assume next that $\|\mathbf{x}\|_\infty = \max_i |x_i| = 1$. Then

$$\|\mathbf{Ax}\|_\infty = \max_i \left| \sum_{j=1}^{n} a_{ij}x_j \right| \leq \max_i \sum_{j=1}^{n} |a_{ij}| \cdot |x_j|$$

$$\leq \max_i \sum_{j=1}^{n} |a_{ij}| \cdot \max_j |x_j| = \max_i \sum_{j=1}^{n} |a_{ij}| .$$

It is easy to show that in this case vector $\mathbf{x}$ with

$$x_j = \begin{cases} 1 \text{ if } a_{i_0 j} \geq 0 \\ -1 \text{ if } a_{i_0 j} < 0 \end{cases}$$

gives equalities everywhere in the above inequality, where i_0 is selected as

$$\max_i \sum_{j=1}^{n} |a_{ij}| = \sum_{j=1}^{n} |a_{i_0 j}| .$$

Assume finally that $\|\mathbf{x}\|_2 = \{\sum_{i=1}^{n} |x_i|^2\}^{1/2} = 1$. Then

$$\|\mathbf{Ax}\|_2^2 = (\mathbf{Ax})^*(\mathbf{Ax}) = \mathbf{x}^*(\mathbf{A}^*\mathbf{A})\mathbf{x} .$$

Since matrix $\mathbf{A}^*\mathbf{A}$ is Hermitian, all eigenvalues are real and the maximum of the above quadratic form is the largest eigenvalue λ^*, and the maximum occurs when $\mathbf{x}$ is selected as an eigenvector associated to λ^* (see, for example, [43]).

Thus, the proof is complete. ∎

DEFINITION 1.10   *A vector norm $\| \cdot \|$ and a matrix norm $\| \cdot \|$ are called compatible if for all vectors and matrices such that $\mathbf{Ax}$ exists,*

$$\|\mathbf{Ax}\| \leq \|\mathbf{A}\| \cdot \|\mathbf{x}\| . \tag{1.22}$$

THEOREM 1.7

Let $\| \cdot \|$ be any vector norm, then it is compatible with the matrix norm that is generated by $\| \cdot \|$.

PROOF   Note first that the norm of the vector $\mathbf{z} = (1/\|\mathbf{x}\|) \cdot \mathbf{x}$ equals 1. Therefore

$$\|\mathbf{A}\mathbf{x}\| = \|\mathbf{x}\| \cdot \left\| \mathbf{A} \cdot \frac{\mathbf{x}}{\|\mathbf{x}\|} \right\| = \|\mathbf{x}\| \cdot \|\mathbf{A}\mathbf{z}\|$$

$$\leq \max_S \|\mathbf{A}\mathbf{z}\| \cdot \|\mathbf{x}\| = \|\mathbf{A}\| \cdot \|\mathbf{x}\| ,$$

which completes the proof.   ∎

REMARK 1.4   It can be easily verified that for every matrix norm satisfying Properties (i), (ii), (iii), and (iv) of Theorem 1.5 there is at least one vector norm with which the matrix norm is compatible. The construction of one of these vector norms is the following. Let $\mathbf{x} = (x_i)$ be an n-dimensional vector. Construct matrix

$$\mathbf{X} = \begin{pmatrix} x_1 & 0 & \cdots & 0 \\ x_2 & 0 & \cdots & 0 \\ \vdots & \vdots & \ddots & \vdots \\ x_n & 0 & \cdots & 0 \end{pmatrix} \tag{1.23}$$

and define the norm of vector $\mathbf{x}$ as the matrix norm of matrix $\mathbf{X}$. Note that a given vector norm may be compatible with more than one matrix norm. Such a case is presented next.   ∎

DEFINITION 1.11   *The Frobenius norm of $n \times n$ real (or complex) matrices is defined as*

$$\|\mathbf{A}\|_F = \left\{ \sum_{i=1}^{n} \sum_{j=1}^{n} |a_{ij}|^2 \right\}^{\frac{1}{2}} . \tag{1.24}$$

By using the triangle inequality and the Cauchy–Schwarz inequality it is easy to see that this matrix norm satisfies all properties of Theorem 1.5 and is compatible with the vector norm $\| \cdot \|_2$.

Example 1.10

Consider vectors

$$\mathbf{x} = \begin{pmatrix} 1 \\ 1 \end{pmatrix}, \qquad \mathbf{y} = \begin{pmatrix} 1 \\ 2 \\ 1 \end{pmatrix}$$

and matrices

$$\mathbf{A} = \begin{pmatrix} 1 & 2 \\ 1 & 1 \end{pmatrix}, \qquad \mathbf{B} = \begin{pmatrix} 1 & 0 & 1 \\ 0 & 1 & 1 \\ 1 & 1 & 0 \end{pmatrix}.$$

The norms of these vectors and matrices can be determined as follows. Simple calculation shows that

$$\|\mathbf{x}\|_1 = 1 + 1 = 2, \qquad \|\mathbf{x}\|_2 = \sqrt{1^2 + 1^2} = \sqrt{2}, \qquad \|\mathbf{x}\|_\infty = \max\{1; 1\} = 1;$$
$$\|\mathbf{y}\|_1 = 1 + 2 + 1 = 4, \ \|\mathbf{y}\|_2 = \sqrt{1^2 + 2^2 + 1^2} = \sqrt{6}, \ \|\mathbf{y}\|_\infty = \max\{1; 2; 1\} = 2;$$

$$\|\mathbf{A}\|_1 = \max\{1 + 1; 2 + 1\} = 3,$$

$$\|\mathbf{A}\|_F = \{1^2 + 2^2 + 1^2 + 1^2\}^{1/2} = \sqrt{7},$$

$$\|\mathbf{A}\|_\infty = \max\{1 + 2; 1 + 1\} = 3,$$

and

$$\|\mathbf{B}\|_1 = \|\mathbf{B}\|_\infty = \max\{1 + 1; 1 + 1; 1 + 1\} = 2,$$

$$\|\mathbf{B}\|_F = \{1^2 + 1^2 + 1^2 + 1^2 + 1^2 + 1^2\}^{1/2} = \sqrt{6}.$$

Note that

$$\mathbf{A}^*\mathbf{A} = \begin{pmatrix} 1 & 1 \\ 2 & 1 \end{pmatrix} \begin{pmatrix} 1 & 2 \\ 1 & 1 \end{pmatrix} = \begin{pmatrix} 2 & 3 \\ 3 & 5 \end{pmatrix}$$

and

$$\mathbf{B}^*\mathbf{B} = \begin{pmatrix} 1 & 0 & 1 \\ 0 & 1 & 1 \\ 1 & 1 & 0 \end{pmatrix} \begin{pmatrix} 1 & 0 & 1 \\ 0 & 1 & 1 \\ 1 & 1 & 0 \end{pmatrix} = \begin{pmatrix} 2 & 1 & 1 \\ 1 & 2 & 1 \\ 1 & 1 & 2 \end{pmatrix}.$$

By using standard software to find eigenvalues of real symmetric matrices, we find that the eigenvalues of $\mathbf{A}^*\mathbf{A}$ and $\mathbf{B}^*\mathbf{B}$ are $(1/2)(7 \pm \sqrt{45})$ and 1, 1, 4, respectively. Hence,

$$\|\mathbf{A}\|_2 = \sqrt{\frac{1}{2}(7 + \sqrt{45})} \approx 2.6180 \qquad \text{and} \qquad \|\mathbf{B}\|_2 = \sqrt{4} = 2.$$

In the next part of this subsection, important relations between matrix norms and the eigenvalues of square matrices are discussed. These results locate the eigenvalues into finite regions which allow us to select appropriate initial approximations of the eigenvalues in applying iteration methods for finding them.

THEOREM 1.8
Let $\| \cdot \|$ be a matrix norm. If λ is an eigenvalue of matrix $\mathbf{A}$, then $|\lambda| \leq \|\mathbf{A}\|$.

PROOF   The eigenvalue equation $\mathbf{A}\mathbf{x} = \lambda\mathbf{x}$ implies that

$$|\lambda| \cdot \|\mathbf{x}\| = \|\lambda\mathbf{x}\| = \|\mathbf{A}\mathbf{x}\| \leq \|\mathbf{A}\| \cdot \|\mathbf{x}\| \,.$$

Dividing by $\|\mathbf{x}\| \neq 0$ yields the assertion.   ∎

The assertion of the theorem is illustrated in the next example.

Example 1.11

Consider the matrix

$$\mathbf{A} = \begin{pmatrix} 1 & 2 \\ 1 & 1 \end{pmatrix} \,.$$

In Example 1.10 we derived that its $p = 1, 2, \infty$, and Frobenius norms are $3, \sqrt{1/2(7 \pm \sqrt{45})}, 3$, and $\sqrt{7}$, respectively. The characteristic polynomial of $\mathbf{A}$ is

$$\varphi(\lambda) = (1 - \lambda)^2 - 2 = \lambda^2 - 2\lambda - 1 \,,$$

therefore, the eigenvalues of $\mathbf{A}$ are

$$\lambda_{1,2} = \frac{2 \pm \sqrt{8}}{2} = 1 \pm \sqrt{2} \,.$$

That is,
$$\lambda_1 = 1 - \sqrt{2} \approx -0.4142 \,,$$

and
$$\lambda_2 = 1 + \sqrt{2} \approx 2.4142 \,.$$

Note that the smallest norm of $\mathbf{A}$ equals

$$\sqrt{\frac{1}{2}(7 + \sqrt{45})} \approx 2.6180 \,,$$

which bounds the absolute values of both eigenvalues.

It is possible to restrict even further the domain $|\lambda| \leq \|\mathbf{A}\|$ for the location of the eigenvalues of $\mathbf{A}$ by the following result, which is known as the Gerschgorin theorem

THEOREM 1.9
[The Gerschgorin Theorem] For $i = 1, 2, \ldots, n$, let

$$r_i = \sum_{\substack{j=1 \\ j \neq i}}^{n} |a_{ij}| \,, \tag{1.25}$$

and let B_i denote the closed ball with center a_{ii} and radius r_i. Then all eigenvalues of $\mathbf{A}$ lie in the domain

$$D = B_1 \cup B_2 \cup \cdots \cup B_n \,. \tag{1.26}$$

PROOF   Let λ be an eigenvalue of $\mathbf{A}$ with associated eigenvector $\mathbf{x} = (x_i)$. Let i_0 be determined by the relation

$$|x_{i_0}| = \max_i |x_i| \,.$$

The eigenvalue equation of matrix $\mathbf{A}$ implies that for all i,

$$\lambda x_i = \sum_{j=1}^{n} a_{ij} x_j \,.$$

Therefore, by selecting $i = i_0$ and subtracting $a_{i_0 i_0} x_{i_0}$ from both sides,

$$(\lambda - a_{i_0 i_0}) x_{i_0} = \sum_{\substack{j=1 \\ j \neq i_0}}^{n} a_{i_0 j} x_j \,,$$

which implies that

$$|(\lambda - a_{i_0 i_0}) x_{i_0}| = \left| \sum_{\substack{j=1 \\ j \neq i_0}}^{n} a_{i_0 j} x_j \right| \leq \sum_{\substack{j=1 \\ j \neq i_0}}^{n} |a_{i_0 j}| \cdot |x_j| \,.$$

Divide by $|x_{i_0}| \neq 0$ to get

$$|\lambda - a_{i_0 i_0}| \leq \sum_{\substack{j=1 \\ j \neq i_0}}^{n} |a_{i_0 j}| \cdot \frac{|x_j|}{|x_{i_0}|} \leq \sum_{\substack{j=1 \\ j \neq i_0}}^{n} |a_{i_0 j}| = r_{i_0} \,.$$

Hence, $\lambda \in B_{i_0}$, which proves the assertion. ∎

This theorem is illustrated next.

Example 1.12

In the case of matrix $\mathbf{A}$ discussed earlier in Example 1.11, we have the domains

$$B_1 = \{\lambda \ \mid \ |\lambda - 1| \leq 2\}$$

and

$$B_2 = \{\lambda \ \mid \ |\lambda - 1| \leq 1\} .$$

Because the first disk contains the second one,

$$B_1 \cup B_2 = B_1 = \{\lambda \ \mid \ |\lambda - 1| \leq 2\} .$$

The resulting region is illustrated in Figure 1.8.

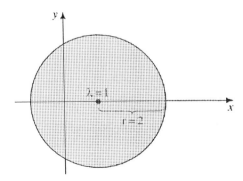

Figure 1.8 Illustration of Example 1.12.

REMARK 1.5 Since the eigenvalues of $\mathbf{A}$ and $\mathbf{A}^T$ are the same, define

$$r_j^T = \sum_{\substack{i=1 \\ i \neq j}}^{n} |a_{ij}| , \tag{1.27}$$

and let B_j^T denote the closed disk with center a_{jj} and radius r_j^T. Then all eigenvalues of $\mathbf{A}$ lie in the domain

$$D^T = B_1^T \cup B_2^T \cup \cdots \cup B_n^T. \tag{1.28}$$

∎

We close this section with an important property of real matrices, which will have important applications in later chapters.

Let $\mathbf{A}$ be an $n \times m$ real matrix. Let $R(\mathbf{A})$ denote the range space of $\mathbf{A}$:

$$R(\mathbf{A}) = \{\mathbf{y} \mid \mathbf{A}\mathbf{x} = \mathbf{y} \text{ with some } \mathbf{x}\} ,$$

and let $N(\mathbf{A}^T)$ denote the null space of $\mathbf{A}^T$:

$$N(\mathbf{A}^T) = \{\mathbf{x} \mid \mathbf{A}^T\mathbf{x} = \mathbf{0}\} .$$

THEOREM 1.10
$R(\mathbf{A})$ and $N(\mathbf{A}^T)$ *are orthogonal complementary subspaces in* $\mathbf{R}^n$, *that is:*

(i)   *If* $\mathbf{u} \in R(\mathbf{A})$ *and* $\mathbf{v} \in N(\mathbf{A}^T)$, *then* $\mathbf{u}^T\mathbf{v} = 0$.

(ii) *If for a vector* $\mathbf{v} \in \mathbf{R}^n$, $\mathbf{u}^T\mathbf{v} = 0$ *with all* $\mathbf{u} \in R(\mathbf{A})$ *then* $\mathbf{v} \in N(\mathbf{A}^T)$.

PROOF

(i)   Assume first that $\mathbf{u} \in R(\mathbf{A})$ and $\mathbf{v} \in N(\mathbf{A}^T)$. Then $\mathbf{u} = \mathbf{A}\mathbf{x}$ with some $\mathbf{x}$, and $\mathbf{A}^T\mathbf{v} = \mathbf{0}$. Therefore,

$$\mathbf{u}^T\mathbf{v} = (\mathbf{x}^T\mathbf{A}^T)\mathbf{v} = \mathbf{x}^T(\mathbf{A}^T\mathbf{v}) = \mathbf{x}^T\mathbf{0} = 0 .$$

(ii) Assume next that for a vector $\mathbf{v}$, $\mathbf{u}^T\mathbf{v} = 0$ with all $\mathbf{u} \in R(\mathbf{A})$. Note that $\mathbf{A}\mathbf{x} \in R(\mathbf{A})$ with $\mathbf{x} = \mathbf{A}^T\mathbf{v}$; therefore we may select $\mathbf{u} = \mathbf{A}\mathbf{x} = \mathbf{A}\mathbf{A}^T\mathbf{v}$. Hence,

$$0 = \mathbf{u}^T\mathbf{v} = \mathbf{v}^T\mathbf{A}\mathbf{A}^T\mathbf{v} = (\mathbf{A}^T\mathbf{v})^T(\mathbf{A}^T\mathbf{v}) = \|\mathbf{A}^T\mathbf{v}\|_2^2 ,$$

which implies that $\mathbf{A}^T\mathbf{v} = \mathbf{0}$, that is, $\mathbf{v} \in N(\mathbf{A}^T)$.

∎

COROLLARY 1.2
Arbitrary $\mathbf{x} \in \mathbf{R}^n$ *can be uniquely represented as* $\mathbf{x} = \mathbf{u} + \mathbf{v}$, *where* $\mathbf{u} \in R(\mathbf{A})$ *and* $\mathbf{v} \in N(\mathbf{A}^T)$.

PROOF   Let $\mathbf{u}_1, \mathbf{u}_2, \ldots, \mathbf{u}_k$ be an orthogonal basis in $R(\mathbf{A})$, and extend it to an orthogonal basis $\mathbf{u}_1, \mathbf{u}_2, \ldots, \mathbf{u}_k, \mathbf{v}_1, \mathbf{v}_2, \ldots, \mathbf{v}_{n-k}$ of $\mathbf{R}^n$. Then $\mathbf{v}_1, \mathbf{v}_2, \ldots, \mathbf{v}_{n-k}$ is a basis of $N(\mathbf{A}^T)$. Therefore, all $\mathbf{x} \in \mathbf{R}^n$ can be represented as

$$\mathbf{x} = \alpha_1 \mathbf{u}_1 + \cdots + \alpha_k \mathbf{u}_k + \beta_1 \mathbf{v}_1 + \cdots + \beta_{n-k} \mathbf{v}_{n-k} \; .$$

Select

$$\mathbf{u} = \alpha_1 \mathbf{u}_1 + \cdots + \alpha_k \mathbf{u}_k \qquad \text{and} \qquad \mathbf{v} = \beta_1 \mathbf{v}_1 + \cdots + \beta_{n-k} \mathbf{v}_{n-k}$$

to obtain the desired representation.

Assume next that there are two representations:

$$\mathbf{x} = \mathbf{u} + \mathbf{v} = \tilde{\mathbf{u}} + \tilde{\mathbf{v}} \; .$$

Then

$$\mathbf{u} - \tilde{\mathbf{u}} = \tilde{\mathbf{v}} - \mathbf{v} \; ,$$

where $\mathbf{u} - \tilde{\mathbf{u}} \in R(\mathbf{A})$ and $\tilde{\mathbf{v}} - \mathbf{v} \in N(\mathbf{A}^T)$. Because these vectors are orthogonal,

$$\|\mathbf{u} - \tilde{\mathbf{u}}\|_2^2 = (\mathbf{u} - \tilde{\mathbf{u}})^T (\mathbf{u} - \tilde{\mathbf{u}}) = (\tilde{\mathbf{v}} - \mathbf{v})^T (\mathbf{u} - \tilde{\mathbf{u}}) = 0 \; .$$

Therefore, $\mathbf{u} = \tilde{\mathbf{u}}$ and $\tilde{\mathbf{v}} = \mathbf{v}$, which completes the proof.   ∎

COROLLARY 1.3
In the previous decomposition, $\|\mathbf{x}\|_2^2 = \|\mathbf{u}\|_2^2 + \|\mathbf{v}\|_2^2$.

PROOF   Since $\mathbf{u}$ and $\mathbf{v}$ are orthogonal,

$$\|\mathbf{x}\|_2^2 = \|\mathbf{u} + \mathbf{v}\|_2^2 = (\mathbf{u} + \mathbf{v})^T (\mathbf{u} + \mathbf{v})$$

$$= \mathbf{u}^T \mathbf{u} + \mathbf{u}^T \mathbf{v} + \mathbf{v}^T \mathbf{u} + \mathbf{v}^T \mathbf{v} = \mathbf{u}^T \mathbf{u} + \mathbf{v}^T \mathbf{v} = \|\mathbf{u}\|_2^2 + \|\mathbf{v}\|_2^2 \; .$$

∎

REMARK 1.6   The above corollary is also known as the theorem of Pythagoras in n-dimensional Euclidean spaces.   ∎

1.3.2   Special Matrix Forms

In the discussion of linear systems, special matrix transformations and certain canonical matrix forms are often applied. They are the subjects of this subsection.

Assume first that an $n \times n$ matrix has n distinct eigenvalues, $\lambda_1, \lambda_2, \ldots, \lambda_n$. Let the associated eigenvectors be denoted by $\mathbf{x}_1, \mathbf{x}_2, \ldots, \mathbf{x}_n$. We remind the reader that a scalar λ and vector $\mathbf{x} \neq \mathbf{0}$ are an eigenvalue and an associated eigenvector of a matrix $\mathbf{A}$, if $\mathbf{A}\mathbf{x} = \lambda\mathbf{x}$. First we prove that these eigenvectors are linearly independent. Assume that there are constants $c_1, c_2, \ldots, c_n$ such that

$$c_1\mathbf{x}_1 + c_2\mathbf{x}_2 + \cdots + c_n\mathbf{x}_n = \mathbf{0}$$

and at least one $c_i \neq 0$. Multiply this equation first by λ_1 and then by matrix $\mathbf{A}$ to get equalities

$$\lambda_1 c_1\mathbf{x}_1 + \lambda_1 c_2\mathbf{x}_2 + \cdots + \lambda_1 c_n\mathbf{x}_n = \mathbf{0}$$

and

$$\lambda_1 c_1\mathbf{x}_1 + \lambda_2 c_2\mathbf{x}_2 + \cdots + \lambda_n c_n\mathbf{x}_n = \mathbf{0} \ ,$$

where we used the fact that $\mathbf{A}\mathbf{x}_k = \lambda_k\mathbf{x}_k$ for $k = 1, 2, \ldots, n$. Subtracting the second equation from the first one we have

$$(\lambda_1 - \lambda_2)c_2\mathbf{x}_2 + \cdots + (\lambda_1 - \lambda_n)c_n\mathbf{x}_n = \mathbf{0} \ .$$

Therefore, we conclude that if vectors $\mathbf{x}_1, \mathbf{x}_2, \ldots, \mathbf{x}_n$ are linearly dependent, then by dropping $\mathbf{x}_1$, the remaining vectors $\mathbf{x}_2, \ldots, \mathbf{x}_n$ are still linearly dependent. If we continue this idea sequentially dropping vectors $\mathbf{x}_2, \ldots, \mathbf{x}_{n-1}$, we will conclude that vector $\mathbf{x}_n$ itself forms a linearly dependent set of vectors. Since $\mathbf{x}_n \neq \mathbf{0}$ (being an eigenvector), this is a contradiction.

This observation implies the following important theorem.

THEOREM 1.11
Assume that the eigenvalues λ_i of the $n \times n$ matrix $\mathbf{A}$ are distinct. Then there is a nonsingular matrix $\mathbf{T}$ such that

$$\mathbf{TAT}^{-1} = diag(\lambda_1, \ldots, \lambda_n) \ , \qquad (1.29)$$

where this notation means a diagonal matrix with diagonal elements $\lambda_1, \ldots, \lambda_n$.

PROOF   Equations

$$\mathbf{A}\mathbf{x}_i = \lambda_i \mathbf{x}_i \qquad (i = 1, 2, \ldots, n)$$

can be summarized as

$$\mathbf{A} \cdot (\mathbf{x}_1, \ldots, \mathbf{x}_n) = (\mathbf{x}_1, \ldots, \mathbf{x}_n) \cdot diag(\lambda_1, \ldots, \lambda_n) ,$$

where $(\mathbf{x}_1, \ldots, \mathbf{x}_n)$ denotes the n-column matrix with column vectors $\mathbf{x}_1, \ldots, \mathbf{x}_n$. Define $\mathbf{T} = (\mathbf{x}_1, \ldots, \mathbf{x}_n)^{-1}$, which exists, since columns $\mathbf{x}_1, \ldots, \mathbf{x}_n$ are linearly independent. Premultiply the above equation by $\mathbf{T}$ to get relation (1.29).   ∎

REMARK 1.7   The proof of the theorem suggests the following diagonalization algorithm:

Step 1   Find the eigenvalues λ_i and associated eigenvectors $\mathbf{x}_i$ of matrix $\mathbf{A}$.

Step 2   Form matrix $(\mathbf{x}_1, \mathbf{x}_2, \ldots, \mathbf{x}_n)$.

Step 3   Invert this matrix to obtain $\mathbf{T}$.

For finding the eigenvalues and eigenvectors of matrix $\mathbf{A}$ and for inverting matrix $\mathbf{T}$, standard computer programs are available.   ∎

REMARK 1.8   The assertion remains true even in the slightly more general case, when the distinct eigenvalues $\lambda_1, \ldots, \lambda_r$ have multiplicities $m_1, \ldots, m_r$, and for each i, there are m_i linearly independent eigenvectors associated to λ_i. However, in the general case, the theorem does not hold, but the matrix can be transformed into a *Jordan canonical* form. That is, there exists a nonsingular matrix $\mathbf{T}$ such that

$$\mathbf{T}\mathbf{A}\mathbf{T}^{-1} = \begin{pmatrix} \mathbf{J}_1 & & & O \\ & \mathbf{J}_2 & & \\ & & \ddots & \\ O & & & \mathbf{J}_s \end{pmatrix} \tag{1.30}$$

with $s \geq r$, and for $j = 1, 2, \ldots, s$,

$$\mathbf{J}_j = \begin{pmatrix} \lambda_i & 1 & & & O \\ & \lambda_i & 1 & & \\ & & \ddots & \ddots & \\ & & & \lambda_i & 1 \\ O & & & & \lambda_i \end{pmatrix}. \tag{1.31}$$

Note that the order of $\mathbf{J}_j$ is not greater than m_i, and in each Jordan block $\mathbf{J}_j$ the same eigenvalue forms the diagonal; however, the same eigenvalue can be found simultaneously in different Jordan blocks.

A Jordan canonical form with 2×2, 3×3, 2×2, and 1×1 blocks is illustrated in Figure 1.9, where all elements not indicated are equal to zero. ∎

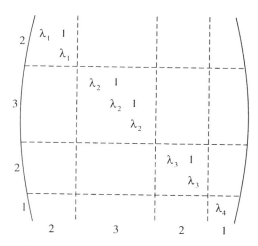

Figure 1.9   A special Jordan canonical form.

In general, matrix transformations $\mathbf{TAT}^{-1}$ with nonsingular matrices $\mathbf{T}$ are called *similarity transformations*, and matrices $\mathbf{A}$ and $\mathbf{TAT}^{-1}$ are called *similar*. It is well known from linear algebra that the characteristic polynomials of similar matrices are the same, therefore they have the same eigenvalues.

An important consequence of the above theorem is known as the Cayley-Hamilton theorem:

THEOREM 1.12

[The Cayley–Hamilton Theorem] Let $\mathbf{A}$ *be an* $n \times n$ *real or complex matrix, and let* φ *denote its characteristic polynomial. Then*

$$\varphi(\mathbf{A}) = \mathbf{O} \,,$$

where $\mathbf{O}$ *is an* $n \times n$ *matrix with all elements equal to zero.*

PROOF   Because the characteristic polynomials and eigenvalues of $n \times n$ matrices depend continuously on the matrix elements, it is sufficient to show that the assertion holds for matrices with distinct eigenvalues. In this case (1.29) implies that

$$\mathbf{A} = \mathbf{T}^{-1} \cdot diag(\lambda_1, \ldots, \lambda_n)\mathbf{T} \,.$$

Therefore, for all $j \geq 1$,

$$\mathbf{A}^j = (\mathbf{T}^{-1} \cdot diag(\lambda_1, \ldots, \lambda_n) \cdot \mathbf{T})(\mathbf{T}^{-1} \cdot diag(\lambda_1, \ldots, \lambda_n) \cdot \mathbf{T})$$

$$\cdots \cdot (\mathbf{T}^{-1} \cdot diag(\lambda_1, \ldots, \lambda_n) \cdot \mathbf{T})$$

$$= \mathbf{T}^{-1} \cdot [diag(\lambda_1, \ldots, \lambda_n)]^j \cdot \mathbf{T} = \mathbf{T}^{-1} \cdot diag(\lambda_1^j, \ldots, \lambda_n^j) \cdot \mathbf{T} \,,$$

which implies that for all finite polynomials

$$p(\lambda) = a_0 + a_1\lambda + a_2\lambda^2 + \cdots + a_k\lambda^k \,,$$

$$p(\mathbf{A}) = \sum_{j=0}^{k} a_j\mathbf{T}^{-1} \cdot diag(\lambda_1^j, \ldots, \lambda_n^j) \cdot \mathbf{T}$$

$$= \mathbf{T}^{-1}\left[\sum_{j=0}^{k} a_j\, diag(\lambda_1^j, \ldots, \lambda_n^j)\right]\mathbf{T}$$

$$= \mathbf{T}^{-1} \cdot diag(p(\lambda_1), \ldots, p(\lambda_n))\mathbf{T} \,.$$

If $p = \varphi$, then $p(\lambda_i) = 0$ $(i = 1, 2, \ldots, n)$, hence $\varphi(\mathbf{A}) = \mathbf{O}$.   ∎

Example 1.13

Consider matrix

$$A = \begin{pmatrix} 0 & \omega \\ -\omega & 0 \end{pmatrix},$$

where $\omega > 0$ is a given constant. Note this matrix has an important role in analyzing harmonic motions (see Example 3.3). The diagonal form of A will now be determined. The characteristic polynomial of A is as follows:

$$\varphi(\lambda) = \det \begin{pmatrix} -\lambda & \omega \\ -\omega & -\lambda \end{pmatrix} = \lambda^2 + \omega^2,$$

where $\det(\cdot)$ denotes the determinants of matrices. Therefore, the eigenvalues are $\lambda_1 = j\omega$, $\lambda_2 = -j\omega$, where $j = \sqrt{-1}$. The eigenvector associated with λ_1 is the solution of the homogeneous linear equation

$$(A - \lambda_1 I)x_1 = \begin{pmatrix} -j\omega & \omega \\ -\omega & -j\omega \end{pmatrix} \begin{pmatrix} x_{11} \\ x_{12} \end{pmatrix} = \begin{pmatrix} 0 \\ 0 \end{pmatrix}.$$

Select $x_{11} = 1$, then $x_{12} = j$. Therefore, $x_1 = (1, j)^T$. One can similarly verify that $x_2 = (1, -j)^T$ is an eigenvector associated with λ_2. Since the eigenvalues are different, Equation (1.29) can be applied, which implies that with $T = (x_1, x_2)^{-1}$,

$$A = T^{-1} \cdot diag(\lambda_1, \lambda_2) \cdot T = \begin{pmatrix} 1 & 1 \\ j & -j \end{pmatrix} \begin{pmatrix} j\omega & 0 \\ 0 & -j\omega \end{pmatrix} \begin{pmatrix} \frac{1}{2} & -\frac{j}{2} \\ \frac{1}{2} & \frac{j}{2} \end{pmatrix}.$$

In this case, the Cayley–Hamilton theorem can be easily demonstrated because

$$\varphi(A) = A^2 + w^2 I = \begin{pmatrix} -\omega^2 & 0 \\ 0 & -\omega^2 \end{pmatrix} + \begin{pmatrix} \omega^2 & 0 \\ 0 & \omega^2 \end{pmatrix} = O.$$

Note that the construction of the diagonal and Jordan canonical forms of square matrices require the computation of the eigenvalues and the eigenvectors. A summary of such methods is presented for example in [42] and in [43].

In the diagonal transformation (1.29) and the Jordan form (1.30), matrix T is nonsingular. In many cases, transformations with more special matrices are important. Such a special case is given next.

THEOREM 1.13

Let $\mathbf{A}$ be an $n \times n$ real or complex matrix. Then there exists a (unitary) matrix $\mathbf{U}$ such that $\mathbf{U}^{-1} = \mathbf{U}^*$, and $\mathbf{U}\mathbf{A}\mathbf{U}^*$ is upper triangular.

PROOF   Finite induction is used for n. If $n = 1$, then $\mathbf{A}$ is a scalar, and therefore $\mathbf{U} = 1$ is satisfactory. Assume next that the assertion holds for $n - 1$. Let $\mathbf{A}$ be an $n \times n$ matrix, and let λ_1 be an eigenvalue of $\mathbf{A}$ with associated eigenvector $\mathbf{x}_1$. Select vectors $\mathbf{x}_2, \ldots, \mathbf{x}_n$ in such a way that $\{\mathbf{x}_1, \mathbf{x}_2, \ldots, \mathbf{x}_n\}$ is an orthonormal system, and define matrix $\mathbf{U}_1 = (\mathbf{x}_1, \mathbf{x}_2, \ldots, \mathbf{x}_n)^*$. Obviously $\mathbf{U}_1$ is a unitary matrix, since

$$\mathbf{U}_1 \mathbf{U}_1^* = \begin{pmatrix} \mathbf{x}_1^* \\ \vdots \\ \mathbf{x}_n^* \end{pmatrix} (\mathbf{x}_1, \ldots, \mathbf{x}_n) = \begin{pmatrix} \mathbf{x}_1^* \mathbf{x}_1 & \ldots & \mathbf{x}_1^* \mathbf{x}_n \\ \vdots & & \vdots \\ \mathbf{x}_n^* \mathbf{x}_1 & \ldots & \mathbf{x}_n^* \mathbf{x}_n \end{pmatrix} = \mathbf{I} \,,$$

where the diagonal elements equal 1 and the off-diagonal elements are all equal to zero. Observe that

$$\mathbf{U}_1 \mathbf{A} \mathbf{U}_1^* = \mathbf{U}_1 \mathbf{A}(\mathbf{x}_1, \mathbf{x}_2, \ldots, \mathbf{x}_n)$$

$$= \mathbf{U}_1(\mathbf{A}\mathbf{x}_1, \mathbf{A}\mathbf{x}_2, \ldots, \mathbf{A}\mathbf{x}_n)$$

$$= \mathbf{U}_1(\lambda_1 \mathbf{x}_1, \mathbf{A}\mathbf{x}_2, \ldots, \mathbf{A}\mathbf{x}_n)$$

$$= \begin{pmatrix} \mathbf{x}_1^* \\ \vdots \\ \mathbf{x}_n^* \end{pmatrix} (\lambda_1 \mathbf{x}_1, \mathbf{A}\mathbf{x}_2, \ldots, \mathbf{A}\mathbf{x}_n) \,.$$

The elements in the first column of the product are $\lambda_1 \mathbf{x}_1^* \mathbf{x}_1 = \lambda_1$, $\lambda_1 \mathbf{x}_2^* \mathbf{x}_1 = 0, \ldots, \lambda_1 \mathbf{x}_n^* \mathbf{x}_1 = 0$. Therefore, the first column is $(\lambda_1, 0, \ldots, 0)^T$, and so

$$\mathbf{U}_1 \mathbf{A} \mathbf{U}_1^* = \begin{pmatrix} \lambda_1 & \mathbf{a}^T \\ \mathbf{0} & \mathbf{A}_1 \end{pmatrix} \,,$$

where $\mathbf{a}^T$ is an $(n-1)$-element row vector, $\mathbf{0}$ is the $(n-1)$-element zero column vector, and $\mathbf{A}_1$ is an $(n-1) \times (n-1)$ matrix. By the inductive hypothesis, there exists an $(n-1)$-order unitary matrix $\mathbf{U}_2$ such that $\mathbf{U}_2 \mathbf{A}_1 \mathbf{U}_2^*$ is upper triangular. Define

$$\mathbf{U}_3 = \begin{pmatrix} 1 & \mathbf{0}^T \\ \mathbf{0} & \mathbf{U}_2 \end{pmatrix} \,,$$

where $\mathbf{0}^T$ is the $(n-1)$-dimensional zero row vector. Note that $\mathbf{U}_3$ is also unitary, since

$$\mathbf{U}_3\mathbf{U}_3^* = \begin{pmatrix} 1 & \mathbf{0}^T \\ \mathbf{0} & \mathbf{U}_2 \end{pmatrix} \begin{pmatrix} 1 & \mathbf{0}^T \\ \mathbf{0} & \mathbf{U}_2^* \end{pmatrix}$$

$$= \begin{pmatrix} 1 & \mathbf{0}^T \\ \mathbf{0} & \mathbf{U}_2\mathbf{U}_2^* \end{pmatrix}$$

$$= \begin{pmatrix} 1 & \mathbf{0}^T \\ \mathbf{0} & \mathbf{I} \end{pmatrix} = \mathbf{I}\,.$$

Finally we show that $\mathbf{U} = \mathbf{U}_3\mathbf{U}_1$ satisfies the assertion of the theorem. Note first that

$$\mathbf{U}\mathbf{U}^* = \mathbf{U}_3\mathbf{U}_1\mathbf{U}_1^*\mathbf{U}_3^* = \mathbf{U}_3\mathbf{I}\mathbf{U}_3^* = \mathbf{U}_3\mathbf{U}_3^* = \mathbf{I}\,,$$

that is, $\mathbf{U}$ is unitary. Furthermore,

$$\mathbf{U}\mathbf{A}\mathbf{U}^* = \mathbf{U}_3\mathbf{U}_1\mathbf{A}\mathbf{U}_1^*\mathbf{U}_3^*$$

$$= \begin{pmatrix} 1 & \mathbf{0}^T \\ \mathbf{0} & \mathbf{U}_2 \end{pmatrix} \begin{pmatrix} \lambda_1 & \mathbf{a}^T \\ \mathbf{0} & \mathbf{A}_1 \end{pmatrix} \begin{pmatrix} 1 & \mathbf{0}^T \\ \mathbf{0} & \mathbf{U}_2^* \end{pmatrix}$$

$$= \begin{pmatrix} \lambda_1 & \mathbf{a}^T\mathbf{U}_2^* \\ \mathbf{0} & \mathbf{U}_2\mathbf{A}_1\mathbf{U}_2^* \end{pmatrix}\,,$$

which is upper triangular.   ∎

In the special case, when $\mathbf{A}$ is real and symmetric, a much stronger result holds.

THEOREM 1.14

Assume that $\mathbf{A}$ is an $n \times n$ real symmetric matrix, then there exists a real orthogonal matrix $\mathbf{Q}$ such that

$$\mathbf{Q}^T = \mathbf{Q}^{-1} \quad \text{and} \quad \mathbf{Q}\mathbf{A}\mathbf{Q}^T = diag(\lambda_1, \ldots, \lambda_n)\,, \tag{1.32}$$

where $\lambda_1, \lambda_2, \ldots, \lambda_n$ are the eigenvalues of $\mathbf{A}$.

PROOF   First we show that all eigenvalues of $\mathbf{A}$ are real. Let λ_1 be an eigenvalue of $\mathbf{A}$ with associated eigenvector $\mathbf{x}_1$, then

$$\mathbf{A}\mathbf{x}_1 = \lambda_1 \mathbf{x}_1 .$$

Premultiplying this equation by $\mathbf{x}_1^*$ and dividing by $\mathbf{x}_1^*\mathbf{x}_1 \neq 0$ yields the relation

$$\lambda_1 = \frac{\mathbf{x}_1^*\mathbf{A}\mathbf{x}_1}{\mathbf{x}_1^*\mathbf{x}_1} .$$

Observe that both the numerator and the denominator are real, since

$$\overline{\mathbf{x}_1^*\mathbf{A}\mathbf{x}_1} = (\mathbf{x}_1^*\mathbf{A}\mathbf{x}_1)^* = \mathbf{x}_1^*\mathbf{A}\mathbf{x}_1$$

and

$$\overline{\mathbf{x}_1^*\mathbf{x}_1} = (\mathbf{x}_1^*\mathbf{x}_1)^* = \mathbf{x}_1^*\mathbf{x}_1 .$$

Hence λ_1 is real, and therefore the associated eigenvector can also be selected as a real vector because the homogeneous equation $(\mathbf{A} - \lambda_1 \mathbf{I})\mathbf{x} = \mathbf{0}$ has real coefficients.

The construction presented in the proof of the previous theorem implies that $\mathbf{U}$ can be selected as real.

We prove finally that the selection $\mathbf{Q} = \mathbf{U}$ satisfies the assertion. Because $\mathbf{Q}$ is real, $\mathbf{Q}^* = \mathbf{Q}^T$. By denoting the upper triangular matrix $\mathbf{Q}\mathbf{A}\mathbf{Q}^T$ by $\mathbf{A}_1$,

$$\mathbf{A}_1^T = (\mathbf{Q}\mathbf{A}\mathbf{Q}^T)^T = \mathbf{Q}\mathbf{A}^T\mathbf{Q}^T = \mathbf{Q}\mathbf{A}\mathbf{Q}^T = \mathbf{A}_1 .$$

Hence $\mathbf{A}_1$ is diagonal, which completes the proof.   ∎

REMARK 1.9   It is known that the eigenvectors $\mathbf{x}_1, \ldots, \mathbf{x}_n$ of a real symmetric $n \times n$ matrix $\mathbf{A}$ can be selected as an orthonormal system. That is, $\mathbf{x}_i^T\mathbf{x}_i = 1$ and $\mathbf{x}_i^T\mathbf{x}_j = 0$ $(i \neq j; i, j = 1, \ldots, n)$. Then we may select $\mathbf{Q} = (\mathbf{x}_1, \ldots, \mathbf{x}_n)^T$. The eigenvalues and eigenvectors can be determined by using standard computer packages.   ∎

COROLLARY 1.4

Assume that $\mathbf{A}$ is a real, symmetric, and positive semidefinite matrix, that is, for arbitrary vector $\mathbf{v}$, $\mathbf{v}^T\mathbf{A}\mathbf{v} \geq 0$. Then there exists a nonsingular matrix $\mathbf{T}$ such that

$$\mathbf{A} = \mathbf{T}^T diag(1, \ldots, 1, 0, \ldots, 0)\mathbf{T} . \tag{1.33}$$

PROOF   Let $\lambda_1, \ldots, \lambda_k$ denote the nonzero eigenvalues of $\mathbf{A}$, then $\lambda_i > 0$ $(i = 1, 2, \ldots, k)$ and

$$\mathbf{A} = \mathbf{Q}^T \cdot diag(\lambda_1, \ldots, \lambda_n) \cdot \mathbf{Q} = \mathbf{Q}^T \cdot diag(\lambda_1, \ldots, \lambda_k, 0, \ldots, 0)\mathbf{Q} \, .$$

Observe that the second factor can be rewritten as

$$diag\left(\sqrt{\lambda_1}, \ldots, \sqrt{\lambda_k}, 1, \ldots, 1\right) \cdot diag(1, \ldots, 1, 0, \ldots, 0)$$

$$\cdot \, diag\left(\sqrt{\lambda_1}, \ldots, \sqrt{\lambda_k}, 1, \ldots, 1\right) \, ;$$

therefore the selection

$$\mathbf{T} = diag\left(\sqrt{\lambda_1}, \ldots, \sqrt{\lambda_k}, 1, \ldots, 1\right)\mathbf{Q}$$

satisfies the assertion.   ∎

Example 1.14

Decomposition (1.33) will now be constructed for matrix

$$\mathbf{A} = \begin{pmatrix} 1 & 1 \\ 1 & 1 \end{pmatrix} .$$

First the method suggested in the proof of the theorem is illustrated.

Method 1. The characteristic polynomial of $\mathbf{A}$ is

$$\varphi(\lambda) = \det \begin{pmatrix} 1 - \lambda & 1 \\ 1 & 1 - \lambda \end{pmatrix} = (1 - \lambda)^2 - 1 = \lambda^2 - 2\lambda \, ;$$

therefore $\lambda_1 = 2$, $\lambda_2 = 0$. Similar to the previous example, one may easily verify that the normalized eigenvectors are

$$\mathbf{x}_1 = \frac{1}{\sqrt{2}}(1, 1)^T \quad \text{and} \quad \mathbf{x}_2 = \frac{1}{\sqrt{2}}(1, -1)^T \, ,$$

which implies that

$$\mathbf{Q} = (\mathbf{x}_1, \mathbf{x}_2)^T = \frac{1}{\sqrt{2}} \begin{pmatrix} 1 & 1 \\ 1 & -1 \end{pmatrix} .$$

From (1.32) we conclude that

$$\mathbf{A} = \mathbf{Q}^T diag(\lambda_1, \ldots, \lambda_n)\mathbf{Q}$$

$$= \frac{1}{\sqrt{2}} \begin{pmatrix} 1 & 1 \\ 1 & -1 \end{pmatrix} \begin{pmatrix} 2 & 0 \\ 0 & 0 \end{pmatrix} \frac{1}{\sqrt{2}} \begin{pmatrix} 1 & 1 \\ 1 & -1 \end{pmatrix}$$

$$= \begin{pmatrix} \frac{1}{\sqrt{2}} & \frac{1}{\sqrt{2}} \\ \frac{1}{\sqrt{2}} & -\frac{1}{\sqrt{2}} \end{pmatrix} \begin{pmatrix} 2 & 0 \\ 0 & 0 \end{pmatrix} \begin{pmatrix} \frac{1}{\sqrt{2}} & \frac{1}{\sqrt{2}} \\ \frac{1}{\sqrt{2}} & -\frac{1}{\sqrt{2}} \end{pmatrix} .$$

The factored form (1.33) of $\mathbf{A}$ can be obtained in the same way as shown in the proof:

$$\mathbf{A} = \begin{pmatrix} \frac{1}{\sqrt{2}} & \frac{1}{\sqrt{2}} \\ \frac{1}{\sqrt{2}} & -\frac{1}{\sqrt{2}} \end{pmatrix} \begin{pmatrix} \sqrt{2} & 0 \\ 0 & 1 \end{pmatrix} \begin{pmatrix} 1 & 0 \\ 0 & 0 \end{pmatrix} \begin{pmatrix} \sqrt{2} & 0 \\ 0 & 1 \end{pmatrix} \begin{pmatrix} \frac{1}{\sqrt{2}} & \frac{1}{\sqrt{2}} \\ \frac{1}{\sqrt{2}} & -\frac{1}{\sqrt{2}} \end{pmatrix}$$

$$= \begin{pmatrix} 1 & \frac{1}{\sqrt{2}} \\ 1 & -\frac{1}{\sqrt{2}} \end{pmatrix} \begin{pmatrix} 1 & 0 \\ 0 & 0 \end{pmatrix} \begin{pmatrix} 1 & 1 \\ \frac{1}{\sqrt{2}} & -\frac{1}{\sqrt{2}} \end{pmatrix} .$$

A special method is illustrated next.

Method 2. For matrices of small order, the computation of the decomposition is reduced to a system of nonlinear equations that can be easily solved in many cases. Find matrix $\mathbf{T}$ as

$$\mathbf{T} = \begin{pmatrix} a & b \\ c & d \end{pmatrix} ,$$

where a, b, c, and d are considered to be unknown. Since $rank(\mathbf{A}) = 1$, the decomposition has the form

$$\mathbf{A} = \begin{pmatrix} 1 & 1 \\ 1 & 1 \end{pmatrix} = \begin{pmatrix} a & c \\ b & d \end{pmatrix} \begin{pmatrix} 1 & 0 \\ 0 & 0 \end{pmatrix} \begin{pmatrix} a & b \\ c & d \end{pmatrix} .$$

Compare the corresponding components on the two sides to get the system of equations

$$a^2 = 1$$

$$ab = 1$$

$$b^2 = 1 .$$

For example, $a = b = 1$ is a solution. The second row of $\mathbf{T}$ is arbitrary. Select for example $c = 1$, $d = -1$ to guarantee that $\mathbf{T}$ is nonsingular.

Hence we obtain the decomposition

$$\begin{pmatrix} 1 & 1 \\ 1 & 1 \end{pmatrix} = \begin{pmatrix} 1 & 1 \\ 1 & -1 \end{pmatrix} \begin{pmatrix} 1 & 0 \\ 0 & 0 \end{pmatrix} \begin{pmatrix} 1 & 1 \\ 1 & -1 \end{pmatrix} .$$

Note that this result differs from the previous decomposition; however, simple calculation shows that it also satisfies Equation (1.33).

Our next result is a special factorization of real (or complex) matrices.

THEOREM 1.15
Let $\mathbf{A}$ be an $m \times n$ matrix with $rank(\mathbf{A}) = r \ (\leq \min\{m, n\})$. Then there exist matrices $\mathbf{A}_1$ and $\mathbf{B}_1$, such that $\mathbf{A}_1$ has r columns and $\mathbf{B}_1$ has r rows; furthermore,

$$\mathbf{A} = \mathbf{A}_1 \mathbf{B}_1 . \tag{1.34}$$

PROOF   Let $\mathbf{A} = (\mathbf{a}_1, \mathbf{a}_2, \dots, \mathbf{a}_n)$ and assume that $\mathbf{a}_{i_1}, \mathbf{a}_{i_2}, \dots, \mathbf{a}_{i_r}$ form a basis of the column space of $\mathbf{A}$. Then for all $j = 1, 2, \dots, n$,

$$\mathbf{a}_j = \sum_{k=1}^{r} \alpha_{kj} \cdot \mathbf{a}_{i_k}$$

with some constants α_{kj}, which implies that

$$\mathbf{A} = \left(\sum_{k=1}^{r} \alpha_{k1} \cdot \mathbf{a}_{i_k}, \dots, \sum_{k=1}^{r} \alpha_{kn} \cdot \mathbf{a}_{i_k} \right)$$

$$= (\mathbf{a}_{i_1}, \dots, \mathbf{a}_{i_r}) \cdot \begin{pmatrix} \alpha_{11} & \cdots & \alpha_{1n} \\ \vdots & & \vdots \\ \alpha_{r1} & \cdots & \alpha_{rn} \end{pmatrix} .$$

Hence the selection

$$\mathbf{A}_1 = (\mathbf{a}_{i_1}, \dots, \mathbf{a}_{i_r}), \qquad \mathbf{B}_1 = (\alpha_{kj})$$

satisfies the assertion.   ∎

Example 1.15

Consider again matrix

$$\mathbf{A} = \begin{pmatrix} 1 & 1 \\ 1 & 1 \end{pmatrix} .$$

Because the first column gives the basis of the column space,

$$\mathbf{A} = (1 \cdot \mathbf{a}_1, 1 \cdot \mathbf{a}_1) = \mathbf{a}_1 \cdot (1 , 1) = \begin{pmatrix} 1 \\ 1 \end{pmatrix} (1 , 1) \ .$$

Therefore, we may select

$$\mathbf{A}_1 = \begin{pmatrix} 1 \\ 1 \end{pmatrix} \quad \text{and} \quad \mathbf{B}_1 = (1 , 1) \ .$$

1.3.3   Matrix functions

In obtaining and analyzing the solutions of linear difference and differential equations, which govern the state transitions of dynamic linear systems, special functions of real matrices have an important role. In particular, the computation of matrix powers and matrix exponentials are used in such investigations. This subsection is devoted to defining and examining matrix functions.

Assume that the complex power series

$$f(z) = a_0 + a_1 z + a_2 z^2 + \cdots + a_k z^k + \cdots$$

is convergent for $|z| < R$.

DEFINITION 1.12   *Let $\mathbf{A}$ be a square matrix, then $f(\mathbf{A})$ is defined as the sum of the series*

$$a_0 \mathbf{I} + a_1 \mathbf{A} + a_2 \mathbf{A}^2 + \cdots + a_k \mathbf{A}^k + \cdots \qquad (1.35)$$

if this matrix series converges.

Example 1.16

Assume that $\mathbf{A}^N = \mathbf{0}$ with some N. Then for all $k \geq N$, $\mathbf{A}^k = \mathbf{O}$. Therefore $f(\mathbf{A})$ exists and

$$f(\mathbf{A}) = a_0 \mathbf{I} + a_1 \mathbf{A} + a_2 \mathbf{A}^2 + \cdots + a_{N-1} \mathbf{A}^{N-1} \ . \qquad (1.36)$$

Hence, $f(\mathbf{A})$ is given in a finite form.

Example 1.17

Assume that $R > 1$ and $\mathbf{A}^2 = \mathbf{A}$. Then for all $k \geq 2$, $\mathbf{A}^k = \mathbf{A}$, and

$$f(\mathbf{A}) = a_0\mathbf{I} + \sum_{k=1}^{\infty} a_k\mathbf{A}^k = a_0\mathbf{I} + \left(\sum_{k=1}^{\infty} a_k\right) \cdot \mathbf{A} = a_0\mathbf{I} + (f(1) - a_0)\mathbf{A}.$$

$$(1.37)$$

This formula is easy to compute.

Example 1.18

Assume that $\mathbf{A} = diag(\lambda_1, \ldots, \lambda_n)$ with $|\lambda_i| < R$ $(i = 1, 2, \ldots, n)$. Then

$$f(\mathbf{A}) = \sum_{k=0}^{\infty} a_k\, diag(\lambda_1^k, \ldots, \lambda_n^k)$$

$$= diag\left(\sum_{k=0}^{\infty} a_k\lambda_1^k, \ldots, \sum_{k=0}^{\infty} a_k\lambda_n^k\right)$$

$$= diag(f(\lambda_1), \ldots, f(\lambda_n)).$$

$$(1.38)$$

Hence, a closed form representation is obtained again.

Example 1.19

Let $\mathbf{A}$ be a ν-order Jordan block

$$\mathbf{A} = \begin{pmatrix} \lambda & 1 & & & O \\ & \lambda & 1 & & \\ & & \ddots & \ddots & \\ & & & \ddots & 1 \\ O & & & & \lambda \end{pmatrix}$$

with $|\lambda| < R$. Introduce matrix

$$\mathbf{N} = \begin{pmatrix} 0 & 1 & & & O \\ & 0 & 1 & & \\ & & \ddots & \ddots & \\ & & & \ddots & 1 \\ O & & & & 0 \end{pmatrix}$$

to have
$$\mathbf{A} = \lambda \mathbf{I} + \mathbf{N} \ .$$

Since $\mathbf{N}^\nu = \mathbf{O}$,

$$\mathbf{A}^k = (\lambda \mathbf{I} + \mathbf{N})^k = \lambda^k \mathbf{I} + \binom{k}{1} \lambda^{k-1} \mathbf{N} + \cdots + \binom{k}{\nu-1} \lambda^{k-\nu+1} \mathbf{N}^{\nu-1}$$

for all $k \geq \nu$. Therefore,

$$f(\mathbf{A}) = \sum_{k=0}^{\infty} a_k \mathbf{A}^k = \sum_{k=0}^{\infty} a_k \sum_{l=0}^{\nu-1} \lambda^{k-l} \cdot \binom{k}{l} \mathbf{N}^l \ ,$$

where we used the fact that for $l > k$,

$$\binom{k}{l} = 0 \ .$$

Therefore,

$$f(\mathbf{A}) = \sum_{l=0}^{\nu-1} \mathbf{N}^l \sum_{k=0}^{\infty} \binom{k}{l} a_k \lambda^{k-l} = \sum_{l=0}^{\nu-1} \mathbf{N}^l \cdot \frac{1}{l!} f^{(l)}(\lambda) \ . \qquad (1.39)$$

The matrix form of this representation is the following:

$$\begin{pmatrix} f(\lambda) & \frac{1}{1!} f'(\lambda) & \frac{1}{2!} f''(\lambda) & \frac{1}{3!} f'''(\lambda) & \cdots & \frac{1}{(\nu-1)!} f^{(\nu-1)}(\lambda) \\ & f(\lambda) & \frac{1}{1!} f'(\lambda) & \frac{1}{2!} f''(\lambda) & \cdots & \frac{1}{(\nu-2)!} f^{(\nu-2)}(\lambda) \\ & & f(\lambda) & \frac{1}{1!} f'(\lambda) & \cdots & \frac{1}{(\nu-3)!} f^{(\nu-3)}(\lambda) \\ & & & f(\lambda) & & \frac{1}{(\nu-4)!} f^{(\nu-4)}(\lambda) \\ & & & & \ddots & \vdots \\ & & & & & \frac{1}{1!} f'(\lambda) \\ O & & & & & f(\lambda) \end{pmatrix} \ .$$

Note that a finite representation is obtained.

Example 1.20

Assume now that $\mathbf{A}$ is a Jordan canonical form

$$\mathbf{A} = \begin{pmatrix} \mathbf{J}_1 & & & O \\ & \mathbf{J}_2 & & \\ & & \ddots & \\ O & & & \mathbf{J}_s \end{pmatrix} \ .$$

Since

$$\mathbf{A}^k = \begin{pmatrix} \mathbf{J}_1^k & & & O \\ & \mathbf{J}_2^k & & \\ & & \ddots & \\ O & & & \mathbf{J}_s^k \end{pmatrix},$$

we have

$$f(\mathbf{A}) = \begin{pmatrix} f(\mathbf{J}_1) & & & O \\ & f(\mathbf{J}_2) & & \\ & & \ddots & \\ O & & & f(\mathbf{J}_s) \end{pmatrix}. \tag{1.40}$$

Each diagonal block of this matrix can be determined by using the method of the previous example.

Matrix functions are usually determined by using special matrix transformations. This principle is based on the following result.

THEOREM 1.16
Assume that $f(\mathbf{A})$ exists, and furthermore $\mathbf{T}$ is a nonsingular matrix of the same order as $\mathbf{A}$. Then

$$f(\mathbf{TAT}^{-1}) = \mathbf{T}f(\mathbf{A})\mathbf{T}^{-1}. \tag{1.41}$$

PROOF   Note first that for $k \geq 1$,

$$(\mathbf{TAT}^{-1})^k = (\mathbf{TAT}^{-1}) \cdot (\mathbf{TAT}^{-1}) \cdot \ldots \cdot (\mathbf{TAT}^{-1}) = \mathbf{TA}^k\mathbf{T}^{-1}.$$

Therefore,

$$f(\mathbf{TAT}^{-1}) = \sum_{k=0}^{\infty} a_k (\mathbf{TAT}^{-1})^k = \sum_{k=0}^{\infty} a_k \mathbf{TA}^k\mathbf{T}^{-1}$$

$$= \mathbf{T}\left(\sum_{k=0}^{\infty} a_k \mathbf{A}^k\right)\mathbf{T}^{-1} = \mathbf{T} \cdot f(\mathbf{A}) \cdot \mathbf{T}^{-1}.$$

∎

COROLLARY 1.5
If $\mathbf{A} = \mathbf{T}^{-1} \cdot diag(\lambda_1, \ldots, \lambda_n) \cdot \mathbf{T}$, then the result of Example 1.18 implies that

$$f(\mathbf{A}) = \mathbf{T}^{-1} \cdot diag(f(\lambda_1), \ldots, f(\lambda_n)) \cdot \mathbf{T}. \tag{1.42}$$

COROLLARY 1.6

In the general case $\mathbf{A} = \mathbf{T}^{-1} \cdot \mathbf{J} \cdot \mathbf{T}$, *where* $\mathbf{J}$ *is a Jordan canonical form given in Example 1.20. Therefore,*

$$f(\mathbf{A}) = \mathbf{T}^{-1} \cdot \begin{pmatrix} f(\mathbf{J}_1) & & & O \\ & f(\mathbf{J}_2) & & \\ & & \ddots & \\ O & & & f(\mathbf{J}_s) \end{pmatrix} \mathbf{T}. \qquad (1.43)$$

The theorem suggests the following algorithm for finding matrix functions:

Step 1 Transform matrix $\mathbf{A}$ to a special form $\tilde{\mathbf{A}} = \mathbf{TAT}^{-1}$.

Step 2 Find $f(\tilde{\mathbf{A}})$.

Step 3 Compute $f(\mathbf{A})$ as $\mathbf{T}^{-1} f(\tilde{\mathbf{A}})\mathbf{T}$.

Note that the first step can be performed by using standard computer packages. Several packages even compute special matrix functions, such as matrix exponentials (see Example 1.22 for definition).

Example 1.21

Let t be a positive integer, and let the eigenvalues of matrix $\mathbf{A}$ be $\lambda_1, \lambda_2, \ldots, \lambda_r$ with multiplicities $m_1, m_2, \ldots, m_r$. We will present a special representation of $\mathbf{A}^t$. The result of this example will be used later in solving linear difference equations with constant coefficients. It is known from relation (1.30) that there exists a nonsingular matrix $\mathbf{T}$ such that

$$\mathbf{A} = \mathbf{T}^{-1} \begin{pmatrix} \mathbf{J}_1 & & & O \\ & \mathbf{J}_2 & & \\ & & \ddots & \\ O & & & \mathbf{J}_s \end{pmatrix} \mathbf{T},$$

where each matrix $\mathbf{J}_j$ is a Jordan block. From Equation (1.41) we conclude that

$$\mathbf{A}^t = \mathbf{T}^{-1} \cdot \begin{pmatrix} \mathbf{J}_1^t & & & O \\ & \mathbf{J}_2^t & & \\ & & \ddots & \\ O & & & \mathbf{J}_s^t \end{pmatrix} \mathbf{T}.$$

Consider first one block $\mathbf{J}_j^t$ with order ν_j. Select function $f(z) = z^t$, then Equation (1.39) implies that

$$\mathbf{J}_j^t = \sum_{l=0}^{v_j-1} \mathbf{N}_j^l \frac{1}{l!} t(t-1)\ldots(t-l+1)\lambda_i^{t-l} = \lambda_i^t \sum_{l=0}^{m_i-1} t^l \mathbf{M}_{jl} ,$$

where λ_i is the eigenvalue in block $\mathbf{J}_j$, and $\mathbf{M}_{jl}$ is a constant matrix. If $v_j < m_i$, then $\mathbf{M}_{jl} = \mathbf{O}$ for $l \geq v_j$. Therefore,

$$\mathbf{A}^t = \mathbf{T}^{-1} \sum_{j=1}^{s} \begin{pmatrix} \mathbf{O} & & & & & O \\ & \ddots & & & & \\ & & \mathbf{O} & & & \\ & & & \mathbf{J}_j^t & & \\ & & & & \mathbf{O} & \\ & & & & & \ddots & \\ O & & & & & \mathbf{O} \end{pmatrix} \mathbf{T}$$

$$= \mathbf{T}^{-1} \sum_{j=1}^{s} \lambda_i^t \sum_{l=0}^{m_i-1} t^l \begin{pmatrix} \mathbf{O} & & & & & O \\ & \ddots & & & & \\ & & \mathbf{O} & & & \\ & & & \mathbf{M}_{jl} & & \\ & & & & \mathbf{O} & \\ & & & & & \ddots & \\ O & & & & & \mathbf{O} \end{pmatrix} \mathbf{T} .$$

Because each eigenvalue appears in at least one Jordan block,

$$\mathbf{A}^t = \sum_{i=1}^{r} \lambda_i^t \sum_{l=0}^{m_i-1} t^l \mathbf{C}_{il} , \tag{1.44}$$

where $\mathbf{C}_{il}$ is a constant matrix. We mention that this representation will have many applications in later chapters.

Example 1.22

Let $\mathbf{A}$ be a square matrix and t be a real constant. In linear systems theory, the matrix exponential $e^{\mathbf{A}t}$ has special importance, since in Section 2.1.2 we will see that the solutions of time-invariant linear differential equations can be easily constructed using these matrix functions. A special representation of this matrix exponential will be introduced next. Our derivation and the final result are analogous to those of the previous example. Select now $f(z) = e^{zt}$, and if $\mathbf{J}_j$ is a Jordan block, then from (1.39) we have

$$e^{\mathbf{J}_j t} = \sum_{l=0}^{v_j-1} \mathbf{N}_j^l \frac{1}{l!} t^l e^{\lambda_i t} = e^{\lambda_i t} \sum_{l=0}^{m_i-1} t^l \mathbf{K}_{jl} ,$$

where $\mathbf{K}_{jl}$ is a constant matrix. A similar argument that was shown in the previous example yields to our final formula:

$$e^{\mathbf{A}t} = \sum_{i=1}^{r} e^{\lambda_i t} \sum_{l=0}^{m_i-1} t^l \mathbf{B}_{il} , \tag{1.45}$$

where $\mathbf{B}_{il}$ is a constant matrix. Like the previous example, this representation will be applied frequently in later chapters.

The derivations presented in Examples 1.21 and 1.22 have only theoretical importance. In practical problems, matrices $e^{\mathbf{A}t}$ and $\mathbf{A}^t$ are computed directly without repeating the above derivations. Such direct methods will be introduced in Example 1.23.

Since matrix exponentials are often used in the theory of continuous systems, we now summarize their basic properties.

THEOREM 1.17
The following relations are true:

(i) $e^{\mathbf{A} \cdot 0} = \mathbf{I}$.

(ii) $\frac{d}{dt} e^{\mathbf{A}t} = \mathbf{A} \cdot e^{\mathbf{A}t}$.

(iii) $e^{\mathbf{A}t} \cdot e^{\mathbf{A}\tau} = e^{\mathbf{A}(t+\tau)}$.

(iv) $(e^{\mathbf{A}t})^{-1} = e^{-\mathbf{A}t}$.

(v)   *If* $\mathbf{AB} = \mathbf{BA}$, *then* $e^{\mathbf{A}t} \cdot e^{\mathbf{B}t} = e^{(\mathbf{A}+\mathbf{B})t}$.

PROOF

(i)   Use the definition of matrix functions to get $e^{\mathbf{A} \cdot 0} = e^{\mathbf{O}} = \mathbf{I} + (1/1!)\mathbf{O} + (1/2!)\mathbf{O}^2 + \cdots = \mathbf{I}$.

(ii)   Simple differentiation shows that

$$\frac{d}{dt} e^{\mathbf{A}t} = \frac{d}{dt} \left(\sum_{k=0}^{\infty} \frac{1}{k!} \mathbf{A}^k t^k \right) = \sum_{k=1}^{\infty} \frac{1}{k!} \mathbf{A}^k k t^{k-1}$$

$$= \mathbf{A} \sum_{k=1}^{\infty} \frac{1}{(k-1)!} \mathbf{A}^{k-1} t^{k-1} = \mathbf{A} \sum_{l=0}^{\infty} \frac{1}{l!} \mathbf{A}^l t^l = \mathbf{A} e^{\mathbf{A}t} \, .$$

(iii,v)   We first prove that if $\mathbf{MN} = \mathbf{NM}$, then $e^{\mathbf{M}} \cdot e^{\mathbf{N}} = e^{\mathbf{M}+\mathbf{N}}$.

Note that the Cauchy product of infinite series implies that

$$e^{\mathbf{M}} \cdot e^{\mathbf{N}} = \left(\sum_{k=0}^{\infty} \frac{1}{k!} \mathbf{M}^k \right) \left(\sum_{l=0}^{\infty} \frac{1}{l!} \mathbf{N}^l \right)$$

$$= \sum_{m=0}^{\infty} \left(\frac{1}{0!} \mathbf{M}^0 \cdot \frac{1}{m!} \mathbf{N}^m + \frac{1}{1!} \mathbf{M}^1 \cdot \frac{1}{(m-1)!} \mathbf{N}^{m-1} + \cdots \right.$$

$$\left. + \frac{1}{m!} \mathbf{M}^m \cdot \frac{1}{0!} \mathbf{N}^0 \right)$$

$$= \sum_{m=0}^{\infty} \sum_{n=0}^{m} \frac{1}{n!(m-n)!} \mathbf{M}^n \mathbf{N}^{m-n}$$

$$= \sum_{m=0}^{\infty} \left(\frac{1}{m!} \sum_{n=0}^{m} \binom{m}{n} \mathbf{M}^n \mathbf{N}^{m-n} \right)$$

$$= \sum_{m=0}^{\infty} \frac{1}{m!} (\mathbf{M} + \mathbf{N})^m = e^{\mathbf{M}+\mathbf{N}} \, .$$

The assumption that matrices $\mathbf{M}, \mathbf{N}$ commute has been used in applying the binomial theorem for matrices. If $\mathbf{M}$ and $\mathbf{N}$ do not commute, then, for example,

$$(\mathbf{M} + \mathbf{N})^2 = (\mathbf{M} + \mathbf{N})(\mathbf{M} + \mathbf{N})$$

$$= \mathbf{M}^2 + \mathbf{MN} + \mathbf{NM} + \mathbf{N}^2$$

$$\neq \mathbf{M}^2 + 2\mathbf{MN} + \mathbf{N}^2 \, .$$

For proving relations (iii) and (v), select

$$\mathbf{M} = \mathbf{A}t, \quad \mathbf{N} = \mathbf{A}\tau \quad \text{and} \quad \mathbf{M} = \mathbf{A}t, \quad \mathbf{N} = \mathbf{B}t \,,$$

respectively.

(iv)   From (i) and (iii) we conclude that

$$e^{\mathbf{A}t} \cdot e^{-\mathbf{A}t} = e^{\mathbf{A}t - \mathbf{A}t} = e^{\mathbf{O}} = \mathbf{I} \,,$$

which implies the assertion.

■

Example 1.23

Consider matrix

$$\mathbf{A} = \begin{pmatrix} 0 & \omega \\ -\omega & 0 \end{pmatrix} \,,$$

which was the subject of our earlier Example 1.13. The matrix $e^{\mathbf{A}t}$ will now be determined. In order to illustrate the methodology, several alternative approaches will be used.

Method 1. First we use the definition of matrix functions. Note first that

$$\mathbf{A} = \omega \cdot \mathbf{B}, \text{ where } \mathbf{B} = \begin{pmatrix} 0 & 1 \\ -1 & 0 \end{pmatrix};$$

$$\mathbf{A}^2 = \begin{pmatrix} -\omega^2 & 0 \\ 0 & -\omega^2 \end{pmatrix} = -\omega^2\mathbf{I}, \mathbf{A}^3 = -\omega^3\mathbf{B}, \mathbf{A}^4 = \omega^4\mathbf{I}, \text{ etc.}$$

In general,

$$\mathbf{A}^k = \begin{cases} (-1)^m\omega^k\mathbf{I} & \text{if } k = 2m \\ (-1)^m\omega^k\mathbf{B} & \text{if } k = 2m + 1 \,. \end{cases}$$

Therefore,

$$e^{\mathbf{A}t} = \sum_{k=0}^{\infty} \frac{1}{k!}\mathbf{A}^k t^k$$

$$= \sum_{\substack{m \\ (k=\text{even})}} (-1)^m\frac{\omega^{2m}t^{2m}}{(2m)!}\mathbf{I} + \sum_{\substack{m \\ (k=\text{odd})}} (-1)^m\frac{\omega^{2m+1}t^{2m+1}}{(2m + 1)!}\mathbf{B}$$

$$= \cos(\omega t)\mathbf{I} + \sin(\omega t)\mathbf{B}$$

$$= \begin{pmatrix} \cos(\omega t) & \sin(\omega t) \\ -\sin(\omega t) & \cos(\omega t) \end{pmatrix} .$$

Method 2. Next we apply the diagonal transformation of $\mathbf{A}$, which was derived in Example 1.13. Combine that result with relation (1.42) to obtain

$$e^{\mathbf{A}t} = \mathbf{T}^{-1} \cdot diag(e^{j\omega t}, e^{-j\omega t})\mathbf{T}$$

$$= \begin{pmatrix} 1 & 1 \\ j & -j \end{pmatrix} \begin{pmatrix} e^{j\omega t} & 0 \\ 0 & e^{-j\omega t} \end{pmatrix} \begin{pmatrix} \frac{1}{2} & -\frac{j}{2} \\ \frac{1}{2} & \frac{j}{2} \end{pmatrix}$$

$$= \begin{pmatrix} \frac{e^{j\omega t} + e^{-j\omega t}}{2} & \frac{e^{j\omega t} - e^{-j\omega t}}{2j} \\ -\frac{e^{j\omega t} - e^{-j\omega t}}{2j} & \frac{e^{j\omega t} + e^{-j\omega t}}{2} \end{pmatrix}$$

$$= \begin{pmatrix} \cos(\omega t) & \sin(\omega t) \\ -\sin(\omega t) & \cos(\omega t) \end{pmatrix} .$$

Method 3. Now we apply the special form (1.45). Because the eigenvalues of $\mathbf{A}$ are distinct, $m_1 = m_2 = 1$. Therefore,

$$e^{\mathbf{A}t} = e^{\lambda_1 t}\mathbf{B}_{10} + e^{\lambda_2 t}\mathbf{B}_{20} , \tag{1.46}$$

where matrices $\mathbf{B}_{10}$ and $\mathbf{B}_{20}$ are to be determined. We can easily formulate two equations for the two unknowns in the following way. First, substitute $t = 0$ into the above equation to get

$$\mathbf{I} = \mathbf{B}_{10} + \mathbf{B}_{20} ,$$

where we used property (i) of Theorem 1.17.

Differentiate Equation (1.46) with respect to t and substitute $t = 0$ into the resulting relation to have

$$\mathbf{A} = \lambda_1 \mathbf{B}_{10} + \lambda_2 \mathbf{B}_{20} = j\omega \mathbf{B}_{10} - j\omega \mathbf{B}_{20} .$$

Here we used property (ii) of Theorem 1.17. From the first equation,

$$\mathbf{B}_{20} = \mathbf{I} - \mathbf{B}_{10} ,$$

and by substituting this relation into the second equation,

$$\mathbf{A} = j\omega \mathbf{B}_{10} - j\omega(\mathbf{I} - \mathbf{B}_{10}) ,$$

which implies that

$$\mathbf{B}_{10} = \frac{1}{2j\omega}(\mathbf{A} + j\omega\mathbf{I}) = \frac{1}{2j\omega}\begin{pmatrix} j\omega & \omega \\ -\omega & j\omega \end{pmatrix} = \begin{pmatrix} \frac{1}{2} & -\frac{j}{2} \\ \frac{j}{2} & \frac{1}{2} \end{pmatrix}$$

and

$$\mathbf{B}_{20} = \mathbf{I} - \mathbf{B}_{10} = \begin{pmatrix} \frac{1}{2} & \frac{j}{2} \\ -\frac{j}{2} & \frac{1}{2} \end{pmatrix}.$$

Hence

$$e^{\mathbf{A}t} = \begin{pmatrix} \frac{1}{2} & -\frac{j}{2} \\ \frac{j}{2} & \frac{1}{2} \end{pmatrix} e^{j\omega t} + \begin{pmatrix} \frac{1}{2} & \frac{j}{2} \\ -\frac{j}{2} & \frac{1}{2} \end{pmatrix} e^{-j\omega t}$$

$$= \begin{pmatrix} \cos(\omega t) & \sin(\omega t) \\ -\sin(\omega t) & \cos(\omega t) \end{pmatrix}.$$

Discrete time-invariant systems are usually analyzed based on the properties of matrix $\mathbf{A}^t$, which are summarized next.

THEOREM 1.18
The following relations are true:

(i) $\mathbf{A}^0 = \mathbf{I}$.

(ii) $\mathbf{A}^{t+1} = \mathbf{A} \cdot \mathbf{A}^t$.

(iii) $\mathbf{A}^t \cdot \mathbf{A}^\tau = \mathbf{A}^{t+\tau}$.

(iv) $(\mathbf{A}^t)^{-1} = (\mathbf{A}^{-1})^t$ *assuming that* $\mathbf{A}$ *is invertible.*

(v)   *If* $\mathbf{AB} = \mathbf{BA}$, *then* $\mathbf{A}^t \cdot \mathbf{B}^t = (\mathbf{AB})^t$.

Because the proof of this theorem follows immediately from the definition of matrix powers, the details are left as an exercise.

Example 1.24

In the previous example, $\mathbf{A}^t$ for matrix

$$\mathbf{A} = \begin{pmatrix} 0 & \omega \\ -\omega & 0 \end{pmatrix}$$

was determined by first observing and then proving the general form for $\mathbf{A}, \mathbf{A}^2, \mathbf{A}^3, \dots, \mathbf{A}^t$. In most cases the general expression for $\mathbf{A}^t$ is complicated, usually it is hard (if not impossible) to guess from some initial terms. In this example we show two systematic methods for finding $\mathbf{A}^t$ for the above matrix.

Method 1. By using the diagonal transformation of $\mathbf{A}$ (which was determined in Example 1.13), we have

$$\mathbf{A}^t = \mathbf{T}^{-1} \cdot diag((j\omega)^t, (-j\omega)^t)\mathbf{T}$$

$$= \begin{pmatrix} 1 & 1 \\ j & -j \end{pmatrix} \begin{pmatrix} (j\omega)^t & 0 \\ 0 & (-j\omega)^t \end{pmatrix} \begin{pmatrix} \frac{1}{2} & -\frac{j}{2} \\ \frac{1}{2} & \frac{j}{2} \end{pmatrix}$$

$$= \begin{pmatrix} \frac{1}{2}[(j\omega)^t + (-j\omega)^t] & \frac{j}{2}[-(j\omega)^t + (-j\omega)^t] \\ \frac{j}{2}[(j\omega)^t - (-j\omega)^t] & \frac{1}{2}[(j\omega)^t + (-j\omega)^t] \end{pmatrix} .$$

If $t = 2m$, then $(j\omega)^t = (-j\omega)^t = (-1)^m\omega^t$, and if $t = 2m+1$, then $(j\omega)^t = j(-1)^m\omega^t$ and $(-j\omega)^t = -j(-1)^m\omega^t$. Hence, if $t = 2m$, then

$$\mathbf{A}^t = \begin{pmatrix} (-1)^m\omega^t & 0 \\ 0 & (-1)^m\omega^t \end{pmatrix} ,$$

and if $t = 2m + 1$, then

$$\mathbf{A}^t = \begin{pmatrix} 0 & (-1)^m\omega^t \\ -(-1)^m\omega^t & 0 \end{pmatrix} .$$

Method 2. Next we use Equation (1.44), which has now the form:

$$\mathbf{A}^t = (j\omega)^t\mathbf{C}_{10} + (-j\omega)^t\mathbf{C}_{20} .$$

By substituting $t = 0$,

$$\mathbf{I} = \mathbf{C}_{10} + \mathbf{C}_{20} ,$$

and by substituting $t = 1$,

$$\mathbf{A} = j\omega\mathbf{C}_{10} + (-j\omega)\mathbf{C}_{20} .$$

The two equations for $\mathbf{C}_{10}$ and $\mathbf{C}_{20}$ have the solution:

$$\mathbf{C}_{10} = \frac{1}{2}\mathbf{I} - \frac{j}{2\omega}\mathbf{A}$$

and

$$\mathbf{C}_{20} = \frac{1}{2}\mathbf{I} + \frac{j}{2\omega}\mathbf{A} .$$

Therefore,

$$\mathbf{A}^t = (j\omega)^t[\frac{1}{2}\mathbf{I} - \frac{j}{2\omega}\mathbf{A}] + (-j\omega)^t[\frac{1}{2}\mathbf{I} + \frac{j}{2\omega}\mathbf{A}]$$

$$= \frac{1}{2}[(j\omega)^t + (-j\omega)^t]\mathbf{I} + \frac{j}{2\omega}[-(j\omega)^t + (-j\omega)^t]\mathbf{A} ,$$

which coincides with the result obtained by using the previous method.

We conclude this section with a special matrix function, which will be applied in later chapters of this book.

Example 1.25

Let $\mathbf{A}$ be a square matrix, and assume that all eigenvalues of $\mathbf{A}$ are inside the unit circle of the complex plane. Consider the function

$$f(z) = \frac{1}{1-z} = 1 + z + z^2 + z^3 + \cdots ,$$

then

$$f(\mathbf{A}) = \mathbf{I} + \mathbf{A} + \mathbf{A}^2 + \cdots .$$

Note first that in the case of the above series, $R = 1$. Therefore, Examples 1.19 and 1.20 imply that the infinite matrix series $f(\mathbf{A})$ is convergent. Hence, $\mathbf{A}^k \to \mathbf{O}$ as $k \to \infty$.
    Next we prove that $f(\mathbf{A}) = (\mathbf{I} - \mathbf{A})^{-1}$. Consider the equation

$$\mathbf{I} - \mathbf{A}^{t+1} = (\mathbf{I} + \mathbf{A} + \mathbf{A}^2 + \cdots + \mathbf{A}^t)(\mathbf{I} - \mathbf{A}), \qquad (t \geq 0)$$

and let $t \to \infty$. Then

$$\mathbf{I} = f(\mathbf{A}) \cdot (\mathbf{I} - \mathbf{A}) ,$$

which implies that $(\mathbf{I} - \mathbf{A})^{-1}$ exists and equals $f(\mathbf{A})$.
    Finally we note that the results of this example have many applications in matrix theory.

Problems

1.   Assume that in Definition 1.1, condition (iii) is modified as follows:

(iii')   For all $x, y, z \in M$, $\rho(z, x) \leq \rho(x, y) + \rho(y, z)$.

Prove that (i) and (iii') imply condition (ii).

2. Prove that conditions (ii) and (iii) of Definition 1.1 and the assumption that $\rho(x, x) \geq 0$ for all $x \in M$ imply that $\rho(x, y) \geq 0$ for all x, $y \geq 0$.

3. Let $M = \mathbf{R}^1$ and $\rho(x, y) = (x - y)^2$. Is (M, ρ) a metric space?

4. Construct a metric space that has closed balls B_1 and B_2 such that B_1 is a proper subset of B_2, but the radius of B_1 is larger than that of B_2.

5. Assume that (M, ρ_1) and (M, ρ_2) are metric spaces. Prove that (M, ρ) is also a metric space with $\rho = \rho_1 + \rho_2$.

6. Let $M = M' = \mathbf{R}^1$ and $A(x) = \ln x$ with $D(A) = (0, \infty)$. If ρ and ρ' are defined as in Example 1.1, is mapping A bounded?

7. Assume that (M, ρ) is a complete metric space, $M_1 \subseteq M$ is closed, $D(A) = M_1$, and $R(A) \subseteq M_1$. Assume furthermore that for all $x, y \in M_1$, $\rho(A(x), A(y)) \leq \rho(x, y)$. Construct an example such that mapping A has no fixed point in M_1.

8. Construct an example such that under the conditions of the previous problem mapping A has infinitely many fixed points.

9. Solve equation $xe^x = 1/2$ by fixed-point iteration.

10. Let $\mathbf{A}$ be an $n \times n$ real matrix such that $\|\mathbf{A}\| < 1$ with some matrix norm. Prove that equation $\mathbf{x} = \mathbf{A}\mathbf{x} + \mathbf{b}$ has a unique solution for all $\mathbf{b} \in \mathbf{R}^n$.

11. Bound the eigenvalues of matrix

$$\mathbf{A} = \begin{pmatrix} 1 & 1 & 2 \\ -1 & 1 & 1 \\ 0 & 2 & 1 \end{pmatrix}$$

by using the $p = 1, \infty$ and Frobenius norms.

12. Apply first Theorem 1.9 for matrix

$$\mathbf{A} = \begin{pmatrix} 1 & 1 & 2 \\ -1 & 1 & 1 \\ 0 & 3 & 1 \end{pmatrix},$$

and then, apply again for $\mathbf{A}^T$. Which case gives better results?

13. Find the null space of matrix

$$A = \begin{pmatrix} 1 & 1 & 2 \\ -1 & 1 & 1 \\ 0 & 2 & 3 \end{pmatrix} .$$

14. Diagonalize and find decomposition (1.33) for matrix

$$A = \begin{pmatrix} 1 & 2 \\ 2 & 4 \end{pmatrix} .$$

15. Find decomposition (1.34) for matrix

$$A = \begin{pmatrix} 1 & 2 \\ 2 & 4 \end{pmatrix} .$$

16. Find $e^{\mathbf{A}t}$ for

$$A = \begin{pmatrix} 1 & 1 \\ 2 & 2 \end{pmatrix} .$$

17. Find $e^{\mathbf{A}t}$ for

$$A = \begin{pmatrix} 2 & 1 \\ 0 & 2 \end{pmatrix} .$$

18. Find $\mathbf{A}^t$ for

$$A = \begin{pmatrix} 1 & 1 \\ 2 & 2 \end{pmatrix} .$$

19. Find $\mathbf{A}^t$ for

$$A = \begin{pmatrix} 2 & 1 \\ 0 & 2 \end{pmatrix} .$$

20.

(i) Prove the triangle inequality for Example 1.3.

(ii) Prove Theorem 1.5 for the Frobenius norm.

(iii) Prove Theorem 1.18.

21. Prove that the vector norms of n-element real (or complex) vectors are equivalent to each others. That is, if $\| \cdots \|$ and $\| \cdots \|'$ are two vector norms, then there are positive constants a_1 and a_2 such that for all vectors x,

$$a_1 \|x\| \leq \|x\|' \leq a_2 \|x\| .$$

22. Discuss an improvement of the Gerschgorin Theorem by using the fact that the eigenvalues of $\mathbf{A}$ and $\mathbf{D}^{-1}\mathbf{A}\mathbf{D}$ are the same for all nonsingular diagonal matrices $\mathbf{D}$.

23.   Using Theorem 1.8 and the fact that the characteristic polynomial of matrix

$$\mathbf{A} = \begin{pmatrix} 0 & 0 & 0 & \cdots & 0 & -a_0 \\ 1 & 0 & 0 & \cdots & 0 & -a_1 \\ 0 & 1 & 0 & \cdots & 0 & -a_2 \\ \vdots & \vdots & \vdots & \ddots & \vdots & \vdots \\ 0 & 0 & 0 & \cdots & 1 & -a_{n-1} \end{pmatrix}$$

is $\varphi(\lambda) = \lambda^n + a_{n-1}\lambda^{n-1} + \cdots + a_1\lambda + a_0$, show that all roots of this polynomial satisfy inequalities

$$|\lambda| \leq 1 + \max_{0 \leq i \leq n-1}\{|a_i|\} \, ,$$

$$|\lambda| \leq max\{1; |a_0| + |a_1| + \cdots + |a_{n-1}|\} \, ,$$

and

$$|\lambda| \leq \sqrt{(n-1) + \sum_{i=0}^{n-1} |a_i|^2} \, .$$

24.   Let $\mathbf{A}$ be an $n \times n$ real symmetric matrix and assume that λ_1 and λ_n are the smallest and largest eigenvalues of $\mathbf{A}$. Prove that

$$\min_{\mathbf{x} \neq 0} \frac{\mathbf{x}^T \mathbf{A} \mathbf{x}}{\mathbf{x}^T \mathbf{x}} = \lambda_1$$

and

$$\max_{\mathbf{x} \neq 0} \frac{\mathbf{x}^T \mathbf{A} \mathbf{x}}{\mathbf{x}^T \mathbf{x}} = \lambda_n \, .$$

25.   Let $\mathbf{A}$ and $\mathbf{B}$ be $n \times n$ real matrices. Show that

$$\|e^{\mathbf{A}} - e^{\mathbf{B}}\| \leq \|\mathbf{A} - \mathbf{B}\| \frac{e^{\alpha} - e^{\beta}}{\alpha - \beta}$$

where $\|\mathbf{A}\| \leq \alpha$ and $\|\mathbf{B}\| \leq \beta$.

chapter two

Mathematics of Dynamic Processes

This chapter introduces conditions for the solvability of ordinary differential and difference equations, which will be fundamental in describing dynamic processes. In addition, the general solutions of such equations will be constructed in linear cases. In the constant coefficient cases, Laplace transforms and Z-transforms serve as the most commonly used solution methods. Their definitions and main properties are also discussed in this chapter.

2.1   Solution of Ordinary Differential Equations

Dynamic systems with continuous time scale are usually modeled by a system of linear or nonlinear differential equations. Without defining the state of the system formally we mention that the unknown of the differential equations is usually the state of the dynamic system under consideration. Therefore, the determination of the state requires the solution of the governing differential equations, and the examination of the properties of the state is based on those of the solution of differential equations. In this section the existence and uniqueness of the solutions of such equations are discussed, and methods will be introduced to find them.

2.1.1   Existence and Uniqueness Theorems

In this subsection, conditions will be developed for the existence and the uniqueness of the solution of initial value problems of ordinary differential equations. Because the solution represents the state of the corresponding dynamic system, the results of this subsection are often used to determine whether or not a given differential equation represents a

dynamic system. If no solution exists, then no state can be defined, and
if the solution is not unique, then there is a multiple state for the system.
In the theory of dynamic systems, we usually assume the existence of
the unique state.

Consider the first-order explicit differential equation

$$\dot{\mathbf{x}} = \mathbf{f}(t, \mathbf{x}) , \tag{2.1}$$

where $\mathbf{x} : I \to \mathbf{R}^n$ is an unknown real-variable, vector-valued function,
and $f : I \times X \to \mathbf{R}^n$, where I is an interval of the real line and $X \subseteq \mathbf{R}^n$.

A function $\mathbf{x}$ is called the solution of the given differential equation
on an interval $I_1 \subseteq I$, if for all $t \in I_1$, $\mathbf{x}(t)$ and $\dot{\mathbf{x}}(t)$ exist, $\mathbf{x}(t) \in X$;
furthermore $\dot{\mathbf{x}}(t) = \mathbf{f}(t, \mathbf{x}(t))$.

In order to have a solution, function $\mathbf{f}$ must satisfy certain conditions,
as the following example illustrates.

Example 2.1

Let $n = 1, I = \mathbf{R}, X = \mathbf{R}$ and

$$f(t, x) = g(t) = \begin{cases} 0 \text{ if } t \text{ is rational} \\ 1 \text{ if } t \text{ is irrational.} \end{cases}$$

We can easily show that Equation (2.1) has no solution. Assume that
there is a solution $x(t)$ on an interval $I_1 = [a, b]$, then the left-hand
side $\dot{x}$ of the differential equation is integrable on I_1. Therefore, the
right-hand side must also be integrable on I_1. But g is not integrable,
which can be shown as follows. Consider the Riemann sum of function
g:

$$R = \sum_{i=1}^{N} g(\xi_i)(t_i - t_{i-1}) ,$$

where $a = t_0 < t_1 < \cdots < t_N = b$, and for all i, $\xi_i \in [t_{i-1}, t_i]$. If ξ_i
is rational for all i, then $g(\xi_i) = 0$, therefore, $R = 0$. If ξ_i is selected
to be irrational for all i, then

$$R = \sum_{i=1}^{N}(t_i - t_{i-1}) = t_N - t_0 = b - a \neq 0 .$$

If $N \to \infty$, we obtain different limits 0 and $b - a$ for R. Hence
$\int_{I_1} g(t)\, dt$ does not exist.

Assume next that $t_0 \in I$, and $\mathbf{x}_0 \in X$ are given. The *initial value
problem* of the differential equation is given as

$$\dot{\mathbf{x}} = \mathbf{f}(t, \mathbf{x}) , \qquad \mathbf{x}(t_0) = \mathbf{x}_0 . \tag{2.2}$$

That is, a solution which passes through the point $(t_0, \mathbf{x}_0)$ is to be determined as illustrated in Figure 2.1. In systems theory, $\mathbf{x}(t)$ usually denotes the state of the system at time t, and $\mathbf{x}_0$ is the initial state when $t_0 = 0$.

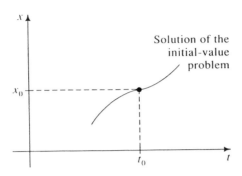

Figure 2.1   Initial value problem.

If $t_0 \neq 0$, then by introducing the new independent variable $t' = t - t_0$, the initial value of t' becomes zero.

The previous example shows that an initial value problem does not need to have a solution. The next example shows even if it has a solution, the solution does not need to be unique.

Example 2.2

Consider the one-dimensional initial value problem

$$\dot{x} = \sqrt{|x|}, \qquad x(1) = 0$$

with $I = X = \mathbf{R}$. Obviously the zero function ($x_1(t) \equiv 0$) satisfies both the differential equation and the initial condition. Consider next the function

$$x_2(t) = \begin{cases} 0 & \text{if } t \leq 1 \\ (\frac{t-1}{2})^2 & \text{if } t > 1 \, . \end{cases}$$

Easy calculation shows that $x_2(1) = 0$, x_2 is differentiable, and

$$\dot{x_2}(t) = \begin{cases} 0 = \sqrt{|x_2|} & \text{if } t \leq 1 \\ 2(\frac{t-1}{2}) \cdot \frac{1}{2} = \frac{t-1}{2} = \sqrt{|x_2|} & \text{if } t > 1 \, . \end{cases}$$

Hence x_2 also solves the initial value problem. Therefore, $x_1(t)$ and $x_2(t)$ are both solutions. That is, the solution is not unique.

Note that in the above example, function $f = \sqrt{|x|}$ is continuous. Therefore, even stronger conditions are needed to guarantee the uniqueness of the solution. In contrast to this example, in dynamic systems theory we always assume that the governing dynamic relations (differential or difference equations) and the initial values uniquely determine the solutions. Otherwise the future behavior of the system is absolutely unpredictable. These additional conditions and the resulting existence and uniqueness theorem are discussed next.

Consider the initial value problem (2.2), where $I = [t_0 - a, t_0 + a]$ and with some norm let

$$X = \{\mathbf{x} \mid \|\mathbf{x} - \mathbf{x}_0\| \le b\}, \qquad (2.3)$$

where $b > 0$ is given. Assume that $\mathbf{f}$ is continuous on $I \times X$, and furthermore it satisfies the Lipschitz condition, that is, there is a constant $L > 0$ such that

$$\|\mathbf{f}(t, \mathbf{x}_1) - \mathbf{f}(t, \mathbf{x}_2)\| \le L \cdot \|\mathbf{x}_1 - \mathbf{x}_2\| \qquad (2.4)$$

for all $t \in I$ and $\mathbf{x}_1, \mathbf{x}_2 \in X$. After proving Theorem 2.1 we will present an easy way to check if the Lipschitz condition holds for a given function $\mathbf{f}$.

Since f is continuous,

$$Q = \max_{(t,x) \in I \times X} \|\mathbf{f}(t, \mathbf{x})\|$$

exists and is finite. Define finally

$$\alpha = \min\left\{ a, \frac{b}{Q}, \frac{1}{L} - \varepsilon \right\},$$

where $\varepsilon > 0$ is a small number. Sets I and X are illustrated in Figure 2.2.

The main result of this section can be now formulated as follows.

THEOREM 2.1

Under the above conditions, the initial value problem has a unique solution on the interval $I_0 = [t_0 - \alpha, t_0 + \alpha]$.

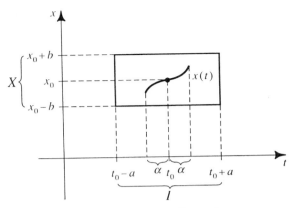

Figure 2.2 Illustration of Theorem 2.1 in one dimension.

PROOF

(a) First we prove that the initial value problem (2.2) is equivalent to the fixed-point problem

$$\mathbf{x}(t) = \mathbf{x}_0 + \int_{t_0}^t \mathbf{f}(\tau, \mathbf{x}(\tau))\, d\tau \ . \tag{2.5}$$

Assume first that $\mathbf{x}$ is the solution of (2.2). Integrate both sides on interval $[t_0, t]$ to get relation

$$\mathbf{x}(t) - \mathbf{x}(t_0) = \int_{t_0}^t \mathbf{f}(\tau, \mathbf{x}(\tau))\, d\tau \ ,$$

and use the initial condition $\mathbf{x}(t_0) = \mathbf{x}_0$ to obtain (2.5). Assume next that $\mathbf{x}$ is a solution of the fixed-point problem (2.5). Differentiate both sides and use the fact that $\mathbf{f}(\tau, \mathbf{x}(\tau))$ is continuous in τ to derive

$$\dot{\mathbf{x}}(t) = \mathbf{f}(t, \mathbf{x}(t)) \ .$$

Now substitute $t = t_0$ into (2.5) to get the initial condition.

(b) Next we will verify that the fixed-point problem (2.5) satisfies all conditions of Theorem 1.3 which implies that there is a unique solution.

Define M as the function space $C(I_0)$ (that is, the set of the continuous functions defined on I_o) with distance $\rho(\mathbf{x}_1, \mathbf{x}_2) = \max_{t \in I_o} \|\mathbf{x}_1(t) - \mathbf{x}_2(t)\|$. Select

$$M_1 = \{\mathbf{x}(t) \mid \mathbf{x} \in M, \mathbf{x}(t_0) = \mathbf{x}_0, \|\mathbf{x}(t) - \mathbf{x}_0\| \le b \text{ for all } t \in I_0\} \ ,$$

and mapping

$$A(\mathbf{x})(t) = \mathbf{x}_0 + \int_{t_0}^{t} \mathbf{f}(\tau, \mathbf{x}(\tau)) \, d\tau \, . \tag{2.6}$$

The right-hand side gives the value of function $A(\mathbf{x})$ at t.
We know from Section 1.2.1 that M is a complete metric space, and because all conditions in defining M_1 are closed (= or $\leq$), M_1 is a closed subset in M. Note that $A(\mathbf{x})$ is defined on the entire M_1, because $\mathbf{f}$ is continuous and therefore, the integral exists in the right-hand side of (2.6). Hence, we may select $D(A) = M_1$.
Next we verify that $R(A) \subseteq M_1$, that is, if $\mathbf{x} \in M_1$, then $A(\mathbf{x}) \in M_1$.
Obviously, $A(\mathbf{x})$ is continuous, since for continuous $\mathbf{f}$ and $\mathbf{x}$ the right-hand side of (2.5) is differentiable with derivative $\mathbf{f}(t, \mathbf{x}(t))$. Simple substitution shows that $A(\mathbf{x})(t_0) = \mathbf{x}_0$; furthermore for all $t \in I_0$,

$$\|A(\mathbf{x})(t) - \mathbf{x}_0\| = \left\| \int_{t_0}^{t} \mathbf{f}(\tau, \mathbf{x}(\tau)) \, d\tau \right\| \leq \int_{t_0}^{t} \|\mathbf{f}(\tau, \mathbf{x}(\tau))\| \, d\tau$$

$$\leq Q \cdot |t - t_0| \leq Q \cdot \alpha \leq b \, .$$

Therefore, $A(\mathbf{x}) \in M_1$.
Finally we show that mapping A is a contraction on M_1. Let $\mathbf{x}_1, \mathbf{x}_2 \in M_1$, then

$$\rho(A(\mathbf{x}_1), A(\mathbf{x}_2)) = \max_{t \in I_0} \left\| \mathbf{x}_0 + \int_{t_0}^{t} \mathbf{f}(\tau, \mathbf{x}_1(\tau)) \, d\tau - \mathbf{x}_0 \right.$$

$$\left. - \int_{t_0}^{t} \mathbf{f}(\tau, \mathbf{x}_2(\tau)) \, d\tau \right\|$$

$$= \max_{t \in I_0} \left\| \int_{t_0}^{t} (\mathbf{f}(\tau, \mathbf{x}_1(\tau)) - \mathbf{f}(\tau, \mathbf{x}_2(\tau))) \, d\tau \right\|$$

$$\leq \max_{t \in I_0} \left| \int_{t_0}^{t} \|\mathbf{f}(\tau, \mathbf{x}_1(\tau)) - \mathbf{f}(\tau, \mathbf{x}_2(\tau))\| \, d\tau \right|$$

$$\leq \max_{t \in I_0} \left| \int_{t_0}^{t} L \cdot \|\mathbf{x}_1(\tau) - \mathbf{x}_2(\tau)\| \, d\tau \right|$$

$$\leq \rho(\mathbf{x}_1, \mathbf{x}_2) \cdot L \cdot \max_{t \in I_0} \left| \int_{t_0}^{t} d\tau \right|$$

$$= \rho(\mathbf{x}_1, \mathbf{x}_2) \cdot L\alpha .$$

Since $L\alpha < 1$, mapping A is a contraction, which completes the proof. ∎

REMARK 2.1 In practical problems we can usually check if condition (2.4) holds. Assume that all partial derivatives of $\mathbf{f}$ with respect to the coordinates of $\mathbf{x}$ are bounded. If they are all continuous on X, then they are also bounded. Then the mean-value theorem of derivatives implies that (2.4) holds (see Problem 2/19). ∎

REMARK 2.2 The proof of this theorem and that of Theorem 1.3 suggest an iteration method for solving the initial value problem (2.2). Select the initial approximation for the solution as $\mathbf{x}_0(t) \equiv \mathbf{x}_0$. Obviously, this constant function belongs to M_1. Construct the iteration sequence by the recursion

$$\mathbf{x}_{k+1}(t) = \mathbf{x}_0 + \int_{t_0}^{t} \mathbf{f}(\tau, \mathbf{x}_k(\tau)) \, d\tau , \tag{2.7}$$

which uniformly converges on I_0 to the unique solution of the initial value problem. ∎

The iteration method (2.7) is illustrated next.

Example 2.3

Consider the one-dimensional initial value problem

$$\dot{x} = x, \qquad x(0) = 1 ,$$

which has the unique solution $x(t) = e^t$. In this case the iteration method (2.7) has the special form

$$x_{k+1}(t) = 1 + \int_{0}^{t} x_k(\tau) \, d\tau .$$

Selecting

$$x_0(t) = 1 ,$$

we get the iteration sequence

$$x_1(t) = 1 + \int_0^t 1 \, d\tau = 1 + t \, ,$$

$$x_2(t) = 1 + \int_0^t (1 + \tau) \, d\tau = t + \frac{t^2}{2} \, ,$$

$$x_3(t) = 1 + \int_0^t \left(1 + \tau + \frac{\tau^2}{2} \right) d\tau = 1 + t + \frac{t^2}{2} + \frac{t^3}{3!} \, ,$$

and so on. It is easy to show by using finite induction that $x_k(t)$ is the kth degree Taylor's polynomial of e^t. Hence, $x_k(t) \rightarrow e^t$ as $t \rightarrow \infty$.

This method, however, has only limited practical value, since in many cases the convergence is very slow and the integration can be performed only by using numerical techniques. Therefore, this method is used only in certain special cases. A summary of modern computer methods for solving initial value problems can be found for example, in [42] and in [36].

Higher order initial value problems can be transformed into higher dimensional first-order initial value problems. Consider the nth order one-dimensional explicit ordinary differential equation

$$x^{(n)} = f(t, x, \dot{x}, \dots, x^{(n-1)}) \tag{2.8}$$

with initial conditions

$$x(t_0) = x_0, \dot{x}(t_0) = \dot{x}_0, \dots, x^{(n-1)}(t_0) = x_0^{(n-1)} \, . \tag{2.9}$$

Here $\dot{x}$ and $\ddot{x}$ are the first and second derivatives, respectively. For $k > 2$, $x^{(k)}$ denotes the kth derivative of x.

Introduce the new variables

$$x_1 = x, x_2 = \dot{x}, x_3 = \ddot{x}, \dots, x_n = x^{(n-1)} \, ,$$

then $x^{(n)} = \dot{x}_n$. Therefore, we obtain the following problem:

$$\dot{x}_1 = x_2$$

$$\dot{x}_2 = x_3$$

$$\vdots$$

$$\dot{x}_{n-1} = x_n$$

$$\dot{x}_n = f(t, x_1, x_2, \ldots, x_n) \tag{2.10}$$

with initial values

$$x_1(t_0) = x_0, x_2(t_0) = \dot{x}_0, \ldots, x_n(t_0) = x_0^{(n-1)} . \tag{2.11}$$

Note that the resulting equations form an n-dimensional first-order initial value problem.

Example 2.4

Consider the second-order initial value problem

$$\ddot{x} = t + x + \dot{x}^2, \qquad x(0) = \dot{x}(0) = 1 .$$

By introducing the variables $x_1 = x$ and $x_2 = \dot{x}$, Equation (2.10) can be written as

$$\dot{x}_1 = x_2$$

$$\dot{x}_2 = t + x_1 + x_2^2 ,$$

and the initial conditions (2.11) are $x_1(0) = x_2(0) = 1$. Hence, the second-order initial value problem is reduced to the initial value problem of a system of first-order differential equations.

2.1.2 *Solution of Linear Differential Equations*

The state of continuous linear systems is obtained by solving the governing linear differential equation, which therefore, is the basis for determining the state and investigating the properties of the state of linear systems. We discuss this problem area in this subsection. First the existence of the unique solution is examined, and then the solution is determined in a closed form, which makes the computation procedure very attractive.

In this section the solutions of linear ordinary differential equations of the form

$$\dot{\mathbf{x}} = \mathbf{A}(t)\mathbf{x} + \mathbf{f}(t) \tag{2.12}$$

are examined, where $\mathbf{A}(t)$ is an $n \times n$ matrix, and $\mathbf{f}(t)$ is an n-element vector. It is assumed that all elements a_{ij} of $\mathbf{A}$ and all elements f_i of $\mathbf{f}$ are continuous functions of t on a closed finite interval $I \subset \mathbf{R}$.

Note that the right-hand side function is continuous, and $\|\mathbf{A}(t)\|$ is bounded, since it is continuous. Furthermore, the right-hand side of the differential equation satisfies the Lipschitz condition:

$$\|(\mathbf{A}(t)\mathbf{x}_1 + \mathbf{f}(t)) - (\mathbf{A}(t)\mathbf{x}_2 + \mathbf{f}(t))\| = \|\mathbf{A}(t)(\mathbf{x}_1 - \mathbf{x}_2)\|$$

$$\leq \|\mathbf{A}(t)\| \cdot \|\mathbf{x}_1 - \mathbf{x}_2\|$$

$$\leq L \cdot \|\mathbf{x}_1 - \mathbf{x}_2\|$$

for all $t \in I$ and $\mathbf{x}_1$, $\mathbf{x}_2 \in \mathbf{R}^n$, where L is an upper bound for $\|\mathbf{A}(t)\|$. Consequently, the initial value problem

$$\dot{\mathbf{x}} = \mathbf{A}(t)\mathbf{x} + \mathbf{f}(t), \qquad \mathbf{x}(t_0) = \mathbf{x}_0 \tag{2.13}$$

has a unique solution for all $t_0 \in I$ and $\mathbf{x}_0 \in \mathbf{R}^n$. By using a slight refinement of the proof of Theorem 2.1 we can verify that the unique solution is defined on the entire interval I.

Consider first the corresponding homogeneous equation:

$$\dot{\mathbf{x}} = \mathbf{A}(t)\mathbf{x} . \tag{2.14}$$

Let $t_0 \in I$, and for $k = 1, 2, \ldots, n$ consider the initial conditions

$$\mathbf{x}(t_0) = \mathbf{e}_k , \tag{2.15}$$

with $\mathbf{e}_k$ being the kth basis vector $(0, \ldots, 0, 1, 0, \ldots, 0)^T$, where the kth element equals 1 and all other elements equal 0. Let $\mathbf{x}_k$ denote the unique solution of this initial value problem with fixed k, and construct the $n \times n$ matrix

$$\phi(t, t_0) = (\mathbf{x}_1(t), \mathbf{x}_2(t), \ldots, \mathbf{x}_n(t)) .$$

THEOREM 2.2
For all $t \in I$, matrix $\phi(t, t_0)$ is nonsingular, and the general solution of the homogeneous Equation (2.14) is given as

$$\mathbf{x}(t) = \phi(t, t_0)\mathbf{c} , \tag{2.16}$$

where $\mathbf{c}$ is a constant vector.

PROOF   First we prove that (2.16) satisfies the homogeneous equation
with arbitrary $c \in R^n$. If $c = (c_k)$, then

$$x(t) = \sum_{k=1}^{n} c_k x_k(t) ,$$

therefore,

$$\dot{x}(t) = \sum_{k=1}^{n} c_k \dot{x}_k(t) = \sum_{k=1}^{n} c_k A(t) x_k(t) = A(t) \sum_{k=1}^{n} c_k x_k(t) = A(t) x(t) .$$

Next we verify that $\phi(t, t_0)$ is nonsingular, that is, its columns $x_1(t), \ldots,$
$x_n(t)$ are linearly independent for all $t \in I$. In contrast to this assertion
assume that there is a $t_1 \in I$ such that

$$\alpha_1 x_1(t_1) + \alpha_2 x_2(t_1) + \cdots + \alpha_n x_n(t_1) = 0 ,$$

where the constants $\alpha_1, \ldots, \alpha_n$ are not all zero. Define function

$$z(t) = \alpha_1 x_1(t) + \alpha_2 x_2(t) + \cdots + \alpha_n x_n(t) ,$$

and note that it satisfies equation (2.14), since it has the form of (2.16)
with $c = (\alpha_k)$. Furthermore $z(t_1) = 0$, that is, function z solves the
initial value problem

$$\dot{z} = A(t)z, \qquad z(t_1) = 0 .$$

The linear independence of the basis vectors e_k ($1 \leq k \leq n$) implies that
$z(t_0) \neq 0$, hence $z(t)$ is not identically zero. Note that function $x(t) \equiv 0$
also solves this initial value problem, and we obtained a contradiction
to the uniqueness of the solution.

Finally we show that any solution $x(t)$ of the homogeneous equation
can be written in the form of (2.16). Let $x(t)$ be a solution with $x(t_0) = x_0$,
then x satisfies the initial value problem

$$\dot{x} = A(t)x, \qquad x(t_0) = x_0 . \tag{2.17}$$

Consider next the function

$$z(t) = \phi(t, t_0)x_0 ,$$

which is a solution of the homogeneous differential equation; furthermore,

$$z(t_0) = (x_1(t_0), \ldots, x_n(t_0))x_0 = (e_1, \ldots, e_n)x_0 = I \cdot x_0 = x_0 .$$

Hence x and z solve the same initial value problem. The uniqueness of the solution implies that $z(t) = x(t)$ for all $t \in I$, which completes the proof. ∎

COROLLARY 2.1

From the end of the proof we conclude that the particular solution of the initial value problem (2.17) is given as

$$x(t) = \phi(t, t_0)x_0 . \tag{2.18}$$

Matrix $\phi(t, t_0)$ is called the *fundamental* (or the *transition*) matrix of equation (2.14).

THEOREM 2.3

The fundamental matrix satisfies the following properties:

(i) $\phi(t_0, t_0) = \mathbf{I}$;

(ii) $\phi(t, t_1)\phi(t_1, t_0) = \phi(t, t_0)$;

(iii) $\phi(t_1, t_0)^{-1} = \phi(t_0, t_1)$;

(iv) $(\partial/\partial t)\phi(t, t_0) = \mathbf{A}(t)\phi(t, t_0)$;

(v) $(\partial/\partial t)\phi(t_0, t) = -\phi(t_0, t)\mathbf{A}(t)$.

PROOF

(i)   The construction of the transition matrix implies that

$$\phi(t_0, t_0) = (x_1(t_0), \dots, x_n(t_0)) = (e_1, \dots, e_n) = \mathbf{I} .$$

(ii) The solution of the initial value problem (2.17) is $x(t) = \phi(t, t_0)x_0$. Denote $x_1 = \phi(t_1, t_0)x_0$, then $x(t)$ obviously satisfies the initial value problem

$$\dot{x} = \mathbf{A}(t)x, \qquad x(t_1) = x_1 .$$

From formula (2.18), however, we know that its solution is

$$\phi(t, t_1)x_1 = \phi(t, t_1)\phi(t_1, t_0)x_0 ;$$

therefore, the uniqueness of the solution implies that this function equals x:

$$\phi(t, t_1)\phi(t_1, t_0)x_0 = \phi(t, t_0)x_0 .$$

Vector $\mathbf{x}_0$ is arbitrary, which implies the assertion.

(iii)   Substitute $t = t_0$ into (ii) and use (i).

(iv)   Simple calculation shows that

$$\frac{\partial}{\partial t}\phi(t, t_0) = \frac{d}{dt}(\mathbf{x}_1(t), \ldots, \mathbf{x}_n(t)) = (\dot{\mathbf{x}}_1(t), \ldots, \dot{\mathbf{x}}_n(t))$$

$$= (\mathbf{A}(t)\mathbf{x}_1(t), \ldots, \mathbf{A}(t)\mathbf{x}_n(t)) = \mathbf{A}(t)(\mathbf{x}_1(t), \ldots, \mathbf{x}_n(t))$$

$$= \mathbf{A}(t)\phi(t, t_0) .$$

(v)   From property (iii) we know that

$$\phi(t_0, t)\phi(t, t_0) = \mathbf{I} .$$

Differentiating both sides with respect to t yields the relation

$$\frac{\partial}{\partial t}\phi(t_0, t)\phi(t, t_0) + \phi(t_0, t)\frac{\partial}{\partial t}\phi(t, t_0) = \mathbf{O} ,$$

which implies that

$$\frac{\partial}{\partial t}\phi(t_0, t) = -\phi(t_0, t)\frac{\partial}{\partial t}\phi(t, t_0)\phi^{-1}(t, t_0)$$

$$= -\phi(t_0, t)\mathbf{A}(t)\phi(t, t_0)\phi^{-1}(t, t_0) = -\phi(t_0, t)\mathbf{A}(t) .$$

∎

REMARK 2.3   Assume that the $n \times n$ matrix $\mathbf{X}(t)$ satisfies the matrix initial value problem

$$\dot{\mathbf{X}}(t) = \mathbf{A}(t) \cdot \mathbf{X}(t), \qquad \mathbf{X}(t_0) = \mathbf{I} . \tag{2.19}$$

Then $\mathbf{X}(t) = \phi(t, t_0)$, which follows from the uniqueness of the solution of the initial value problems (2.14) and (2.15).   ∎

Example 2.5

As a particular numerical example, we now solve the initial value problem

$$\dot{\mathbf{x}} = \begin{pmatrix} t & t \\ 0 & 2t \end{pmatrix} \mathbf{x}, \qquad \mathbf{x}(0) = \begin{pmatrix} 1 \\ 1 \end{pmatrix}$$

using the above results.

First the fundamental matrix will be determined by solving the initial-value problem (2.14) and (2.15).

For $k = 1$, we have the problem

$$\dot{\mathbf{x}} = \begin{pmatrix} t & t \\ 0 & 2t \end{pmatrix} \mathbf{x}, \qquad \mathbf{x}(t_0) = \begin{pmatrix} 1 \\ 0 \end{pmatrix}.$$

If x_1 and x_2 denote the components of $\mathbf{x}$, then this equation can be rewritten as

$$\dot{x}_1 = tx_1 + tx_2, \quad x_1(t_0) = 1$$

$$\dot{x}_2 = 2tx_2, \qquad x_2(t_0) = 0.$$

The second equation is separable:

$$\frac{dx_2}{dt} = 2tx_2,$$

from which we have

$$\frac{dx_2}{x_2} = 2t\,dt,$$

and by integration

$$\log x_2 = t^2 + \log C,$$

where the integration constant is $\log C$. Hence,

$$x_2 = Ce^{t^2}.$$

The initial condition implies equality

$$0 = x_2(t_0) = Ce^{t_0},$$

therefore, $C = 0$, and $x_2(t) \equiv 0$. Then the first equation simplifies as

$$\dot{x}_1 = tx_1, \quad x_1(t_0) = 1.$$

By separating the variables we get

$$\frac{dx_1}{dt} = tx_1,$$

or
$$\frac{dx_1}{x_1} = t\,dt\ .$$

By integration
$$\log x_1 = \frac{t^2}{2} + \log C\ ,$$

so
$$x_1(t) = Ce^{\frac{t^2}{2}}\ .$$

The initial condition implies that
$$1 = x_1(t_0) = Ce^{\frac{t_0^2}{2}}\ ,$$

therefore,
$$C = e^{\frac{-t_0^2}{2}}\ ,$$

and hence
$$x_1(t) = e^{\frac{t^2 - t_0^2}{2}}\ .$$

For $k = 2$, we have the similar problem
$$\dot{\mathbf{x}} = \begin{pmatrix} t & t \\ 0 & 2t \end{pmatrix} \mathbf{x}, \qquad \mathbf{x}(t_0) = \begin{pmatrix} 0 \\ 1 \end{pmatrix}\ ,$$

which can be rewritten as
$$\dot{x}_1 = tx_1 + tx_2, \qquad x_1(t_0) = 0$$
$$\dot{x}_2 = 2tx_2, \qquad\qquad x_2(t_0) = 1\ .$$

We have already derived the general solution of the second equation, and this new initial condition implies that
$$1 = x_2(t_0) = Ce^{t_0^2}\ ,$$

so
$$C = e^{-t_0^2}\ ,$$

and hence
$$x_2(t) = e^{t^2 - t_0^2}\ .$$

Substituting this function into the first equation, an inhomogeneous linear equation is obtained:
$$\dot{x}_1 = tx_1 + te^{t^2 - t_0^2}, \qquad x_1(t_0) = 0\ .$$

From the previous case we know that the general solution of the corresponding homogeneous equation is

$$x_1(t) = Ce^{\frac{t^2}{2}}.$$

Assume now that C also depends on t, and substitute this formula to the inhomogeneous equation:

$$\dot{C}e^{\frac{t^2}{2}} + Cte^{\frac{t^2}{2}} = tCe^{\frac{t^2}{2}} + te^{t^2 - t_0^2},$$

which implies that

$$\dot{C} = te^{\frac{t^2}{2} - t_0^2},$$

therefore,

$$C = e^{\frac{t^2}{2} - t_0^2} + k$$

with some constant k. So,

$$x_1(t) = (e^{\frac{t^2}{2} - t_0^2} + k)e^{\frac{t^2}{2}} = e^{t^2 - t_0^2} + ke^{\frac{t^2}{2}}.$$

The initial condition $x_1(t_0) = 0$ implies that $k = -e^{-\frac{t_0^2}{2}}$, and hence

$$x_1(t) = e^{t^2 - t_0^2} - e^{\frac{t^2 - t_0^2}{2}}.$$

The columns of $\phi(t, t_0)$ are the above solution vectors:

$$\phi(t, t_0) = \begin{pmatrix} e^{\frac{t^2 - t_0^2}{2}} & -e^{\frac{t^2 - t_0^2}{2}} + e^{t^2 - t_0^2} \\ 0 & e^{t^2 - t_0^2} \end{pmatrix}.$$

If $t_0 = 0$, then

$$\phi(t, 0) = \begin{pmatrix} e^{\frac{t^2}{2}} & -e^{\frac{t^2}{2}} + e^{t^2} \\ 0 & e^{t^2} \end{pmatrix}.$$

We can easily show that this matrix as $\mathbf{X}(t)$ satisfies relations (2.19) with $t_0 = 0$:

$$\frac{\partial}{\partial t}\phi(t, 0) = \begin{pmatrix} te^{\frac{t^2}{2}} & -te^{\frac{t^2}{2}} + 2te^{t^2} \\ 0 & 2te^{t^2} \end{pmatrix}$$

$$= \begin{pmatrix} t & t \\ 0 & 2t \end{pmatrix}\begin{pmatrix} e^{\frac{t^2}{2}} & -e^{\frac{t^2}{2}} + e^{t^2} \\ 0 & e^{t^2} \end{pmatrix} = \begin{pmatrix} t & t \\ 0 & 2t \end{pmatrix}\phi(t, 0),$$

and

$$\phi(0,0) = \begin{pmatrix} 1 & -1 + 1 \\ 0 & 1 \end{pmatrix} = \mathbf{I} .$$

Then the solution formula (2.18) implies that

$$\mathbf{x}(t) = \begin{pmatrix} e^{\frac{t^2}{2}} & -e^{\frac{t^2}{2}} + e^{t^2} \\ 0 & e^{t^2} \end{pmatrix} \begin{pmatrix} 1 \\ 1 \end{pmatrix} = \begin{pmatrix} e^{t^2} \\ e^{t^2} \end{pmatrix} .$$

The inhomogeneous equation (2.12) will be solved next. Look for the solution in the form

$$\mathbf{x}(t) = \phi(t, t_0)\mathbf{k}(t) , \tag{2.20}$$

which is a modification of (2.16), where $\mathbf{k}$ now depends on t. Since $\phi(t, t_0)$ is nonsingular, an arbitrary vector-valued function can be represented in this form. Simple substitution shows that $\mathbf{x}(t)$ is a solution if and only if

$$\frac{\partial}{\partial t}\phi(t, t_0)\mathbf{k}(t) + \phi(t, t_0)\dot{\mathbf{k}}(t) = \mathbf{A}(t)\phi(t, t_0)\mathbf{k}(t) + \mathbf{f}(t) .$$

By using properties (iii) and (iv) of Theorem 2.3, we conclude that this equation is equivalent to relation

$$\dot{\mathbf{k}}(t) = \phi(t, t_0)^{-1}\mathbf{f}(t) = \phi(t_0, t)\mathbf{f}(t) .$$

If t_0 is any point in I, then

$$\mathbf{k}(t) = \int_{t_0}^{t} \phi(t_0, \tau)\mathbf{f}(\tau) \, d\tau + \mathbf{k}_1 ,$$

where $\mathbf{k}_1$ is a constant vector. Substituting this expression into (2.20),

$$\mathbf{x}(t) = \phi(t, t_0)\mathbf{k}_1 + \int_{t_0}^{t} \phi(t, t_0)\phi(t_0, \tau)\mathbf{f}(\tau) \, d\tau ,$$

that is,

$$\mathbf{x}(t) = \phi(t, t_0)\mathbf{k}_1 + \int_{t_0}^{t} \phi(t, \tau)\mathbf{f}(\tau) \, d\tau , \tag{2.21}$$

where we used property (ii) of Theorem 2.3. This formula gives the *general solution* of the inhomogeneous equation.

The solution of the initial value problem

$$\dot{\mathbf{x}} = \mathbf{A}(t)\mathbf{x} + \mathbf{f}(t) , \qquad \mathbf{x}(t_0) = \mathbf{x}_0 \tag{2.22}$$

can be obtained by substituting the general solution formula (2.21) into the initial condition. Since

$$\mathbf{x}(t_0) = \phi(t_0, t_0)\mathbf{k}_1 + \int_{t_0}^{t_0} \phi(t_0, \tau)\mathbf{f}(\tau)\,d\tau = \mathbf{I}\mathbf{k}_1 = \mathbf{k}_1\ ,$$

the initial condition is satisfied if and only if $\mathbf{k}_1 = \mathbf{x}_0$. Hence the *particular solution* of (2.22) has the form

$$\mathbf{x}(t) = \phi(t, t_0)\mathbf{x}_0 + \int_{t_0}^{t} \phi(t, \tau)\mathbf{f}(\tau)\,d\tau\ . \tag{2.23}$$

In many applications, the coefficient matrix is periodic, that is, there is a $T > 0$ such that for all $t \geq 0$, $\mathbf{A}(t + T) = \mathbf{A}(t)$. It can be proven that the fundamental matrix for a periodic matrix $\mathbf{A}(t)$ can be written in the form

$$\phi(t, t_0) = \mathbf{P}(t)e^{\mathbf{R}(t-t_0)}\mathbf{P}^{-1}(t_0)\ ,$$

where $\mathbf{R}$ is a constant (possible complex) matrix having the same size as $\mathbf{A}(t)$, and $\mathbf{P}(t)$ is a continuously differentiable matrix function that also has the same size and period as $\mathbf{A}(t)$ and is invertible for all t. This form is known as the **Floquet decomposition**. We mention here that the computation of this decomposition is rather complicated and involves computing the natural logarithm of an invertible matrix. For details see, for example, [32], and for a special case, see Problem 2/25.

Consider next the special case when $\mathbf{A}(t) \equiv \mathbf{A}$, that is, when the coefficient matrix is constant. Then the fundamental matrix can be easily obtained as it is given by the following theorem.

THEOREM 2.4
If $\mathbf{A}(t) \equiv \mathbf{A}$, *then*

$$\phi(t, t_0) = e^{\mathbf{A}\cdot(t-t_0)}\ . \tag{2.24}$$

PROOF It is sufficient to show that this matrix satisfies conditions (2.19). Use properties (i), (ii), and (iii) of Theorem 1.17 to conclude that

$$\phi(t_0, t_0) = e^{\mathbf{A}\cdot 0} = \mathbf{I}$$

and

$$\frac{\partial}{\partial t}\phi(t, t_0) = \frac{d}{dt}e^{\mathbf{A}\cdot t} \cdot e^{-\mathbf{A}\cdot t_0} = \mathbf{A}e^{\mathbf{A}\cdot t} \cdot e^{-\mathbf{A}\cdot t_0}$$

$$= \mathbf{A}e^{\mathbf{A}\cdot(t-t_0)} = \mathbf{A}\phi(t, t_0)\ .$$

These equations complete the proof. ∎

Example 2.6

Now we solve the initial value problem

$$\dot{\mathbf{x}} = \begin{pmatrix} 0 & \omega \\ -\omega & 0 \end{pmatrix} \mathbf{x} + \begin{pmatrix} 0 \\ 1 \end{pmatrix}, \qquad \mathbf{x}(0) = \begin{pmatrix} 1 \\ 0 \end{pmatrix}.$$

In Example 1.23 we computed $e^{\mathbf{A}t}$ for this particular coefficient matrix. Therefore, (2.24) implies that

$$\phi(t, t_0) = \begin{pmatrix} \cos\omega(t - t_0) & \sin\omega(t - t_0) \\ -\sin\omega(t - t_0) & \cos\omega(t - t_0) \end{pmatrix},$$

and from (2.23) we conclude that

$$\mathbf{x}(t) = \begin{pmatrix} \cos\omega(t - 0) & \sin\omega(t - 0) \\ -\sin\omega(t - 0) & \cos\omega(t - 0) \end{pmatrix} \begin{pmatrix} 1 \\ 0 \end{pmatrix}$$

$$+ \int_0^t \begin{pmatrix} \cos\omega(t - \tau) & \sin\omega(t - \tau) \\ -\sin\omega(t - \tau) & \cos\omega(t - \tau) \end{pmatrix} \begin{pmatrix} 0 \\ 1 \end{pmatrix} d\tau$$

$$= \begin{pmatrix} \cos\omega t \\ -\sin\omega t \end{pmatrix} + \int_0^t \begin{pmatrix} \sin\omega(t - \tau) \\ \cos\omega(t - \tau) \end{pmatrix} d\tau$$

$$= \begin{pmatrix} \cos\omega t \\ -\sin\omega t \end{pmatrix} + \frac{1}{\omega} \begin{pmatrix} 1 - \cos\omega t \\ \sin\omega t \end{pmatrix}$$

$$= \frac{1}{\omega} \begin{pmatrix} 1 + (\omega - 1)\cos\omega t \\ -(\omega - 1)\sin\omega t \end{pmatrix}.$$

Hence the solution of the initial value problem is determined.

Finally we mention that matrix exponentials (and therefore, the fundamental matrix) can be determined by using standard computer packages.

2.1.3 Laplace Transform

In this subsection we introduce a very useful function transformation, known as *Laplace transform*. It allows us to reduce the solution of linear differential equations with constant coefficients to the solution of algebraic equations, which are easier to solve.

Laplace transforms are defined as follows.

DEFINITION 2.1   *Let $x(t)$ be a real- or vector-valued function defined for $t \geq 0$. Then the Laplace transform of x is defined as the improper integral*

$$X(s) = \int_0^\infty x(t)e^{-ts}\, dt \ , \tag{2.25}$$

where s is a complex variable.

Note that Laplace transforms are mappings from a function space into another function space. In the above notation $X(s)$ denotes the value of the image function at s.

The *abscissa of convergence* is the smallest real number σ such that the integral (2.25) exists for all s such that $Re\, s > \sigma$, where Re means the real part of a complex number.

In the first part of this subsection the fundamental properties of the Laplace transform are summarized and then the Laplace transforms of some well-known functions are determined. The application of Laplace transforms for solving differential equations will be discussed in the third part of this section.

We start our analysis with an important existence theorem.

THEOREM 2.5
Assume that x is integrable and

$$|x(t)| \leq ke^{\sigma t} \qquad (t \geq 0) \tag{2.26}$$

with some real constants $k > 0$ and σ. Then $X(s)$ exists for all $Re\, s > \sigma$.

PROOF   Let $s = s_1 + js_2$, then

$$\int_0^\infty |x(t)e^{-ts}|\, dt \leq \int_0^\infty |ke^{\sigma t} \cdot e^{-ts}|\, dt = k \int_0^\infty e^{(\sigma-s_1)t} \cdot |e^{-js_2 t}|\, dt$$

$$= k \int_0^\infty e^{(\sigma-s_1)t}\, dt \ ,$$

which is finite for all $s_1 > \sigma$.   ∎

The most used properties of Laplace transforms are summarized as follows.

THEOREM 2.6

The following relations are true:

(i)   If $x(t) = x_1(t) + x_2(t)$, then $X(s) = X_1(s) + X_2(s)$.

(ii)   If $x(t) = \alpha x_1(t)$, then $X(s) = \alpha X_1(s)$.

(iii)   If $x(t) = x_1(t - \triangle)$ with some $\triangle > 0$, where $x_1(t) = 0$ for $t < 0$, then

$$X(s) = e^{-s\triangle} X_1(s) .$$

(iv)   If $x(t) = x_1(t)e^{\alpha t}$ with some real constant α, then

$$X(s) = X_1(s - \alpha) .$$

(v)   If $x(t) = x_1(\frac{t}{T})$ $(T > 0)$, then

$$X(s) = T \cdot X_1(sT) .$$

(vi)   If $x(t) = \dot{x}_1(t)$, then

$$X(s) = -x_1(0) + sX_1(s) ,$$

and in general, if $x(t) = x_1^{(n)}(t)$, then

$$X(s) = -x_1^{(n-1)}(0) - sx_1^{(n-2)}(0) - \cdots - s^{n-1}x_1(0) + s^n X_1(s) .$$

(vii)   If $x(t) = \int_0^t x_1(\tau)\,d\tau$, then

$$X(s) = \frac{1}{s}X_1(s) ,$$

and in general, if $x(t) = \int_0^t \int_0^{\tau_1} \cdots \int_0^{\tau_{n-1}} x_1(\tau_n)\,d\tau_n\,d\tau_{n-1} \cdots d\tau_1$, then

$$X(s) = \frac{1}{s^n}X_1(s) .$$

(viii)   If x is the convolution of x_1 and x_2, that is, if

$$x(t) = (x_1 * x_2)(t) = \int_0^\infty x_1(t - \tau)x_2(\tau)\,d\tau ,$$

then

$$X(s) = X_1(s) \cdot X_2(s) .$$

PROOF   Properties (i) and (ii) are the simple consequences of the linearity of the integral. Relations (iii) through (vii) can be shown by simple calculations based on the elementary facts of integral calculus:

(iii) $X(s) = \int_0^\infty x(t)e^{-st} = \int_0^\infty x_1(t - \triangle)e^{-st}\, dt$.
Introduce the new variable $\tau = t - \triangle$, then this integral equals

$$\int_{-\triangle}^\infty x_1(\tau)e^{-s\tau} \cdot e^{-s\triangle}\, d\tau = e^{-s\triangle}\int_0^\infty x_1(\tau)e^{-s\tau}\, d\tau = e^{-s\triangle}X_1(s)\,,$$

where we used the fact that $x_1(\tau) = 0$ for $\tau < 0$.

(iv) $X(s) = \int_0^\infty x_1(t)e^{\alpha t}e^{-st}\, dt = \int_0^\infty x_1(t)e^{-(s-\alpha)t}\, dt = X_1(s - \alpha)$.

(v) $X(s) = \int_0^\infty x_1(\frac{t}{T})e^{-st}\, dt = \int_0^\infty T \cdot x_1(\tau)e^{-sT\tau}\, d\tau = T \cdot X_1(sT)$,
where we introduced the new variable $\tau = t/T$.

(vi)   Integrating by parts,

$$X(s) = \int_0^\infty \dot{x}_1(t)e^{-st}\, dt = [x_1(t)e^{-st}]_0^\infty - \int_0^\infty x_1(t)(-s)e^{-st}\, dt$$

$$= -x_1(0) + sX_1(s)\,,$$

and the general case can be proven by the repeated application of the above formula. Let $x_{1k}(t)$ denote the kth derivative of $x_1(t)$. Then

$$X(s) = X_{1n}(s) = -x_1^{(n-1)}(0) + s \cdot X_{1,n-1}(s)$$

$$= -x_1^{(n-1)}(0) + s \cdot [-x_1^{(n-2)}(0) + s \cdot X_{1,n-2}(s)]$$

$$= -x_1^{(n-1)}(0) - s \cdot x_1^{(n-2)}(0) + s^2 X_{1,n-2}(s) = \cdots$$

$$= -x_1^{(n-1)}(0) - s \cdot x_1^{(n-2)}(0) - s^2 \cdot x_1^{(n-3)}(0) - \cdots$$

$$- s^{n-1} \cdot x_1(0) + s^n \cdot X_1(s)\,.$$

(vii) Using the previous identities and observing that $x_1 = \dot{x}$ (or $x_1 = x^{(n)}$), furthermore $x(0) = 0$ (or $x(0) = \dot{x}(0) = \cdots = x^{(n-1)}(0) = 0$), we obtain identities

$$X_1(s) = s \cdot X(s) \qquad (\text{or } X_1(s) = s^n \cdot X(s)) \,,$$

from which the assertions follow immediately.

(viii)   Interchange the integrals and use identity (iii) to have

$$X(s) = \int_0^\infty \left(\int_0^\infty x_1(t-\tau)x_2(\tau)\, d\tau \right) e^{-st}\, dt$$

$$= \int_0^\infty \left(\int_0^\infty x_1(t-\tau)e^{-st}\, dt \right) x_2(\tau)\, d\tau$$

$$= \int_0^\infty X_1(s)e^{-s\tau} \cdot x_2(\tau)\, d\tau$$

$$= X_1(s) \cdot X_2(s) \, . \quad \blacksquare$$

The Laplace transforms of the most commonly used functions are summarized in Table 2.1. These relations are verified next.

Table 2.1   Laplace Transforms of Common Functions

No.	$x(t)$	$X(s)$
1	e^{at}	$\frac{1}{s-a}$
2	unit step $= \begin{cases} 1 \text{ if } t > 0 \\ 0 \text{ otherwise} \end{cases}$	$\frac{1}{s}$
3	t^n	$\frac{n!}{s^{n+1}}$
4	$t^n e^{at}$	$\frac{n!}{(s-a)^{n+1}}$
5	$\cos \omega t$	$\frac{s}{s^2+\omega^2}$
6	$\sin \omega t$	$\frac{\omega}{s^2+\omega^2}$
7	$e^{at}\cos \omega t$	$\frac{s-\alpha}{(s-\alpha)^2+\omega^2}$
8	$e^{at}\sin \omega t$	$\frac{\omega}{(s-\alpha)^2+\omega^2}$
9	$\delta_\varepsilon(t) = \begin{cases} 1/\varepsilon \text{ if } 0 < t \le \varepsilon \\ 0 \quad \text{otherwise} \end{cases}$	$\frac{1}{s\varepsilon}(1 - e^{-s\varepsilon})$
10	unit impulse $\delta(t) = \lim_{\varepsilon \to 0} \delta_\varepsilon(t)$	1

Numbers 1, 3, and 9 are proven by simple integration:

$$\int_0^\infty e^{at} \cdot e^{-st} \, dt = \int_0^\infty e^{(a-s)t} \, dt = \frac{1}{s-a} \qquad (Res > a),$$

$$\int_0^\infty t^n e^{-st} \, dt = \left[t^n \cdot \frac{e^{-st}}{-s} \right]_0^\infty - \int_0^\infty n t^{n-1} \cdot \frac{e^{-st}}{-s} \, dt$$

$$= \frac{n}{s} \int_0^\infty t^{n-1} e^{-st} \, dt = \frac{n(n-1)}{s^2} \int_0^\infty t^{n-2} e^{-st} \, dt \dots$$

$$\vdots$$

$$= \frac{n!}{s^n} \int_0^\infty t^0 e^{-st} \, dt = \frac{n!}{s^{n+1}} \ ,$$

and

$$\int_0^\infty \delta_\varepsilon(t) e^{-st} \, dt = \int_0^\varepsilon \frac{1}{\varepsilon} e^{-st} \, dt = \frac{1}{s\varepsilon}(1 - e^{-s\varepsilon}) \ .$$

Select $a = 0$ in No. 1 to prove identity 2. Use No. 3 and Property (iv) of Theorem 2.6 to verify No. 4. Note that No. 5 and No. 6 are implied by identities

$$\cos \omega t = \frac{e^{j\omega t} + e^{-j\omega t}}{2} \ , \qquad \sin \omega t = \frac{e^{j\omega t} - e^{-j\omega t}}{2j}$$

and the linearity of the Laplace transform. Use again property (iv) of Theorem 2.6 and the previous cases to show No. 7 and 8. The last row of the table is verified by letting $\varepsilon \to 0$ in the result of No. 9:

$$\lim_{\varepsilon \to 0} \frac{1}{s\varepsilon}(1 - e^{-s\varepsilon}) = \lim_{\varepsilon \to 0} \frac{se^{-s\varepsilon}}{s} = 1 \ ,$$

where we used the L'Hospital rule.

As an example, Figure 2.3 illustrates function t^n and its Laplace transform for $n = 2$. For the sake of simplicity, only nonnegative values of t and s are considered.

Note that function δ_ε is very seldom used in practical cases. However, it has a great theoretical importance because the unit impulse function is defined as its limit for $\varepsilon \to 0$.

Laplace transforms are very useful in solving linear ordinary differential equations with constant coefficients. The main idea is based on

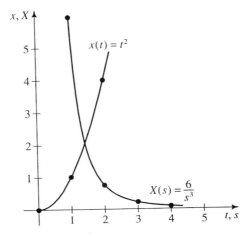

Figure 2.3   A function and its Laplace transform.

Property (vi) of Theorem 2.6 reducing differential equations to algebraic equations, which are easy to solve. Consider first an nth order initial value problem

$$x^{(n)} + a_{n-1}x^{(n-1)} + \cdots + a_1\dot{x} + a_0x = f(t) ,$$

$$x(0) = x_0, \dot{x}(0) = \dot{x}_0, \ldots, x^{(n-1)}(0) = x_0^{(n-1)} . \qquad (2.27)$$

Apply the Laplace transform on both sides, and use Property (vi) of Theorem 2.6 to obtain equality

$$[-x^{(n-1)}(0) - sx^{(n-2)}(0) - \cdots - s^{n-1}x(0) + s^n X(s)] +$$

$$+ a_{n-1}[-x^{(n-2)}(0) - sx^{(n-3)}(0) - \cdots - s^{n-2}x(0) + s^{n-1}X(s)]$$

$$+ \cdots + a_1[-x(0) + sX(s)] + a_0 X(s) = F(s) ,$$

where $X(s)$ and $F(s)$ are the Laplace transforms of $x(t)$ and $f(t)$, respectively. This equation can be rewritten as

$$X(s) = \frac{q(s)}{p(s)} + \frac{F(s)}{p(s)} , \qquad (2.28)$$

where

$$p(s) = s^n + a_{n-1}s^{n-1} + \cdots + a_1s + a_0 ,$$

and

$$q(s) = s^{n-1}x(0) + s^{n-2}[\dot{x}(0) + a_{n-1}x(0)] + \cdots$$

$$+ s[x^{(n-2)}(0) + a_{n-1}x^{(n-3)}(0)$$

$$+ \cdots + a_2 x(0)] + [x^{(n-1)}(0) + a_{n-1}x^{(n-2)}(0) + \cdots + a_1 x(0)] .$$

Note that in the special case when all initial values $x(0) = \dot{x}(0) = \cdots = x^{(n-1)}(0) = 0$, $q(s) = 0$. That is, the first term becomes zero. After $X(s)$ is computed, $x(t)$ has to be determined. In special cases we can use the results of Table 2.1. As an example, assume that $F(s)$ is a rational function. Then $X(s)$ is also rational. Assume that $X(s)$ is strictly proper and the roots $\lambda_1, \ldots, \lambda_r$ of its denominator are distinct. Then the *partial fraction expansion* of $X(s)$ is as follows:

$$X(s) = \frac{R_1}{s - \lambda_1} + \frac{R_2}{s - \lambda_2} + \cdots + \frac{R_r}{s - \lambda_r} , \qquad (2.29)$$

where $R_1, \ldots, R_r$ are constants. Use No. 1 of Table 2.1 to conclude that

$$x(t) = R_1 e^{\lambda_1 t} + R_2 e^{\lambda_2 t} + \cdots + R_r e^{\lambda_r t} . \qquad (2.30)$$

Assume next that the roots $\lambda_1, \ldots, \lambda_r$ have multiplicities $m_1, \ldots, m_r$. Then the *partial fraction expansion* of $X(s)$ can be written as

$$X(s) = \sum_{i=1}^{r} \left[\frac{R_{i1}}{s - \lambda_i} + \frac{R_{i2}}{(s - \lambda_i)^2} + \cdots + \frac{R_{im_i}}{(s - \lambda_i)^{m_i}} \right] , \qquad (2.31)$$

where the numbers R_{ij} can be determined by using elementary methods, which will be illustrated in Example 2.7. To recover $x(t)$, use No. 4 of Table 2.1:

$$x(t) = \sum_{i=1}^{r} e^{\lambda_i t} \left[R_{i1} + \frac{R_{i2} t}{1!} + \cdots + \frac{R_{im_i} t^{m_i - 1}}{(m_i - 1)!} \right] . \qquad (2.32)$$

This procedure can be summarized as follows:

Step 1   Apply Laplace transform on both sides of the differential equation.

Step 2   Solve the resulting algebraic equation.

Step 3   Apply inverse transform to recover the solution of the original problem.

Figure 2.4 illustrates this procedure.

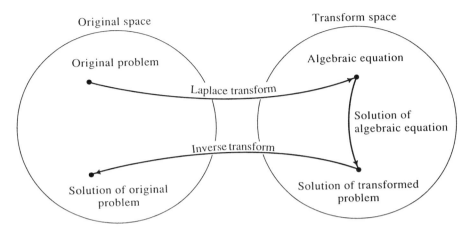

Figure 2.4 Application of Laplace transforms in solving differential equations.

Example 2.7

First we solve the initial value problem

$$\ddot{x} - 3\dot{x} + 2x = 4, \qquad x(0) = 0, \qquad \dot{x}(0) = 1 .$$

By using Laplace transform on both sides of the equation we have

$$-\dot{x}(0) - sx(0) + s^2 X(s) - 3(-x(0) + sX(s)) + 2X(s) = \frac{4}{s} ,$$

which implies that

$$X(s)(s^2 - 3s + 2) = \frac{4}{s} + 1 .$$

Therefore,

$$X(s) = \frac{1 + \frac{4}{s}}{(s-1)(s-2)} = \frac{s+4}{s(s-1)(s-2)} .$$

Because the denominator has distinct roots, relations (2.29) and (2.30) can be used. The unknown coefficients of the partial fraction expansion

$$\frac{s+4}{s(s-1)(s-2)} = \frac{R_1}{s} + \frac{R_2}{s-1} + \frac{R_3}{s-2} \qquad (2.33)$$

can be obtained by applying one of the following methods.

Method 1 (Algebraic approach). Multiply both sides by $s(s-1)(s-2)$ and rearrange the terms to get

$$s + 4 = s^2(R_1 + R_2 + R_3) + s(-3R_1 - 2R_2 - R_3) + (2R_1) .$$

Comparing the like coefficients, the linear equations

$$R_1 + R_2 + R_3 = 0$$

$$-3R_1 - 2R_2 - R_3 = 1$$

$$2R_1 = 4$$

are obtained; therefore, the solution is

$$R_1 = 2, \quad R_2 = -5, \quad R_3 = 3 .$$

Thus, we conclude that

$$X(s) = \frac{2}{s} - \frac{5}{s-1} + \frac{3}{s-2} ,$$

and (2.30) implies that the solution is as follows:

$$x(t) = 2e^{0 \cdot t} - 5e^{1 \cdot t} + 3e^{2 \cdot t} = 2 - 5e^t + 3e^{2t} .$$

Method 2 (Residue method). First multiply both sides of (2.33) by s and then evaluate the resulting equation at $s = 0$ to get

$$\frac{s+4}{(s-1)(s-2)}\Big|_{s=0} = R_1 + s\left(\frac{R_2}{s-1} + \frac{R_3}{s-2}\right)\Big|_{s=0} ,$$

which implies that

$$R_1 = \frac{4}{(-1)(-2)} = 2 .$$

Next multiply both sides of (2.33) by $(s - 1)$ and evaluate at $s = 1$:

$$\frac{s+4}{s(s-2)}\Big|_{s=1} = R_2 + (s-1)\left(\frac{R_1}{s} + \frac{R_3}{s-2}\right)\Big|_{s=1} ;$$

therefore,

$$R_2 = \frac{1+4}{1\cdot(-1)} = -5 .$$

And finally, multiply both sides of (2.33) by $(s - 2)$ and evaluate at $s = 2$. The resulting relation is

$$\frac{s+4}{s(s-1)}\Big|_{s=2} = R_3 + (s-2)\left(\frac{R_1}{s} + \frac{R_2}{s-1}\right)\Big|_{s=2} ,$$

that is,

$$R_3 = \frac{2+4}{2\cdot1} = 3 .$$

Note that this is the same result we derived before.

Example 2.8

Next we solve the initial value problem

$$\ddot{x} - \dot{x} = 1, \; x(0) = 0, \; \dot{x}(0) = 1 .$$

By using Laplace transform on both sides of the equation we have

$$-\dot{x}(0) - sx(0) + s^2 X(s) - (-x(0) + sX(s)) = \frac{1}{s} ,$$

which can be rewritten as

$$X(s)(s^2 - s) = \frac{1}{s} + 1 .$$

Therefore,

$$X(s) = \frac{1 + \frac{1}{s}}{s(s-1)} = \frac{s+1}{s^2(s-1)} .$$

Note that the roots of the denominator are $\lambda_1 = 0$ and $\lambda_2 = 1$ with multiplicities $m_1 = 2$ and $m_2 = 1$. Therefore, $X(s)$ can be written in the more general form of (2.31), which specializes to the following:

$$\frac{s+1}{s^2(s-1)} = \frac{R_{11}}{s} + \frac{R_{12}}{s^2} + \frac{R_{21}}{s-1} . \tag{2.34}$$

The unknown constants R_{11}, R_{12}, and R_{21} can be determined by using either of the methods shown in the previous example.

Method 1. Multiply both sides by $s^2(s-1)$ and rearrange the terms:

$$s + 1 = s^2(R_{11} + R_{21}) + s(-R_{11} + R_{12}) + (-R_{12}) .$$

Compare the like terms to get the linear equations

$$R_{11} + R_{21} = 0$$

$$-R_{11} + R_{12} = 1$$

$$-R_{12} = 1 .$$

The solution is

$$R_{11} = -2, \qquad R_{12} = -1, \qquad R_{21} = 2 .$$

Method 2. Multiply both sides of (2.34) by s^2 to get equation

$$\frac{s+1}{s-1} = R_{11}s + R_{12} + s^2 \cdot \frac{R_{21}}{s-1} .$$

Substitute $s = 0$ to conclude that $R_{12} = -1$. Because R_{11} is multiplied by s, it cancels at $s = 0$. That is, R_{11} cannot be determined directly. However, by differentiating both sides, multiplier s disappears. The resulting equation becomes

$$\frac{(s-1) - (s+1)}{(s-1)^2} = R_{11} + s^2 \cdot \frac{-R_{21}}{(s-1)^2} + 2s \cdot \frac{R_{21}}{s-1} .$$

Evaluate this equation at $s = 0$ to conclude that

$$R_{11} = -2 .$$

Multiply next both sides of (2.34) by $s - 1$ and evaluate at $s = 1$:

$$\frac{s+1}{s^2}\Big|_{s=1} = R_{21} + (s-1)\left(\frac{R_{11}}{s} + \frac{R_{12}}{s^2}\right)\Big|_{s=1} ,$$

which implies that $R_{21} = 2$.

Therefore,

$$X(s) = -\frac{2}{s} - \frac{1}{s^2} + \frac{2}{s-1} ,$$

and from relation (2.32) we have the solution:

$$x(t) = e^{0 \cdot t}(-2 - t) + e^{1 \cdot t} \cdot 2 = -2 - t + 2e^t \ .$$

In the general case, when a root λ_i has multiplicity m_i, multiply first Equation (2.31) by $(s - \lambda_i)^{m_i}$, then differentiate the resulting equation $0, 1, 2, \ldots, m_i - 1$ times and substitute $s = \lambda_i$ to these equations to recover the constants $R_{i1}, R_{i2}, \ldots, R_{im_i}$.

Our next example illustrates a case when $X(s)$ is rational but not strictly proper.

Example 2.9

Assume that the Laplace transform of a function is

$$X(s) = \frac{s^2 + s - 1}{s(s - 1)} \ .$$

Then it can be rewritten as the sum of a polynomial and a strictly proper rational function:

$$X(s) = X_1(s) + X_2(s)$$

with

$$X_1(s) = 1 \text{ and } X_2(s) = \frac{2s - 1}{s(s - 1)} \ .$$

Simple calculation shows that

$$X_2(s) = \frac{1}{s} + \frac{1}{s - 1} \ .$$

Then the last row of Table 2.1 and relation (2.30) imply that

$$x(t) = x_1(t) + x_2(t) = \delta(t) + e^{0 \cdot t} \cdot 1 + e^{1 \cdot t} \cdot 1$$

$$= \delta(t) + 1 + e^t \ .$$

Consider next the system

$$\dot{\mathbf{x}} = \mathbf{A}\mathbf{x} + \mathbf{f}(t) \ , \qquad \mathbf{x}(0) = \mathbf{x}_0 \ , \tag{2.35}$$

where $\mathbf{A}$ is an $n \times n$ constant matrix and $\mathbf{f}$ is an n-dimensional function of t. By applying the Laplace transform on both sides, we have equation

$$-\mathbf{x}(0) + s\mathbf{X}(s) = \mathbf{A} \cdot \mathbf{X}(s) + \mathbf{F}(s) \ ,$$

that is,

$$(s\mathbf{I} - \mathbf{A})\mathbf{X}(s) = \mathbf{x}_0 + \mathbf{F}(s) \ .$$

Assuming that s is not an eigenvalue of $\mathbf{A}$, $s\mathbf{I} - \mathbf{A}$ is invertible and

$$\mathbf{X}(s) = (s\mathbf{I} - \mathbf{A})^{-1}(\mathbf{x}_0 + \mathbf{F}(s)) \ . \tag{2.36}$$

The components of $\mathbf{x}(t)$ can then be recovered by applying the above methods.

Example 2.10

Consider again the initial value problem

$$\dot{\mathbf{x}} = \begin{pmatrix} 0 & \omega \\ -\omega & 0 \end{pmatrix} \mathbf{x} + \begin{pmatrix} 0 \\ 1 \end{pmatrix}, \qquad \mathbf{x}(0) = \begin{pmatrix} 1 \\ 0 \end{pmatrix}$$

of Example 2.6. In our case,

$$\mathbf{A} = \begin{pmatrix} 0 & \omega \\ -\omega & 0 \end{pmatrix}, \qquad \mathbf{f}(t) = \begin{pmatrix} 0 \\ 1 \end{pmatrix}, \text{ and } \qquad \mathbf{x}_0 = \begin{pmatrix} 1 \\ 0 \end{pmatrix} \ .$$

By using No. 2 of Table 2.1 we have

$$\mathbf{X}(s) = \begin{pmatrix} s & -\omega \\ \omega & s \end{pmatrix}^{-1} \begin{pmatrix} 1 \\ 1/s \end{pmatrix}$$

$$= \frac{1}{s^2 + \omega^2} \begin{pmatrix} s & \omega \\ -\omega & s \end{pmatrix} \begin{pmatrix} 1 \\ 1/s \end{pmatrix}$$

$$= \begin{pmatrix} \frac{\omega + s^2}{s(s^2 + \omega^2)} \\ \frac{-\omega + 1}{s^2 + \omega^2} \end{pmatrix} = \begin{pmatrix} \frac{1/\omega}{s} + \frac{(1 - 1/\omega)s}{s^2 + \omega^2} \\ \frac{-\omega + 1}{s^2 + \omega^2} \end{pmatrix},$$

and from Nos. 5 and 6 of Table 2.1 we have

$$\mathbf{x}(t) = \begin{pmatrix} \frac{1}{\omega} + (1 - \frac{1}{\omega}) \cos \omega t \\ \frac{-\omega + 1}{\omega} \sin \omega t \end{pmatrix} = \frac{1}{\omega} \begin{pmatrix} 1 + (\omega - 1) \cos \omega t \\ -(w - 1) \sin \omega t \end{pmatrix} \ .$$

Hence, the solution is determined.

In cases when we cannot recognize $x(t)$ from $X(s)$ by using Table 2.1 and/or elementary methods, the following general result may be useful.

THEOREM 2.7
Assume that σ is the abscissa of convergence in the Laplace transform of $x(t)$. Then $x(t)$ can be determined with the inversion formula

$$x(t) = \frac{1}{2\pi j} \int_{\sigma_0 - j\infty}^{\sigma_0 + j\infty} X(s)e^{st}\, ds, \qquad (2.37)$$

where $\sigma_0 > \sigma$ is arbitrary.

REMARK 2.4   Note that the integration domain may be any vertical line in the complex plane that lies in the region of convergence.
   The proof of this result can be found for example in [26]. We note that the integral (2.37) is usually difficult to compute, therefore, the direct use of the inversion formula is not an easy task. However, one may apply numerical integration and standard computer programs to determine the values of the integral (2.37) for given values of t.   ∎

2.2   *Solution of Difference Equations*

Dynamic systems with discrete time scale are usually modeled by a system of difference equations. The solution of the difference equations represent the state of the dynamic system. Therefore, the determination of the state requires the solution of the governing difference equations. Similar to the differential equation case discussed earlier in Section 2.1.2, this section gives a summary of the existence and solution methodology of nonlinear and linear difference equations.

2.2.1   *General Solutions*

In this subsection, the existence of the unique solution of the initial value problem of difference equations is discussed. Similar to the differential equation case, the results of this subsection are useful in determining whether or not a given difference equation really represents a dynamic system with discrete time scale.
   Consider the first-order explicit difference equation

$$\mathbf{x}(t + 1) = \mathbf{f}(t, \mathbf{x}(t)) \qquad (2.38)$$

where $\mathbf{x} : \mathbf{N} \to \mathbf{R}^n$ is an integer variable vector-valued unknown function (here $\mathbf{N} = \{0, 1, 2, \ldots\}$) and $\mathbf{f} : \mathbf{N} \times X \to \mathbf{R}^n$, where $X \subseteq \mathbf{R}^n$. A function $\mathbf{x}$ is called the solution of the above difference equation if for all $t \in \mathbf{N}$, $\mathbf{x}(t) \in X$ and Equation (2.38) is satisfied. In the case of difference equations, we will always assume that the initial time is zero.

This assumption will make all further formulations easier without losing the essence of the problems. If the initial time $t_0 \neq 0$, then we have to introduce a new time-variable $t' = t - t_0$. Note furthermore that there is an obvious analogy between the difference equations (2.38) and the iteration process (1.18) for solving fixed-point problems.

Note first that function $\mathbf{f}$ and set X should satisfy certain conditions in order to have a solution, as illustrated in the following example. If no solution exists, Equation (2.38) cannot describe the behavior of dynamic systems.

Example 2.11

Consider equation

$$x(t+1) = \sqrt{-|x(t)|} - 1 .$$

Note first that function

$$f(t, x) = \sqrt{-|x|} - 1$$

is defined only for $x = 0$. Therefore, we must select $x(0) = 0$ in order to start the solution. Then

$$x(1) = \sqrt{-|x(0)|} - 1 = -1 ,$$

which cannot be substituted into function f. Hence no solution exists.

Our main existence result can be summarized as follows.

THEOREM 2.8
Assume that for all $\mathbf{x} \in X$ and $t \in \mathbf{N}$, $\mathbf{f}(t, \mathbf{x}) \in X$. Then there is a unique solution in X starting from any arbitrary initial value $\mathbf{x}(0) = \mathbf{x}_0 \in X$.

PROOF   If $\mathbf{x}(0) \in X$, then $\mathbf{f}(0, \mathbf{x}(0))$ exists and is in X. Therefore, $\mathbf{x}(1) \in X$. By induction, assume that $\mathbf{x}(t) \in X$. Then $f(t, \mathbf{x}(t))$ exists and is in X, that is, $\mathbf{x}(t + 1) \in X$. Hence, the proof is completed.   ∎

Higher order difference equations are given by the recursion

$$x(t+n) = f(t, x(t), x(t+1), \ldots, x(t+n-1)) , \tag{2.39}$$

and in order to guarantee the uniqueness of the solution, the initial values

$$x(0) = x_0, x(1) = x_1, \ldots, x(n-1) = x_{n-1} \tag{2.40}$$

are given. As in the case of higher order ordinary differential equations, it is possible to transform Equation (2.39) into a first-order system of the form (2.38). This transformation is based on the introduction of the new variables

$$x_1(t) = x(t), x_2(t) = x(t+1), \ldots, x_n(t) = x(t+n-1) .$$

Then we have the first-order system

$$x_1(t+1) = x_2(t)$$

$$x_2(t+1) = x_3(t)$$

$$\vdots$$

$$x_{n-1}(t+1) = x_n(t)$$

$$x_n(t+1) = f(t, x_1(t), \ldots, x_n(t)) \tag{2.41}$$

of difference equations with initial values

$$x_1(0) = x_0, x_2(0) = x_1, \ldots, x_n(0) = x_{n-1} . \tag{2.42}$$

Example 2.12

Consider the second-order initial value problem

$$x(t+2) = t + x(t) + x(t+1)^2, \qquad x(0) = x(1) = 1 .$$

Introduce the new variables $x_1(t) = x(t)$ and $x_2(t) = x(t+1)$, then Equation (2.41) can be rewritten as

$$x_1(t+1) = x_2(t)$$

$$x_2(t+1) = t + x_1(t) + x_2(t)^2 ,$$

and the initial conditions (2.42) are

$$x_1(0) = x_2(0) = 1 .$$

2.2.2   Solution of Linear Difference Equations

The state of discrete linear systems is obtained as the solution of the governing linear difference equation, which therefore, is the basis for determining the state and investigating the state properties. In this subsection the solution of linear difference equations is examined; the closed form solution introduced next is often applied in solving practical problems.

Consider the linear difference equations of the form

$$\mathbf{x}(t+1) = \mathbf{A}(t)\mathbf{x}(t) + \mathbf{f}(t) \tag{2.43}$$

where $\mathbf{A}(t)$ is an $n \times n$ matrix and $\mathbf{f}(t)$ is an n-vector defined for all $t \in \mathbf{N}$. Obviously this equation satisfies the conditions of Theorem 2.8 with $X = \mathbf{R}^n$; therefore, for all $\mathbf{x}_0 \in \mathbf{R}^n$, there is a unique solution satisfying the initial condition $\mathbf{x}(0) = \mathbf{x}_0$.

First, a general formula will be derived for the solution of Equation (2.43). The repeated application of Equation (2.43) gives the solutions:

$$\mathbf{x}(1) = \mathbf{A}(0)\mathbf{x}_0 + \mathbf{f}(0) \ ,$$

$$\mathbf{x}(2) = \mathbf{A}(1)\mathbf{x}(1) + \mathbf{f}(1)$$

$$= \mathbf{A}(1)\mathbf{A}(0)\mathbf{x}_0 + \mathbf{A}(1)\mathbf{f}(0) + \mathbf{f}(1) \ ,$$

$$\mathbf{x}(3) = \mathbf{A}(2)\mathbf{x}(2) + \mathbf{f}(2)$$

$$= \mathbf{A}(2)\mathbf{A}(1)\mathbf{A}(0)\mathbf{x}_0 + \mathbf{A}(2)\mathbf{A}(1)\mathbf{f}(0) + \mathbf{A}(2)\mathbf{f}(1) + \mathbf{f}(2) \ ,$$

and so on. By using finite induction, it is easy to see that, in general,

$$\mathbf{x}(t) = \mathbf{A}(t-1)\mathbf{A}(t-2)\dots\mathbf{A}(1)\mathbf{A}(0)\mathbf{x}_0 +$$

$$\sum_{\tau=0}^{t-2} \mathbf{A}(t-1)\mathbf{A}(t-2)\dots\mathbf{A}(\tau+1)\mathbf{f}(\tau) + \mathbf{f}(t-1) \ .$$

The third term, $\mathbf{f}(t-1)$, can be considered as the extension of the summation for $\tau = t-1$, because in this case, $\tau + 1 > t - 1$, so $\mathbf{A}(t-1)\mathbf{A}(t-2)\cdots\mathbf{A}(\tau+1)$ represents an "empty" product with identity matrix value. Introduce the notation

$$\phi(t, \tau) = \mathbf{A}(t-1)\mathbf{A}(t-2)\cdot\ldots\cdot\mathbf{A}(\tau)$$

to obtain the general formula

$$\mathbf{x}(t) = \phi(t,0)\mathbf{x}_0 + \sum_{\tau=0}^{t-1} \phi(t,\tau+1)\mathbf{f}(\tau) \ . \tag{2.44}$$

Note that this equation is analogous to the general solution (2.23) of linear inhomogeneous first-order differential equations. In that case, integral substitutes the summation.

Consider next the homogeneous case when $\mathbf{f}(t) = \mathbf{0}$ for all $t \in \mathbf{N}$. Then the second term of (2.44) equals zero, and hence, the general solution of the resulting homogeneous equation is as follows:

$$\mathbf{x}(t) = \phi(t,0)\mathbf{x}_0 \ , \tag{2.45}$$

which is exactly the same formula as (2.18) for the solution of the corresponding differential equations.

Example 2.13

We now solve the one-dimensional equation

$$x(t+1) = (t+1)^2 \cdot x(t) + 1$$

with initial condition $x(0) = 1$. In this case $n = 1$, $\mathbf{A}(t) = (t+1)^2$, and $\mathbf{f}(t) = 1$. Therefore,

$$\phi(t,\tau) = t^2 \cdot (t-1)^2 \cdot \ldots \cdot (\tau+1)^2 = \left(\frac{t!}{\tau!}\right)^2 ,$$

and so

$$x(t) = \left(\frac{t!}{0!}\right)^2 \cdot 1 + \sum_{\tau=0}^{t-1} \left(\frac{t!}{(\tau+1)!}\right)^2 \cdot 1$$

$$= (t!)^2 + (t!)^2 \sum_{\tau=0}^{t-1} \frac{1}{((\tau+1)!)^2}$$

$$= (t!)^2 \sum_{\tau=0}^{t} \frac{1}{(\tau!)^2} \ .$$

Similarly to the case of linear differential equations, $\phi(t,\tau)$ is called the *fundamental* (or *transition*) matrix of Equation (2.43).

Consider next the special case, when $\mathbf{A}(t) = \mathbf{A}$ and $\mathbf{f}(t) = \mathbf{f}$, that is, when the equation is time-invariant. Then the transition matrix has the special form

$$\phi(t, \tau) = \mathbf{A}^{t-\tau} , \qquad (2.46)$$

and therefore, the solution of the initial value problem

$$\mathbf{x}(t + 1) = \mathbf{A} \cdot \mathbf{x}(t) + \mathbf{f}, \qquad \mathbf{x}(0) = \mathbf{x}_0$$

can be written as

$$\mathbf{x}(t) = \mathbf{A}^t \cdot \mathbf{x}_0 + \left(\sum_{\tau=0}^{t-1} \mathbf{A}^{t-\tau-1} \right) \cdot \mathbf{f} . \qquad (2.47)$$

Note that $\mathbf{A}^0 = \mathbf{I}$ in the term of $\tau = t - 1$. We mention that the matrix operations needed to implement this solution formula can be performed by using standard program packages.

This formula is illustrated next.

Example 2.14

Consider the initial value problem

$$\mathbf{x}(t + 1) = \begin{pmatrix} 1 & 1 \\ 0 & 1 \end{pmatrix} \mathbf{x}(t) + \begin{pmatrix} 0 \\ 1 \end{pmatrix}, \qquad \mathbf{x}(0) = \begin{pmatrix} 1 \\ 0 \end{pmatrix} .$$

In this case $n = 2$,

$$\mathbf{A} = \begin{pmatrix} 1 & 1 \\ 0 & 1 \end{pmatrix}, \qquad \mathbf{f} = \begin{pmatrix} 0 \\ 1 \end{pmatrix}, \qquad \text{and} \qquad \mathbf{x}_0 = \begin{pmatrix} 1 \\ 0 \end{pmatrix} .$$

Note first that

$$\mathbf{A}^2 = \begin{pmatrix} 1 & 1 \\ 0 & 1 \end{pmatrix} \begin{pmatrix} 1 & 1 \\ 0 & 1 \end{pmatrix} = \begin{pmatrix} 1 & 2 \\ 0 & 1 \end{pmatrix} ,$$

$$\mathbf{A}^3 = \begin{pmatrix} 1 & 2 \\ 0 & 1 \end{pmatrix} \begin{pmatrix} 1 & 1 \\ 0 & 1 \end{pmatrix} = \begin{pmatrix} 1 & 3 \\ 0 & 1 \end{pmatrix} ,$$

and by finite induction one may easily verify that

$$\mathbf{A}^t = \begin{pmatrix} 1 & t \\ 0 & 1 \end{pmatrix} .$$

Therefore, relation (2.47) implies that

$$\mathbf{x}(t) = \begin{pmatrix} 1 & t \\ 0 & 1 \end{pmatrix} \cdot \begin{pmatrix} 1 \\ 0 \end{pmatrix} + \sum_{\tau=0}^{t-1} \begin{pmatrix} 1 & t - \tau - 1 \\ 0 & 1 \end{pmatrix} \begin{pmatrix} 0 \\ 1 \end{pmatrix}$$

$$= \begin{pmatrix} 1 \\ 0 \end{pmatrix} + \sum_{\tau=0}^{t-1} \begin{pmatrix} t - \tau - 1 \\ 1 \end{pmatrix}$$

$$= \begin{pmatrix} 1 + \sum_{\tau=0}^{t-1} \tau \\ t \end{pmatrix} = \begin{pmatrix} \frac{t^2 - t + 2}{2} \\ t \end{pmatrix} .$$

Hence, the solution of the initial value problem is determined.

Finally, we remark that Theorem 2.3 can be modified for difference equations, and the details are left to the reader.

2.2.3   Z-transform

In this section we introduce a very useful function transformation, which reduces the solution of linear difference equations with constant coefficients to the solution of algebraic equations. This transformation is called Z-transform, and is considered the discrete time counterpart of Laplace transforms.

After an existence theorem the fundamental properties of the Z-transform are outlined, and the Z-transforms of the most frequently applied functions are derived. At the end of this section the application of Z-transforms to solve difference equations is discussed.

DEFINITION 2.2   *Let $x(t)$ be a real- or vector-valued function defined for all $t \in \mathbf{N}$. Then the Z-transform of x is defined by the infinite series*

$$X(z) = \sum_{t=0}^{\infty} \frac{x(t)}{z^t} . \tag{2.48}$$

Similar to Laplace transform, the Z-transform maps a function space into another function space. Here $X(z)$ is the value of the transformed function at z.

Our first result is an existence theorem.

THEOREM 2.9
Assume that

$$|x(t)| \leq k \cdot \sigma^t (t = 0, 1, 2, \ldots) \tag{2.49}$$

with some real constants $k > 0$ and $\sigma \geq 0$. Then $X(z)$ exists for all $|z| > \sigma$.

PROOF   Note that under the assumption of the theorem,

$$\left| \frac{x(t)}{z^t} \right| \leq \frac{k\sigma^t}{|z|^t} = k \cdot \left(\frac{\sigma}{|z|} \right)^t,$$

and therefore, series (2.48) is majorized by the convergent series

$$k \sum_{t=0}^{\infty} \left(\frac{\sigma}{|z|} \right)^t.$$

∎

The *radius of convergence* is the smallest real number $\sigma \geq 0$ such that series (2.48) converges for all z such that $|z| > \sigma$.

The most frequently used properties of Z-transforms are given next.

THEOREM 2.10
The following relations are true:

(i)   *If $x(t) = x_1(t) + x_2(t)$, then $X(z) = X_1(z) + X_2(z)$.*

(ii)   *If $x(t) = \alpha x_1(t)$, then $X(z) = \alpha X_1(z)$.*

(iii)   *If $x(t) = x_1(t)a^t$, where a is a constant, then $X(z) = X_1(z/a)$.*

(iv)   *If $x(t) = x_1(t+1)$, then*

$$X(z) = z \cdot X_1(z) - z \cdot x_1(0) \, ,$$

and in general, if $x(t) = x_1(t+n)$ with n being a positive integer, then

$$X(z) = z^n \cdot X_1(z) - z^n \cdot x_1(0) - z^{n-1} \cdot x_1(1) - \cdots - z \cdot x_1(n-1) \, .$$

(v)   *If $x(t) = x_1(t-1)$ then with $x_1(\tau) = 0$ ($\tau < 0$),*

$$X(z) = \frac{1}{z} X_1(z) \, ,$$

and in general, if $x(t) = x_1(t - n)$ with some positive integer n with $x_1(\tau) = 0$ ($\tau < 0$), then

$$X(z) = \frac{1}{z^n} X_1(z) \,.$$

PROOF   Properties (i) and (ii) are obvious, (iii) can be shown as follows:

$$X(z) = \sum_{t=0}^{\infty} \frac{x_1(t) a^t}{z^t} = \sum_{t=0}^{\infty} \frac{x_1(t)}{\left(\frac{z}{a}\right)^t} = X_1\left(\frac{z}{a}\right) \,;$$

(iv)   Simple calculation shows that in the general case,

$$X(z) = \sum_{t=0}^{\infty} \frac{x_1(t + n)}{z^t} = z^n \sum_{t=0}^{\infty} \frac{x_1(t + n)}{z^{t+n}} = z^n \sum_{\tau=n}^{\infty} \frac{x_1(\tau)}{z^\tau}$$

$$= z^n \left(\sum_{\tau=0}^{\infty} \frac{x_1(\tau)}{z^\tau} - \frac{x_1(0)}{z^0} - \frac{x_1(1)}{z^1} - \cdots - \frac{x_1(n - 1)}{z^{n-1}} \right)$$

$$= z^n X_1(z) - z^n \cdot x_1(0) - z^{n-1} \cdot x_1(1) - \cdots - z \cdot x_1(n - 1) \,.$$

Select $n = 1$ as a special case to have the first identity.

(v)   Since in the general case $x_1(t) = x(t + n)$ $(t \geq n)$ and $x(\tau) = 0$ $(\tau < 0)$, the previous property implies that

$$X_1(z) = z^n X(z) \,.$$

Divide both sides by z^n to obtain the assertion. The first identity follows by selecting $n = 1$.   ∎

The Z-transforms of the most commonly used functions are summarized in Table 2.2. These relations are proven next.
Numbers 1 and 7 are proven by simple calculation:

$$\sum_{t=0}^{\infty} \frac{1}{z^t} = \frac{1}{1 - \frac{1}{z}} = \frac{z}{1 - z} \,,$$

and

$$\sum_{t=0}^{N-1} \frac{1}{z^t} = 1 \cdot \frac{1 - \left(\frac{1}{z}\right)^N}{1 - \frac{1}{z}} = \frac{1 - z^N}{z^{N-1}(1 - z)} \,.$$

Table 2.2 Z-Transforms of Common Functions

No.	$x(t)$	$X(z)$
1	1	$\frac{z}{z-1}$
2	a^t	$\frac{z}{z-a}$
3	ta^{t-1}	$\frac{z}{(z-a)^2}$
4	t	$\frac{z}{(z-1)^2}$
5	$\begin{cases} a^{t-1} & \text{if } t \geq 1 \\ 0 & \text{if } t = 0 \end{cases}$	$\frac{1}{z-a}$
6	$\begin{cases} \binom{t-1}{k-1} a^{t-k} & \text{if } t \geq 1 \\ 0 & \text{if } t = 0 \end{cases}$	$\frac{1}{(z-a)^k}$
7	$\begin{cases} 1 \text{ if } 0 \leq t < N \\ 0 \text{ otherwise} \end{cases}$	$\frac{1-z^N}{z^{N-1}(1-z)}$
8	$\begin{cases} 1 \text{ if } t = 0 \\ 0 \text{ otherwise} \end{cases}$	1

Number 2 is the consequence of No. 1 and Item (iii) of Theorem 2.10. Number 3 is obtained by simple differentiation from No. 2:

$$\sum_{t=0}^{\infty} \frac{ta^{t-1}}{z^t} = \frac{d}{da} \sum_{t=0}^{\infty} \frac{a^t}{z^t} = \frac{d}{da} \cdot \frac{z}{z-a} = \frac{z}{(z-a)^2} ,$$

and No. 4 is derived from the previous identity by selecting $a = 1$. Number 5 is implied by No. 2 and Property (v) of Theorem 2.10. Number 6 can be proven by using No. 5 and the $(k-1)$th derivative with respect to a:

$$\sum_{t=1}^{\infty} \binom{t-1}{k-1} \frac{a^{t-k}}{z^t} = \frac{1}{(k-1)!} \frac{d^{k-1}}{da^{k-1}} \sum_{t=1}^{\infty} \frac{a^{t-1}}{z^t}$$

$$= \frac{1}{(k-1)!} \frac{d^{k-1}}{da^{k-1}} \frac{1}{z-a} = \frac{1}{(z-a)^k} .$$

And finally, No. 8 is obtained from No. 7 by selecting $N = 1$.

As an example, Figure 2.5 illustrates the discrete function a^t ($t = 0, 1, 2, \ldots$) and its Z-transform for $a = 1.5$. For the sake of simplicity, only nonnegative values of z are considered.

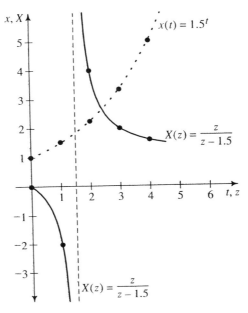

Figure 2.5 A discrete function and its Z-transform.

Z-transforms are very useful in solving linear difference equations with constant coefficients. The methodology is very similar to that used in the case of differential equations.

Consider first the nth order problem

$$x(t+n) + a_{n-1}x(t+n-1) + \cdots + a_1x(t+1) + a_0x(t) = f(t) \quad (2.50)$$

with initial values

$$x(0) = x_0, x(1) = x_1, \ldots, x(n-1) = x_{n-1} \ .$$

Apply Z-transform on both sides of the equation and use Property (iv) of Theorem 2.10 to obtain equality

$$[z^n X(z) - z^n x(0) - z^{n-1}x(1) - \cdots - zx(n-1)]$$

$$+ a_{n-1}[z^{n-1}X(z) - z^{n-1}x(0) - z^{n-2}x(1) - \cdots - zx(n-2)] + \cdots$$

$$+ a_1[zX(z) - zx(0)] + a_0X(z) = F(z) \ ,$$

where $X(z)$ and $F(z)$ denote the Z-transforms of $x(t)$ and $f(t)$, respectively. Rearranging the terms

$$X(z) = \frac{q(z)}{p(z)} + \frac{F(z)}{p(z)} \ , \tag{2.51}$$

where

$$p(z) = z^n + a_{n-1}z^{n-1} + \cdots + a_1 z + a_0 \ ,$$

$$q(z) = z[a_1 x(0) + \cdots + a_{n-1}x(n-2) + x(n-1)] + z^2[a_2 x(0)$$

$$+ \cdots + x(n-2)] + \cdots + z^{n-1}[a_{n-1}x(0) + x(1)] + z^n x(0) \ .$$

In the special case, when all initial values $x_0, \ldots, x_{n-1}$ are equal to zero, $q(z) = 0$. Therefore, the first term of (2.51) is zero. After $X(z)$ is determined, the solution $x(t)$ is found by using the Z-transforms of known functions given in Table 2.2. As an important special case, assume that $X(z)$ is a strictly proper rational function and the roots $\lambda_1, \ldots, \lambda_r$ of the denominator of $X(z)$ are distinct. Similarly to the case of differential equations, let

$$X(z) = \frac{R_1}{z - \lambda_1} + \frac{R_2}{z - \lambda_2} + \cdots + \frac{R_r}{z - \lambda_r}$$

be the partial fraction expansion of $X(z)$. Then use No. 5 of Table 2.2 to verify that

$$x(t) = \begin{cases} 0 & \text{if } t = 0 \\ R_1 \lambda_1^{t-1} + R_2 \lambda_2^{t-1} + \cdots + R_r \lambda_r^{t-1} & \text{if } t \geq 1 \ . \end{cases} \tag{2.52}$$

Assume next that the roots $\lambda_1, \ldots, \lambda_r$ have multiplicities $m_1, \ldots, m_r$. In this more general case,

$$X(z) = \sum_{i=1}^{r} \left[\frac{R_{i1}}{z - \lambda_i} + \frac{R_{i2}}{(z - \lambda_i)^2} + \cdots + \frac{R_{im_i}}{(z - \lambda_i)^{m_i}} \right] \ ,$$

and use No. 6 of Table 2.2 to obtain the solution

$$x(t) = \sum_{i=1}^{r} \left[R_{i1} \lambda_i^{t-1} + R_{i2} \binom{t-1}{1} \lambda_i^{t-2} + \cdots \right.$$

$$\left. + R_{im_i} \binom{t-1}{m_i - 1} \lambda_i^{t-m_i} \right] \tag{2.53}$$

for $t \geq 1$ and $x(0) = 0$.

Similar to the application of Laplace transforms, this procedure can be summarized as follows:

Step 1   Apply Z-transform on both sides of the difference equation.

Step 2   Solve the resulting algebraic equation.

Step 3   Apply inverse transform to recover the solution of the original problem.

Note that Figure 2.4 illustrates this procedure, when Laplace transforms are replaced by Z-transforms.

Example 2.15

The Fibonacci numbers are defined by recursion

$$x(t + 2) = x(t + 1) + x(t), \quad x(0) = 0, \quad x(1) = 1 .$$

By applying Z-transforms on both sides, we have

$$z^2 X(z) - z^2 x(0) - z x(1) = z X(z) - z x(0) + X(z) ,$$

that is,

$$X(z) = \frac{z}{z^2 - z - 1} = \frac{z}{\left(z - \frac{1+\sqrt{5}}{2}\right)\left(z - \frac{1-\sqrt{5}}{2}\right)} .$$

A calculation similar to that presented in Example 2.7 shows that the partial fraction expansion of $X(z)$ is given as

$$X(z) = \frac{\frac{1+\sqrt{5}}{2\sqrt{5}}}{z - \frac{1+\sqrt{5}}{2}} - \frac{\frac{1-\sqrt{5}}{2\sqrt{5}}}{z - \frac{1-\sqrt{5}}{2}} ,$$

and finally No. 5 of Table 2.2 shows that

$$x(t) = \frac{1 + \sqrt{5}}{2\sqrt{5}} \left(\frac{1 + \sqrt{5}}{2}\right)^{t-1} - \frac{1 - \sqrt{5}}{2\sqrt{5}} \left(\frac{1 - \sqrt{5}}{2}\right)^{t-1}$$

$$= \frac{1}{\sqrt{5}} \left[\left(\frac{1 + \sqrt{5}}{2}\right)^{t} - \left(\frac{1 - \sqrt{5}}{2}\right)^{t} \right] .$$

Hence, the solution of the above second-order initial value problem is determined.

Example 2.16

Next we solve the initial value problem

$$x(t+1) + 2x(t) = 4^t, \qquad x(0) = 0 .$$

Applying Z-transforms on both sides yields to equation

$$zX(z) - zx(0) + 2X(z) = \frac{z}{z-4} ,$$

where we used No. 2 of Table 2.2 with $a = 4$. Solve this equation for $X(z)$:

$$X(z) = \frac{z}{(z+2)(z-4)} .$$

It is easy to verify that the partial fraction expansion is as follows:

$$X(z) = \frac{\frac{1}{3}}{z+2} + \frac{\frac{2}{3}}{z-4} ,$$

and therefore, No. 5 of Table 2.2 implies that $x(0) = 0$ and

$$x(t) = \frac{1}{3}(-2)^{t-1} + \frac{2}{3}4^{t-1} \qquad (t \geq 1) .$$

Hence, the solution of the above second-order initial value problem is determined.

Consider next the system

$$\mathbf{x}(t+1) = \mathbf{A}\mathbf{x}(t) + \mathbf{f}(t), \qquad \mathbf{x}(0) = \mathbf{x}_0 , \tag{2.54}$$

where $\mathbf{A}$ is an $n \times n$ constant matrix and $\mathbf{f}$ is an n-dimensional function of the nonnegative integer t. Apply Z-transforms on both sides to get equation

$$z\mathbf{X}(z) - z\mathbf{x}(0) = \mathbf{A}\mathbf{X}(z) + \mathbf{F}(z) ,$$

where $\mathbf{X}$ and $\mathbf{F}$ are the Z-transforms of $\mathbf{x}$ and $\mathbf{f}$. Assuming that z is not an eigenvalue of $\mathbf{A}$,

$$\mathbf{X}(z) = (z\mathbf{I} - \mathbf{A})^{-1}(z\mathbf{x}_0 + \mathbf{F}(z)) . \tag{2.55}$$

Function $\mathbf{x}(t)$ can then be recovered by applying the previous method for each component of $\mathbf{X}(z)$.

Example 2.17

The second-order equation of the Fibonacci numbers presented in Example 2.15 can be transformed into the following two-dimensional first-order equation:

$$x(t+1) = \begin{pmatrix} 0 & 1 \\ 1 & 1 \end{pmatrix} x(t), \qquad x(0) = \begin{pmatrix} 0 \\ 1 \end{pmatrix}.$$

Here $x_1(t) = x(t)$ and $x_2(t) = x(t+1)$ are the new variables. In this case,

$$A = \begin{pmatrix} 0 & 1 \\ 1 & 1 \end{pmatrix}, \qquad f(t) = \begin{pmatrix} 0 \\ 0 \end{pmatrix}, \qquad \text{and} \qquad x_0 = \begin{pmatrix} 0 \\ 1 \end{pmatrix};$$

therefore, (2.55) implies that

$$X(z) = \begin{pmatrix} z & -1 \\ -1 & z-1 \end{pmatrix}^{-1} \begin{pmatrix} 0 \\ z \end{pmatrix}$$

$$= \frac{1}{z^2 - z - 1} \begin{pmatrix} z-1 & 1 \\ 1 & z \end{pmatrix} \begin{pmatrix} 0 \\ z \end{pmatrix} = \begin{pmatrix} \frac{z}{z^2-z-1} \\ \frac{z^2}{z^2-z-1} \end{pmatrix}.$$

Since $x(t) = x_1(t)$, we conclude that

$$X(z) = \frac{z}{z^2 - z - 1},$$

and $x(t)$ can be determined in the same way as demonstrated in Example 2.15.

Problems

1. Does problem $\dot{x} = t \cdot \sin x$, $x(0) = 1$ have a unique solution?

2. Find the fundamental matrix for differential equation

$$\dot{x} = \begin{pmatrix} \frac{1}{t} & 0 \\ 0 & \frac{1}{t} \end{pmatrix} x, \qquad x(1) = \begin{pmatrix} 1 \\ 1 \end{pmatrix}.$$

3. Apply the iteration method (2.7) to solve the initial value problem

$$\dot{x} = tx, \qquad x(0) = 1.$$

4. Solve the initial value problem

$$\dot{x} = \begin{pmatrix} 1 & 1 \\ 2 & 2 \end{pmatrix} x + \begin{pmatrix} 1 \\ 0 \end{pmatrix}, \qquad x(0) = \begin{pmatrix} 1 \\ 1 \end{pmatrix}$$

by using formula (2.23).

5. Solve the initial value problem

$$\dot{x} = \begin{pmatrix} 2 & 1 \\ 0 & 2 \end{pmatrix} x + \begin{pmatrix} 1 \\ 1 \end{pmatrix}, \qquad x(0) = \begin{pmatrix} 1 \\ 0 \end{pmatrix}$$

by using formula (2.23).

6. Solve the initial value problem

$$x(t+1) = \begin{pmatrix} 1 & 1 \\ 2 & 2 \end{pmatrix} x(t) + \begin{pmatrix} 1 \\ 0 \end{pmatrix}, \qquad x(0) = \begin{pmatrix} 1 \\ 1 \end{pmatrix}$$

by using formula (2.44).

7. Solve the initial value problem

$$x(t+1) = \begin{pmatrix} 2 & 1 \\ 0 & 2 \end{pmatrix} x(t) + \begin{pmatrix} 1 \\ 1 \end{pmatrix}, \qquad x(0) = \begin{pmatrix} 1 \\ 0 \end{pmatrix}$$

by using formula (2.44).

8. Solve the initial value problem

$$\dot{x} = \begin{pmatrix} 1 & 1 \\ 2 & 2 \end{pmatrix} x + \begin{pmatrix} 1 \\ 0 \end{pmatrix}, \qquad x(0) = \begin{pmatrix} 1 \\ 1 \end{pmatrix}$$

by Laplace transform.

9. Solve the initial value problem

$$\dot{x} = \begin{pmatrix} 2 & 1 \\ 0 & 2 \end{pmatrix} x + \begin{pmatrix} 1 \\ 1 \end{pmatrix}, \qquad x(0) = \begin{pmatrix} 1 \\ 0 \end{pmatrix}$$

by Laplace transform.

10. Solve the initial value problem

$$x(t+1) = \begin{pmatrix} 1 & 1 \\ 2 & 2 \end{pmatrix} x(t) + \begin{pmatrix} 1 \\ 0 \end{pmatrix}, \qquad x(0) = \begin{pmatrix} 1 \\ 1 \end{pmatrix}$$

by Z-transform.

11.   Solve the initial value problem

$$\mathbf{x}(t+1) = \begin{pmatrix} 2 & 1 \\ 0 & 2 \end{pmatrix} \mathbf{x}(t) + \begin{pmatrix} 1 \\ 1 \end{pmatrix}, \qquad \mathbf{x}(0) = \begin{pmatrix} 1 \\ 0 \end{pmatrix}$$

by Z-transform.

12.   Use Laplace transform to solve equation

$$\ddot{x} + 3\dot{x} + 2x = 1, \qquad x(0) = \dot{x}(0) = 0.$$

13.   Rewrite the second-order equation

$$\ddot{x} + 3\dot{x} + 2x = 1, \qquad x(0) = \dot{x}(0) = 0$$

as a system of first-order equations.

14.   Use Z-transform to solve equation

$$x(t+2) + 3x(t+1) + 2x(t) = 1, \qquad x(0) = x(1) = 0.$$

15.   Rewrite equation

$$x(t+2) + 3x(t+1) + 2x(t) = 1, \qquad x(0) = x(1) = 0$$

as a system of first-order equations.

16.   Solve the resulting equation of Problem 13 by Laplace transform, and compare your results to that of Problem 12.

17.   Solve the resulting equation of Problem 15 by Z-transform, and compare your results to that of Problem 14.

18.   Which initial condition should be selected at $t_0 = 0$ so that the trajectory of equation

$$\dot{\mathbf{x}} = \begin{pmatrix} 1 & 1 \\ 2 & 2 \end{pmatrix} \mathbf{x} + \begin{pmatrix} 1 \\ 0 \end{pmatrix}$$

passes through the point

$$\begin{pmatrix} 1 \\ 1 \end{pmatrix}$$

at $t = 1$?

19.   Assume that $\mathbf{f} = (f_i)$, $\mathbf{x} = (x_j)$, and all partial derivatives $\partial f_i / \partial x_j$ are bounded in the neighborhood of the initial point. Prove that there exists a $L > 0$ such that inequality (2.4) holds in this neighborhood.

20. Formulate and verify Theorem 2.3 for linear difference equations.

21. Assume that $\mathbf{A}(t)$ is an invertible matrix for all $t \in [a, b]$. Show that

$$\frac{d}{dt}\mathbf{A}^{-1}(t) = -\mathbf{A}^{-1}(t)\dot{\mathbf{A}}(t)\mathbf{A}^{-1}(t) \ .$$

22. By substitution show that $\mathbf{x}(t)$ given in Equation (2.23) is the solution of the initial value problem (2.13).

23. Prove that the solution of the initial value problem

$$\dot{x}(t) = \mathbf{A}(t)x(t), \qquad x(t_0) = x_0$$

satisfies the inequality

$$\|x(t)\|_2 \leq \|x_0\|_2 e^{\int_{t_0}^{t} \|\mathbf{A}(\tau)\|_2 d\tau}$$

for $t \geq t_0$.

24. Let $\mathbf{A}$ be a constant $n \times n$ matrix. Show that

$$e^{\mathbf{A}t} = \mathbf{I} + \mathbf{A}\int_0^t e^{\mathbf{A}\tau} d\tau \ .$$

25. Let $\mathbf{B}$ be a constant $n \times n$ real matrix. Show that if the eigenvalues of $\mathbf{B}$ are distinct and positive, then there is a real $n \times n$ matrix $\mathbf{A}$ such that $\mathbf{B} = e^{\mathbf{A}}$.

chapter three

Characterization of Systems

This chapter first introduces the mathematical concept of dynamic systems, and then methods for their solutions are presented. As we have seen in the previous chapter, linear differential and difference equations are easy to solve; therefore, the main method for solving nonlinear systems is based on linearization and the numerical solution of the resulting linear equations. Special linear methods are also discussed in this chapter, and decompositions will be introduced to reduce specially structured high-dimensional problems to smaller dimensional ones.

3.1   The Concept of Dynamic Systems

Many situations in applied sciences can be modeled by dynamic equations. The term dynamic refers to phenomena that produce time-changing patterns and the characteristics of the pattern at one time being interrelated with those of other times.

If the time scale is assumed to be continuous, then the direction of the change in the characteristics is usually described by a differential equation because derivatives represent these directions. In the case of a discrete time scale, the characteristics of consecutive time periods are interrelated by difference equations. In dynamic systems theory, the systems characteristics are usually divided into three classes:

1.   All effects arriving into the system from the outside world form the *input* of the system.

2.   The internal variables are summarized as the *state* of the system.

3.   The *output* either comprises that portion of the system's state which can be directly determined by external measurements, or summarizes the response of the system to the input.

Let $\mathbf{u}(t)$, $\mathbf{x}(t)$, and $\mathbf{y}(t)$ denote the input, state, and output of a system at time period t. Then the system is represented as the block diagram shown in Figure 3.1.

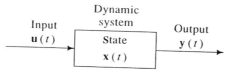

Figure 3.1   Block diagram representation of dynamic systems.

A *dynamic system with continuous time scale* and state–space description is presented as

$$\dot{\mathbf{x}}(t) = \mathbf{f}(t, \mathbf{x}(t), \mathbf{u}(t)) \tag{3.1}$$

$$\mathbf{y}(t) = \mathbf{g}(t, \mathbf{x}(t)) \; , \tag{3.2}$$

where the first equation is known as the *state transition equation* and the second relation is known as the *output equation*. Some authors allow function g to depend also on the input. In this chapter we discuss only the form (3.2). It is assumed that for all $t \geq 0$,

$$\mathbf{x}(t) \in X, \qquad \mathbf{u}(t) \in U \; ,$$

where $X \subseteq \mathbf{R}^n$ and $U \subseteq \mathbf{R}^m$ are called the *state space* and *input space* of the system; furthermore,

$$\mathbf{f} : [0, \infty) \times X \times U \to \mathbf{R}^n \qquad \text{and} \qquad \mathbf{g} : [0, \infty) \times X \to \mathbf{R}^p \; .$$

This notation means that for all $t \in [0, \infty)$ and $\mathbf{x}(t) \in X$ and $\mathbf{u}(t) \in U$, the function value of $\mathbf{f}$ is in $\mathbf{R}^n$ and the value of function $\mathbf{g}$ is in $\mathbf{R}^p$. Note that the dimensions of the input, state, and output are m, n, and p, respectively. It will usually be assumed that for all input functions $\mathbf{u}(t)$ and all initial states $\mathbf{x}_0$, the initial value problem

$$\dot{\mathbf{x}} = \mathbf{f}(t, \mathbf{x}, \mathbf{u}), \qquad \mathbf{x}(0) = \mathbf{x}_0$$

satisfies the conditions of Theorem 2.1, that is, there is a unique state function $\mathbf{x}(t)$ $(t \geq 0)$.

A *dynamic system with discrete time scale* is presented as

$$\mathbf{x}(t + 1) = \mathbf{f}(t, \mathbf{x}(t), \mathbf{u}(t)) \tag{3.3}$$

$$\mathbf{y}(t) = \mathbf{g}(t, \mathbf{x}(t)) \ . \tag{3.4}$$

Analogous to the continuous time scale case, the first equation is called the *state transition equation* and the second relation is known as the *output equation*. It is now assumed that for all $t = 0, 1, 2, \ldots$,

$$\mathbf{x}(t) \in X \qquad \text{and} \qquad \mathbf{u}(t) \in U \ ,$$

where X and U are the same as before; furthermore,

$$\mathbf{f} : \mathbf{N} \times X \times U \to X \qquad \text{and} \qquad \mathbf{g} : \mathbf{N} \times X \to \mathbf{R}^p,$$

where $\mathbf{N} = \{0, 1, 2, \ldots\}$. Observe that for all initial vectors $\mathbf{x}_0 \in X$ and input functions $\mathbf{u}$, the initial value problem

$$\mathbf{x}(t + 1) = \mathbf{f}(t, \mathbf{x}(t), \mathbf{u}(t)), \qquad \mathbf{x}(0) = \mathbf{x}_0$$

satisfies the conditions of Theorem 2.8; therefore, the solution $\mathbf{x}(t)$ exists and is unique.

The above concepts are illustrated next.

Example 3.1

Consider a point mass m in the presence of an inverse square force field $-k/r^2$, such as gravity. Assume that the mass is equipped with the ability to exert a thrust u_1 in the radial direction and a thrust u_2 in the tangential direction. This situation is illustrated in Figure 3.2 and is usually referred as the *satellite problem* because it describes the dynamics of an orbiting satellite that has no friction effects. First the equations of motion of this system will be derived.

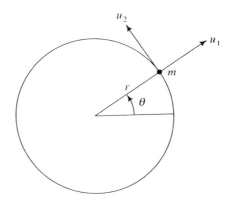

Figure 3.2   Illustration of the satellite problem.

The total energy of the system is called the Lagrangian, which is $L = K - P$, where

$$K = \frac{1}{2}m(\dot{r}^2 + r^2\dot{\theta}^2)$$

is the kinetic energy and

$$P = -\frac{k}{r}$$

is the potential energy. It is known that Lagrange's equations in the coordinate q read

$$\frac{d}{dt}\left(\frac{\partial L}{\partial \dot{q}}\right) - \frac{\partial L}{\partial q} = F,$$

where F is the external force in the q direction. Therefore, we obtain equations

$$m\ddot{r} - mr\dot{\theta}^2 + \frac{k}{r^2} = u_1$$

and

$$2r\dot{r}\dot{\theta}m + r^2\ddot{\theta}m = ru_2.$$

For the sake of simplicity, select $m = 1$, and solve these equations for $\ddot{r}$ and $\ddot{\theta}$ to get

$$\ddot{r} = r\dot{\theta}^2 - \frac{k}{r^2} + u_1$$

$$\ddot{\theta} = -\frac{2\dot{\theta}\dot{r}}{r} + \frac{1}{r}u_2.$$

If we introduce the notation $r_1 = r$, $r_2 = \dot{r}$, $\theta_1 = \theta$, and $\theta_2 = \dot{\theta}$, then these equations can be rewritten as follows:

$$\dot{r}_1 = r_2$$

$$\dot{r}_2 = r_1\theta_2^2 - \frac{k}{r_1^2} + u_1$$

$$\dot{\theta}_1 = \theta_2$$

$$\dot{\theta}_2 = -\frac{2\theta_2 r_2}{r_1} + \frac{1}{r_1}u_2. \qquad (3.5)$$

Assume finally that the position parameters r and θ form the output, then the output equations are

$$y_1 = r_1 \quad \text{and} \quad y_2 = \theta_1 \,. \tag{3.6}$$

The resulting equations coincide with the general formulation of dynamic systems with continuous time scale, where u_1, u_2 are the inputs; $r_1, r_2, \theta_1, \theta_2$ are the state variables; and y_1 and y_2 are the outputs.

Example 3.2

The simple models of national economic dynamics are usually based on the variables

$$Y(t) = \text{national income (or national product)}$$

$$C(t) = \text{consumption}$$

$$I(t) = \text{investment}$$

$$G(t) = \text{government expenditure}\,.$$

Variable $Y(t)$ is the total amount earned during a period by all individuals (or the total value of goods and services produced) in the economy, $C(t)$ is the total amount spent by individuals for goods and services, and $I(t)$ is the total amount invested in time period t. Variable $G(t)$ is the total amount spent by the government in the same time period. Obviously,

$$Y(t) = C(t) + I(t) + G(t)\,. \tag{3.7}$$

Two additional assumptions are made:

$$C(t) = mY(t) \tag{3.8}$$

and

$$Y(t+1) - Y(t) = rI(t)\,, \tag{3.9}$$

where

$$m = \text{marginal propensity to consume } (0 < m < 1)$$

$$r = \text{growth factor } (r > 0)\,.$$

Note that m is the fraction of the consumption and the national income, and r is the fraction of the national income increase and investment.

Combine the above equations to get the dynamic relation

$$Y(t+1) = Y(t) + rI(t) = Y(t) + r[Y(t) - C(t) - G(t)]$$

$$= Y(t) + r[Y(t) - mY(t) - G(t)] \,,$$

which can be simplified as

$$Y(t+1) = [1 + r - rm]Y(t) - rG(t) \,. \tag{3.10}$$

Consider Y as a state variable and G as an input. Then this relation is a special case of the general formulation (3.3), and, for example, (3.8) can serve as output equation, when consumption is considered to be the most important variable to measure. Note that Equation (3.10) is known as the Harrod-type model.

3.2   *Equilibrium and Linearization*

In many applications, the natural rest points of a dynamic system are much more interesting than the mechanism of change. The rest points are known as equilibrium points and they are defined as follows:

DEFINITION 3.1   *A vector $\bar{x}$ is an equilibrium of a dynamic system with an input function u if it has the property that once the state reaches $\bar{x}$ it remains at $\bar{x}$ for all future time.*

In particular, if a system is described by the dynamic equation

$$\dot{x}(t) = f(t, x(t), u(t)) \,,$$

then an equilibrium is a state $\bar{x}$ satisfying

$$f(t, \bar{x}, u(t)) = 0 \tag{3.11}$$

for all $t \geq 0$. If the system is described by the difference equation

$$x(t+1) = f(t, x(t), u(t)),$$

then an equilibrium is a state $\bar{x}$ that satisfies

$$\bar{x} = f(t, \bar{x}, u(t)) \tag{3.12}$$

for all $t = 0, 1, 2, \ldots$.

Note that the existence of the equilibrium states and numerical methods for determining them can be discussed based on the general theory of Section 1.2.3. The details are not given here.

Example 3.3

Consider the dynamic equations of harmonic motion of a unit mass with position $\theta(t)$ and velocity $v(t)$:

$$\dot{\theta} = v$$

$$\dot{v} = -\omega^2\theta + u ,$$

where a force input is assumed. Introduce the new state variables $x_1 = \omega\theta$ and $x_2 = v$. Then

$$\dot{x}_1 = \omega\dot{\theta} = \omega v = \omega x_2$$

and

$$\dot{x}_2 = \dot{v} = -\omega^2\theta + u = -\omega(\omega\theta) + u = -\omega x_1 + u .$$

By introducing matrices

$$\mathbf{A} = \begin{pmatrix} 0 & \omega \\ -\omega & 0 \end{pmatrix}$$

and

$$\mathbf{b} = \begin{pmatrix} 0 \\ 1 \end{pmatrix} ,$$

this system can be written in the standard notation

$$\dot{\mathbf{x}} = \mathbf{A}\mathbf{x} + \mathbf{b}u ,$$

which now has the particular form

$$\dot{\mathbf{x}} = \begin{pmatrix} 0 & \omega \\ -\omega & 0 \end{pmatrix}\mathbf{x} + \begin{pmatrix} 0 \\ 1 \end{pmatrix} u ;$$

here we assume that $u(t) = 1$ for all $t \geq 0$. The equilibrium is the solution of the algebraic equation

$$\begin{pmatrix} 0 & \omega \\ -\omega & 0 \end{pmatrix} \bar{\mathbf{x}} + \begin{pmatrix} 0 \\ 1 \end{pmatrix} = \begin{pmatrix} 0 \\ 0 \end{pmatrix} .$$

The solution of this equation is

$$\bar{\mathbf{x}} = - \begin{pmatrix} 0 & \omega \\ -\omega & 0 \end{pmatrix}^{-1} \begin{pmatrix} 0 \\ 1 \end{pmatrix} = - \begin{pmatrix} 0 & -\frac{1}{\omega} \\ \frac{1}{\omega} & 0 \end{pmatrix} \begin{pmatrix} 0 \\ 1 \end{pmatrix} = \begin{pmatrix} \frac{1}{\omega} \\ 0 \end{pmatrix} .$$

Hence at the equilibrium state, $\bar{x}_1 = \frac{1}{\omega}$ and $\bar{x}_2 = 0$, that is, the position is $\bar{\theta} = \frac{1}{\omega}\bar{x}_1 = \frac{1}{\omega^2}$ and the velocity is zero.

Example 3.4

Consider next the dynamic economic system introduced in Example 3.2. If $\bar{Y}$ is an equilibrium state of the system, then for all $t = 0, 1, 2, \ldots,$

$$\bar{Y} = [1 + r - rm]\bar{Y} - rG(t),$$

which can be satisfied if $G(t)$ is a constant G_0, since $r \neq 0$. Therefore, $\bar{Y}$ can be obtained by solving this equation for $\bar{Y}$:

$$\bar{Y} = \frac{1}{1-m}G_0 .$$

The values of the other variables at the equilibrium are obtained from Equations (3.8) and (3.9)

$$\bar{C} = \frac{m}{1-m}G_0 \quad \text{and} \quad \bar{I} = 0 .$$

Hence, the equilibrium of the system is given by vector $(1/(1-m)G_0, m/(1-m)G_0, 0)$.

Example 3.5

In this example we show that the satellite problem has no equilibrium point with zero inputs $u_1(t) = u_2(t) = 0$ ($t \geq 0$). From (3.5) and (3.11) we conclude that if there is an equilibrium, it must satisfy the equations

$$r_2 = 0$$

$$r_1\theta_2^2 - \frac{k}{r_1^2} = 0$$

$$\theta_2 = 0$$

$$-\frac{2\theta_2 r_2}{r_1} = 0 .$$

If $k \neq 0$, then the second and third equations contradict each other. However, one may easily verify that with certain nonzero inputs u_1 and u_2, the satellite problem does have equilibrium points.

The *linearization of a nonlinear system* is based on the approximation of the nonlinear functions $\mathbf{f}$ and $\mathbf{g}$ by linear Taylor's polynomials around the equilibrium point. Let $\bar{\mathbf{x}}$ denote an equilibrium with input function $\bar{\mathbf{u}}(t)$. Introduce matrices

$$
\mathbf{J}_x(t, \mathbf{x}, \mathbf{u}) = \begin{pmatrix}
\frac{\partial f_1}{\partial x_1}(t, \mathbf{x}, \mathbf{u}) & \frac{\partial f_1}{\partial x_2}(t, \mathbf{x}, \mathbf{u}) & \cdots & \frac{\partial f_1}{\partial x_n}(t, \mathbf{x}, \mathbf{u}) \\
\frac{\partial f_2}{\partial x_1}(t, \mathbf{x}, \mathbf{u}) & \frac{\partial f_2}{\partial x_2}(t, \mathbf{x}, \mathbf{u}) & \cdots & \frac{\partial f_2}{\partial x_n}(t, \mathbf{x}, \mathbf{u}) \\
\vdots & \vdots & \ddots & \vdots \\
\frac{\partial f_n}{\partial x_1}(t, \mathbf{x}, \mathbf{u}) & \frac{\partial f_n}{\partial x_2}(t, \mathbf{x}, \mathbf{u}) & \cdots & \frac{\partial f_n}{\partial x_n}(t, \mathbf{x}, \mathbf{u})
\end{pmatrix}
$$

and

$$
\mathbf{J}_u(t, \mathbf{x}, \mathbf{u}) = \begin{pmatrix}
\frac{\partial f_1}{\partial u_1}(t, \mathbf{x}, \mathbf{u}) & \frac{\partial f_1}{\partial u_2}(t, \mathbf{x}, \mathbf{u}) & \cdots & \frac{\partial f_1}{\partial u_m}(t, \mathbf{x}, \mathbf{u}) \\
\frac{\partial f_2}{\partial u_1}(t, \mathbf{x}, \mathbf{u}) & \frac{\partial f_2}{\partial u_2}(t, \mathbf{x}, \mathbf{u}) & \cdots & \frac{\partial f_2}{\partial u_m}(t, \mathbf{x}, \mathbf{u}) \\
\vdots & \vdots & \ddots & \vdots \\
\frac{\partial f_n}{\partial u_1}(t, \mathbf{x}, \mathbf{u}) & \frac{\partial f_n}{\partial u_2}(t, \mathbf{x}, \mathbf{u}) & \cdots & \frac{\partial f_n}{\partial u_m}(t, \mathbf{x}, \mathbf{u})
\end{pmatrix},
$$

where $\mathbf{f} = (f_i)$, $\mathbf{x} = (x_j)$, and $\mathbf{u} = (u_k)$. Note that $\mathbf{J}_x(t, \mathbf{x}, \mathbf{u})$ is an $n \times n$ matrix and $\mathbf{J}_u(t, \mathbf{x}, \mathbf{u})$ is an $n \times m$ matrix. Note that they are the Jacobian matrices of $\mathbf{f}$ with respect to $\mathbf{x}$ and $\mathbf{u}$, respectively, because the matrix elements are the partial derivatives of the components of $\mathbf{f}$ with respect to the elements of $\mathbf{x}$ and $\mathbf{u}$. The linear Taylor's polynomial approximation of $\mathbf{f}$ is then given as

$$
\mathbf{f}(t, \mathbf{x}, \mathbf{u}) \approx \mathbf{f}(t, \bar{\mathbf{x}}, \bar{\mathbf{u}}) + \mathbf{J}_x(t, \bar{\mathbf{x}}, \bar{\mathbf{u}}) \cdot (\mathbf{x} - \bar{\mathbf{x}}) + \mathbf{J}_u(t, \bar{\mathbf{x}}, \bar{\mathbf{u}}) \cdot (\mathbf{u} - \bar{\mathbf{u}}) .
$$

Assume first that the system is described by the differential equation (3.1). The above derivation implies that the linearized equation can be written as follows:

$$
\dot{\mathbf{x}}_\delta = \mathbf{A}(t)\mathbf{x}_\delta + \mathbf{B}(t)\mathbf{u}_\delta , \tag{3.13}
$$

where

$$
\mathbf{x}_\delta = \mathbf{x} - \bar{\mathbf{x}} \qquad \text{and} \qquad \mathbf{u}_\delta = \mathbf{u} - \bar{\mathbf{u}} ;
$$

furthermore

$$
\mathbf{A}(t) = \mathbf{J}_x(t, \bar{\mathbf{x}}, \bar{\mathbf{u}}(t)) \qquad \text{and} \qquad \mathbf{B}(t) = \mathbf{J}_u(t, \bar{\mathbf{x}}, \bar{\mathbf{u}}(t)) .
$$

Assume next that the system is described by the difference Equation (3.3). Then the linearized equation has the form

$$\mathbf{x}_\delta(t+1) = \mathbf{A}(t)\mathbf{x}_\delta(t) + \mathbf{B}(t)\mathbf{u}_\delta(t) , \qquad (3.14)$$

where $\mathbf{x}_\delta$, $\mathbf{u}_\delta$, $\mathbf{A}(t)$, and $\mathbf{B}(t)$ are the same as before.

In cases when the output relation

$$\mathbf{y} = \mathbf{g}(t, \mathbf{x})$$

is nonlinear, it also has to be linearized in order to get a system with only linear relations. Let $\bar{\mathbf{x}}$ denote the equilibrium, and $\bar{\mathbf{y}} = \mathbf{g}(t, \bar{\mathbf{x}})$. Linearize the function $\mathbf{g}$ about $\bar{\mathbf{x}}$:

$$\mathbf{y} \approx \mathbf{g}(t, \bar{\mathbf{x}}) + \mathbf{J}_g(t, \bar{\mathbf{x}})(\mathbf{x} - \bar{\mathbf{x}}) ,$$

where $\mathbf{J}_g$ is the Jacobian of $\mathbf{g}$ with respect to $\mathbf{x}$. By introducing the new output function $\mathbf{y}_\delta = \mathbf{y} - \bar{\mathbf{y}}$, the linearized output relation has the form:

$$\mathbf{y}_\delta(t) = \mathbf{C}(t)\mathbf{x}_\delta(t) \qquad (3.15)$$

with

$$\mathbf{C}(t) = \mathbf{J}_g(t, \bar{\mathbf{x}}) .$$

In the general case, where the system does not have an equilibrium state or we do not know what it is, we must proceed as follows. Select first an input function $\bar{\mathbf{u}}(t)$, and solve the corresponding differential or difference equation. Let the solution be $\mathbf{x}_0(t)$. Introduce the new variable $\mathbf{x}_\delta(t) = \mathbf{x}(t) - \mathbf{x}_0(t)$ in the equation, then $\bar{\mathbf{x}}_\delta(t) \equiv 0$ satisfies the resulting equation (that is, $\bar{\mathbf{x}}_\delta = 0$ is an equilibrium point); then, apply the above linearization method for this equation and zero equilibrium.

Example 3.6

The linearization process is applied now to the satellite problem, which was introduced in Example 3.1. For the sake of simplicity, select $\sigma = 1$, $k = \sigma^3 w^2$, and zero input $\bar{u}_1(t) = \bar{u}_2(t) = 0$, and observe that functions

$$r_1(t) = \sigma, \quad r_2(t) = 0, \quad \theta_1(t) = \omega t, \quad \theta_2(t) = \omega$$

solve Equation (3.5). By using the above relations, we therefore have

$$\mathbf{x}_0(t) = (\sigma, 0, \omega t, \omega)^T .$$

Introduce the new variable $\mathbf{x}_\delta = (x_1, x_2, x_3, x_4)^T$ with

$$x_1 = r_1 - \sigma$$

$$x_2 = r_2$$

$$x_3 = \theta_1 - \omega t$$

$$x_4 = \theta_2 - \omega ,$$

then these functions satisfy the differential equations

$$\dot{x}_1 = x_2$$

$$\dot{x}_2 = (x_1 + 1)(x_4 + \omega)^2 - \frac{k}{(x_1 + 1)^2} + u_1$$

$$\dot{x}_3 = x_4$$

$$\dot{x}_4 = -\frac{2(x_4 + \omega)x_2}{x_1 + 1} + \frac{1}{x_1 + 1} u_2 .$$

We know from the construction of these equations that with zero input, vector $\bar{\mathbf{x}}_\delta \equiv \mathbf{0}$ forms an equilibrium. The elements of the Jacobian matrices $\mathbf{J}_x$ and $\mathbf{J}_u$ are determined next at this equilibrium. Let f_1, f_2, f_3, f_4 denote the right-hand side functions. Note that only the second and fourth equations are nonlinear; therefore, the partial derivatives of only f_2 and f_4 have to be determined and only f_2 and f_4 are to be linearized. At the zero equilibrium state, these partial derivatives are as follows:

$$\frac{\partial f_2}{\partial x_1} = (x_4 + \omega)^2 + \frac{2k}{(x_1 + 1)^3} = \omega^2 + 2k = \omega^2 + 2\omega^2 = 3\omega^2,$$

$$\frac{\partial f_2}{\partial x_2} = \frac{\partial f_2}{\partial x_3} = 0,$$

$$\frac{\partial f_2}{\partial x_4} = 2(x_1 + 1)(x_4 + \omega) = 2\omega,$$

$$\frac{\partial f_2}{\partial u_1} = 1,$$

$$\frac{\partial f_2}{\partial u_2} = 0,$$

$$\frac{\partial f_4}{\partial x_1} = \frac{2(x_4 + w)x_2}{(x_1 + 1)^2} - \frac{1}{(x_1 + 1)^2}u_2 = 0,$$

$$\frac{\partial f_4}{\partial x_2} = \frac{-2(x_4 + w)}{x_1 + 1} = -2w,$$

$$\frac{\partial f_4}{\partial x_3} = 0,$$

$$\frac{\partial f_4}{\partial x_4} = \frac{-2x_2}{x_1 + 1} = 0,$$

$$\frac{\partial f_4}{\partial u_1} = 0,$$

$$\frac{\partial f_4}{\partial u_2} = \frac{1}{x_1 + 1} = 1.$$

Hence, using Equation (3.13), the linearized equations are summarized as

$$\begin{pmatrix} \dot{x}_1 \\ \dot{x}_2 \\ \dot{x}_3 \\ \dot{x}_4 \end{pmatrix} = \begin{pmatrix} 0 & 1 & 0 & 0 \\ 3w^2 & 0 & 0 & 2w \\ 0 & 0 & 0 & 1 \\ 0 & -2w & 0 & 0 \end{pmatrix} \begin{pmatrix} x_1 \\ x_2 \\ x_3 \\ x_4 \end{pmatrix} + \begin{pmatrix} 0 & 0 \\ 1 & 0 \\ 0 & 0 \\ 0 & 1 \end{pmatrix} \begin{pmatrix} u_1 \\ u_2 \end{pmatrix}.$$

A similar but more complicated calculation than the one that was shown in Example 1.23 gives the fundamental matrix of the system:

$$\phi(t, t_0) =$$

$$\begin{pmatrix} 4 - 3\cos w(t - t_0) & \sin w(t - t_0)/w & 0 & 2(1 - \cos w(t - t_0))/w \\ 3w \sin w(t - t_0) & \cos w(t - t_0) & 0 & 2\sin w(t - t_0) \\ 6(-w(t - t_0) + \sin w(t - t_0)) & -2(w - \cos w(t - t_0))/w & 1 & (-3w(t - t_0) + 4\sin w(t - t_0))/w \\ 6w(-1 + \cos w(t - t_0)) & -2\sin w(t - t_0) & 0 & -3 + 4\cos w(t - t_0) \end{pmatrix}$$

This matrix is the basis for predicting the future states of the system.

3.3   *Continuous Linear Systems*

In this section, dynamic systems of the form

$$\dot{\mathbf{x}}(t) = \mathbf{A}(t)\mathbf{x}(t) + \mathbf{B}(t)\mathbf{u}(t), \qquad \mathbf{x}(t_0) = \mathbf{x}_0 \qquad (3.16)$$

$$\mathbf{y}(t) = \mathbf{C}(t)\mathbf{x}(t) \qquad (3.17)$$

will be reexamined in order to find their solutions in particular forms. The differential equation is a special case of the general first-order inhomogeneous linear equation (2.12) with

$$\mathbf{f}(t) = \mathbf{B}(t)\mathbf{u}(t) . \qquad (3.18)$$

Therefore, the entire solution methodology discussed in Sections 2.1.2 and 2.1.3 is now applicable. Our discussions are divided into two parts. In the first part, methods based on the fundamental matrix are presented. Because the solution is obtained directly using the state vector, this methodology is called the *state–space approach*. The other method is based on Laplace transforms, and it is called the *transfer function approach*, because — as we will see later — it is based on a special relation between the Laplace transforms of the input and output functions, which is known as the transfer function. Notice that the transfer function approach can be used only in the case of constant matrices $\mathbf{A}, \mathbf{B}$ and $\mathbf{C}$.

3.3.1   *State–Space Approach*

The general solution for functions $\mathbf{x}$ and $\mathbf{y}$ is obtained directly from relation (2.23). The resulting equations can be formulated as follows.

THEOREM 3.1
The general solution of system (3.16) and (3.17) is given by relations

$$\mathbf{x}(t) = \phi(t, t_0)\mathbf{x}_0 + \int_{t_0}^{t} \phi(t, \tau)\mathbf{B}(\tau)\mathbf{u}(\tau)\, d\tau \qquad (3.19)$$

and

$$\mathbf{y}(t) = \mathbf{C}(t)\mathbf{x}(t) = \mathbf{C}(t)\phi(t, t_0)\mathbf{x}_0 + \int_{t_0}^{t} \mathbf{C}(t)\phi(t, \tau)\mathbf{B}(\tau)\mathbf{u}(\tau)\, d\tau . \quad (3.20)$$

These solution formulas are illustrated in the following example.

Example 3.7

Now we give the solution of the dynamic system

$$\dot{\mathbf{x}} = \begin{pmatrix} 0 & \omega \\ -\omega & 0 \end{pmatrix}\mathbf{x} + \begin{pmatrix} 0 \\ 1 \end{pmatrix}u, \qquad \mathbf{x}(0) = \begin{pmatrix} 1 \\ 0 \end{pmatrix}$$

$$y = (1,1)\mathbf{x},$$

which was introduced earlier in Example 1.13. Here $(1,1)$ is a row vector.

In Example 1.23 we derived the fundamental matrix:

$$\phi(t,t_0) = \begin{pmatrix} \cos\omega(t-t_0) & \sin\omega(t-t_0) \\ -\sin\omega(t-t_0) & \cos\omega(t-t_0) \end{pmatrix}.$$

Therefore, the state variable is

$$\mathbf{x}(t) = \begin{pmatrix} \cos\omega t & \sin\omega t \\ -\sin\omega t & \cos\omega t \end{pmatrix}\begin{pmatrix} 1 \\ 0 \end{pmatrix}$$

$$+ \int_0^t \begin{pmatrix} \cos\omega(t-\tau) & \sin\omega(t-\tau) \\ -\sin\omega(t-\tau) & \cos\omega(t-\tau) \end{pmatrix}\begin{pmatrix} 0 \\ 1 \end{pmatrix} u(\tau)\,d\tau$$

$$= \begin{pmatrix} \cos\omega t \\ -\sin\omega t \end{pmatrix} + \int_0^t \begin{pmatrix} \sin\omega(t-\tau) \\ \cos\omega(t-\tau) \end{pmatrix} u(\tau)\,d\tau.$$

As a special case, assume that $u(t) \equiv 1$, then the calculations coincide with those of Example 2.6, and the state vector becomes

$$\mathbf{x}(t) = \frac{1}{w}\begin{pmatrix} 1+(w-1)\cos\omega t \\ -(w-1)\sin\omega t \end{pmatrix}.$$

The direct input–output relation can be derived as follows:

$$y(t) = (1,1)\mathbf{x}(t) = (\cos\omega t - \sin\omega t) + \int_0^t (\sin\omega(t-\tau)+\cos\omega(t-\tau))u(\tau)\,d\tau.$$

In the special case of $u(t) \equiv 1$ we have

$$y(t) = (1,1)\frac{1}{w}\begin{pmatrix} 1+(\omega-1)\cos\omega t \\ -(\omega-1)\sin\omega t \end{pmatrix} = \frac{1}{w}(1+(\omega-1)(\cos\omega t - \sin\omega t)).$$

Equations (3.19) and (3.20) have high practical significance, because the state and/or output can be directly computed at any future time period t with any arbitrary initial state $\mathbf{x}_0$ and input function $\mathbf{u}(t)$.

The application of Equations (3.19) and (3.20) consists of the following steps:

Step 1   Determine the fundamental matrix $\phi(t, \tau)$.

Step 2   For the designated values of t, apply Equations (3.19) and (3.20) to get $\mathbf{x}(t)$ and/or $\mathbf{y}(t)$.

Note that Equations (3.19) and (3.20) have a special structure. The first terms depend only on the initial value $\mathbf{x}_0$, and the second terms depend only on the input. This property can be applied as follows. If the system has to be solved repeatedly with the same input but with several variants for the initial state, then the second terms have to be computed only once because only the first terms change. Similarly, if $\mathbf{x}_0$ is fixed, but the input $\mathbf{u}$ changes, then the first terms are fixed and only the second terms have to be recalculated. Finally we remark that in many cases the integrals in (3.19) and (3.20) cannot be determined analytically. In such cases, numerical integration methods are used. A summary of such algorithms can be found, for example, in [42] and in [44].

3.3.2 *Transfer Functions*

In this subsection we assume that matrices $\mathbf{A}$, $\mathbf{B}$, and $\mathbf{C}$ of the dynamic system

$$\dot{\mathbf{x}} = \mathbf{A}\mathbf{x} + \mathbf{B}\mathbf{u}, \qquad \mathbf{x}(0) = \mathbf{x}_0 \qquad (3.21)$$

$$\mathbf{y} = \mathbf{C}\mathbf{x} \qquad (3.22)$$

are constants. This system is called *time invariant*, which refers to the time-independence of the coefficient matrices. The methodology of the previous section can be used for solving this system without any limitation. An alternative method is based on the Laplace transform, which is the subject of this section.

Note that Equation (3.21) is a special case of (2.35) with $\mathbf{f}(t) = \mathbf{B}\mathbf{u}(t)$. Then relation (2.36) and the linearity of Laplace transforms imply the following result.

THEOREM 3.2
The general solution of system (3.21) and (3.22) can be given as

$$\mathbf{X}(s) = \mathbf{R}(s)\mathbf{x}_0 + \mathbf{R}(s)\mathbf{B}\mathbf{U}(s) \tag{3.23}$$

and

$$\mathbf{Y}(s) = \mathbf{C}\mathbf{R}(s)\mathbf{x}_0 + \mathbf{H}(s)\mathbf{U}(s) \, , \tag{3.24}$$

where

$$\mathbf{R}(s) = (s\mathbf{I} - \mathbf{A})^{-1} \quad and \quad \mathbf{H}(s) = \mathbf{C}(s\mathbf{I} - \mathbf{A})^{-1}\mathbf{B} \, . \tag{3.25}$$

Matrix $\mathbf{R}(s)$ is called the *resolvent matrix*, and $\mathbf{H}(s)$ is called the *transfer function*.

Note that in (3.23) and (3.24) the first terms depend only on the initial state $\mathbf{x}_0$, and the second terms depend only on the input. Hence, if the initial state $\mathbf{x}_0$ is zero, then the first term drops out, and if the input is zero, then the second term cancels. The comment made in the last paragraph of the previous section also applies in this case.

In the special case of a single-input, single-output system with zero initial state, Equation (3.24) reduces to the relation

$$Y(s) = H(s)U(s) \, ,$$

where H, U, Y are scalars. In this case, the transfer function is the fraction of $Y(s)$ and $U(s)$:

$$H(s) = \frac{Y(s)}{U(s)} \, .$$

The most important properties of $\mathbf{R}(s)$ and $\mathbf{H}(s)$ are discussed next.

THEOREM 3.3
$\mathbf{R}(s)$ *is the Laplace transform of* $e^{\mathbf{A}t}$.

PROOF   Select $\mathbf{u}(t) = \mathbf{0}$ for all $t \geq 0$, then from (3.19) and (3.23) we know that

$$\mathbf{x}(t) = e^{\mathbf{A}t}\mathbf{x}_0 \quad and \quad \mathbf{X}(s) = \mathbf{R}(s)\mathbf{x}_0 \, .$$

Applying Laplace transform on both sides of the first equation and comparing the resulting equality to the second equation yield to the identity

$$\mathbf{X}(s) = \mathbf{E}(s)\mathbf{x}_0 = \mathbf{R}(s)\mathbf{x}_0 \, ,$$

where $\mathbf{E}(s)$ denotes the Laplace transform of $e^{\mathbf{A}t}$. Because $\mathbf{x}_0$ is arbitrary, the assertion follows.   ∎

COROLLARY 3.1
$\mathbf{H}(s)$ *is the Laplace transform of* $\mathbf{C}e^{\mathbf{A}t}\mathbf{B}$, *which is a simple consequence of the theorem, the linearity of Laplace transforms, and relation* $\mathbf{H}(s) = \mathbf{C}\mathbf{R}(s)\mathbf{B}$.

In the first chapter we saw that by using appropriate matrix transformations, matrices can be transformed into special forms. These transformations are equivalent to introducing the new variable

$$\tilde{\mathbf{x}} = \mathbf{T}\mathbf{x}$$

in system (3.21) and (3.22), where $\mathbf{T}$ is a nonsingular matrix. Then

$$\dot{\tilde{\mathbf{x}}} = \mathbf{T}\dot{\mathbf{x}} = \mathbf{T}\mathbf{A}\mathbf{x} + \mathbf{T}\mathbf{B}\mathbf{u} = \mathbf{T}\mathbf{A}\mathbf{T}^{-1}\tilde{\mathbf{x}} + \mathbf{T}\mathbf{B}\mathbf{u}$$

and

$$\mathbf{y} = \mathbf{C}\mathbf{T}^{-1}\tilde{\mathbf{x}} \ .$$

That is, the new system has the form

$$\dot{\tilde{\mathbf{x}}} = \tilde{\mathbf{A}}\mathbf{x} + \tilde{\mathbf{B}}\mathbf{u} \tag{3.26}$$

$$\mathbf{y} = \tilde{\mathbf{C}}\tilde{\mathbf{x}} \tag{3.27}$$

with

$$\tilde{\mathbf{A}} = \mathbf{T}\mathbf{A}\mathbf{T}^{-1}, \qquad \tilde{\mathbf{B}} = \mathbf{T}\mathbf{B}, \qquad \text{and} \qquad \tilde{\mathbf{C}} = \mathbf{C}\mathbf{T}^{-1} \ .$$

THEOREM 3.4
Let $\mathbf{H}(s)$ *and* $\tilde{\mathbf{H}}(s)$ *denote the transfer functions of system (3.21)–(3.22) and (3.26)–(3.27), respectively. Then*

$$\mathbf{H}(s) = \tilde{\mathbf{H}}(s) \ .$$

PROOF   Simple calculation shows that

$$\tilde{\mathbf{H}}(s) = \tilde{\mathbf{C}}(s\mathbf{I} - \tilde{\mathbf{A}})^{-1}\tilde{\mathbf{B}}$$

$$= \mathbf{C}\mathbf{T}^{-1}[s\mathbf{T}\mathbf{T}^{-1} - \mathbf{T}\mathbf{A}\mathbf{T}^{-1}]^{-1}\mathbf{T}\mathbf{B}$$

$$= \mathbf{C}\mathbf{T}^{-1}[\mathbf{T}(s\mathbf{I} - \mathbf{A})\mathbf{T}^{-1}]^{-1}\mathbf{T}\mathbf{B}$$

$$= \mathbf{CT}^{-1}\mathbf{T}(s\mathbf{I} - \mathbf{A})^{-1}\mathbf{T}^{-1}\mathbf{TB}$$

$$= \mathbf{C}(s\mathbf{I} - \mathbf{A})^{-1}\mathbf{B} = \mathbf{H}(s) . \quad \blacksquare$$

The constructions of the resolvent matrix and the transfer function are illustrated next.

Example 3.8

Consider again the system

$$\dot{\mathbf{x}} = \begin{pmatrix} 0 & \omega \\ -\omega & 0 \end{pmatrix} \mathbf{x} + \begin{pmatrix} 0 \\ 1 \end{pmatrix} u, \qquad \mathbf{x}(0) = \begin{pmatrix} 1 \\ 0 \end{pmatrix}$$

$$y = (1,1)\mathbf{x} ,$$

which was the subject of our earlier Examples 3.3 and 3.7. In this case,

$$\mathbf{A} = \begin{pmatrix} 0 & \omega \\ -\omega & 0 \end{pmatrix}, \quad \mathbf{B} = \begin{pmatrix} 0 \\ 1 \end{pmatrix}, \quad \mathbf{x}_0 = \begin{pmatrix} 1 \\ 0 \end{pmatrix}, \quad \text{and} \quad \mathbf{C} = (1,1) .$$

The resolvent matrix is

$$\mathbf{R}(s) = (s\mathbf{I} - \mathbf{A})^{-1} = \begin{pmatrix} s & -\omega \\ \omega & s \end{pmatrix}^{-1} = \frac{1}{s^2 + \omega^2} \begin{pmatrix} s & \omega \\ -\omega & s \end{pmatrix} ,$$

where the inversion can be verified by simple multiplication. Therefore, the transfer function is obtained by simple algebra:

$$\mathbf{H}(s) = \mathbf{C}(s\mathbf{I} - \mathbf{A})^{-1}\mathbf{B} = \frac{1}{s^2 + \omega^2}(1,1) \begin{pmatrix} s & \omega \\ -\omega & s \end{pmatrix} \begin{pmatrix} 0 \\ 1 \end{pmatrix} = \frac{s + \omega}{s^2 + \omega^2} .$$

These results can be substituted into (3.23) to derive the input-state relation:

$$\mathbf{X}(s) = \frac{1}{s^2 + \omega^2} \begin{pmatrix} s & \omega \\ -\omega & s \end{pmatrix} \begin{pmatrix} 1 \\ 0 \end{pmatrix} + \frac{1}{s^2 + \omega^2} \begin{pmatrix} s & \omega \\ -\omega & s \end{pmatrix} \begin{pmatrix} 0 \\ 1 \end{pmatrix} U(s)$$

$$= \frac{1}{s^2 + \omega^2} \left[\begin{pmatrix} s \\ -\omega \end{pmatrix} + \begin{pmatrix} \omega \\ s \end{pmatrix} U(s) \right] ,$$

and from (3.24) we have the input–output relation:

$$Y(s) = \frac{1}{s^2 + \omega^2}(1,1) \begin{pmatrix} s & \omega \\ -\omega & s \end{pmatrix} \begin{pmatrix} 1 \\ 0 \end{pmatrix} + \frac{s + \omega}{s^2 + \omega^2} U(s)$$

$$= \frac{s - \omega}{s^2 + \omega^2} + \frac{s + \omega}{s^2 + \omega^2} U(s) \, .$$

As a particular case, assume that the input is selected as $u(t) = 1$ for all $t \geq 0$. Then item No. 2 of Table 2.1 implies that $U(s) = 1/s$. Therefore,

$$\mathbf{X}(s) = \frac{1}{s^2 + \omega^2} \begin{pmatrix} s + \frac{\omega}{s} \\ -\omega + 1 \end{pmatrix} = \begin{pmatrix} \frac{s^2 + \omega}{s(s^2 + \omega^2)} \\ \frac{-\omega + 1}{s^2 + \omega^2} \end{pmatrix} ,$$

which coincides with the result obtained earlier in Example 2.10. Similarly,

$$Y(s) = \frac{1}{s^2 + \omega^2} (s - \omega + \frac{\omega}{s} + 1) = \frac{s^2 + s(-\omega + 1) + \omega}{s(s^2 + \omega^2)} \, .$$

The output $y(t)$ can now be determined in the same way as it was earlier demonstrated in Example 2.7. Note first that the partial fraction expansion of $Y(s)$ is given as

$$Y(s) = \frac{\frac{1}{\omega}}{s} + \frac{\frac{\omega - 1}{\omega} \cdot s}{s^2 + \omega^2} - \frac{\frac{\omega - 1}{\omega} \cdot \omega}{s^2 + \omega^2} \, .$$

Therefore, Nos. 2, 5, and 6 of Table 2.1 imply that

$$y(t) = \frac{1}{\omega} (1 + (\omega - 1)(\cos \omega t - \sin \omega t)) \, ,$$

which coincides with the result of Example 3.7.

Consider finally the special case when the initial state is zero. Then the solution formula (3.24) reduces to

$$\mathbf{Y}(s) = \mathbf{H}(s)\mathbf{U}(s) \, .$$

This simplified formula is illustrated in the modified block diagram representation shown in Figure 3.3, which is essentially the same as the one presented earlier in Figure 3.1, but the input and output are replaced now by their Laplace transforms and the state variable is replaced by the transfer function.

3.3.3 *Equations in Input–Output Form*

In this section, a special linear, time-variant system is discussed, which has a single input and a single output. It is assumed that the input $u(t)$

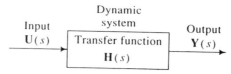

Figure 3.3   Modified block diagram representation of dynamic systems.

and output $y(t)$ are interrelated by equation

$$y^{(n)}(t) + p_{n-1}y^{(n-1)}(t) + \cdots + p_1\dot{y}(t) + p_0y(t)$$

$$= q_{n-1}u^{(n-1)}(t) + \cdots + q_1\dot{u}(t) + q_0u(t) , \quad (3.28)$$

where p_i and q_i $(0 \le i \le n-1)$ are constants and u and y are real-variable real-valued functions.

First we show that the above representation is equivalent to a first-order n-dimensional system. This representation is usually called the *phase variable form* in the systems theory literature. We also note that a controllability canonical form of continuous time-invariant systems (which will be discussed later in Chapter 7) has the same form.

THEOREM 3.5

The input and output of system

$$\dot{\mathbf{x}} = \begin{pmatrix} 0 & 1 & 0 & \cdots & 0 & 0 \\ 0 & 0 & 1 & \cdots & 0 & 0 \\ \vdots & \vdots & \vdots & \cdots & \vdots & \vdots \\ 0 & 0 & 0 & \cdots & 1 & 0 \\ 0 & 0 & 0 & \cdots & 0 & 1 \\ -p_0 & -p_1 & -p_2 & \cdots & -p_{n-2} & -p_{n-1} \end{pmatrix} \mathbf{x} + \begin{pmatrix} 0 \\ 0 \\ \vdots \\ 0 \\ 0 \\ 1 \end{pmatrix} u \quad (3.29)$$

with output equation

$$y = (q_0, q_1, \ldots, q_{n-1})\mathbf{x} \quad (3.30)$$

satisfy the input–output relation (3.28).

PROOF   Let x_i denote the components of $\mathbf{x}$ for $i = 1, 2, \ldots, n$. Then Equation (3.29) implies that

$$\dot{x}_1 = x_2$$

$$\dot{x}_2 = x_3$$

$$\vdots$$

$$\dot{x}_{n-1} = x_n$$

$$\dot{x}_n = -p_0 x_1 - \cdots - p_{n-1} x_n + u .$$

From the first $n - 1$ equations we conclude that

$$x_2 = \dot{x}_1, x_3 = \ddot{x}_1, \ldots, x_n = x_1^{(n-1)} ,$$

and the nth equation implies that

$$x_1^{(n)} + p_{n-1} x_1^{(n-1)} + \cdots + p_1 \dot{x}_1 + p_0 x_1 = u . \qquad (3.31)$$

Introduce mapping

$$A(x_1) = x_1^{(n)} + p_{n-1} x_1^{(n-1)} + \cdots + p_1 \dot{x}_1 + p_0 x_1 ,$$

and note that the linearity of A implies that

$$\frac{d}{dt} A(x_1) = A(\dot{x}_1), \ldots, \frac{d^{n-1}}{dt^{n-1}} A(x_1) = A\left(x_1^{(n-1)}\right) .$$

Because from (3.30) we have

$$y = q_0 x_1 + q_1 \dot{x}_1 + \cdots + q_{n-1} x_1^{(n-1)} , \qquad (3.32)$$

simple calculation shows that

$$A(y) = q_0 A(x_1) + q_1 A(\dot{x}_1) + \cdots + q_{n-1} A\left(x_1^{(n-1)}\right)$$

$$= q_0 A(x_1) + q_1 \frac{d}{dt} A(x_1) + \cdots + q_{n-1} \frac{d}{dt^{n-1}} A(x_1) .$$

The definition of mapping A implies that $A(x_1) = u$, therefore,

$$A(y) = q_0 u + q_1 \dot{u} + \cdots + q_{n-1} u^{(n-1)} ,$$

which completes the proof. ∎

REMARK 3.1 The simple trick which constructs the output y by using the state x_1 of the more simple system (3.31) is called *superposition*. ∎

Example 3.9

Consider the input–output form

$$\ddot{y} + 5\dot{y} + 4y = 2\dot{u} + u .$$

In this case, $n = 2$, $p_1 = 5$, $p_0 = 4$, $q_1 = 2$, $q_0 = 1$. Therefore, system (3.29) and (3.30) can be written as

$$\dot{x} = \begin{pmatrix} 0 & 1 \\ -4 & -5 \end{pmatrix} x + \begin{pmatrix} 0 \\ 1 \end{pmatrix} u$$

$$y = (1, 2)x .$$

This theorem allows us to transform input–output forms into the usual systems model and then use any of the methods of the previous sections. However, the direct use of Laplace transform on the input–output form is more efficient.

Apply Laplace transform on both sides of Equation (3.28) to get

$$[s^n Y(s) - s^{n-1}y(0) - \cdots - sy^{(n-2)}(0) - y^{(n-1)}(0)] + p_{n-1}[s^{n-1}Y(s)$$

$$- s^{n-2}y(0) - \cdots - sy^{(n-3)}(0) - y^{(n-2)}(0)]$$

$$+ \cdots + p_1[sY(s) - y(0)] + p_0 Y(s)$$

$$= q_{n-1}[s^{n-1}U(s) - s^{n-2}u(0) - \cdots - u^{(n-2)}(0)]$$

$$+ \cdots + q_1[sU(s) - u(0)] + q_0 U(s),$$

and solve this equation for $Y(s)$:

$$Y(s) = \frac{q(s)}{p(s)}U(s) + \frac{r(s)}{p(s)} , \qquad (3.33)$$

where

$$p(s) = s^n + p_{n-1}s^{n-1} + \cdots + p_1 s + p_0,$$

$$q(s) = q_{n-1}s^{n-1} + \cdots + q_1 s + q_0,$$

and $r(s)$ is a polynomial of degree not greater than $n - 1$ such that it becomes zero if all initial values $u(0), \ldots, u^{(n-2)}(0), y(0), \ldots, y^{(n-1)}(0)$ are equal to zero.

Note that the coefficient of $U(s)$ in Equation (3.33) is the transfer function:

$$H(s) = \frac{q(s)}{p(s)} \ . \tag{3.34}$$

Example 3.10

Consider again the input–output form

$$\ddot{y} + 5\dot{y} + 4y = 2\dot{u} + u \ ,$$

which was discussed in the previous example. Apply Laplace transform on both sides of the equation to get

$$[s^2 Y(s) - sy(0) - \dot{y}(0)] + 5[sY(s) - y(0)] + 4Y(s) = 2[sU(s) - u(0)] + U(s) \ .$$

Solve for $Y(s)$:

$$Y(s) = \frac{2s + 1}{s^2 + 5s + 4} U(s) + \frac{(s + 5)y(0) + \dot{y}(0) - 2u(0)}{s^2 + 5s + 4} \ .$$

As a particular example, assume that all initial values of the input and output are zero, then

$$Y(s) = \frac{2s + 1}{s^2 + 5s + 4} U(s) \ .$$

3.3.4   Combinations

Transfer functions and their main properties allow us to rewrite the transfer functions of a large structured system in terms of the transfer functions of individual subsystems, which reduces a high-order calculation to a sequence of smaller order calculations.

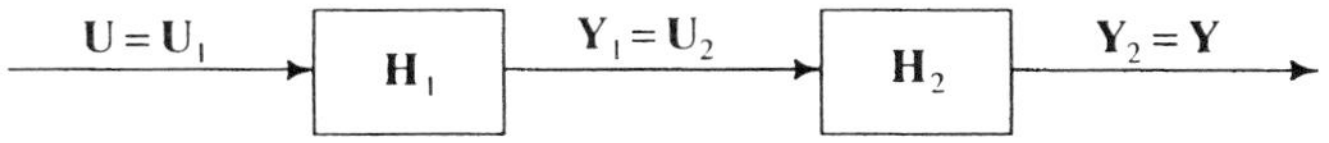

Figure 3.4   Series combination of systems.

Consider first a *series* combination of two systems, shown in Figure 3.4. If the initial states are zero, then

$$\mathbf{Y}_1(s) = \mathbf{H}_1(s)\mathbf{U}_1(s)$$

and

$$\mathbf{Y}_2(s) = \mathbf{H}_2(s)\mathbf{U}_2(s) \ .$$

Since
$$U_2(s) = Y_1(s) \ ,$$

combine the above relations to get

$$Y_2(s) = H_2(s)H_1(s)U_1(s) \ .$$

Since
$$U = U_1 \qquad \text{and} \qquad Y = Y_2 \ ,$$

we conclude that
$$Y(s) = H_2(s)H_1(s)U(s) \ .$$

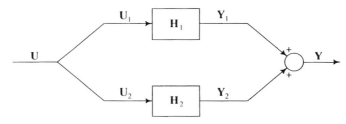

Figure 3.5   Parallel combination of systems.

Hence, the transfer function of *series* combination of systems is the *product* of the transfer functions of the subsystems:

$$H(s) = H_2(s)H_1(s) \ . \tag{3.35}$$

Consider next a *parallel* combination of two systems, shown in Figure 3.5. Because $U_1 = U_2 = U$ and $Y = Y_1 + Y_2$,

$$Y(s) = Y_1(s) + Y_2(s) = H_1(s)U_1(s) + H_2(s)U_2(s) \ ,$$

that is,

$$Y(s) = (H_1(s) + H_2(s))U(s) \ .$$

Hence, the transfer function of *parallel* combination of systems is the *sum* of the transfer functions of the subsystems:

$$H(s) = H_1(s) + H_2(s) \ . \tag{3.36}$$

Consider now the *feedback* structure shown in Figure 3.6. Then

$$\mathbf{Y}(s) = \mathbf{Y}_1(s) = \mathbf{H}_1(s)\mathbf{U}_1(s) = \mathbf{H}_1(s) \cdot [\mathbf{U}(s) + \mathbf{Y}_2(s)]$$

$$= \mathbf{H}_1(s)[\mathbf{U}(s) + \mathbf{H}_2(s)\mathbf{U}_2(s)] = \mathbf{H}_1(s)[\mathbf{U}(s) + \mathbf{H}_2(s)\mathbf{Y}(s)] \ ,$$

that is,

$$\mathbf{Y}(s) = \mathbf{H}_1(s)\mathbf{U}(s) + \mathbf{H}_1(s)\mathbf{H}_2(s)\mathbf{Y}(s) \ .$$

Solve this equation for $\mathbf{Y}(s)$:

$$\mathbf{Y}(s) = [\mathbf{I} - \mathbf{H}_1(s)\mathbf{H}_2(s)]^{-1}\mathbf{H}_1(s)\mathbf{U}(s) \ .$$

Hence, the transfer function of the *feedback* structure is given by relation

$$\mathbf{H}(s) = [\mathbf{I} - \mathbf{H}_1(s)\mathbf{H}_2(s)]^{-1}\mathbf{H}_1(s) \ . \tag{3.37}$$

The above combinations enable us to quickly compute the transfer functions of complex arrangements, and in addition, complex transfer functions can be represented as combinations of subsystems having only simple transfer functions.

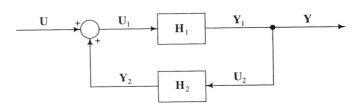

Figure 3.6   Feedback structure of systems.

Example 3.11

Consider the structure shown in Figure 3.7 with the repeated feedback. The inner feedback loop can be substituted by a system with transfer function
$$\mathbf{H}_I(s) = [\mathbf{I} - \mathbf{H}_1(s)\mathbf{H}_2(s)]^{-1}\mathbf{H}_1(s) \ .$$

This system has a series combination with the subsystem having transfer function $\mathbf{H}_3(s)$. Therefore, the transfer function of the upper part of the outer feedback loop is

$$\mathbf{H}_U(s) = \mathbf{H}_3(s)\mathbf{H}_I(s) = \mathbf{H}_3(s)[\mathbf{I} - \mathbf{H}_1(s)\mathbf{H}_2(s)]^{-1}\mathbf{H}_1(s) \ .$$

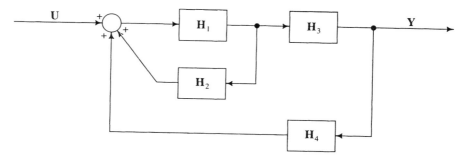

Figure 3.7 A structured combination of systems.

Finally, use (3.37) again to get the transfer function of the entire combination:

$$\mathbf{H}(s) = [\mathbf{I} - \mathbf{H}_U(s)\mathbf{H}_4(s)]^{-1}\mathbf{H}_U(s)$$

$$= [\mathbf{I} - \mathbf{H}_3(s)(\mathbf{I} - \mathbf{H}_1(s)\mathbf{H}_2(s))^{-1}\mathbf{H}_1(s)\mathbf{H}_4(s)]^{-1}\mathbf{H}_3(s)$$

$$\times [\mathbf{I} - \mathbf{H}_1(s)\mathbf{H}_2(s)]^{-1}\mathbf{H}_1(s) .$$

As a numerical example, assume that

$$\mathbf{H}_1(s) = \frac{1}{s}, \ \mathbf{H}_2(s) = \frac{1}{s-1}, \ \mathbf{H}_3(s) = \frac{1}{s-1}, \ \mathbf{H}_4(s) = \frac{2}{s} .$$

Then

$$\mathbf{H}_I(s) = (1 - \frac{1}{s}\frac{1}{s-1})^{-1}\frac{1}{s} = \frac{\frac{1}{s}}{1 - \frac{1}{s(s-1)}} = \frac{s-1}{s^2 - s - 1} ,$$

$$\mathbf{H}_U(s) = \frac{1}{s-1}\frac{s-1}{s^2 - s - 1} = \frac{1}{s^2 - s - 1} ,$$

and

$$\mathbf{H}(s) = (1 - \frac{1}{s^2 - s - 1}\frac{2}{s})^{-1}\frac{1}{s^2 - s - 1} = \frac{\frac{1}{s^2-s-1}}{1 - \frac{2}{s(s^2-s-1)}} = \frac{s}{s^3 - s^2 - s - 2} .$$

Example 3.12

Consider next an input structure with a strictly proper transfer function

$$\mathbf{H}(s) = \frac{q(s)}{p(s)} .$$

Let the distinct roots of p be $\lambda_1, \ldots, \lambda_r$ with multiplicities $m_1, \ldots, m_r$. Then the partial fraction expansion of $\mathbf{H}(s)$ is as follows:

$$\mathbf{H}(s) = \sum_{i=1}^{r} \left[\frac{R_{i1}}{s - \lambda_i} + \frac{R_{i2}}{(s - \lambda_i)^2} + \cdots + \frac{R_{im_i}}{(s - \lambda_i)^{m_i}} \right] .$$

Introduce the transfer functions

$$\mathbf{H}_{ij} = \frac{R_{ij}}{(s - \lambda_i)^j} \qquad (1 \le i \le r, 1 \le j \le m_i) .$$

Because $\mathbf{H}(s)$ is additive, the system can be represented as the parallel combination of the subsystems having the transfer functions $\mathbf{H}_{ij}$.

As a particular case, consider transfer function

$$\mathbf{H}(s) = \frac{2s}{s^2 - 1} .$$

Because

$$\mathbf{H}(s) = \frac{1}{s - 1} + \frac{1}{s + 1} ,$$

we may select

$$\mathbf{H}_{11}(s) = \frac{1}{s - 1} , \qquad \mathbf{H}_{21}(s) = \frac{1}{s + 1} ,$$

and therefore, the parallel combination representation is the one shown in Figure 3.8. Hence the original system with transfer function $\mathbf{H}(s)$ can be represented as a combination of the systems with the transfer functions $\mathbf{H}_{11}(s)$ and $\mathbf{H}_{21}(s)$ of more simple structures.

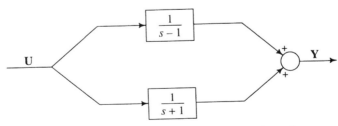

Figure 3.8   Parallel combination representation of Example 3.12.

Analog computers are based on certain realizations of the system equations. The above example shows such a realization, which is known as the *diagonal-form* representation. In the conclusion of this section, some other representations are introduced.

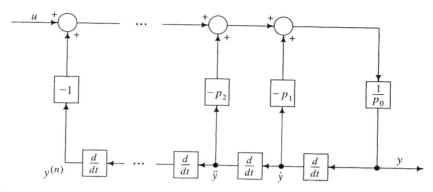

Figure 3.9 Analog model with differentiators.

Consider first the simple input–output form

$$y^{(n)} + p_{n-1}y^{(n-1)} + \cdots + p_1\dot{y} + p_0 y = u , \qquad (3.38)$$

where the right-hand side has no derivative of the input. Assume that a differentiator is available. Then the circuit shown in Figure 3.9 is a possible implementation. Here we assume that $p_0 \neq 0$. If in general $p_0 = p_1 = \cdots = p_k = 0$ with $p_{k+1} \neq 0$ $(0 < k < n)$, then introduce the new variable $z = y^{(k+1)}$ which guarantees that the coefficient of the new variable becomes nonzero. This kind of system implementation, however, has only limited practical importance. Signals are usually corrupted by noise, and the differentiation of noisy signals produces large errors. It is also well known (see, for example, [42]) that the integration of noisy signals can be performed in the accuracy of the noise itself. Therefore, in practical cases, differentiators are replaced by integrators. Note that the integrator is a block with transfer function $1/s$. Assuming that the highest-order derivative of y is available, use integrators successively to obtain all lower order derivatives as shown in Figure 3.10.

In the case of the general input–output form

$$y^{(n)} + p_{n-1}y^{(n-1)} + \cdots + p_1\dot{y} + p_0 y = q_{n-1}u^{(n-1)} + \cdots + q_1\dot{u} + q_0 u , \quad (3.39)$$

we can use superposition as we did in proving Theorem 3.5. If x_1 now

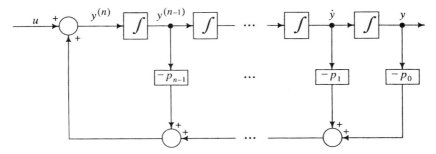

Figure 3.10   Simple implementation with integrators.

denotes the output of system (3.38), then

$$y = q_0 x_1 + q_1 \dot{x}_1 + \cdots + q_{n-1} x_1^{(n-1)} .$$

Note that from x_1 the output y can be obtained by successive application of differentiators, as shown in Figure 3.11.

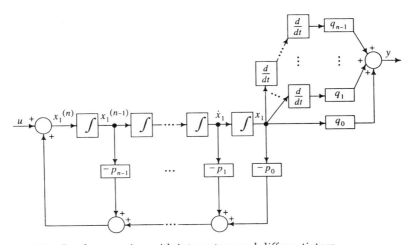

Figure 3.11   Implementation with integrators and differentiators.

   As the next step, we can use a simple trick to eliminate the need for differentiators by noticing that we have integrators and differentiators in series. The only thing we have to do is move the lines with differentiators over the requisite number of integrators. This construct is shown in Figure 3.12, and is called a *controllability-form* representation.
   An analogous representation can be obtained by equating the left- and right-hand sides of Equation (3.39). In the left-hand side, the representa-

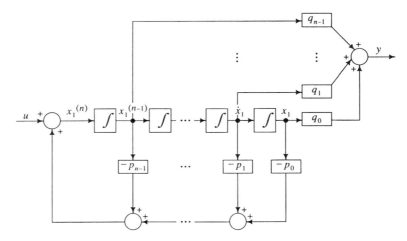

Figure 3.12   A controllability-form representation.

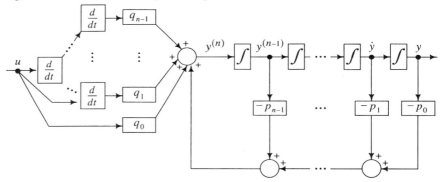

Figure 3.13   An alternative implementation with integrators and differentiators.

tion shown in Figure 3.10 is applied, and in the first step, differentiators are used in the right-hand side. This construct is illustrated in Figure 3.13. Similar to the previous case, we have to eliminate the need for differentiators by using again the simple fact that serial connection of an integrator cancels the effect of a differentiator. This idea leads us to the implementation shown in Figure 3.14, where the constants β_0, $\beta_1, \ldots, \beta_{n-1}$ are unknown. It is easy to see that the selection $\beta_i = q_i$ $(0 \leq i \leq n-1)$ is not satisfactory, since the feedback lines from the top affect the quantities in between the integrators. Note that in Figure 3.12, no feedback of the input was used, so the above difficulty did not arise. The unknown β_i values can be determined as follows. Simple calculation shows that

$$x_0 = y$$

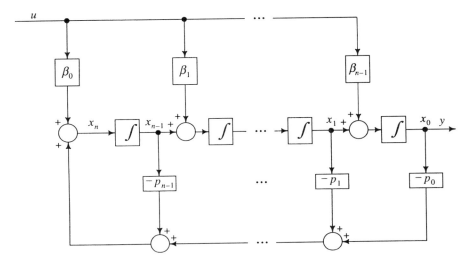

Figure 3.14   An observability-form representation.

and

$$\dot{x}_0 = x_1 + \beta_{n-1} u \ ,$$

$$\dot{x}_1 = x_2 + \beta_{n-2} u \ ,$$

$$\vdots$$

$$\dot{x}_{n-2} = x_{n-1} + \beta_1 u \ ,$$

$$\dot{x}_{n-1} = x_n \ ;$$

furthermore,

$$x_n = \beta_0 u - p_0 x_0 - p_1 x_1 - \cdots - p_{n-1} x_{n-1} \ . \qquad (3.40)$$

These relations imply

$$\ddot{x}_0 = \dot{x}_1 + \beta_{n-1} \dot{u} = x_2 + \beta_{n-2} u + \beta_{n-1} \dot{u}$$

$$x_0^{(3)} = \dot{x}_2 + \beta_{n-1} \ddot{u} + \beta_{n-2} \dot{u} = x_3 + \beta_{n-3} u + \beta_{n-2} \dot{u} + \beta_{n-1} \ddot{u}$$

$$\vdots$$

$$x_0^{(n-1)} = x_{n-1} + \beta_1 u + \beta_2 \dot{u} + \cdots + \beta_{n-1} u^{(n-2)} \ ,$$

$$x_0^{(n)} = x_n + \beta_1 \dot{u} + \cdots + \beta_{n-1} u^{(n-1)} \ ,$$

that is, for $k = 1, 2, \ldots, n$,

$$x_k = y^{(k)} - \beta_{n-k} u - \beta_{n-k+1} \dot{u} - \cdots - \beta_{n-1} u^{(k-1)} \ .$$

Substitute this equation into relation (3.40) to obtain

$$y^{(n)} - \beta_0 u - \beta_1 \dot{u} - \cdots - \beta_{n-1} u^{(n-1)}$$

$$= -p_0 y - p_1 (\dot{y} - \beta_{n-1} u) - p_2 (\ddot{y} - \beta_{n-2} u - \beta_{n-1} \dot{u})$$

$$- \cdots - p_{n-1} (y^{(n-1)} - \beta_1 u - \beta_2 \dot{u} - \cdots - \beta_{n-1} u^{(n-2)}) \ ,$$

which can be simplified as

$$y^{(n)} + p_{n-1} y^{(n-1)} + \cdots + p_1 \dot{y} + p_0 y$$

$$= \beta_{n-1} u^{(n-1)} + (\beta_{n-2} + p_{n-1} \beta_{n-1}) u^{(n-2)}$$

$$+ \cdots + (\beta_1 + p_{n-1} \beta_2 + p_{n-2} \beta_3 + \cdots + p_2 \beta_{n-1}) \dot{u}$$

$$+ (\beta_0 + p_{n-1} \beta_1 + p_{n-2} \beta_2 + \cdots + p_1 \beta_{n-1}) u \ .$$

Compare the right-hand side of this equation to that of Equation (3.39) to see that the β_i values should be selected so that

$$\beta_{n-1} = q_{n-1}$$

$$\beta_{n-2} + p_{n-1} \beta_{n-1} = q_{n-2}$$

$$\vdots$$

$$\beta_1 + p_{n-1} \beta_2 + \cdots + p_3 \beta_{n-2} + p_2 \beta_{n-1} = q_1$$

$$\beta_0 + p_{n-1} \beta_1 + \cdots + p_2 \beta_{n-2} + p_1 \beta_{n-1} = q_0 \ .$$

These relations can be summarized in matrix form as

$$
\begin{pmatrix}
1 & 0 & \cdots & 0 & 0 \\
p_{n-1} & 1 & \cdots & 0 & 0 \\
\vdots & \vdots & \ddots & \vdots & \vdots \\
p_2 & p_3 & \cdots & 1 & 0 \\
p_1 & p_2 & \cdots & p_{n-1} & 1
\end{pmatrix}
\begin{pmatrix}
\beta_{n-1} \\
\beta_{n-2} \\
\vdots \\
\beta_1 \\
\beta_0
\end{pmatrix}
=
\begin{pmatrix}
q_{n-1} \\
q_{n-2} \\
\vdots \\
q_1 \\
q_0
\end{pmatrix} .
\tag{3.41}
$$

We note that the implementation shown in Figure 3.14 with the β_i values obtained from Equation (3.41) is called an *observability-form* representation.

A different approach is based on the transfer function

$$
\mathbf{H}(s) = \frac{q_{n-1}s^{n-1} + \cdots + q_1 s + q_0}{s^n + p_{n-1}s^{n-1} + \cdots + p_1 s + p_0}
$$

of the system (3.39), which implies that

$$
(s^n + p_{n-1}s^{n-1} + \cdots + p_1 s + p_0)Y(s) = (q_{n-1}s^{n-1} + \cdots + q_1 s + q_0)U(s) .
$$

Divide both sides by s^n and rearrange the terms as follows:

$$
Y(s) = \frac{1}{s^n}\left(-p_0 Y(s) + q_0 U(s)\right) + \frac{1}{s^{n-1}}(-p_1 Y(s) + q_1 U(s))
$$

$$
+ \cdots + \frac{1}{s}(-p_{n-1}Y(s) + q_{n-1}U(s)) ,
$$

and observe that for $k = 1, 2, \ldots, n$, $1/s^k$ is the operation of integration k times in succession. This idea is realized in the implementation shown in Figure 3.15, which is also called an *observability-form* representation.

Similar to Figure 3.14, the above implementation can be modified as shown in Figure 3.16, which is also called a *controllability-form* representation.

Finally, we note that the diagonal canonical-form representation corresponds to the diagonal canonical form to be discussed in Section 7.1, the controllability form representations correspond to the controllability canonical forms to be introduced in Section 7.2, and the observability form representations correspond to the observability canonical forms to be discussed in Section 7.3.

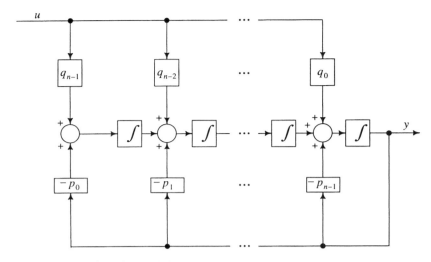

Figure 3.15   Another observability-form representation.

3.3.5   Adjoint and Dual Systems

Consider the time-variant linear system

$$\dot{\mathbf{x}}(t) = \mathbf{A}(t)\mathbf{x}(t) + \mathbf{B}(t)\mathbf{u}(t)$$

$$\mathbf{y}(t) = \mathbf{C}(t)\mathbf{x}(t) \ . \tag{3.42}$$

DEFINITION 3.2   *The adjoint of the above system is defined as*

$$\dot{\mathbf{x}}_a(t) = \mathbf{A}_a(t)\mathbf{x}_a(t) + \mathbf{B}_a(t)\mathbf{u}_a(t)$$

$$\mathbf{y}_a(t) = \mathbf{C}_a(t)\mathbf{x}_a(t) \ , \tag{3.43}$$

where

$$\mathbf{A}_a(t) = -\mathbf{A}^T(t), \quad \mathbf{B}_a(t) = \mathbf{C}^T(t), \quad and \quad \mathbf{C}_a(t) = \mathbf{B}^T(t) \ .$$

The main relations between the adjoint and original systems are given by the following results.

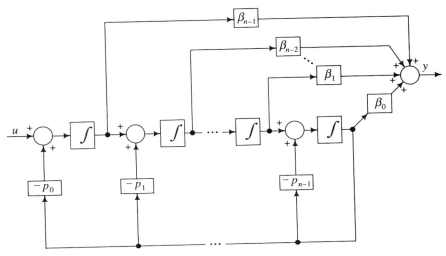

Figure 3.16   Another controllability-form representation.

THEOREM 3.6
The fundamental matrix of the adjoint system is given as

$$\phi_a(t, t_0) = \phi^T(t_0, t) ,$$

where $\phi(t, t_0)$ is the fundamental matrix of the original system.

PROOF   Property (v) of Theorem 2.3 implies that

$$\frac{\partial}{\partial t}\phi^T(t_0, t) = -\mathbf{A}^T(t)\phi^T(t_0, t) ,$$

and from Property (i) of the same theorem we know that

$$\phi^T(t_0, t_0) = \mathbf{I} .$$

The assertion then follows from Equation (2.19).   ■

THEOREM 3.7
If system (3.42) is time-invariant, then the transfer function of its adjoint is

$$\mathbf{H}_a(s) = -\mathbf{H}^T(-s) ,$$

where $\mathbf{H}(s)$ is the transfer function of the original system.

PROOF   Note that

$$\mathbf{H}_a(s) = \mathbf{C}_a(s\mathbf{I} - \mathbf{A}_a)^{-1}\mathbf{B}_a = \mathbf{B}^T(s\mathbf{I} + \mathbf{A}^T)^{-1}\mathbf{C}^T$$

$$= [-\mathbf{C}(-s\mathbf{I} - \mathbf{A})^{-1}\mathbf{B}]^T = -[\mathbf{C}(-s\mathbf{I} - \mathbf{A})^{-1}\mathbf{B}]^T$$

$$= -\mathbf{H}^T(-s) .   \quad \blacksquare$$

The dual of a linear system is defined next.

DEFINITION 3.3   *The dual of system (3.42) is defined as*

$$\dot{\mathbf{x}}_d(t) = \mathbf{A}_d(t)\mathbf{x}_d(t) + \mathbf{B}_d(t)\mathbf{u}_d(t)$$

$$\mathbf{y}_d(t) = \mathbf{C}_d(t)\mathbf{x}_d(t) , \tag{3.44}$$

where

$$\mathbf{A}_d(t) = \mathbf{A}^T(-t), \quad \mathbf{B}_d(t) = \mathbf{C}^T(t), \quad and \quad \mathbf{C}_d(t) = \mathbf{B}^T(t) .$$

The fundamental matrix as well as the transfer function of a dual system can be easily obtained from those of the original system.

THEOREM 3.8
The fundamental matrix of the dual is given as

$$\phi_d(t, t_0) = \phi^T(-t_0, -t) ,$$

where $\phi(t, t_0)$ is the fundamental matrix of the original system.

PROOF   Simple calculation shows that

$$\frac{d}{dt}\phi^T(-t_0, -t) = \left[\frac{d}{dt}\phi(-t_0, -t)\right]^T = [\phi(-t_0, -t)\mathbf{A}(-t)]^T$$

$$= \mathbf{A}^T(-t)\phi^T(-t_0, -t) ,$$

where we used Properties (iii) and (v) of Theorem 2.3.
Furthermore,

$$\phi^T(-t_0, -t_0) = \mathbf{I}^T = \mathbf{I} ,$$

and therefore, matrix $\phi^T(-t_0, -t)$ satisfies Equation (2.19) with coefficient matrix $\mathbf{A}^T(-t)$. Hence, $\phi^T(-t_0, -t)$ is the fundamental matrix of the dual.   ∎

THEOREM 3.9
If system (3.42) is time-invariant, then the transfer function of its dual is

$$\mathbf{H}_d(s) = \mathbf{H}^T(s) \, ,$$

where $\mathbf{H}(s)$ is the transfer function of the original system.

PROOF   By definition,

$$\mathbf{H}_d(s) = \mathbf{C}_d(s\mathbf{I} - \mathbf{A}_d)^{-1}\mathbf{B}_d = \mathbf{B}^T\left(s\mathbf{I} - \mathbf{A}^T\right)^{-1}\mathbf{C}^T$$

$$= [\mathbf{C}(s\mathbf{I} - \mathbf{A})^{-1}\mathbf{B}]^T = \mathbf{H}^T(s) \, . \quad ∎$$

The above concepts and results are illustrated in the following example.

Example 3.13

Consider the system of our earlier Example 3.8:

$$\dot{\mathbf{x}} = \begin{pmatrix} 0 & \omega \\ -\omega & 0 \end{pmatrix} \mathbf{x} + \begin{pmatrix} 0 \\ 1 \end{pmatrix} u$$

$$y = (1, 1)\mathbf{x} \, .$$

The adjoint and dual systems are given as

$$\dot{\mathbf{x}}_a = \begin{pmatrix} 0 & \omega \\ -\omega & 0 \end{pmatrix} \mathbf{x}_a + \begin{pmatrix} 1 \\ 1 \end{pmatrix} u_a$$

$$y_a = (0, 1)\mathbf{x}_a$$

and

$$\dot{\mathbf{x}}_d = \begin{pmatrix} 0 & -\omega \\ \omega & 0 \end{pmatrix} \mathbf{x}_d + \begin{pmatrix} 1 \\ 1 \end{pmatrix} u_d$$

$$y_d = (0, 1)\mathbf{x}_d,$$

respectively. We also know from Example 3.7 that

$$\phi(t, t_0) = \begin{pmatrix} \cos\omega(t-t_0) & \sin\omega(t-t_0) \\ -\sin\omega(t-t_0) & \cos\omega(t-t_0) \end{pmatrix} .$$

Therefore,

$$\phi_a(t, t_0) = \phi^T(t_0, t) = \phi(t, t_0) ,$$

because the cosine function is even and the sine function is odd. Similarly,

$$\phi_d(t, t_0) = \phi^T(-t_0, -t) = \begin{pmatrix} \cos\omega(t-t_0) & \sin\omega(t-t_0) \\ -\sin\omega(t-t_0) & \cos\omega(t-t_0) \end{pmatrix}^T$$

$$= \begin{pmatrix} \cos\omega(t-t_0) & -\sin\omega(t-t_0) \\ \sin\omega(t-t_0) & \cos\omega(t-t_0) \end{pmatrix} .$$

It is also known from Example 3.8 that the transfer function of the original system is

$$\mathbf{H}(s) = \frac{s+\omega}{s^2+\omega^2} ,$$

therefore,

$$\mathbf{H}_a(s) = -\frac{-s+\omega}{s^2+\omega^2} = \frac{s-\omega}{s^2+\omega^2} ,$$

and

$$\mathbf{H}_d(s) = \frac{s+\omega}{s^2+\omega^2} .$$

We mention here that dual and adjoint systems are often applied in systems theory. For example, they will be used in establishing the observability of linear systems based on controllability conditions, and observability canonical forms will be derived by using dual systems.

3.4   Discrete Systems

This section is devoted to the solution of discrete dynamic systems of the form

$$\mathbf{x}(t+1) = \mathbf{A}(t)\mathbf{x}(t) + \mathbf{B}(t)\mathbf{u}(t), \qquad \mathbf{x}(0) = \mathbf{x}_0 \qquad (3.45)$$

$$\mathbf{y}(t) = \mathbf{C}(t)\mathbf{x}(t) . \qquad (3.46)$$

Note that difference Equation (3.45) is a special case of the general linear difference equation model (2.43) with

$$\mathbf{f}(t) = \mathbf{B}(t)\mathbf{u}(t) , \tag{3.47}$$

therefore, the methods discussed earlier in Sections 2.2.2 and 2.2.3 are applicable without limitations.

The *state–space approach* is based on the general solution formula (2.44) and can be given as follows.

THEOREM 3.10
The general solution of system (3.45) and (3.46) is given as

$$\mathbf{x}(t) = \phi(t,0)\mathbf{x}_0 + \sum_{\tau=0}^{t-1} \phi(t,\tau+1)\mathbf{B}(\tau)\mathbf{u}(\tau) \tag{3.48}$$

and

$$\mathbf{y}(t) = \mathbf{C}(t)\mathbf{x}(t) = \mathbf{C}(t)\phi(t,0)\mathbf{x}_0 + \sum_{\tau=0}^{t-1} \mathbf{C}(t)\phi(t,\tau+1)\mathbf{B}(\tau)\mathbf{u}(\tau) . \tag{3.49}$$

The algorithm to compute $\mathbf{x}(t)$ and/or $\mathbf{y}(t)$ is the same as it has been shown for continuous systems. As an illustration consider the following example.

Example 3.14

We now give the solution of the system

$$\mathbf{x}(t+1) = \begin{pmatrix} 1 & 1 \\ 0 & 1 \end{pmatrix} \mathbf{x}(t) + \begin{pmatrix} 0 \\ 1 \end{pmatrix} u(t), \qquad \mathbf{x}(0) = \begin{pmatrix} 1 \\ 0 \end{pmatrix}$$

$$y(t) = (1,1)\mathbf{x}(t) .$$

In Example 2.14 we derived that

$$\phi(t,\tau) = \begin{pmatrix} 1 & t-\tau \\ 0 & 1 \end{pmatrix} ,$$

therefore, (3.48) implies that

$$\mathbf{x}(t) = \begin{pmatrix} 1 & t-0 \\ 0 & 1 \end{pmatrix} \begin{pmatrix} 1 \\ 0 \end{pmatrix} + \sum_{\tau=0}^{t-1} \begin{pmatrix} 1 & t-\tau-1 \\ 0 & 1 \end{pmatrix} \begin{pmatrix} 0 \\ 1 \end{pmatrix} u(\tau)$$

$$= \begin{pmatrix} 1 \\ 0 \end{pmatrix} + \sum_{\tau=0}^{t-1} \begin{pmatrix} t - \tau - 1 \\ 1 \end{pmatrix} u(\tau) \,,$$

and from (3.49) we conclude that the output is given as

$$y(t) = (1,1)\mathbf{x}(t) = 1 + \sum_{\tau=0}^{t-1}(t - \tau)u(\tau) \,.$$

In the particular case when $u(t) \equiv 1$, the calculations coincide with those of Example 2.14 and the state vector is

$$\mathbf{x}(t) = \begin{pmatrix} 1 \\ 0 \end{pmatrix} + \sum_{\tau=0}^{t-1} \begin{pmatrix} t - \tau - 1 \\ 1 \end{pmatrix} = \begin{pmatrix} \frac{t^2-t+2}{2} \\ t \end{pmatrix} \,.$$

This state function is illustrated in Figure 3.17, where the horizontal axis is $x_2(t) = t$, and the vertical axis represents $x_1(t)$.

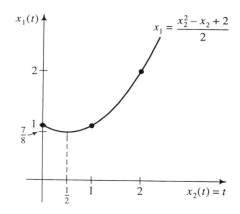

Figure 3.17   The state function of Example 3.14.

Note that Equations (3.48) and (3.49) are analogous to the continuous counterparts (Equations (3.19) and (3.20)); therefore, their applications and main properties are also similar.

In the *time invariant* case, when $\mathbf{A}(t)$, $\mathbf{B}(t)$ and $\mathbf{C}(t)$ do not depend on t, the application of Z transforms is very attractive. The general solution formula (2.55) implies the following result.

THEOREM 3.11

The general solution of the time-invariant system (3.45) and (3.46) can be given as

$$\mathbf{X}(z) = \mathbf{R}(z) \cdot z\mathbf{x}_0 + \mathbf{R}(z)\mathbf{B}U(z) \qquad (3.50)$$

and

$$\mathbf{Y}(z) = \mathbf{CR}(z)z\mathbf{x}_0 + \mathbf{H}(z)U(z) , \qquad (3.51)$$

where

$$\mathbf{R}(z) = (z\mathbf{I} - \mathbf{A})^{-1} \qquad and \qquad \mathbf{H}(z) = \mathbf{C}(z\mathbf{I} - \mathbf{A})^{-1}\mathbf{B} . \qquad (3.52)$$

Similar to the continuous case, $\mathbf{R}(z)$ is called the *resolvent matrix* and $\mathbf{H}(z)$ is called the *transfer function*. Analogously to Theorem 3.3, one may prove that $z \cdot \mathbf{R}(z)$ is the z transform of $\mathbf{A}^t$ and $\mathbf{H}(z)$ does not change if a new variable $\bar{\mathbf{x}} = \mathbf{Tx}$ is introduced with a nonsingular matrix $\mathbf{T}$. That is, Theorem 3.4 remains valid for discrete systems.

Example 3.15

In the case of the system discussed in the previous example,

$$\mathbf{R}(z) = \begin{pmatrix} z-1 & -1 \\ 0 & z-1 \end{pmatrix}^{-1} = \frac{1}{(z-1)^2} \begin{pmatrix} z-1 & 1 \\ 0 & z-1 \end{pmatrix}$$

and

$$\mathbf{H}(z) = (1,1)\frac{1}{(z-1)^2} \begin{pmatrix} z-1 & 1 \\ 0 & z-1 \end{pmatrix} \begin{pmatrix} 0 \\ 1 \end{pmatrix} = \frac{z}{(z-1)^2} .$$

Substitute these results into (3.50) to get the input-state relation:

$$\mathbf{X}(z) = \frac{1}{(z-1)^2} \begin{pmatrix} z-1 & 1 \\ 0 & z-1 \end{pmatrix} z \begin{pmatrix} 1 \\ 0 \end{pmatrix}$$

$$+ \frac{1}{(z-1)^2} \begin{pmatrix} z-1 & 1 \\ 0 & z-1 \end{pmatrix} \begin{pmatrix} 0 \\ 1 \end{pmatrix} U(z)$$

$$= \frac{1}{(z-1)^2} \left[\begin{pmatrix} z^2 - z \\ 0 \end{pmatrix} + \begin{pmatrix} 1 \\ z-1 \end{pmatrix} U(z) \right] ,$$

and from (3.51) the input–output relation is derived:

$$\mathbf{Y}(z) = \frac{1}{(z-1)^2}(1,1)\begin{pmatrix} z-1 & 1 \\ 0 & z-1 \end{pmatrix} z \begin{pmatrix} 1 \\ 0 \end{pmatrix} + \frac{z}{(z-1)^2}U(z)$$

$$= \frac{z^2 - z}{(z-1)^2} + \frac{z}{(z-1)^2}U(z) = \frac{z}{z-1} + \frac{z}{(z-1)^2}U(z).$$

As a particular case, assume that the input $u(t) = 1$ for all $t = 0, 1, 2, \ldots$. Then item No. 1 of Table 2.2 implies that $U(z) = z/(z-1)$. Therefore,

$$\mathbf{X}(z) = \frac{1}{(z-1)^2}\left(\frac{z^2 - z + \frac{z}{z-1}}{(z-1)\frac{z}{z-1}}\right) = \frac{1}{(z-1)^3}\left(\frac{z^3 - 2z^2 + 2z}{z^2 - z}\right)$$

and

$$Y(z) = (1,1)\mathbf{X}(z) = \frac{z^3 - z^2 + z}{(z-1)^3}.$$

The partial fraction representation of $Y(z)$ is as follows:

$$Y(z) = 1 + \frac{2}{z-1} + \frac{2}{(z-1)^2} + \frac{1}{(z-1)^3}.$$

Use Nos. 8 and 6 of Table 2.2 to conclude that the output is given as

$$y(t) = \begin{cases} 1 + 0 + 0 + 0 = 1 & \text{if } t = 0 \\ 0 + 2\begin{pmatrix} t-1 \\ 0 \end{pmatrix} + 2\begin{pmatrix} t-1 \\ 1 \end{pmatrix} + \begin{pmatrix} t-1 \\ 2 \end{pmatrix} = \frac{t^2+t+2}{2} & \text{if } t > 0. \end{cases}$$

Because at $t = 0$ the two parts coincide,

$$y(t) = \frac{t^2 + t + 2}{2} \quad \text{for all} \quad t \geq 0.$$

Hence, the output of the system is determined.

Note that the block diagram representation of discrete systems has the same form as shown in Figure 3.3, where variable s is replaced by z. Discrete systems in input–output form can be discussed in the same way as was demonstrated in Section 3.3.3 for continuous systems. In addition, combinations and representations, as well as adjoint and dual systems are also defined in the same way as they were introduced earlier for continuous systems. The details are left to the reader as simple exercises.

3.5   Applications

In this section some particular nonlinear and linear dynamic systems are introduced from different fields of applied sciences. In the first part, some applications in engineering are outlined, and in the second part, case studies from the social sciences are reported. Pick and choose: study only the ones you like. In this chapter these examples illustrate the concepts of an equilibrium point, in some cases linearization, the fundamental matrix and the transfer function. In subsequent chapters, these same examples will be used to illustrate principles of stability, controllability and observability, and other concepts in systems theory.

3.5.1   Dynamic Systems in Engineering

1. Our first system models *harmonic motion*. If we apply a force to a point mass attached to an ideal spring in a frictionless environment, the mass will oscillate sinusoidally. The input could be either a position command; a velocity command, which could be supplied by a velocity-servo system; or both combined.

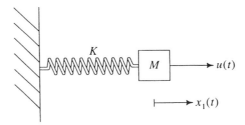

Figure 3.18   An undamped, spring-mass system.

Let θ be the position and v be the velocity of the mass M shown in Figure 3.18. It is well known that

$$\ddot{\theta} = -K\theta + u \ ,$$

if the input is a force applied on the mass when we assume that $M = 1$. In Example 3.3 we derived that with the new parameter $\omega = \sqrt{K}$, the system equations can be summarized as

$$\dot{\mathbf{x}} = \begin{pmatrix} 0 & \omega \\ -\omega & 0 \end{pmatrix} \mathbf{x} + \begin{pmatrix} 0 \\ 1 \end{pmatrix} u \ . \tag{3.53}$$

This is one real-world system that could produce the equations we used in Examples 1.13, 2.6, and 3.3. As shown in Example 3.3, the equilibrium point of this system for a step input is

$$\bar{\mathbf{x}} = \begin{pmatrix} \frac{1}{\omega} \\ 0 \end{pmatrix} . \tag{3.54}$$

This means if we apply a step input of force, the system will come to an equilibrium where the spring is stretched and the velocity is zero. We note that the complete solution of this system has been elaborated earlier in the examples.

2. *A linear second-order mechanical system* may be the most common and most intuitive model of physical systems. In this example we investigate many different properties of linear second-order systems.

The Newtonian equation for the spring-mass-dashpot system of Figure 3.19 is

$$f(t) = M\ddot{\theta} + B\dot{\theta} + K\theta , \tag{3.55}$$

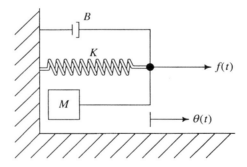

Figure 3.19   A simple damped spring-mass-dashpot system.

where M represents the mass of the object, B the viscosity, and K the elasticity. Notice that the case of $M = 1$ and $B = 0$ corresponds to the previous application, when we select $u = f$. This second-order equation can be reduced to a two-dimensional system of first-order equations by introducing the variables $x_1 = \theta$ and $x_2 = \dot{\theta}$. Then we have

$$\frac{d}{dt} \begin{pmatrix} x_1 \\ x_2 \end{pmatrix} = \begin{pmatrix} 0 & 1 \\ -\frac{K}{M} & -\frac{B}{M} \end{pmatrix} \begin{pmatrix} x_1 \\ x_2 \end{pmatrix} + \begin{pmatrix} 0 \\ \frac{1}{M} \end{pmatrix} u , \tag{3.56}$$

where $u = f$ is the input of the system. Hence, in this case $n = 2$,

$$\mathbf{A} = \begin{pmatrix} 0 & 1 \\ -\frac{K}{M} & -\frac{B}{M} \end{pmatrix}, \qquad \mathbf{b} = \begin{pmatrix} 0 \\ \frac{1}{M} \end{pmatrix} . \qquad (3.57)$$

The equilibrium state of the system with constant force f_0 is the solution of the equation

$$\begin{pmatrix} 0 & 1 \\ -\frac{K}{M} & -\frac{B}{M} \end{pmatrix} \begin{pmatrix} \bar{x}_1 \\ \bar{x}_2 \end{pmatrix} + \begin{pmatrix} 0 \\ \frac{1}{M} \end{pmatrix} f_0 = \begin{pmatrix} 0 \\ 0 \end{pmatrix} ,$$

which is as follows:

$$\bar{x}_1 = \frac{1}{K} f_0 \qquad \text{and} \qquad \bar{x}_2 = 0 . \qquad (3.58)$$

That is, at the equilibrium state, the position is $1/K$ times the force and the velocity is zero.

The characteristic polynomial of matrix $\mathbf{A}$ is given as

$$\varphi(\lambda) = det \begin{pmatrix} -\lambda & 1 \\ -\frac{K}{M} & -\frac{B}{M} - \lambda \end{pmatrix} = \lambda^2 + \frac{B}{M}\lambda + \frac{K}{M} , \qquad (3.59)$$

therefore, the eigenvalues are

$$\lambda_{1,2} = \frac{-B \pm \sqrt{B^2 - 4MK}}{2M} . \qquad (3.60)$$

This formula makes the computation of the fundamental matrix rather complicated. Therefore, the transfer function approach is more attractive. The transfer function is

$$\frac{\Theta(s)}{F(s)} = \frac{1}{Ms^2 + Bs + K} . \qquad (3.61)$$

We now define two parameters ζ and ω_n because they have physical significance and create mathematical simplicity:

$$\zeta = \frac{B}{2\sqrt{KM}} \qquad (3.62)$$

and

$$\omega_n = \sqrt{\frac{K}{M}} . \qquad (3.63)$$

The undamped natural frequency, ω_n, is the frequency at which the system would oscillate if the damping, B, were zero. When we substitute these new parameters, the transfer function becomes

$$\frac{\Theta(s)}{F(s)} = \frac{1}{K}\frac{\omega_n^2}{s^2 + 2\zeta\omega_n s + \omega_n^2} . \tag{3.64}$$

This system exhibits four different types of behavior; they are defined uniquely by the value of the damping ratio, ζ. Figure 3.20 summarizes these responses. The mathematical details are omitted.

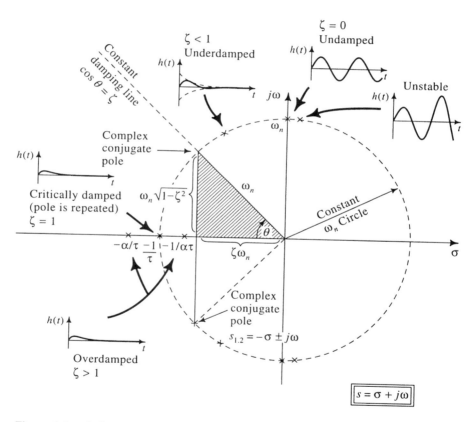

Figure 3.20 Pole-zero diagrams and impulse responses for second-order systems.

The expression under the square root of Equation (3.60) is called the *discriminant*. It has special significance because its square root is imaginary when $B^2 < 4MK$. Applying the quadratic formula to the denominator of Equation (3.64) shows that the two roots are

$$s_{1,2} = -\zeta\omega_n \pm \omega_n \sqrt{\zeta^2 - 1} \, . \qquad (3.65)$$

The roots are called the poles of the system. The ζ and ω_n parameters are most significant when $\zeta < 1$ and the discriminant is negative; therefore, it is preferable to rewrite Equation (3.65) in a way that directly illustrates the real and imaginary components:

$$s_{1,2} = -\zeta\omega_n \pm j\omega_n \sqrt{1 - \zeta^2} = -\zeta\omega_n \pm j\omega_d \, , \qquad (3.66)$$

where w_d is a new parameter called the *damped natural frequency*. It is the frequency of the oscillation in response to, say, a step input, when the damping, B, is not zero.

This important relation demonstrates that ζ determines whether the discriminant is positive or negative. In particular, the roots are

1. negative real when $\zeta > 1$,

2. repeated real when $\zeta = 1$,

3. complex when $0 < \zeta < 1$,

4. purely imaginary when $\zeta = 0$.

To summarize, the potentially oscillatory behavior of the second-order system is characterized mathematically by whether or not the poles are complex. First-order poles can only be real; therefore, a second-order denominator polynomial is the minimum order that allows complex poles to exist. However, the poles of high-order systems can be real or complex, depending on the numerical values of the system's parameters.

When possible, transfer functions are expressed in factored form like this:

$$H(s) = \frac{(s + z_1)(s + z_2)\ldots(s + z_m)}{(s + p_1)(s + p_2)\ldots(s + p_n)} \, , \qquad (3.67)$$

where p_i represents the ith pole and z_k represents the kth zero. It is of conceptual value to plot the poles and zeros of a system on the complex plane because the characteristic patterns of dynamic response in different regions are readily remembered. This plotting also provides the basis for analyzing stability. The details will be given in Chapter 4.

The Laplace transform variable s exhibits the properties of a complex variable. It is a complex frequency variable defined by

$$s = \sigma + j\omega$$

where σ and ω are, respectively, the real and imaginary parts.

The four pole configurations of the basic second-order system as functions of ζ are shown in Figure 3.20, along with a sketched impulse response for each case. The right-triangle relationship between the real and imaginary parts and the undamped natural frequency of the system ω_n should be noted. In particular, the damped natural frequency of the system ω_d is specified by the imaginary part,

$$\omega_d = \omega_n \sqrt{1 - \zeta^2} .$$

Similarly, the real part $-\zeta\omega_n$ corresponds to the inverse exponential decay time constant for the impulse response's envelope. A radial line from the origin is called a *constant damping line* because the angle from the negative real axis is given by

$$\zeta = \cos\theta .$$

So, if ζ is constant, then the angle is constant as well. A circle about the origin is at constant ω_n. For the overdamped case, the poles are arranged on either side of $-1/\tau$ (which also equals $-\zeta\omega_n$ for the $\zeta = 1$ condition).

As a brief review of the mathematical techniques used in analyzing system responses, we will now derive the time response of a linear MBK system with critical damping. The reader is encouraged to perform derivations of the impulse and step responses for the over- and underdamped systems.

Find the step response for the MBK system of Figure 3.19. Let $k = K$, then

$$\frac{\Theta(s)}{F(s)} = \frac{1}{k} \frac{\omega_n^2}{s^2 + 2\omega_n s + \omega_n^2} \tag{3.68}$$

for the particular case where ζ is unity (critically damped). If $f(t)$ is a unit step, then

$$\Theta(s) = \frac{1}{sk} \left(\frac{\omega_n^2}{s^2 + 2\omega_n s + \omega_n^2} \right) = \frac{1}{k} \frac{\omega_n^2}{s(s + \omega_n)^2} . \tag{3.69}$$

We will evaluate this by the method of partial fractions. In this case,

$$\Theta(s) = \frac{1}{k} \frac{\omega_n^2}{s(s + \omega_n)^2} = \frac{A}{s} + \frac{B}{s + \omega_n} + \frac{C}{(s + \omega_n)^2} . \tag{3.70}$$

To find A, multiply both sides of Equation (3.70) by the denominator of the A term (in this case s), then let s take on a value that would make the denominator equal zero (in this instance, 0). We find that

$$A = \frac{1}{k} \ .$$

To find C, multiply Equation (3.70) by $(s + \omega_n)^2$ and then let $s = -\omega_n$ to see that

$$C = \frac{-\omega_n}{k} \ .$$

If we try the same trick for B, we will get

$$B = \frac{0}{0}$$

which is indeterminate. So we must apply differentiation as we did earlier in Example 2.8. Multiply Equation (3.70) by $(s + \omega_n)^2$:

$$\frac{1}{k}\frac{\omega_n^2}{s} = (s + \omega_n)^2\frac{A}{s} + (s + \omega_n)B + C \ . \tag{3.71}$$

Take the derivative with respect to s:

$$\frac{1}{k}\frac{-\omega_n^2}{s^2} = \frac{s^2 - \omega_n^2}{s^2}A + B \ . \tag{3.72}$$

And now evaluate this at $s = -\omega_n$:

$$B = \frac{-1}{k} \ .$$

Therefore, in response to a unit step of force, the position becomes

$$\Theta(s) = \frac{1}{ks} + \frac{-1}{k(s + \omega_n)} + \frac{-\omega_n}{k(s + \omega_n)^2} \ . \tag{3.73}$$

From the table of Laplace transforms (Table 2.1), we find the resulting time response to be

$$\theta(t) = \frac{1}{k} + \frac{-1}{k}e^{-\omega_n t} + \frac{-1}{k}\omega_n te^{-\omega_n t} \qquad (t > 0) \ ,$$

which can be simplified as

$$\theta(t) = \frac{1}{k}[1 - (1 + \omega_n t)e^{-\omega_n t}] \ . \tag{3.74}$$

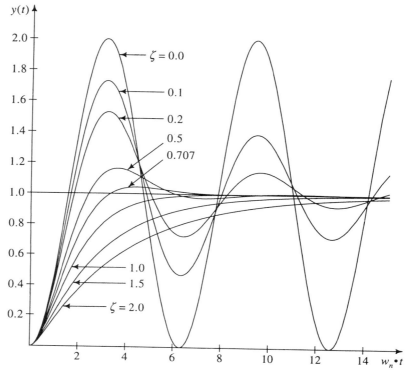

Figure 3.21   Step responses of linear second-order systems.

Similar results for over- and underdamped systems are presented in Figures 3.21 and 3.22 and in Table 3.1.

3. An *electrical system* is discussed next. Assume we apply an ideal voltage source to the ideal resistors, capacitor, and inductor shown in Figure 3.23.

From the theory of simple electric circuits we know that

$$v_s = i_L R_1 + \frac{d}{dt} i_L L + v_C \tag{3.75}$$

and

$$i_C = i_L - i_{R_2} = i_L - \frac{v_C}{R_2} \; ; \tag{3.76}$$

furthermore

$$v_C = \frac{1}{C} \int i_C \, dt \; . \tag{3.77}$$

It is often useful to let the state variables be associated with the energy storage elements. So, let us set $x_1 = i_L$, $x_2 = v_C$, and $u = v_s$, then

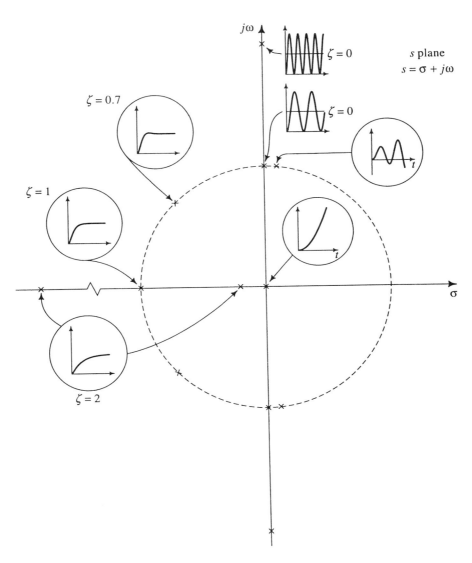

Figure 3.22 Step responses and pole locations of linear second-order systems.

Table 3.1 Step Responses of Second-Order Systems

Damping ratio range	Transfer function	Frequency-domain step response	Time-domain step response
$\zeta = 0$ Undamped	$\frac{\omega_n^2}{s^2+\omega_n^2}$	$\frac{\omega_n^2}{s(s^2+\omega_n^2)}$	$1 - \cos\omega_n t$
$0 < \zeta < 1$ Underdamped	$\frac{\omega_n^2}{s^2+2\zeta\omega_n s+\omega_n^2}$	$\frac{\omega_n^2}{s(s^2+2\zeta\omega_n s+\omega_n^2)}$	$1 - \frac{e^{-\zeta\omega_n t}}{\sqrt{1-\zeta^2}}\sin\left(\omega_n\sqrt{1-\zeta^2}t+\varphi\right)$ where $\varphi = \text{Arc tan }\frac{\sqrt{1-\zeta^2}}{\zeta}$
$\zeta = 1$ Critically damped	$\frac{\omega_n^2}{(s+\omega_n)^2}$	$\frac{\omega_n^2}{s(s+\omega_n)^2}$	$1 - (1 + \omega_n t)e^{-\omega_n t}$
$\zeta > 1$ Overdamped $\zeta \equiv \frac{1+\alpha^2}{2\alpha}$	$\frac{\omega_n^2}{(s+\omega_n/\alpha)(s+\alpha\omega_n)}$	$\frac{\omega_n^2}{s(s+\omega_n/\alpha)(s+\alpha\omega_n)}$	$1 + \frac{1}{\alpha^2-1}(e^{-\alpha\omega_n t} - \alpha^2 e^{-\omega_n t/\alpha})$

from (3.75) we have

$$\dot{x}_1 = -\frac{R_1}{L}x_1 - \frac{1}{L}x_2 + \frac{1}{L}u\ ,$$

and from (3.76) and (3.77),

$$\dot{x}_2 = \frac{1}{C}x_1 - \frac{1}{CR_2}x_2\ .$$

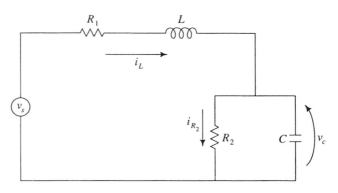

Figure 3.23   A simple electrical system.

Introduce the following notation:

$$\mathbf{A} = \begin{pmatrix} -\frac{R_1}{L} & -\frac{1}{L} \\ \frac{1}{C} & -\frac{1}{CR_2} \end{pmatrix}, \qquad \mathbf{b} = \begin{pmatrix} \frac{1}{L} \\ 0 \end{pmatrix} . \tag{3.78}$$

Then the above equations can be summarized as

$$\dot{\mathbf{x}} = \mathbf{A}\mathbf{x} + \mathbf{b}u .$$

If the voltage across the capacitor is the output of interest, then the output is given as $y = (0,1)\mathbf{x}$, that is, $\mathbf{c}^T = (0,1)$.

For a unit-step input the equilibrium point is the solution of equation

$$\mathbf{A} \cdot \bar{\mathbf{x}} + \mathbf{b} \cdot 1 = \mathbf{0} ,$$

which is

$$\bar{\mathbf{x}} = \begin{pmatrix} \frac{1}{R_1+R_2} \\ \frac{R_2}{R_1+R_2} \end{pmatrix} . \tag{3.79}$$

Note that a step input applied to this system will produce a constant but nonzero steady-state voltage across the capacitor and a constant but nonzero current through the inductor.

The solution of the system can be given in both state–space form and by the transfer function approach. The transfer function method is easy to apply, and easy calculation shows that in this case

$$H(s) = (0,1) \begin{pmatrix} s + \frac{R_1}{L} & \frac{1}{L} \\ -\frac{1}{C} & s + \frac{1}{CR_2} \end{pmatrix}^{-1} \begin{pmatrix} \frac{1}{L} \\ 0 \end{pmatrix}$$

$$= \frac{1}{(s + \frac{R_1}{L})(s + \frac{1}{CR_2}) + \frac{1}{LC}} (0,1) \begin{pmatrix} s + \frac{1}{CR_2} & -\frac{1}{L} \\ \frac{1}{C} & s + \frac{R_1}{L} \end{pmatrix} \begin{pmatrix} \frac{1}{L} \\ 0 \end{pmatrix}$$

$$= \frac{\frac{1}{LC}}{s^2 + s(\frac{R_1}{L} + \frac{1}{CR_2}) + (\frac{R_1+R_2}{LCR_2})} .$$

The solution in the state–space form is left as an easy exercise.

4. A simple *transistor circuit* can be modeled as shown in Figure 3.24. In the output circuit $h_{fe}i_b$ is a dependent current source. This time let us relate the state variables to the input and output of the circuit. Let

the base current, i_b, be x_1 and the output voltage, v_{out}, be x_2, then

$$\dot{\mathbf{x}} = \begin{pmatrix} -\frac{h_{ie}}{L} & 0 \\ \frac{h_{fe}}{C} & 0 \end{pmatrix} \mathbf{x} + \begin{pmatrix} \frac{1}{L} \\ 0 \end{pmatrix} e_s$$

and

$$\mathbf{c}^T = (0, 1) \ . \tag{3.80}$$

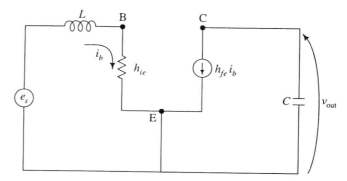

Figure 3.24   A mid-frequency model for a simple transistor circuit.

The **A** matrix looks strange with a column of all zeros, and indeed, the circuit does exhibit odd behavior. For example, as we will show, there is no equilibrium state for a unit step input of e_s. But this is reasonable because the model is for mid-frequencies, and a unit step does not qualify. In response to a unit step, the output voltage will increase linearly until the model is no longer valid.

If e_s is considered to be the input, then the system is

$$\dot{\mathbf{x}} = \begin{pmatrix} -\frac{h_{ie}}{L} & 0 \\ \frac{h_{fe}}{C} & 0 \end{pmatrix} \mathbf{x} + \begin{pmatrix} \frac{1}{L} \\ 0 \end{pmatrix} u \ . \tag{3.81}$$

If $u(t) \equiv 1$, then at the equilibrium state:

$$\begin{pmatrix} -\frac{h_{ie}}{L} & 0 \\ \frac{h_{fe}}{C} & 0 \end{pmatrix} \begin{pmatrix} \bar{x}_1 \\ \bar{x}_2 \end{pmatrix} + \begin{pmatrix} \frac{1}{L} \\ 0 \end{pmatrix} = \begin{pmatrix} 0 \\ 0 \end{pmatrix} \ . \tag{3.82}$$

That is,

$$-\frac{h_{ie}}{L} \bar{x}_1 + \frac{1}{L} = 0$$

$$\frac{h_{fe}}{C}\bar{x}_1 = 0 \ . \tag{3.83}$$

Since $h_{fe}/C \neq 0$, the second equation implies that $\bar{x}_1 = 0$, and by substituting this value into the first equation we get the obvious contradiction $1/L = 0$. Hence, with nonzero constant input *no* equilibrium state exists.

The fundamental matrix of this system can be determined easily. By introducing the notation $\alpha = -h_{ie}/L$ and $\beta = h_{fe}/C$,

$$\mathbf{A} = \begin{pmatrix} \alpha & 0 \\ \beta & 0 \end{pmatrix} \ .$$

Note first that

$$\mathbf{A}^2 = \begin{pmatrix} \alpha & 0 \\ \beta & 0 \end{pmatrix}\begin{pmatrix} \alpha & 0 \\ \beta & 0 \end{pmatrix} = \begin{pmatrix} \alpha^2 & 0 \\ \alpha\beta & 0 \end{pmatrix} = \alpha\mathbf{A} \ ,$$

therefore,

$$\mathbf{A}^3 = \mathbf{A}\cdot\mathbf{A}^2 = \mathbf{A}\cdot\alpha\mathbf{A} = \alpha\mathbf{A}^2 = \alpha\cdot\alpha\mathbf{A} = \alpha^2\mathbf{A} \ ,$$

and so on. Finite induction shows that

$$\mathbf{A}^k = \alpha^{k-1}\cdot\mathbf{A} \ .$$

Hence,

$$e^{\mathbf{A}t} = \mathbf{I} + \sum_{k=1}^{\infty}\frac{\alpha^{k-1}\mathbf{A}t^k}{k!} = \mathbf{I} + \mathbf{A}\cdot\frac{1}{\alpha}\cdot\sum_{k=1}^{\infty}\frac{\alpha^k t^k}{k!}$$

$$= \mathbf{I} + \mathbf{A}\cdot\frac{1}{\alpha}(e^{\alpha t} - 1) = \begin{pmatrix} e^{\alpha t} & 0 \\ \frac{\beta(e^{\alpha t}-1)}{\alpha} & 1 \end{pmatrix} \ .$$

The transfer function of the above system can also be easily determined:

$$H(s) = (0,1)\begin{pmatrix} s-\alpha & 0 \\ -\beta & s \end{pmatrix}^{-1}\begin{pmatrix} \frac{1}{L} \\ 0 \end{pmatrix}$$

$$= (0,1)\frac{1}{s^2 - \alpha s}\begin{pmatrix} s & 0 \\ \beta & s-\alpha \end{pmatrix}\begin{pmatrix} \frac{1}{L} \\ 0 \end{pmatrix}$$

$$= \frac{1}{s^2 - \alpha s} \left(\beta, \, s - \alpha \right) \begin{pmatrix} \frac{1}{L} \\ 0 \end{pmatrix} = \frac{\frac{\beta}{L}}{s^2 - \alpha s} \, .$$

The state of the system can be then determined easily by using the above results for $e^{\mathbf{A}t}$ or $H(s)$.

5. A *hydraulic system* is now presented. For the two-tank liquid reservoir system shown in Figure 3.25, if we ignore the effects of fluid inertia and assume that the system elements are linear, we can write the mass-continuity equations in terms of the liquid levels h_1 and h_2:

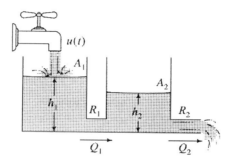

Figure 3.25   A two-tank hydraulic system.

$$A_1 \frac{dh_1}{dt} = -\frac{h_1 - h_2}{R_1} + u(t) \, ,$$

$$A_2 \frac{dh_2}{dt} = \frac{h_1 - h_2}{R_1} - \frac{h_2}{R_2} \, , \tag{3.84}$$

where A_1 and A_2 are the cross-sectional areas of tanks 1 and 2. Rearranging terms, we have

$$\frac{d}{dt} h_1 = -\frac{1}{R_1 A_1} h_1 + \frac{1}{R_1 A_1} h_2 + \frac{1}{A_1} u(t) \, ,$$

$$\frac{d}{dt} h_2 = \frac{1}{R_1 A_2} h_1 - \left(\frac{1}{R_1 A_2} + \frac{1}{R_2 A_2} \right) h_2 \, . \tag{3.85}$$

We find that in our general formulation

$$\mathbf{x}(t) = \begin{pmatrix} h_1(t) \\ h_2(t) \end{pmatrix} , \qquad \mathbf{A} = \begin{pmatrix} -\frac{1}{R_1 A_1} & \frac{1}{R_1 A_1} \\ \frac{1}{R_1 A_2} & -\left(\frac{1}{R_1 A_2} + \frac{1}{R_2 A_2} \right) \end{pmatrix} , \tag{3.86}$$

$$\mathbf{b} = \begin{pmatrix} \frac{1}{A_1} \\ 0 \end{pmatrix} . \tag{3.87}$$

If the output of interest is the flow Q_1 through the connecting pipe, then we have

$$Q_1 = \frac{h_1 - h_2}{R_1} . \tag{3.88}$$

In standard matrix form, the output is expressed as a function of the state vector as

$$Q_1 = \left(\frac{1}{R_1}, -\frac{1}{R_1} \right) \cdot \begin{pmatrix} h_1 \\ h_2 \end{pmatrix} . \tag{3.89}$$

Therefore, in our standard notation,

$$\mathbf{c}^T = \left(\frac{1}{R_1}, -\frac{1}{R_1} \right) . \tag{3.90}$$

In response to a unit step increase in water flow, the system will arrive at the following equilibrium point:

$$\bar{x} = \begin{pmatrix} R_1 + R_2 \\ R_2 \end{pmatrix} . \tag{3.91}$$

That is, the height of the water in each tank will have changed.

This system can be solved by using both state space approach and transfer functions. The details are omitted.

6. In our earlier Example 3, we have examined a single input, single output electrical system. In this example, that earlier model will be extended in order to illustrate a *multiple input electrical system* model. Figure 3.26 shows an electrical network with two voltage sources. Single electric circuit theory implies that

$$v_1 = Ri_1 + L_1 \frac{di_1}{dt} + v_C \tag{3.92}$$

$$v_2 = L_2 \frac{di_2}{dt} + v_C \tag{3.93}$$

and

$$v_C = \frac{1}{C} \int (i_1 + i_2) dt . \tag{3.94}$$

Introduce the state variables $x_1 = i_1, x_2 = i_2, x_3 = v_C$, and the input variables $u_1 = v_1, u_2 = v_2$ to obtain the following differential equations for the state variables:

$$\dot{x}_1 = -\frac{R}{L_1} x_1 - \frac{1}{L_1} x_3 + \frac{1}{L_1} u_1$$

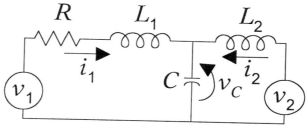

Figure 3.26 A multiple input electrical system.

$$\dot{x}_2 = -\frac{1}{L_2}x_3 + \frac{1}{L_2}u_2$$

$$\dot{x}_3 = \frac{1}{C}x_1 + \frac{1}{C}x_2 \ .$$

Assume furthermore that the output of the system is $y = v_C$. Introduce the following notation:

$$\mathbf{A} = \begin{pmatrix} -\frac{R}{L_1} & 0 & -\frac{1}{L_1} \\ 0 & 0 & -\frac{1}{L_2} \\ \frac{1}{C} & \frac{1}{C} & 0 \end{pmatrix}, \qquad \mathbf{B} = \begin{pmatrix} \frac{1}{L_1} & 0 \\ 0 & \frac{1}{L_2} \\ 0 & 0 \end{pmatrix},$$

and

$$\mathbf{C} = (0, 0, 1) \ .$$

Then the above differential equations can be summarized as

$$\dot{\mathbf{x}} = \mathbf{A}\mathbf{x} + \mathbf{B}\mathbf{u}$$

$$y = \mathbf{C}\mathbf{x} \ . \tag{3.95}$$

As a particular example, we will next determine the equilibrium state of this system with constant unit-step inputs $u_1 = u_2 = 1$. The equilibrium states are the solutions of equation

$$\begin{pmatrix} -\frac{R}{L_1} & 0 & -\frac{1}{L_1} \\ 0 & 0 & -\frac{1}{L_2} \\ \frac{1}{C} & \frac{1}{C} & 0 \end{pmatrix} \begin{pmatrix} \bar{x}_1 \\ \bar{x}_2 \\ \bar{x}_3 \end{pmatrix} + \begin{pmatrix} \frac{1}{L_1} & 0 \\ 0 & \frac{1}{L_2} \\ 0 & 0 \end{pmatrix} \begin{pmatrix} 1 \\ 1 \end{pmatrix} = \begin{pmatrix} 0 \\ 0 \\ 0 \end{pmatrix},$$

which can be rewritten as

$$-\frac{R}{L_1}\bar{x}_1 - \frac{1}{L_1}\bar{x}_3 + \frac{1}{L_1} = 0$$

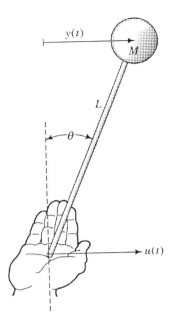

Figure 3.27   An inverted pendulum.

$$-\frac{1}{L_2}\bar{x}_3 + \frac{1}{L_2} = 0$$

$$\frac{1}{C}\bar{x}_1 + \frac{1}{C}\bar{x}_2 = 0 \ .$$

The second equation implies that $\bar{x}_3 = 1$ and by substituting this value into the first equation we see that $\bar{x}_1 = 0$, and the third equation implies that $\bar{x}_2 = 0$. Hence,

$$\bar{\mathbf{x}} = \begin{pmatrix} 0 \\ 0 \\ 1 \end{pmatrix} \tag{3.96}$$

is the unique equilibrium, which can be interpreted by noticing that after the capacitor reaches the voltage level of the inputs, no current will flow in the two loops.

7. There are several mechanical problems — such as orbiting a satellite and controlling a rocket — that have the character of complex *balancing problems*. As a simple version of this type of problem, let us consider balancing a stick on your hand, as illustrated in Figure 3.27.

For simplicity, consider balancing a stick of length L with all of its mass M concentrated on the top. Let the input, $u(t)$, be the position of the person's hand (i.e., the bottom of the stick). The position of the top of the stick will be

$$y(t) = u(t) + L\sin\theta(t) . \tag{3.97}$$

Next we will write the equation for the sum of the moments about the pivot point of the stick:

$$MgL\sin\theta(t) = ML^2\ddot{\theta}(t) + \ddot{u}(t)ML\cos\theta(t) . \tag{3.98}$$

The first term comes from gravity acting on the mass, M. The second term is due to the rotational inertial of the mass on the stick, and the last term shifts the inertial term down to the pivot point.

Now this is a nonlinear system, and nonlinear systems are hard to analyze. So let's make our life easier by linearizing the system. If the stick is nearly at rest in the vertical position (where θ is small), we can say $\cos\theta \approx 1$ and $\sin\theta \approx \theta$. After making these substitutions we can eliminate θ from the two equations to get

$$\ddot{y}(t) = \frac{g}{L}[y(t) - u(t)] . \tag{3.99}$$

For ease of notation let us set $L = 1$. Then, defining the velocity $v(t) = \dot{y}(t)$, the system has the state space representation

$$\begin{pmatrix} \dot{y}(t) \\ \dot{v}(t) \end{pmatrix} = \begin{pmatrix} 0 & 1 \\ g & 0 \end{pmatrix} \begin{pmatrix} y(t) \\ v(t) \end{pmatrix} + \begin{pmatrix} 0 \\ -g \end{pmatrix} u(t) . \tag{3.100}$$

There is only one equilibrium position:

$$\bar{y}(t) = u_0 \qquad \text{and} \qquad \bar{v}(t) = 0 ,$$

where u_0 is a step input. To arrive at this equilibrium state, the top of the stick $y(t)$ had to be moved over the same distance as the bottom, u_0. In this equilibrium state, the angle θ and the velocity will be zero.

Note that with zero input, the general solution of this system is the same as it will be in the case of the warfare model to be introduced in the next section, when we select $h_1 = -g$ and $h_2 = -1$. For the actual solution, see that case study.

8. Our next model deals with a *cart* of mass M that has *two sticks* on it of lengths L_1 and L_2 and masses of M_1 and M_2, as shown in Figure 3.28. If we assume small angles θ_1 and θ_2, we can make the same linearization as before. Please note that in this problem, the input, $u(t)$, is a force, not a position as in previous application. Let $v(t)$ be the velocity of the cart. We can sum horizontal forces acting on the cart to get

$$M\dot{v} = -M_1 g\theta_1 - M_2 g\theta_2 + u .$$

Next we can sum torques about each pivot point to get

$$M_1(\dot{v} + L_1\ddot{\theta}_1) = M_1 g \theta_1$$

$$M_2(\dot{v} + L_2\ddot{\theta}_2) = M_2 g \theta_2 . \tag{3.101}$$

For simplicity, let $M_1 = M_2$. Let us now define our state variables as

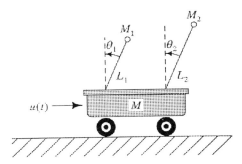

Figure 3.28 A cart with two sticks.

$$x_1 = \theta_1 \quad x_2 = \theta_2 \quad x_3 = \dot{\theta}_1 \quad x_4 = \dot{\theta}_2 \tag{3.102}$$

to obtain (after eliminating $\dot{v}$ from the equations of motion) the state equations $\dot{\mathbf{x}} = \mathbf{A}\mathbf{x} + \mathbf{b}u$ with

$$\mathbf{A} = \begin{pmatrix} 0 & 0 & 1 & 0 \\ 0 & 0 & 0 & 1 \\ a_1 & a_2 & 0 & 0 \\ a_3 & a_4 & 0 & 0 \end{pmatrix}, \qquad \mathbf{b} = \begin{pmatrix} 0 \\ 0 \\ -\frac{1}{ML_1} \\ -\frac{1}{ML_2} \end{pmatrix}, \tag{3.103}$$

where

$$a_1 = \frac{(M + M_2)g}{ML_1}, \qquad a_2 = \frac{M_2 g}{ML_1},$$

$$a_3 = \frac{M_2 g}{ML_2}, \qquad a_4 = \frac{(M + M_2)g}{ML_2} . \tag{3.104}$$

Now what happens if we apply a unit step to the system, i.e., push the cart? It is easy to see that at the equilibrium state, $\bar{x}_1$ and $\bar{x}_2$ are nonzero and $\bar{x}_3 = \bar{x}_4 = 0$, meaning the levers must not be moving, but they are tipped.

9. An *electrical heating system* will be next examined. A temperature-controlled oven is illustrated in Figure 3.29, where the oven temperature is controlled by the heat input u into the jacket. Introduce the following notation:

$$A_1 = \text{inside jacket surface}$$

$$A_2 = \text{outside jacket surface}$$

$$C_1 = \text{heat capacity of inside space}$$

$$C_2 = \text{heat capacity of jacket}$$

$$h_1 = \text{film coefficient for inside surfaces}$$

$$h_2 = \text{film coefficient for outside surfaces}$$

$$T_0 = \text{outside temperature}$$

$$T_1 = \text{inside temperature}$$

$$T_2 = \text{jacket temperature}$$

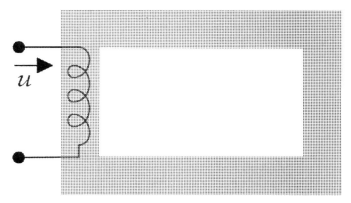

Figure 3.29   An electrical heating system.

Similarly to our earlier hydraulic system, we can easily formulate the heat balance equations for inside space and the jacket as follows:

$$C_1 \dot{T}_1 = A_1 h_1 (T_2 - T_1)$$

$$C_2 \dot{T}_2 = A_2 h_2 (T_0 - T_2) + A_1 h_1 (T_1 - T_2) + u \ .$$

Introduce the state variables $x_1 = T_1 - T_0$ and $x_2 = T_2 - T_0$, and assume that the output of the system is $y = T_1 - T_0$. Then by assuming that the outside temperature T_0 is constant, the above equations can be rewritten as

$$\dot{x}_1 = -\frac{A_1 h_1}{C_1} x_1 + \frac{A_1 h_1}{C_1} x_2$$

$$\dot{x}_2 = \frac{A_1 h_1}{C_2} x_1 - \frac{A_1 h_1 + A_2 h_2}{C_2} x_2 + \frac{1}{C_2} u$$

$$y = x_1 \ .$$

These equations can be summarized as

$$\dot{\mathbf{x}} = \begin{pmatrix} -\frac{A_1 h_1}{C_1} & \frac{A_1 h_1}{C_1} \\ \frac{A_1 h_1}{C_2} & -\frac{A_1 h_1 + A_2 h_2}{C_2} \end{pmatrix} \mathbf{x} + \begin{pmatrix} 0 \\ \frac{1}{C_2} \end{pmatrix} u \qquad (3.105)$$

with output equation

$$y = (1, 0)\mathbf{x} \ .$$

10. The following lumped-parameter model for a *nuclear reactor* is based on [21]. Nuclear fission reactors are described by the same basic dynamic principles, whether they are thermal reactors or fast reactors and whether the nuclear fuel is U^{235}, Pu^{239}, or U^{233}. The essential phenomenon is neutron-induced fission of these isotopes, with the accompanying release of other neutrons, usually two or three per fission event, thus making a self-sustaining neutron chain reaction possible.

The basic concepts of reactor dynamics, common to all types of fission reactors, are reactivity, neutron generation time, and delayed neutrons.

Reactivity is how much the neutron reproduction factor k differs from unity. It is an integral property of the entire reactor. The lumped-parameter (point-reactor) model is satisfactory only when k is near unity, when the reactor is almost critical. The reactivity depends on the size of the reactor, the relative amounts and densities of various materials, and the neutron cross-sections for scattering, absorption, and fission. Because all of these are affected by temperature, pressure, and other effects of fission (arising primarily from the dissipation of kinetic energy of the fission fragments), the reactivity depends on the power history of the reactor. The computation of this reactivity feedback is one of the central problems of reactor dynamics. Furthermore, because the dynamic equations contain the product of reactivity and instantaneous power, the equations are generally nonlinear.

The *neutron generation time* is the mean time for neutron reproduction. It is also an integral property of the entire reactor. It may be as short as 10^{-8} sec for a fast reactor or as long as 10^{-3} sec for a thermal reactor, where neutrons slow down considerably and subsequently diffuse at thermal energies before causing fission. The generation time depends primarily on the number of scattering collisions that a typical neutron undergoes before it escapes from the reactor or disappears in a nuclear reaction. These phenomena are called the leakage and absorption, respectively.

Delayed neutrons, although representing less than one percent of the neutron production in fission, are extremely important in determining the time scales in reactor dynamics. These neutrons are released in certain nuclear transitions that occur in a few types of highly excited fission fragments, and the relevant processes have half-lives of the order of a few seconds.

When the reproduction factor is sufficiently large that the neutron chain reaction would be self-sustaining with only the prompt neutrons (neutrons released immediately in fission), the neutron generation time is dominant in determining the time scale. When the reactor is not too far from critical, in the regime where prompt neutrons alone would be insufficient to sustain a chain reaction, the relatively large delay times of the delayed neutrons are dominant, even though the delayed-neutron fraction is small. If all neutrons were prompt, it would be extremely difficult to control a reactor by conventional mechanical means, such as movement of fuel, neutron absorbers, or neutron reflectors, because of the high frequency response required to compensate for the short neutron generation time.

The point-reactor equations are

$$\frac{dD}{dt} = \frac{\rho - \beta}{l}D + \sum_{i=1}^{m} \lambda_i D_i + q$$

and

$$\frac{dD_i}{dt} = \frac{\beta_i}{l}D - \lambda_i D_i , \qquad (3.106)$$

where

D = neutron density (or power, etc.)

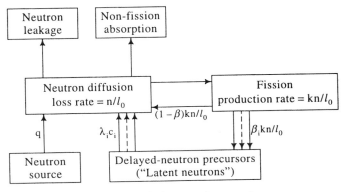

Figure 3.30 Simplified neutron cycle for a nuclear reactor.

D_i = precursor density (latent-neutron density or latent power, etc.

same units as D)

t = time

ρ = reactivity $(k - 1)/k$, the fractional change in neutron reproduction

factor

β = delayed-neutron factor $(\sum_{i=1}^{m} \beta_i)$

l = neutron generation time

λ_i = decay constant for precursor decay

q = effective source strength (same unit as dD/dt)

m = the number of delayed-neutron groups

The parameters β_i, λ_i, and l are assumed constant. The point-reactor model thus consists of $m + 1$ coupled first-order differential equations together with the specification of the functions $\rho(t)$ and $q(t)$. In general, ρ is a function of D and the system is nonlinear; however, in the absence of reactivity feedback, $\rho(t)$ is an explicit function of time and the system is linear.

Using our usual vector-matrix notation, this system's model can be reformulated as follows. Assume that the state vector is given as $\mathbf{x} = (D, D_1, D_2, \ldots, D_m)^T$ and the input is the effective source strength, that is, $u = q$. Then we have

$$\dot{\mathbf{x}} = \mathbf{A}\mathbf{x} + \mathbf{b}u ,\tag{3.107}$$

where

$$\mathbf{A} = \begin{pmatrix} \frac{\rho-\beta}{l} & \lambda_1 & \lambda_2 & \lambda_3 & \cdots & \lambda_{m-1} & \lambda_m \\ \frac{\beta_1}{l} & -\lambda_1 & 0 & 0 & \cdots & 0 & 0 \\ \vdots & \vdots & \vdots & \vdots & \ddots & \vdots & \vdots \\ \frac{\beta_m}{l} & 0 & 0 & 0 & \cdots & 0 & -\lambda_m \end{pmatrix}\tag{3.108}$$

is an $(m+1) \times (m+1)$ size matrix, and $\mathbf{b} = (1, 0, 0, \ldots, 0)^T$ is an $(m+1)$-dimensional vector.

The equilibrium of the system (with constant $q = q_0$) can be obtained by solving equation

$$\mathbf{A}\mathbf{x} + \mathbf{b}q_0 = \mathbf{0} ,\tag{3.109}$$

which is an $(m+1)$ dimensional system of linear equations. Summing up the equations of (3.108) we get the relation

$$\left(\frac{\rho - \beta}{l} + \frac{1}{l} \sum_{i=1}^{m} \beta_i \right) D + q_0 = 0 ,\tag{3.110}$$

which reduces to

$$\frac{\rho}{l} D + q_0 = 0 .$$

Hence,

$$\bar{D} = -\frac{lq_0}{\rho} ,$$

and from the $(i+1)$st Equation of (3.106) we conclude that

$$\frac{\beta_i}{l}\bar{D} - \lambda_i \bar{D}_i = 0 ,$$

that is,

$$\bar{D}_i = \frac{\beta_i}{l\lambda_i}\bar{D} .$$

Criticality, defined as $k = 1$ ($\rho = 0$), is, strictly speaking, a non-equilibrium situation; in the presence of a source, the critical reactor is divergent. A neutron source, inserted prior to reactor startup to provide adequate detector readings, may be withdrawn as criticality is approached; nevertheless, neutrons from spontaneous fission and cosmic rays will always represent sources. In consequence, an operating reactor at steady power is always slightly subcritical, although the reactivity as given by Equation (3.106) is usually undetectably small. This will be the case when $\bar{D}/l$ is large, and if the magnitude of the reactivity at power is very much smaller than the magnitude of the shutdown reactivity, then the source may be neglected in further calculations.

3.5.2 Dynamic Systems in Social Sciences

1. Our first example is known as the two-dimensional *predator–prey* model. Imagine an island populated primarily by goats and wolves. The goats survive by eating the abundant vegetation of the island, and the wolves survive by eating the goats. Let $G(t)$ and $W(t)$ denote the goat and wolf populations at time t. The predator–prey model has the form

$$\dot{G}(t) = aG(t) - bG(t)W(t) \tag{3.111}$$

$$\dot{W}(t) = -cW(t) + dG(t)W(t) \tag{3.112}$$

where a, b, c, and d are positive constants. Equation (3.111) tells us that the prey population growth per unit time is proportional to the prey population, and the decrease rate is proportional to the product of the populations of prey and predator. The second equation implies that the predator growth is negatively proportional to the predator population (reflecting competition) and is positively proportional to the product of the two populations. If $W(t) \equiv 0$, then the first equation reflects the *exponential growth* model

$$\dot{G}(t) = aG(t) ,$$

which has the solution

$$G(t) = e^{at} \cdot G(0) .$$

In the general case, it is known that the solution is periodic (see for example, [17], Section 48). Figure 3.31 shows the solution in the special case, when

$$a = 0.25, \quad b = 0.01, \quad c = 1.0, \quad \text{and} \quad d = 0.01 ;$$

furthermore, the initial values are given as

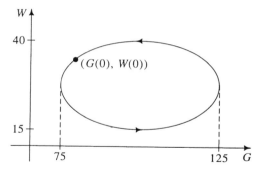

Figure 3.31   Predator–prey trajectory.

$$G(0) = 80 \text{ and } W(0) = 30 \ .$$

The above model description does not have formal input. However, improved weather conditions can result in the increase of the growth rate of the goat population, since more vegetation means more food for the goats. Therefore, parameter a can be assumed to be controlled in this way by the nature. In this case, Equation (3.111) can be replaced by

$$\dot{G}(t) = (a + u(t))G(t) - bG(t)W(t) \ . \tag{3.113}$$

The equilibrium $(\bar{G}, \bar{W})$ of system (3.111) and (3.112) is obtained by solving equation

$$a\bar{G} - b\bar{G}\bar{W} = 0$$

$$-c\bar{W} + d\bar{G}\bar{W} = 0 \ .$$

It is easy to verify that the two solutions are

$$\bar{G} = 0, \quad \bar{W} = 0 \quad \text{and} \quad \bar{G} = \frac{c}{d}, \quad \bar{W} = \frac{a}{b} \ .$$

Therefore, we have two equilibrium points. In the case of the first equilibrium, both populations are zero, and in the case of the second equilibrium, both populations are positive. The first equilibrium is trivial, so we will only discuss the case of the second equilibrium.

We apply the linearization procedure to find the linear approximation of the system around the positive equilibrium. The practical value of linearization is the fact that the theory of linear systems can be applied to the linearized model.

The Jacobian of the right-hand side functions of (3.111) and (3.112) has the form

$$\mathbf{J}(G, W) = \begin{pmatrix} a - bW & -bG \\ dW & -c + dG \end{pmatrix} ,$$

therefore, at the positive equilibrium

$$\mathbf{J}(\bar{G}, \bar{W}) = \begin{pmatrix} 0 & -\frac{bc}{d} \\ \frac{ad}{b} & 0 \end{pmatrix} ,$$

and consequently, with new variables $G_\delta = G - c/d$ and $W_\delta = W - a/b$ from Equation (3.13) we obtain the linearized model

$$\dot{G}_\delta = -\frac{bc}{d} W_\delta$$

$$\dot{W}_\delta = \frac{ad}{b} G_\delta .$$

These equations can be easily rewritten in order to have the original variables:

$$\dot{G} = -\frac{bc}{d} \left(W - \frac{a}{b} \right) = -\frac{bc}{d} W + \frac{ac}{d}$$

$$\dot{W} = \frac{ad}{b} \left(G - \frac{c}{d} \right) = \frac{ad}{b} G - \frac{ac}{b} . \tag{3.114}$$

Note that the structure of this model is very similar to that shown in Example 3.7. More details on interacting populations can be found, for example, in [28]. Finally, we note that the warfare model and epidemics, which will be discussed later in this section, are mathematically special cases of this model.

2. Our next model is known as the *cohort population* model. Assume that the population of a country is divided into age groups (or cohorts). Let n denote the number of age groups, and let $P_i(t)$ denote the population of group i at time period t. We assume that this system is described by difference equations

$$P_{i+1}(t + 1) = a_i P_i(t) \qquad (i = 1, \ldots, n - 1)$$

and

$$P_1(t + 1) = b_1 P_1(t) + \cdots + b_n P_n(t) .$$

The first equation is interpreted as the surviving portion of group i simply moving up to the $(i + 1)$st age group after one time period.

The second equation gives the number of individuals born during the last time period, where the coefficients $b_1, \ldots, b_n$ represent the birth rates in the different age groups. Assume furthermore that at time t, $u_i(t)$ individuals join the population of age group i from outside. Therefore, this situation can be modeled by the linear time-invariant discrete system

$$\mathbf{p}(t+1) = \mathbf{A} \cdot \mathbf{p}(t) + \mathbf{I} \cdot \mathbf{u}(t) , \qquad (3.115)$$

where

$$\mathbf{p} = \begin{pmatrix} P_1 \\ P_2 \\ P_3 \\ \vdots \\ P_n \end{pmatrix} , \qquad \mathbf{A} = \begin{pmatrix} b_1 & b_2 & b_3 & \cdots & b_{n-1} & b_n \\ a_1 & & & & & O \\ & a_2 & & & & \\ & & a_3 & & & \\ & & & \ddots & & \\ O & & & & a_{n-1} & 0 \end{pmatrix} , \qquad \mathbf{u} = \begin{pmatrix} u_1 \\ u_2 \\ u_3 \\ \vdots \\ u_n \end{pmatrix} .$$

If we are interested only in the total population, then we may select the output

$$P_0(t) = P_1(t) + \cdots + P_n(t) = (1, 1, \ldots, 1)\mathbf{p}(t) .$$

This model was originally introduced by Leslie [27], and matrix $\mathbf{A}$ in (3.115) is usually called the Leslie-matrix. The population structure can be predicted for any future time by solving the governing difference equation. Some properties of this system will be analyzed in later chapters.

3. In the two-nation *arms race* model, let $X(t)$ and $Y(t)$ denote the armament levels of the two nations at time t. The well-known Richardson's model [38] can be written as follows:

$$\dot{X}(t) = aY(t) - bX(t) + \alpha \qquad (3.116)$$

$$\dot{Y}(t) = cX(t) - dY(t) + \beta . \qquad (3.117)$$

This model shows that the arms race of each nation is negatively proportional to its armament level and positively proportional to the armament level of the other nation. Constants a, b, c, and d largely depend on the overall relations of the two nations. It is also interesting to note that Chestnut [7] describes cooperative security systems between the two nations based on systems theoretical concepts and methods. By introducing the notation

$$\mathbf{x} = \begin{pmatrix} X \\ Y \end{pmatrix} , \qquad \mathbf{A} = \begin{pmatrix} -b & a \\ c & -d \end{pmatrix} , \qquad \mathbf{f} = \begin{pmatrix} \alpha \\ \beta \end{pmatrix} ,$$

this system can be rewritten in our general form

$$\dot{\mathbf{x}} = \mathbf{A}\mathbf{x} + \mathbf{f} \ . \tag{3.118}$$

This system has no formal input in this formulation. Assume next that improved relations between the two nations can decrease the arms race, and the worsening of their relations may increase it. If $u(t)$ denotes a measure of the "goodness" of the relation of the two nations, then the above model can be extended as

$$\dot{\mathbf{x}}(t) = \begin{pmatrix} -b & a \\ c & -d \end{pmatrix} \mathbf{x}(t) + \begin{pmatrix} \alpha \\ \beta \end{pmatrix} u(t) \ , \tag{3.119}$$

where $u(t)$ is a formal input. The original Richardson's model is the special case of this formulation by selecting the constant input $u(t) \equiv 1$. In this case the equilibrium can be determined by solving equations

$$a\bar{Y} - b\bar{X} + \alpha = 0$$

$$c\bar{X} - d\bar{Y} + \beta = 0 \ .$$

Assume that $bd \neq ac$, then the solution is

$$\bar{X} = \frac{a\beta + \alpha d}{bd - ac} \qquad \text{and} \qquad \bar{Y} = \frac{c\alpha + \beta b}{bd - ac} \ .$$

Assume that the output is defined as the armament level of the first nation. That is,

$$y = (1, 0)\mathbf{x} \ .$$

Then the transfer function of the system is the following:

$$H(s) = (1, 0) \begin{pmatrix} s + b & -a \\ -c & s + d \end{pmatrix}^{-1} \begin{pmatrix} \alpha \\ \beta \end{pmatrix}$$

$$= (1, 0) \frac{1}{s^2 + (b + d)s + (bd - ac)} \begin{pmatrix} s + d & a \\ c & s + b \end{pmatrix} \begin{pmatrix} \alpha \\ \beta \end{pmatrix}$$

$$= \frac{1}{s^2 + (b + d)s + (bd - ac)} (s + d, a) \begin{pmatrix} \alpha \\ \beta \end{pmatrix}$$

$$= \frac{s\alpha + (\alpha d + \beta a)}{s^2 + (b + d)s + (bd - ac)} \ .$$

This transfer function allows us to determine the output of the system of any future time period by using the solution formula (3.24).

A natural generalization of this model to the *multi-nation* case can be given as follows:

$$\dot{X}_i(t) = \sum_{j=1}^{n} \alpha_{ij} X_j(t) + \beta_i \qquad (1 \leq i \leq n) \qquad (3.120)$$

where $X_i(t)$ is the armament level of nation i in time period t; furthermore $\alpha_{ii} < 0$ and $\alpha_{ij} > 0$ for $j \neq i$. Introduce notation $\mathbf{x} = (X_i)$, $\mathbf{A} = (\alpha_{ij})$, and $\mathbf{f} = (\beta_i)$, then this model has the usual form (3.118). The equilibrium is the solution of equation

$$\mathbf{A}\bar{\mathbf{x}} + \mathbf{f} = \mathbf{0} .$$

If $\mathbf{A}^{-1}$ exists, then vector

$$\bar{\mathbf{x}} = -\mathbf{A}^{-1}\mathbf{f}$$

gives the equilibrium.

4. A *warfare* model can be formulated as follows. Assume that two forces are engaged in a war. Let X_1 and X_2 denote the numbers of units in the two forces. The members of the fighting forces are characterized by their "hitting powers," which are the numbers of casualties per unit time that one member can inflict on the enemy. The hitting powers h_1 and h_2 are determined by military technology. By assuming that the hitting power of each force is directed uniformly against all units of the enemy, we obtain the following relations:

$$\dot{X}_1 = -h_2 X_2$$

$$\dot{X}_2 = -h_1 X_1 . \qquad (3.121)$$

Use the notation

$$\mathbf{x} = \begin{pmatrix} X_1 \\ X_2 \end{pmatrix} \qquad \text{and} \qquad \mathbf{A} = \begin{pmatrix} 0 & -h_2 \\ -h_1 & 0 \end{pmatrix}$$

to get the formulation

$$\dot{\mathbf{x}} = \mathbf{A}\mathbf{x} .$$

Note that this system is a special case of the linearized predator–prey model by selecting

$$\frac{bc}{d} = h_2 \qquad \text{and} \qquad \frac{ad}{b} = -h_1$$

with zero constant terms.

This first-order system can be easily rewritten to a single second-order equation. Differentiate the first equation of (3.121) to see that

$$\ddot{X}_1 = -h_2 \dot{X}_2 = h_1 h_2 X_1 \;,$$

that is,

$$\ddot{X}_1 - h_1 h_2 X_1 = 0. \tag{3.122}$$

The characteristic polynomial of this equation is

$$\lambda^2 - h_1 h_2 = 0$$

with eigenvalues $\lambda_{1,2} = \pm\sqrt{h_1 h_2}$. Therefore, the general solution of Equation (3.122) is given as

$$X_1(t) = c_1 e^{t\sqrt{h_1 h_2}} + c_2 e^{-t\sqrt{h_1 h_2}} \;,$$

and from the first equation of (3.121) we conclude that

$$X_2(t) = -\frac{1}{h_2}\dot{X}_1 = -c_1\sqrt{\frac{h_1}{h_2}}e^{t\sqrt{h_1 h_2}} + c_2\sqrt{\frac{h_1}{h_2}}e^{-t\sqrt{h_1 h_2}} \;.$$

The coefficients c_1 and c_2 can be determined from the initial conditions:

$$X_1(0) = c_1 + c_2 = x_{10}$$

$$X_2(0) = \sqrt{\frac{h_1}{h_2}}(-c_1 + c_2) = x_{20} \;,$$

and the results are as follows:

$$c_1 = \frac{1}{2}\left(x_{10} - \sqrt{\frac{h_2}{h_1}}x_{20}\right) \;, \qquad c_2 = \frac{1}{2}\left(x_{10} + \sqrt{\frac{h_2}{h_1}}x_{20}\right) \;.$$

Hence,

$$X_1(t) = \frac{1}{2}\left(x_{10} - \sqrt{\frac{h_2}{h_1}}x_{20}\right)e^{t\sqrt{h_1 h_2}} + \frac{1}{2}\left(x_{10} + \sqrt{\frac{h_2}{h_1}}x_{20}\right)e^{-t\sqrt{h_1 h_2}}$$

$$= x_{10}\cosh\left(t\sqrt{h_1 h_2}\right) - x_{20}\sqrt{\frac{h_2}{h_1}}\sinh\left(t\sqrt{h_1 h_2}\right) \;,$$

and

$$X_2(t) = -\frac{1}{2}\left(x_{10} - \sqrt{\frac{h_2}{h_1}}x_{20}\right)\sqrt{\frac{h_1}{h_2}}e^{t\sqrt{h_1 h_2}}$$

$$+ \frac{1}{2}\left(x_{10} + \sqrt{\frac{h_2}{h_1}}x_{20}\right)\sqrt{\frac{h_1}{h_2}}e^{-t\sqrt{h_1 h_2}}$$

$$= -x_{10}\sqrt{\frac{h_1}{h_2}}\sinh\left(t\sqrt{h_1 h_2}\right) + x_{20}\cosh\left(t\sqrt{h_1 h_2}\right) ,$$

where cosh and sinh are the well-known hyperbolic functions.

An alternative solution method is based on computing $e^{\mathbf{A}t}$ and then applying the solution formula (2.18). This method is illustrated next. Since

$$\mathbf{A}^2 = \begin{pmatrix} h_1 h_2 & 0 \\ 0 & h_1 h_2 \end{pmatrix} = h_1 h_2 \mathbf{I} ,$$

we have

$$\mathbf{A}^3 = \mathbf{A}^2 \cdot \mathbf{A} = h_1 h_2 \mathbf{A},$$

$$\mathbf{A}^4 = \mathbf{A}^2 \cdot \mathbf{A}^2 = (h_1 h_2)^2 \mathbf{I} ,$$

$$\mathbf{A}^5 = \mathbf{A}^3 \cdot \mathbf{A}^2 = (h_1 h_2)^2 \mathbf{A} ,$$

and so on. Finite induction can be applied to show that

$$\mathbf{A}^k = \begin{cases} (h_1 h_2)^m \mathbf{I} & \text{if } k = 2m \\ (h_1 h_2)^m \mathbf{A} & \text{if } k = 2m + 1 . \end{cases}$$

Therefore,

$$e^{\mathbf{A}t} = \sum_{k=0}^{\infty} \frac{1}{k!} \mathbf{A}^k t^k = \sum_{m_{k=\text{even}}} \frac{(h_1 h_2)^m t^{2m}}{(2m)!} \mathbf{I} + \sum_{m_{k=\text{odd}}} \frac{(h_1 h_2)^m t^{2m+1}}{(2m+1)!} \mathbf{A}$$

$$= \sum_m \frac{(\sqrt{h_1 h_2})^{2m} t^{2m}}{(2m)!} \mathbf{I} + \frac{1}{\sqrt{h_1 h_2}} \sum_m \frac{(\sqrt{h_1 h_2})^{2m+1} t^{2m+1}}{(2m+1)!} \mathbf{A}$$

$$= \cosh(t\sqrt{h_1 h_2})\mathbf{I} + \frac{1}{\sqrt{h_1 h_2}} \sinh(t\sqrt{h_1 h_2})\mathbf{A}$$

$$= \begin{pmatrix} \cosh(t\sqrt{h_1 h_2}) & -\sqrt{\frac{h_2}{h_1}} \sinh(t\sqrt{h_1 h_2}) \\ -\sqrt{\frac{h_1}{h_2}} \sinh(t\sqrt{h_1 h_2}) & \cosh(t\sqrt{h_1 h_2}) \end{pmatrix}.$$

Hence,

$$\mathbf{x}(t) = e^{\mathbf{A}t} \begin{pmatrix} x_{10} \\ x_{20} \end{pmatrix}$$

$$= \begin{pmatrix} x_{10} \cosh(t\sqrt{h_1 h_2}) - x_{20}\sqrt{\frac{h_2}{h_1}} \sinh(t\sqrt{h_1 h_2}) \\ -x_{10}\sqrt{\frac{h_1}{h_2}} \sinh(t\sqrt{h_1 h_2}) + x_{20} \cosh(t\sqrt{h_1 h_2}) \end{pmatrix}, \quad (3.123)$$

which coincides with the result obtained earlier.

For more details of this model, see, for example, [25] and [40].

5. Examining *epidemics* of disease in human population is a very important application of dynamic systems theory. Consider a population of individuals, and assume that a disease spreads by contact between individuals. It is assumed that infected individuals either die, become isolated, or recover and become immune. Therefore, in any time period t the population consists of $x(t)$ susceptible individuals, $y(t)$ infected and circulating individuals, and $z(t)$ further individuals who either have been removed (died or isolated) or are immune. The dynamics of the system are described by differential equations

$$\dot{x} = -\alpha x y$$

$$\dot{y} = \alpha x y - \beta y$$

$$\dot{z} = \beta y .$$

The first equation shows that the decline in the number of susceptible individuals is proportional to the product of the numbers of susceptible and infected (but still circulating) individuals. The first term of the right-hand side of the second equation shows the number of newly infected individuals, and the second term gives the number of individuals who have been removed. The third equation means that these removed individuals increase the number of the third group. Note that z does not appear in the first two equations, therefore, it is sufficient to consider the system consisting of only the first two equations:

$$\dot{x} = -\alpha xy$$

$$\dot{y} = \alpha xy - \beta y . \tag{3.124}$$

Mathematically, this system is a special case of the predator–prey model (3.111) and (3.112) by selecting $a = 0$ and $b = d$. This analogy is expected, since when goats are eaten by wolves and susceptible individuals are infected, they are simply removed from their populations. A comprehensive summary on deterministic and stochastic models of epidemics can be found in [6].

6. In Example 3.2 we have introduced the *Harrod-type national economy* model

$$Y(t + 1) = (1 + r - rm)Y(t) - rG(t) ,$$

where $Y(t)$ is the national income and $G(t)$ is the government expenditure.

By using the notation of Section 3.4, in our case $n = 1$,

$$\mathbf{A}(t) = 1 + r - rm, \qquad \mathbf{B}(t) = -r, \qquad \text{and} \qquad \mathbf{u}(t) = G(t) ,$$

when $G(t)$ is considered as the input of the system. The solution of the system can be obtained by using Equations (3.48) and (2.47):

$$Y(t) = (1 + r - rm)^t Y_0 - \sum_{\tau=0}^{t-1} (1 + r - rm)^{t-\tau-1} \cdot rG(\tau)$$

where $Y_0 = Y(0)$. Hence, the national income can be directly computed for any future time period $t > 0$.

7. Our next model is concerned with the supply and demand of a single commodity. Assume that the demand function $d(p)$ gives the demand of the commodity as a function of the price p. Function d is assumed to

be decreasing. The amount $s(p)$ of the commodity that will be supplied by the producers also depends on the price. It is now assumed that function s is increasing. The above properties of functions d and s reflect the economic facts that higher market price results in the decline of the demand of the commodity, but it leads to higher producers' profit, and therefore, the producers increase their production levels.

Let $p(t)$ denote the price in time period t. The producer makes his decision on his production level based on the current price, but the resulting supply is available only in the next time period. Therefore, the supply $S(t)$ satisfies equation

$$S(t + 1) = s(p(t)) .$$

When this supply shows up on the market, its price is determined by the demand function by adjusting it so that the entire supply is sold. Therefore, the demand $D(t + 1)$ at time period $t + 1$ satisfies relation

$$D(t + 1) = d(p(t + 1)) .$$

It is also assumed that the market is in equilibrium in time period $t + 1$, which means that the supply equals the demand:

$$S(t + 1) = D(t + 1) .$$

That is,

$$s(p(t)) = d(p(t + 1)) .$$

Let d^{-1} denote the inverse of function d; then this equation can be rewritten as

$$p(t + 1) = d^{-1}(s(p(t))) . \tag{3.125}$$

Hence, the price function is a solution of this nonlinear difference equation. The solution is made unique by specifying the initial price $p(0)$.

The equilibrium of Equation (3.125) is the solution of the nonlinear equation

$$\bar{p} = d^{-1}(s(\bar{p})) ,$$

which is the fixed-point problem of function $d^{-1}(s(.))$. If this function satisfies the conditions of Theorem 1.3, then process (3.125) converges to the unique equilibrium. Sufficient conditions for these properties are, for example, the following:

(a) $d^{-1}(s(p))$ exists for all $p \geq 0$;

(b) $d^{-1}(s(p)) \geq 0$ for all $p \geq 0$;

(c) $(d/dp)d^{-1}(s(p))$ exists and $|(d/dp)d^{-1}(s(p))| \leq K$ for all $p \geq 0$,
where $K \in [0,1)$ is a constant.

Here we used the results of Example 1.5. The solution of the governing
difference equation (3.125) is illustrated in Figure 3.32. The resulting
rectangular spiral resembles a cobweb. Therefore, in the literature this
model is usually called the *cobweb model*.

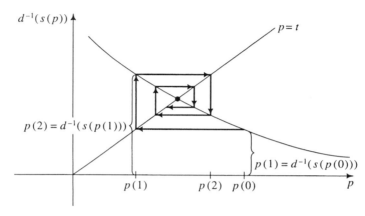

Figure 3.32   Illustration of the cobweb model.

As a special case assume that

$$d(p) = ap + a_0, \qquad s(p) = bp + b_0 .$$

The equilibrium price $\bar{p}$ is the solution of the equation

$$a\bar{p} + a_0 = b\bar{p} + b_0 ,$$

which is

$$\bar{p} = \frac{b_0 - a_0}{a - b} . \qquad (3.126)$$

Furthermore, Equation (3.125) now has the form

$$ap(t + 1) + a_0 = bp(t) + b_0 ,$$

that is, the governing difference equation is the following:

$$p(t + 1) = \frac{b}{a}p(t) + \frac{b_0 - a_0}{a} . \qquad (3.127)$$

If the initial price $p(0) = p_0$, then from Equation (2.47) we conclude that

$$p(t) = \left(\frac{b}{a}\right)^t p_0 + \sum_{\tau=0}^{t-1} \left(\frac{b}{a}\right)^{t-\tau-1} \cdot \frac{b_0 - a_0}{a}$$

$$= \left(\frac{b}{a}\right)^t p_0 + \frac{(\frac{b}{a})^t - 1}{\frac{b}{a} - 1} \cdot \frac{b_0 - a_0}{a} \ .$$

That is, the price can be computed easily for any future time period $t > 0$. Note that the sequence $p(t)$ is convergent if and only if $|b| < |a|$, and in this case,

$$\lim_{t \to \infty} p(t) = \frac{-1}{\frac{b}{a} - 1} \cdot \frac{b_0 - a_0}{a} = \frac{b_0 - a_0}{a - b} \ ,$$

which is the equilibrium price (see Equation (3.126)).

This model is investigated in more detail in [29].

8. Our next application is a continuous system of *interrelated markets*. Assume there are n commodities, and let p_i $(i = 1, 2, \ldots, n)$ denote the price of commodity i. If $s_i(p_1, \ldots, p_n)$ is the supply from commodity i, then it is usually assumed that

$$\frac{\partial s_i}{\partial p_i} > 0, \quad \frac{\partial s_i}{\partial p_j} \leq 0 \quad (j \neq i) \ .$$

These conditions reflect the tendency that manufacturers produce those items that give them higher profit. It is also assumed that the demands for the different commodities are interrelated by the demand functions $d_i(p_1, \ldots, p_n)$, where we assume that

$$\frac{\partial d_i}{\partial p_i} < 0 \quad \text{and} \quad \frac{\partial d_i}{\partial p_j} \geq 0 \quad (j \neq i) \ .$$

At each time period the difference $d_i - s_i$ is the shortage if it is positive, and a surplus if it is negative. It is also assumed that for each commodity the market price moves as directed by the shortage, rising if the shortage is positive, decreasing if negative, and does not change if zero. Therefore, the dynamics are governed by differential equations

$$\dot{p}_i = K_i(d_i(p_1, \ldots, p_n) - s_i(p_1, \ldots, p_n)) \tag{3.128}$$

for $i = 1, 2, \ldots, n$, where K_i is a strictly increasing function.

As a special case, assume that functions K_i, d_i, and s_i are all linear, that is,

$$K_i(d_i - s_i) = k_i \cdot (d_i - s_i), \qquad d_i(p_1, \ldots, p_n)$$

$$= \sum_j a_{ij} p_j + a_{i0}, \qquad s_i(p_1, \ldots, p_n)$$

$$= \sum_j b_{ij} p_j + b_{i0} \ ,$$

where

$$k_i > 0, \quad a_{ii} < 0, \quad a_{ij} \geq 0 \quad (j \neq i), \quad b_{ii} > 0, \text{ and } b_{ij} \leq 0 \quad (j \neq i) \ .$$
$$(3.129)$$

Introduce the notation

$$\mathbf{p} = (p_i) \ , \qquad \mathbf{K} = diag(k_1, \ldots, k_n) \ ,$$

$$\mathbf{A} = (a_{ij}) \ , \qquad \mathbf{B} = (b_{ij}) \ ,$$

$$\mathbf{a}_0 = (a_{i0}) \ , \qquad \mathbf{b}_0 = (b_{i0}) \ ;$$

then Equation (3.128) can be rewritten as

$$\dot{\mathbf{p}} = \mathbf{K} \cdot ((\mathbf{A} - \mathbf{B})\mathbf{p} + \mathbf{a}_0 - \mathbf{b}_0) \ . \qquad (3.130)$$

The equilibrium prices $\bar{\mathbf{p}}$ of this system can be determined by solving equation

$$\mathbf{K}((\mathbf{A} - \mathbf{B})\bar{\mathbf{p}} + \mathbf{a}_0 - \mathbf{b}_0) = \mathbf{0} \ .$$

If $(\mathbf{A} - \mathbf{B})^{-1}$ exists, then vector

$$\bar{\mathbf{p}} = -(\mathbf{A} - \mathbf{B})^{-1}(\mathbf{a}_0 - \mathbf{b}_0)$$

gives the equilibrium. The mathematical properties of this model are examined in [3]. We mention that this model will be further examined in later chapters.

9. *Oligopoly models* have a very important role in economic theory. A simple version of the classical Cournot model is now introduced.

Assume that N firms produce a homogeneous good and sell it on the same market. Let $d(p)$ denote the market demand function. Assume $d(p)$ is strictly decreasing; it then has an inverse $p(d)$. Let $x_1, \ldots, x_N$

denote the output of the firms and let $C_k(x_k)$ be the cost function of firm k, for $k = 1, \ldots, N$. The profit of firm k is given as the difference of its revenue and cost:

$$\varphi_k(x_1, \ldots, x_N) = x_k p \left(\sum_{l=1}^{N} x_l \right) - C_k(x_k) . \qquad (3.131)$$

For the sake of simplicity, assume that p and C_k $(k = 1, 2, \ldots, N)$ are linear functions:

$$p(s) = as + b \qquad \left(s = \sum_{l=1}^{N} x_l, \quad a < 0 \right) \qquad (3.132)$$

$$C_k(x_k) = b_k x_k + c_k \qquad (x_k \geq 0, \quad b_k > 0) .$$

First a *discrete* dynamic model is introduced. Let $x_1(0), \ldots, x_N(0)$ denote the production levels of the firms in the initial time period $t = 0$. At any further time period $t + 1$ $(t \geq 0)$, the output selection of each firm is obtained by maximizing its profit $\varphi_k(x_1(t), \ldots, x_{k-1}(t), x_k, x_{k+1}(t), \ldots, x_N(t))$ by assuming that all other firms will select again the same outputs that they have selected in the preceding time period. This assumption is called the *Cournot expectation*. That is, $x_k(t + 1)$ is the solution of the optimization problem:

$$\text{maximize } x_k \left(a \left(\sum_{l \neq k} x_l(t) + x_k \right) + b \right) - (b_k x_k + c_k)$$

$$\text{subject to } x_k \geq 0 .$$

Because $a < 0$, the objective function has a unique maximizer:

$$x_k(t+1) = -\frac{1}{2} \sum_{l \neq k} x_l(t) + \frac{b_k - b}{2a} \qquad (k = 1, \ldots, N) , \qquad (3.133)$$

by assuming that these values are nonnegative. Summarize these equations as the system

$$\mathbf{x}(t + 1) = \mathbf{A}_c \mathbf{x}(t) + \mathbf{f}_c , \qquad (3.134)$$

where

$$\mathbf{x} = \begin{pmatrix} x_1 \\ x_2 \\ \cdots \\ x_N \end{pmatrix}, \quad \mathbf{A}_c = \begin{pmatrix} 0 & -\frac{1}{2} & -\frac{1}{2} & \cdots & -\frac{1}{2} \\ -\frac{1}{2} & 0 & -\frac{1}{2} & \cdots & -\frac{1}{2} \\ \vdots & \vdots & \vdots & \ddots & \vdots \\ -\frac{1}{2} & -\frac{1}{2} & -\frac{1}{2} & \cdots & 0 \end{pmatrix},$$

and

$$\mathbf{f}_c = \frac{1}{2a} \begin{pmatrix} b_1 - b \\ b_2 - b \\ \cdots \\ b_N - b \end{pmatrix}.$$

The equilibrium of this system can be obtained by solving the linear equations

$$\bar{x}_k = -\frac{1}{2} \sum_{l \neq k} \bar{x}_l + \frac{b_k - b}{2a} \qquad (k = 1, 2, \ldots, N) . \tag{3.135}$$

Add these relations for $k = 1, 2, \ldots, N$ to obtain

$$\bar{s} = -\frac{1}{2}(N - 1)\bar{s} + \frac{B - bN}{2a} ,$$

where $\bar{s} = \sum_{k=1}^{N} \bar{x}_k$ and $B = \sum_{k=1}^{N} b_k$. Therefore,

$$\bar{s} = \frac{B - bN}{(N + 1)a} ,$$

and from (3.135) we have

$$\bar{x}_k = -\frac{1}{2}(\bar{s} - \bar{x}_k) + \frac{b_k - b}{2a} ,$$

which has the solution

$$\bar{x}_k = -\bar{s} + \frac{b_k - b}{a} = \frac{-B - b + (N + 1)b_k}{(N + 1)a} \qquad (k = 1, 2, \ldots, N) .$$

A modified version of the above model is based on the assumption that at each time period, each firm forms *expectations adaptively* on the output $s_k = \sum_{l \neq k} x_l$ of the rest of the industry according to the rule

$$s_k^E(t + 1) = s_k^E(t) + m_k \left(\sum_{l \neq k} x_l(t) - s_k^E(t) \right) , \tag{3.136}$$

where m_k is a positive constant. Then each firm maximizes its expected profit

$$x_k \left(a \left(s_k^E(t+1) + x_k \right) + b \right) - (b_k x_k + c_k) \, .$$

It is easy to verify that the profit maximizing output is

$$x_k(t+1) = -\frac{1}{2} s_k^E(t+1) + \frac{b_k - b}{2a}$$

$$= -\frac{1}{2} \left[s_k^E(t) + m_k \left(\sum_{l \neq k} x_l(t) - s_k^E(t) \right) \right] + \frac{b_k - b}{2a}$$

$$= -\frac{m_k}{2} \sum_{l \neq k} x_l(t) - \frac{1 - m_k}{2} s_k^E(t) + \frac{b_k - b}{2a} \qquad (3.137)$$

by assuming again that these values are nonnegative. Note first that by selecting $m_k = 1$ $(k = 1, 2, \ldots, N)$, Equation (3.137) reduces to (3.133). In the general case, Equations (3.136) and (3.137) can be summarized as

$$\begin{pmatrix} \mathbf{x}(t+1) \\ \mathbf{s}^E(t+1) \end{pmatrix} = \mathbf{A}_a \begin{pmatrix} \mathbf{x}(t) \\ \mathbf{s}^E(t) \end{pmatrix} + \mathbf{f}_a, \qquad (3.138)$$

where $\mathbf{x} = (x_k)$, $\mathbf{s}^E = (s_k^E)$,

$$\mathbf{A}_a = \begin{pmatrix}
0 & -\frac{m_1}{2} & \cdots & -\frac{m_1}{2} & -\frac{1-m_1}{2} & & \\
-\frac{m_2}{2} & 0 & \cdots & -\frac{m_2}{2} & & -\frac{1-m_2}{2} & O \\
& & \ddots & & & \ddots & \\
-\frac{m_N}{2} & -\frac{m_N}{2} & \cdots & 0 & O & & -\frac{1-m_N}{2} \\
0 & m_1 & \cdots & m_1 & 1-m_1 & & \\
m_2 & 0 & \cdots & m_2 & & 1-m_2 & O \\
& & \ddots & & & \ddots & \\
m_N & m_N & \cdots & 0 & O & & 1-m_N
\end{pmatrix}$$

and

$$\mathbf{f}_a = \frac{1}{2a} \begin{pmatrix} b_1 - b \\ b_2 - b \\ \ldots \\ b_N - b \\ 0 \\ 0 \\ \ldots \\ 0 \end{pmatrix} .$$

The *continuous* counterpart of the above models can be introduced as follows. Assume again that the initial outputs $x_1(0), \ldots, x_N(0)$ of the firms are known. At each time period $t \geq 0$, each firm adjusts its output proportionally to its marginal profit. Since the marginal profit of firm k is the derivative of φ_k with respect to x_k, the system is driven by the differential equations

$$\dot{x}_k(t) = m_k \cdot \left[2ax_k(t) + a \sum_{l \neq k} x_l(t) + b - b_k \right] \qquad (k = 1, \ldots, N) ,$$

which can be summarized as

$$\dot{\mathbf{x}}(t) = \mathbf{M} \cdot \mathbf{A} \cdot \mathbf{x}(t) + \mathbf{f} , \qquad\qquad (3.139)$$

where $\mathbf{M} = diag(m_1, \ldots, m_N)$ with positive diagonal elements,

$$\mathbf{A} = \begin{pmatrix} 2a & a & \cdots & a \\ a & 2a & \cdots & a \\ \multicolumn{4}{c}{\dotfill} \\ a & a & \cdots & 2a \end{pmatrix} \qquad \text{and} \qquad \mathbf{f} = \begin{pmatrix} m_1(b - b_1) \\ m_2(b - b_2) \\ \ldots \\ m_N(b - b_N) \end{pmatrix} .$$

The equilibrium of this continuous system is the solution of equation

$$\mathbf{MA\bar{x}} + \mathbf{f} = \mathbf{0} ,$$

which is

$$\bar{\mathbf{x}} = -(\mathbf{MA})^{-1}\mathbf{f} = -\mathbf{A}^{-1}\mathbf{M}^{-1}\mathbf{f} .$$

Here we used the fact that if $m_k > 0$ $(k = 1, 2, \ldots, N)$ and $a < 0$, then both matrices $\mathbf{A}$ and $\mathbf{M}$ are invertible.

Finally we remark that a summary of discrete and continuous oligopolies is presented in [35].

Problems

1. Give the solutions (3.19) and (3.20) of the system

$$\dot{x} = \begin{pmatrix} \frac{1}{t} & 0 \\ 0 & \frac{1}{t} \end{pmatrix} x + \begin{pmatrix} 1 \\ 1 \end{pmatrix} u,$$

$$y = (1,1)x \, .$$

2. Derive the state solution (3.19) of system

$$\dot{x} = \begin{pmatrix} 1 & 1 \\ 2 & 2 \end{pmatrix} x + \begin{pmatrix} 1 \\ 0 \end{pmatrix} u, \qquad x(0) = \begin{pmatrix} 1 \\ 1 \end{pmatrix} \, .$$

What is your result in the special case of $u(t) \equiv 1$? Compare the result to that of Problem 2.4.

3. Find the state solution (3.19) for system

$$\dot{x} = \begin{pmatrix} 2 & 1 \\ 0 & 2 \end{pmatrix} x + \begin{pmatrix} 1 \\ 1 \end{pmatrix} u, \qquad x(0) = \begin{pmatrix} 1 \\ 0 \end{pmatrix},$$

and in the special case of $u(t) \equiv 1$, compare the results to those of Problem 2.5.

4. Find the transfer function for system

$$\dot{x} = \begin{pmatrix} 1 & 1 \\ 2 & 2 \end{pmatrix} x + \begin{pmatrix} 1 \\ 0 \end{pmatrix} u, \qquad x(0) = \begin{pmatrix} 1 \\ 1 \end{pmatrix},$$

if the output is $y = (1,1)x$.

5. Find the transfer function for system

$$\dot{x} = \begin{pmatrix} 2 & 1 \\ 0 & 2 \end{pmatrix} x + \begin{pmatrix} 1 \\ 1 \end{pmatrix} u, \qquad x(0) = \begin{pmatrix} 1 \\ 0 \end{pmatrix},$$

if the output is $y = (0,1)x$.

6. Determine the transfer function for a system

$$\dot{x} = Ax + Bu$$

$$y = Cx + Du \, .$$

7.   The simple mechanical system shown in Figure 3.33 is described by equation

$$\frac{d}{dt}\begin{pmatrix} x_1 \\ x_2 \end{pmatrix} = \begin{pmatrix} 0 & 1 \\ 0 & -6 \end{pmatrix}\begin{pmatrix} x_1 \\ x_2 \end{pmatrix} + \begin{pmatrix} 0 \\ 2 \end{pmatrix}u,$$

$$y = (1,0)\begin{pmatrix} x_1 \\ x_2 \end{pmatrix},$$

where $x_1 = y$ and $x_2 = v$.

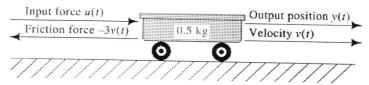

Figure 3.33   Illustration of Problem 7.

Derive the solution (3.19) and (3.20) of this system.

8.   Find the transfer function for problem

$$\dot{x} = \begin{pmatrix} 0 & 1 \\ 0 & -6 \end{pmatrix}x + \begin{pmatrix} 0 \\ 2 \end{pmatrix}u,$$

$$y = (1,0)x.$$

9.   Derive the state solution (3.48) of the system

$$x(t+1) = \begin{pmatrix} 1 & 1 \\ 2 & 2 \end{pmatrix}x(t) + \begin{pmatrix} 1 \\ 0 \end{pmatrix}u(t), \qquad x(0) = \begin{pmatrix} 1 \\ 1 \end{pmatrix}.$$

What is your result for $u(t) \equiv 1$? Compare the result to that of Problem 2.6.

10.   Derive the state solution (3.48) of the system

$$x(t+1) = \begin{pmatrix} 2 & 1 \\ 0 & 2 \end{pmatrix}x(t) + \begin{pmatrix} 1 \\ 1 \end{pmatrix}u(t), \qquad x(0) = \begin{pmatrix} 1 \\ 0 \end{pmatrix},$$

and in the case of $u(t) \equiv 1$, compare the result to that of Problem 2.7.

11. Find the transfer function for system

$$x(t+1) = \begin{pmatrix} 1 & 1 \\ 2 & 2 \end{pmatrix} x(t) + \begin{pmatrix} 1 \\ 0 \end{pmatrix} u(t), \qquad x(0) = \begin{pmatrix} 1 \\ 1 \end{pmatrix},$$

if the output is $y = (1,1)x$.

12. Find the transfer function for system

$$x(t+1) = \begin{pmatrix} 2 & 1 \\ 0 & 2 \end{pmatrix} x(t) + \begin{pmatrix} 1 \\ 1 \end{pmatrix} u(t), \qquad x(0) = \begin{pmatrix} 1 \\ 0 \end{pmatrix},$$

if the output is $y = (0,1)x$.

13. The electric circuit shown in Figure 3.34 is described by the first-order equation

$$L\frac{di(t)}{dt} + (R_1 + R_2)i(t) = u(t)$$

$$y(t) = R_2 i(t)$$

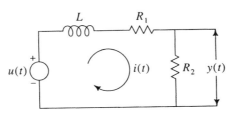

Figure 3.34 Illustration of Problem 13.

Find the transfer function and the solution of the system in the state–space form. If $u(t) \equiv 1$, does this system have an equilibrium state?

14. Find the equilibrium state of the nonlinear system

$$\dot{x}_1 = (x_1 + x_2)^2 ,$$

$$\dot{x}_2 = e^{x_1 - x_2} - 1 ,$$

$$y = x_1^2 + 2x_2 ,$$

and linearize it around the equilibrium.

15.   Solve the equation

$$\ddot{y} + 5\dot{y} + 6y = \dot{u} + 3u \, ,$$

which is given in input–output form. Apply Theorem 3.5 to obtain an equivalent first-order system.

16.   Solve equation

$$\ddot{y} + 6\dot{y} + 4y = \dot{u} + 8u$$

and apply Theorem 3.5 to obtain an equivalent first-order system.

17.   Derive the transfer function for system

$$\ddot{y} + 5\dot{y} + 6y = \dot{u} + 3u \, .$$

18.   Derive the transfer function for system

$$\ddot{y} + 6\dot{y} + 4y = \dot{u} + 8u \, .$$

19.   Find the transfer function for Figure 3.35:

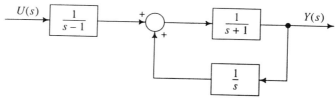

Figure 3.35   Illustration of Problem 19.

20.   Find the transfer function for Figure 3.36:

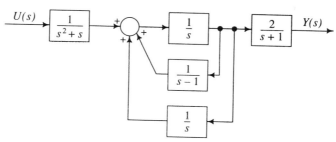

Figure 3.36   Illustration of Problem 20.

21.   Derive a simple formula for $\frac{d}{ds}\mathbf{H}(s)$, where $\mathbf{H}(s)$ is the transfer function of a time-invariant linear system.

22.   Let $\mathbf{a}, \mathbf{b} \in \mathbf{R}^n$ and $\mathbf{A}$ be an $n \times n$ real matrix. By using the simple relation

$$(\mathbf{A} + \mathbf{a}\mathbf{b}^T)^{-1} = \mathbf{A}^{-1} - \frac{\mathbf{A}^{-1}\mathbf{a}\mathbf{b}^T\mathbf{A}^{-1}}{1 + \mathbf{b}^T\mathbf{A}^{-1}\mathbf{a}}$$

derive an updating formula for the transfer function of system

$$\dot{\mathbf{x}} = (\mathbf{A} + \mathbf{a}\mathbf{b}^T)\mathbf{x} + \mathbf{B}\mathbf{u}$$

$$\mathbf{y} = \mathbf{C}\mathbf{x} .$$

23.   Consider systems

$$\dot{\mathbf{x}}(t) = \mathbf{A}(t)\mathbf{x}(t) + \mathbf{B}(t)\mathbf{u}(t), \quad \mathbf{x}(t_0) = \mathbf{x}_0$$

and

$$\dot{\mathbf{z}}(t) = \mathbf{A}(t)\mathbf{z}(t) + \bar{\mathbf{B}}(t)\mathbf{u}(t), \quad \mathbf{z}(t_0) = \mathbf{x}_0 ,$$

where $\bar{\mathbf{B}}(t)$ is an approximate of $\mathbf{B}(t)$. Derive an upper bound for $\|\mathbf{x}(t) - \mathbf{z}(t)\|$.

24.   Let $\mathbf{H}(s)$ and $\bar{\mathbf{H}}(s)$ denote the transfer functions of systems

$$\dot{\mathbf{x}} = \mathbf{A}\mathbf{x} + \mathbf{B}\mathbf{u}$$

$$\mathbf{y} = \mathbf{C}\mathbf{x}$$

and

$$\dot{\mathbf{z}} = \bar{\mathbf{A}}\mathbf{z} + \mathbf{B}\mathbf{u}$$

$$\mathbf{y} = \mathbf{C}\mathbf{z}$$

where $\bar{\mathbf{A}}$ is an approximation of $\mathbf{A}$. By using the bound

$$\|\mathbf{X}^{-1} - \mathbf{Y}^{-1}\| \leq \frac{\|\mathbf{X} - \mathbf{Y}\| \cdot \|\mathbf{X}^{-1}\|^2}{1 - \|\mathbf{X} - \mathbf{Y}\| \cdot \|\mathbf{X}^{-1}\|} ,$$

(which holds for all real $n \times n$ matrices $\mathbf{X}$ and $\mathbf{Y}$ such that $\|\mathbf{X} - \mathbf{Y}\| < \|\mathbf{X}^{-1}\|^{-1}$, see for example, [44]) derive an upper bound for $\|\mathbf{H}(s) - \bar{\mathbf{H}}(s)\|$.

25.   Redo Problem 23 for discrete systems.

chapter four

Stability Analysis

In applied sciences, the term stability has a very broad meaning. However, in the theory of dynamic systems, stability is usually defined with respect to a given equilibrium. If the initial state $\mathbf{x}_0$ is selected as an equilibrium state $\bar{\mathbf{x}}$ of the system, then the state will remain at $\bar{\mathbf{x}}$ for all future time. When the initial state is selected close to the equilibrium state, the system might remain close to the equilibrium or it might move away. In the first section of this chapter we will introduce conditions that guarantee whenever the system starts near an equilibrium state, it remains near it, perhaps even converging to the equilibrium state as the time increases. These kinds of stability are called the Lyapunov-stability and asymptotical stability, respectively. In the first part of this chapter we will introduce the *Lyapunov stability theory* to examine Lyapunov stability and asymptotical stability of linear and nonlinear systems.

In many applications we have to guarantee that the state of linear systems remains bounded, even converging to zero with a certain convergence rate as $t \to \infty$ if zero input is applied. These kinds of properties are defined as uniform and uniform exponential stability, which will be also discussed in the first part of this chapter.

Notice that Lyapunov stability, asymptotical stability, uniform and uniform exponential stability represent properties of the state of the system. Therefore, they are called *internal* stability concepts. In the second part of this chapter, *external* stability will be introduced and investigated for linear systems, when we will find conditions that with zero initial state a bounded input always evokes a bounded output. In this case we will not be interested in the behavior of the state, only the input–output relation is considered.

4.1   The Elements of the Lyapunov Stability Theory

In this section only time-invariant systems will be considered. Continuous time-invariant systems have the form

$$\dot{\mathbf{x}}(t) = \mathbf{f}(\mathbf{x}(t)) , \tag{4.1}$$

and discrete time-invariant systems are modeled by the difference equation

$$\mathbf{x}(t+1) = \mathbf{f}(\mathbf{x}(t)) . \tag{4.2}$$

Here we assume that $\mathbf{f} : X \rightarrow \mathbf{R}^n$, where $X \subseteq \mathbf{R}^n$ is the state space. We also assume that function $\mathbf{f}$ is continuous; furthermore, for arbitrary initial state $\mathbf{x}_0 \in X$ there is a unique solution of the corresponding initial value problem $\mathbf{x}(t_0) = \mathbf{x}_0$, and the entire trajectory $\mathbf{x}(t)$ is in X. Assume furthermore that t_0 denotes the initial time period of the system.

It is also known from the previous chapter that a vector $\bar{\mathbf{x}} \in X$ is an equilibrium state of the continuous system (4.1) if and only if $\mathbf{f}(\bar{\mathbf{x}}) = \mathbf{0}$, and it is an equilibrium state of the discrete system (4.2) if and only if $\bar{\mathbf{x}} = \mathbf{f}(\bar{\mathbf{x}})$. In this chapter the equilibrium of a system will always mean the equilibrium *state*, if it is not specified otherwise. In analyzing the dependence of the state trajectory $\mathbf{x}(t)$ on the selection of the initial state $\mathbf{x}_0$ nearby the equilibrium, the following stability types are considered.

DEFINITION 4.1

(i)   *An equilibrium point $\bar{\mathbf{x}}$ is stable if there is an $\varepsilon_0 > 0$ with the following property: For all ε_1, $0 < \varepsilon_1 < \varepsilon_0$, there is an $\varepsilon > 0$ such that if $\|\bar{\mathbf{x}} - \mathbf{x}_0\| < \varepsilon$, then $\|\bar{\mathbf{x}} - \mathbf{x}(t)\| < \varepsilon_1$ for all $t > t_0$.*

(ii)   *An equilibrium point $\bar{\mathbf{x}}$ is asymptotically stable if it is stable and there is an $\varepsilon > 0$ such that whenever $\|\bar{\mathbf{x}} - \mathbf{x}_0\| < \varepsilon$, then $\mathbf{x}(t) \rightarrow \bar{\mathbf{x}}$ as $t \rightarrow \infty$.*

(iii)   *An equilibrium point $\bar{\mathbf{x}}$ is globally asymptotically stable if it is stable and with arbitrary initial state $\mathbf{x}_0 \in X$, $\mathbf{x}(t) \rightarrow \bar{\mathbf{x}}$ as $t \rightarrow \infty$.*

The first definition says an equilibrium $\bar{\mathbf{x}}$ is stable if the entire trajectory $\mathbf{x}(t)$ is closer to the equilibrium than any small ε_1, if the initial state $\mathbf{x}_0$ is selected close enough to the equilibrium. In the case of asymptotic stability $\mathbf{x}(t)$ has to converge, in addition, to the equilibrium as $t \rightarrow \infty$. If an equilibrium is globally asymptotically stable, then $\mathbf{x}(t)$ converges to the equilibrium regardless of how the initial state $\mathbf{x}_0$ is selected. These concepts are illustrated in Figure 4.1.

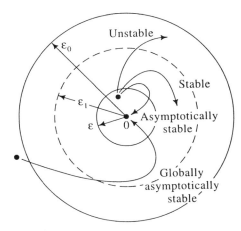

Figure 4.1   Stability concepts.

In the systems theory literature our stability concept is sometimes called marginal stability, and asymptotic stability is called stability. In this book we will always use our terminologies.

4.1.1   Lyapunov Functions

Assume that $\bar{x}$ is an equilibrium state of a dynamic (continuous or discrete) system, and let Ω denote a subset of the state space X such that $\bar{x} \in \Omega$.

DEFINITION 4.2   *A real-valued function V defined on Ω is called a Lyapunov function, if*

   (i)   *V is continuous;*

   (ii) *V has a unique global minimum at $\bar{x}$ with respect to all other points in Ω;*

   (iii) *for any state trajectory $x(t)$ contained in Ω, $V(x(t))$ is nonincreasing in t.*

The Lyapunov function can be interpreted as the generalization of the energy function in mechanical systems. The first requirement simply means that the graph of V has no breaks. The second requirement means that the graph of V has its lowest point at the equilibrium, and the third requirement generalizes the well-known fact of mechanical systems, that the energy of a free mechanical system with friction always decreases, unless the system is at rest.

THEOREM 4.1

Assume that there exists a Lyapunov function V on the spherical region

$$\Omega = \{\mathbf{x} \mid \|\mathbf{x} - \bar{\mathbf{x}}\| < \varepsilon_0\}, \qquad (4.3)$$

where $\varepsilon_0 > 0$ is given, furthermore $\Omega \subseteq X$. Then the equilibrium is stable.

PROOF   We present here the proof only for the discrete case, the continuous counterpart can be proven in the same way.

Let $\varepsilon_1 \in (0, \varepsilon_0)$ be arbitrary. Select $\varepsilon_2 \in (0, \varepsilon_1)$ so that if $\|\mathbf{x} - \bar{\mathbf{x}}\| < \varepsilon_2$ then $\|\mathbf{f}(\mathbf{x}) - \bar{\mathbf{x}}\| < \varepsilon_0$. Such ε_2 exists, since $\mathbf{f}(\bar{\mathbf{x}}) = \bar{\mathbf{x}}$ and $\mathbf{f}$ is continuous. Therefore, if $\|\mathbf{x}(t) - \bar{\mathbf{x}}\| < \varepsilon_2$ for some $t \geq 0$, then $\|\mathbf{x}(t+1) - \bar{\mathbf{x}}\| < \varepsilon_0$. Define now

$$m = \min\{V(\mathbf{x}) \mid \varepsilon_2 \leq \|\mathbf{x} - \bar{\mathbf{x}}\| \leq \varepsilon_0\},$$

which exists, since V is continuous and the set

$$\{\mathbf{x} \mid \varepsilon_2 \leq \|\mathbf{x} - \bar{\mathbf{x}}\| \leq \varepsilon_0\}$$

is compact. Note that this set does not contain the equilibrium state $\bar{\mathbf{x}}$; therefore, $m > V(\bar{\mathbf{x}})$. The continuity of function V implies that there exists an $\varepsilon \in (0, \varepsilon_2)$ such that $V(\mathbf{x}) < m$ as $\|\mathbf{x} - \bar{\mathbf{x}}\| < \varepsilon$.

Finally we show that this $\varepsilon > 0$ satisfies the conditions of Part (i) of Definition 4.1. Assume now that $\|\mathbf{x}(0) - \bar{\mathbf{x}}\| < \varepsilon$, then $V(\mathbf{x}(0)) < m$. Since $V(\mathbf{x}(t))$ is decreasing, for all $t \geq 0$, $V(\mathbf{x}(t)) < m$. Therefore, the definition of m implies that for all t, $\|\mathbf{x}(t) - \bar{\mathbf{x}}\| < \varepsilon_2 < \varepsilon_1$, which completes the proof.   ∎

THEOREM 4.2

Assume that in addition to the conditions of Theorem 4.1, the Lyapunov function $V(\mathbf{x}(t))$ is strictly decreasing in t, unless $\mathbf{x}(t) = \bar{\mathbf{x}}$. Then the stability is asymptotic.

PROOF   We present again the proof only for the discrete case, because the continuous counterpart can be discussed in a similar way.

Select the initial state as $\|\mathbf{x}(0) - \bar{\mathbf{x}}\| < \varepsilon$, where ε is defined in the proof of the previous theorem. We shall prove that $\mathbf{x}(t) \to \bar{\mathbf{x}}$ as $t \to \infty$.

Assume that this limit relation does not hold. Since for all t, $\|\mathbf{x}(t) - \bar{\mathbf{x}}\| < \varepsilon_2$, we conclude that sequence $\mathbf{x}(t)$ $(t = 0, 1, 2, \ldots)$ must have a convergent subsequence such that $\mathbf{x}(t_k) \to \mathbf{x}^* \neq \bar{\mathbf{x}}$ as $k \to \infty$. Since sequence $\mathbf{x}(t_k + 1)$ $(k = 0, 1, 2, \ldots)$ is also bounded, it must also have a convergent subsequence $\mathbf{x}(t_{k_i} + 1) \to \mathbf{x}^{**}$. The strict monotonicity of the Lyapunov function implies that for all $i = 0, 1, 2, \ldots,$

$$V(\mathbf{x}(t_{k_{i+1}})) \leq V(\mathbf{x}(t_{k_i} + 1)) < V(\mathbf{x}(t_{k_i})), \tag{4.4}$$

since $t_{k_{i+1}} \leq t_{k_i} + 1 < t_{k_i}$.

By letting $i \to \infty$ and using the continuity of the Lyapunov function we have

$$V(\mathbf{x}^*) \leq V(\mathbf{x}^{**}) \leq V(\mathbf{x}^*),$$

which implies that

$$V(\mathbf{x}^{**}) = V(\mathbf{x}^*). \tag{4.5}$$

Note that since $\mathbf{f}$ is continuous,

$$\mathbf{x}^{**} = \lim_{i \to \infty} \mathbf{x}(t_{k_i} + 1) = \lim_{i \to \infty} \mathbf{f}(\mathbf{x}(t_{k_i}))$$

$$= \mathbf{f}\left(\lim_{i \to \infty} \mathbf{x}(t_{k_i})\right) = \mathbf{f}(\mathbf{x}^*).$$

This relation and (4.5) contradict the strict monotonicity of the Lyapunov function.

Thus, the proof is completed.   ∎

THEOREM 4.3
Assume that the Lyapunov function is defined on the entire state space X, $V(\mathbf{x}(t))$ is strictly decreasing in t unless $\mathbf{x}(t) = \bar{\mathbf{x}}$, furthermore $V(\mathbf{x})$ tends to infinity as any component of $\mathbf{x}$ gets arbitrarily large in magnitude. Then the stability is globally asymptotic.

PROOF   Only the discrete case is shown, since the continuous case is similar. Let $\mathbf{x}_0 \in X$ be arbitrary. The monotonicity of the Lyapunov function implies that $V(\mathbf{x}(t)) \leq V(\mathbf{x}_0)$ for all $t \geq 0$. Therefore, sequence $\mathbf{x}(t)$ is bounded. The rest of the proof is the same as it was given for the previous theorem.   ∎

REMARK 4.1   This Lyapunov theory provides an alternative way to prove the convergence of the iteration sequence introduced in Theorem 1.3 by selecting the Lyapunov function $V(x) = \rho(x, x^*)$ for the difference equation $x(t+1) = A(x(t))$.   ∎

Example 4.1

Consider the differential equation

$$\dot{\mathbf{x}} = \begin{pmatrix} 0 & \omega \\ -\omega & 0 \end{pmatrix} \mathbf{x} + \begin{pmatrix} 0 \\ 1 \end{pmatrix},$$

which was earlier investigated in Examples 1.23 and 2.6. In Example 3.3 we verified that the equilibrium is given as $\bar{\mathbf{x}} = (1/\omega, 0)^T$, and from Example 2.6 we know that

$$\phi(t, t_0) = \begin{pmatrix} \cos \omega(t - t_0) & \sin \omega(t - t_0) \\ -\sin \omega(t - t_0) & \cos \omega(t - t_0) \end{pmatrix}.$$

Assume that the initial state is selected from the neighborhood of the equilibrium, that is,

$$\mathbf{x}(0) = \begin{pmatrix} 1/\omega + \alpha \\ \beta \end{pmatrix},$$

where α and β are small in magnitude. The general solution formula (2.23) implies that

$$\mathbf{x}(t) = \begin{pmatrix} \cos \omega t & \sin \omega t \\ -\sin \omega t & \cos \omega t \end{pmatrix} \begin{pmatrix} 1/\omega + \alpha \\ \beta \end{pmatrix}$$

$$+ \int_0^t \begin{pmatrix} \cos \omega(t - \tau) & \sin \omega(t - \tau) \\ -\sin \omega(t - \tau) & \cos \omega(t - \tau) \end{pmatrix} \begin{pmatrix} 0 \\ 1 \end{pmatrix} d\tau$$

$$= \begin{pmatrix} (\frac{1}{\omega} + \alpha) \cos \omega t + \beta \sin \omega t \\ -(\frac{1}{\omega} + \alpha) \sin \omega t + \beta \cos \omega t \end{pmatrix} + \begin{pmatrix} \frac{1}{\omega} - \frac{1}{\omega} \cos \omega t \\ \frac{1}{\omega} \sin \omega t \end{pmatrix},$$

where we used some results from Example 2.6. Hence,

$$\mathbf{x}(t) = \begin{pmatrix} \frac{1}{\omega} + \alpha \cos \omega t + \beta \sin \omega t \\ -\alpha \sin \omega t + \beta \cos \omega t \end{pmatrix},$$

and therefore,

$$\mathbf{x}(t) - \bar{\mathbf{x}} = \begin{pmatrix} \alpha \cos \omega t + \beta \sin \omega t \\ -\alpha \sin \omega t + \beta \cos \omega t \end{pmatrix}.$$

This formula can be obtained also by observing that $\mathbf{x}(t) - \bar{\mathbf{x}}$ solves the homogeneous equation. Simple calculation shows that

$$\|\mathbf{x}(t) - \bar{\mathbf{x}}\|_2 = \sqrt{\alpha^2 + \beta^2}\,,$$

which implies that if α and β are sufficiently small in magnitude, then $\|\mathbf{x}(t) - \bar{\mathbf{x}}\| < \varepsilon$ for any positive $\varepsilon > 0$. Thus, the equilibrium is stable. From this relation we also conclude that for $t \to \infty$, $\mathbf{x}(t) \not\to \bar{\mathbf{x}}$. Hence, the stability is not asymptotic.

The stability of the equilibrium can also be verified directly by using Theorem 4.1 without computing the solution. Select the Lyapunov function

$$V(\mathbf{x}) = (\mathbf{x} - \bar{\mathbf{x}})^T (\mathbf{x} - \bar{\mathbf{x}}) = \|\mathbf{x} - \bar{\mathbf{x}}\|_2^2\,.$$

This is continuous in $\mathbf{x}$; furthermore, it has its minimal (zero) value at $\mathbf{x} = \bar{\mathbf{x}}$. Therefore, to establish the stability of the equilibrium we have to show only that $V(\mathbf{x}(t))$ is decreasing. Simple differentiation shows that

$$\frac{d}{dt}V(\mathbf{x}(t)) = 2(\mathbf{x} - \bar{\mathbf{x}})^T \cdot \dot{\mathbf{x}} = 2(\mathbf{x} - \bar{\mathbf{x}})^T (\mathbf{A}\mathbf{x} + \mathbf{b})$$

with

$$\mathbf{A} = \begin{pmatrix} 0 & \omega \\ -\omega & 0 \end{pmatrix} \quad \text{and} \quad \mathbf{b} = \begin{pmatrix} 0 \\ 1 \end{pmatrix}.$$

That is, with $\mathbf{x} = (x_1, x_2)^T$,

$$\frac{d}{dt}V(\mathbf{x}(t)) = 2\left(x_1 - \frac{1}{\omega}, x_2\right)\begin{pmatrix} \omega x_2 \\ -\omega x_1 + 1 \end{pmatrix}$$

$$= 2(\omega x_1 x_2 - x_2 - \omega x_1 x_2 + x_2) = 0\,.$$

Therefore, function $V(\mathbf{x}(t))$ is a constant, which is a (not strictly) decreasing function. That is, all conditions of Theorem 4.1 are satisfied, which imply the stability of the equilibrium.

Theorems 4.1, 4.2, and 4.3 guarantee the stability, asymptotical stability, and global asymptotical stability of the equilibrium, if a Lyapunov function is found. Note that failure in finding such Lyapunov functions does not imply that the system is unstable or that the stability is not asymptotical or globally asymptotical.

In the previous discussions the stability, asymptotical and global asymptotical stability of the equilibrium were examined. We can easily extend these concepts to any particular solution $\mathbf{x}_0(t)$ of a linear or nonlinear system. Assume that $\mathbf{x}_0(t)$ is a solution of Equation (4.1) or (4.2), and introduce the new variable

$$\mathbf{x}(t) = \mathbf{x}_0(t) + \mathbf{z}(t) \, ,$$

then Equation (4.1) can be modified as

$$\dot{\mathbf{z}}(t) = \mathbf{f}(\mathbf{x}_0(t) + \mathbf{z}(t)) - \dot{\mathbf{x}}_0(t) \, ,$$

and Equation (4.2) has the equivalent modified form

$$\mathbf{z}(t + 1) = \mathbf{f}(\mathbf{x}_0(t) + \mathbf{z}(t)) - \mathbf{x}_0(t + 1) \, .$$

Notice that in both cases $\mathbf{z}(t) \equiv \mathbf{0}$ is an equilibrium. Therefore, the stability, asymptotical stability and global asymptotical stability of a trajectory $\mathbf{x}_0(t)$ can be defined as the same concept applied to the new equilibrium of the modified continuous or discrete system. Therefore, the methods introduced above can be used in the more general case as well.

The asymptotic stability of nonlinear systems can be examined via linearization as follows. Consider the time-invariant continuous and discrete systems

$$\dot{\mathbf{x}}(t) = \mathbf{f}(\mathbf{x}(t))$$

and

$$\mathbf{x}(t + 1) = \mathbf{f}(\mathbf{x}(t)) \, .$$

Let $\mathbf{J}(\mathbf{x})$ denote the Jacobian of $\mathbf{f}(\mathbf{x})$, and let $\bar{\mathbf{x}}$ be an equilibrium of the system. It is known from Section 3.2 that the method of linearization around the equilibrium results in the time-invariant linear systems

$$\dot{\mathbf{x}}_\delta(t) = \mathbf{J}(\bar{\mathbf{x}})\mathbf{x}_\delta(t) \, ,$$

and

$$\mathbf{x}_\delta(t + 1) = \mathbf{J}(\bar{\mathbf{x}})\mathbf{x}_\delta(t) \, ,$$

where $\mathbf{x}_\delta(t) = \mathbf{x}(t) - \bar{\mathbf{x}}$. It is also known from the theory of ordinary differential equations (see, for example, [18]) that the asymptotic stability of the zero vector in the linearized system implies the asymptotic stability of the equilibrium $\bar{\mathbf{x}}$ in the original nonlinear system. Hence, the analysis of the asymptotical stability of nonlinear systems can be reduced to that of time-invariant linear systems, which is the topic of the next section.

4.1.2   *The stability of time-variant linear systems*

Consider first the time-variant continuous linear system

$$\dot{\mathbf{x}}(t) = \mathbf{A}(t)\mathbf{x}(t) + \mathbf{b}(t) \ , \tag{4.6}$$

where $\mathbf{A}(t)$ is an $n \times n$ matrix and $\mathbf{b}(t)$ is an n-dimensional vector. It is assumed that all components of matrix $\mathbf{A}(t)$ and $\mathbf{b}(t)$ are continuous functions for $t \geq t_0$. Let $\bar{\mathbf{x}}$ be an equilibrium of this system, and let $\mathbf{x}(t)$ be any trajectory of the system.

Then

$$\dot{\mathbf{x}}(t) = (\mathbf{A}(t)\mathbf{x}(t) + \mathbf{b}(t)) - (\mathbf{A}(t)\bar{\mathbf{x}} + \mathbf{b}(t)) = \mathbf{A}(t)(\mathbf{x}(t) - \bar{\mathbf{x}}) \ ;$$

therefore, $\mathbf{x}_\delta(t) = \mathbf{x}(t) - \bar{\mathbf{x}}$ satisfies the homogeneous equation

$$\dot{\mathbf{x}}_\delta = \mathbf{A}(t)\mathbf{x}_\delta \ .$$

From Equation (2.18) we then know that

$$\mathbf{x}(t) - \bar{\mathbf{x}} = \phi(t, t_0)(\mathbf{x}_0 - \bar{\mathbf{x}}) \ , \tag{4.7}$$

where $\mathbf{x}_0 = \mathbf{x}(t_0)$ and $\phi(t, t_0)$ is the fundamental matrix. Hence the difference of the state and the equilibrium is given in a closed form.

Consider next the time-invariant discrete linear system

$$\mathbf{x}(t + 1) = \mathbf{A}(t)\mathbf{x}(t) + \mathbf{b}(t) \ , \tag{4.8}$$

and assume again that $\bar{\mathbf{x}}$ is an equilibrium. Then

$$\mathbf{x}(t + 1) - \bar{\mathbf{x}} = (\mathbf{A}(t)\mathbf{x}(t) + \mathbf{b}(t)) - (\mathbf{A}(t)\bar{\mathbf{x}} + \mathbf{b}(t)) = \mathbf{A}(t)(\mathbf{x}(t) - \bar{\mathbf{x}}) \ ,$$

therefore, $\mathbf{x}_\delta(t) = \mathbf{x}(t) - \bar{\mathbf{x}}$ satisfies the homogeneous equation

$$\mathbf{x}_\delta(t + 1) = \mathbf{A}(t)\mathbf{x}_\delta(t) \ .$$

Then, similar to the continuous case, from the solution formula (2.45) we know that

$$\mathbf{x}(t) - \bar{\mathbf{x}} = \phi(t, 0)(\mathbf{x}_0 - \bar{\mathbf{x}}) \ , \tag{4.9}$$

where $\phi(t, 0)$ is the fundamental matrix. Hence the difference of the state vector and the equilibrium is given again in a closed form. From Equations (4.7) and (4.9) we have the following result.

THEOREM 4.4

1. *The equilibrium $\bar{\mathbf{x}}$ is stable if and only if $\phi(t, t_0)$ in the continuous case (or $\phi(t, 0)$ in the discrete case) is bounded for $t \geq t_0$ (or $t \geq 0$).*

2. *The equilibrium $\bar{\mathbf{x}}$ is asymptotically stable if and only if $\phi(t, t_0)$ (or $\phi(t, 0)$) is bounded and tends to zero as $t \to \infty$*

COROLLARY 4.1

For systems (4.6) and (4.8) asymptotical stability and global asymptotical stability are equivalent, if the state space is defined as the entire $\mathbf{R}^n$.

We mention here that Theorem 4.4 can also be used to check the stability and asymptotical stability of time-invariant linear systems as well, as it is illustrated in the following example.

Example 4.2

Consider again the system of the previous example. Since each component of the fundamental matrix

$$\phi(t, t_0) = \begin{pmatrix} \cos \omega(t - t_0) & \sin \omega(t - t_0) \\ -\sin \omega(t - t_0) & \cos \omega(t - t_0) \end{pmatrix}$$

is bounded, the system is stable. If $t \to \infty$, then the elements do not converge to zero, therefore, the stability is not asymptotical.

In the systems theory literature a zero-input time-variant linear system is called *uniform stable*, if $\phi(t, \tau)$ is bounded for all $t \geq \tau \geq t_0$. Similarly to Theorem 4.4 one can easily prove that uniform stability is equivalent to the condition that $\|\mathbf{x}(t)\| \leq Q \cdot \|\mathbf{x}_0\|$ for all $t \geq \tau \geq t_0$ with $\mathbf{x}(\tau) = \mathbf{x}_0$ and some finite positive constant Q. (The equivalence of the norms of n-element real or complex vectors implies that any vector norm can be used in this condition.) A zero-input time-variant linear system is called *uniform exponentially stable* if there exist finite positive constants Q and P such that for all elements of the fundamental matrix, $|\phi_{ij}(t, \tau)| \leq Pe^{-Q(t-\tau)}$. One can also prove that uniform exponential stability is equivalent to the assumption that with some finite positive constants Q' and P', $\|\mathbf{x}(t)\| \leq P'e^{-Q'(t-\tau)}\|\mathbf{x}_0\|$ for any $t \geq \tau \geq t_0$ and $\mathbf{x}_0$ where $\mathbf{x}(\tau) = \mathbf{x}_0$.

4.1.3   The Stability of Time-Invariant Linear Systems

This section is divided into two parts. In the first part the stability of linear time-invariant systems given in state–space form is analyzed. In the second part, methods based on transfer functions are discussed.

In Chapter 2 we have derived closed formulas for the fundamental matrices $\phi(t, t_0)$ and $\phi(t, 0)$ of continuous and discrete systems.

Assuming that $\lambda_1, \lambda_2, \ldots, \lambda_r$ are the distinct eigenvalues of $\mathbf{A}$ with multiplicities $m_1, m_2, \ldots, m_r$, we conclude from Equations (1.45) and (1.44) that for continuous systems

$$\phi(t, t_0) = e^{\mathbf{A}(t - t_0)} = \sum_{i=1}^{r} e^{\lambda_i(t - t_0)} \sum_{l=0}^{m_i - 1} (t - t_0)^l \mathbf{B}_{il} , \qquad (4.10)$$

and for discrete systems

$$\phi(t, 0) = \mathbf{A}^t = \sum_{i=1}^{r} \lambda_i^t \sum_{l=0}^{m_i - 1} t^l \mathbf{C}_{il} , \qquad (4.11)$$

where $\mathbf{B}_{il}$ and $\mathbf{C}_{il}$ are constant matrices.

These formulas and Theorem 4.4 imply the following stability conditions.

THEOREM 4.5

(i)   *Assume that for all eigenvalues of $\mathbf{A}$, Re $\lambda_i \leq 0$ in the continuous case (or $|\lambda_i| \leq 1$ in the discrete case), and all eigenvalues with the property Re $\lambda_i = 0$ (or $|\lambda_i| = 1$) have single multiplicity; then the equilibrium is stable.*

(ii) *The stability is asymptotic if and only if for all $i = 1, 2, \ldots, r$, Re $\lambda_i < 0$ (or $|\lambda_i| < 1$).*

PROOF

(i)   If Re $\lambda_i < 0$ (or $|\lambda_i| < 1$), then the ith term of (4.10) (or (4.11)) is bounded. If $m_i = 1$, then this term reduces to

$$e^{\lambda_i(t - t_0)} \mathbf{B}_{io} (\text{or } \lambda_i^t \mathbf{C}_{io}) ,$$

which is bounded even if Re $\lambda_i = 0$ (or $|\lambda_i| = 1$).

(ii) Assume first that Re $\lambda_i < 0$ (or $|\lambda_i| < 1$) for all i. Then in Equation (4.10) (or (4.11)) each term of the right-hand side tends to zero as $t \to \infty$. The eigenvalue conditions are also necessary, since the derivation of relations (1.45) and (1.44) implies that all eigenvalues λ_i show up in at least one term.

∎

REMARK 4.2   Note that Part (i) gives only sufficient conditions for the stability of the equilibrium. As the following example shows, these conditions are not necessary.   ∎

Example 4.3

Consider first the continuous system $\dot{\mathbf{x}} = \mathbf{O}\mathbf{x}$, where $\mathbf{O}$ is the zero matrix. Note that all constant functions $\mathbf{x}(t) \equiv \bar{\mathbf{x}}$ are solutions and also equilibriums. Since

$$\phi(t, t_0) = e^{\mathbf{O}(t-t_0)} = \mathbf{I}$$

is bounded (being independent of t), all equilibriums are stable but $\mathbf{O}$ has only one eigenvalue $\lambda_1 = 0$ with zero real part and multiplicity n, where n is the order of the system.

Consider next the discrete systems $\mathbf{x}(t + 1) = \mathbf{I}\mathbf{x}(t)$, when all constant functions $\mathbf{x}(t) \equiv \bar{\mathbf{x}}$ are also solutions and equilibriums. Furthermore,

$$\phi(t, 0) = \mathbf{A}^t = \mathbf{I}^t = \mathbf{I},$$

which is obviously bounded. Therefore, all equilibriums are stable, but the condition of Part (i) of the theorem is violated again.

Based on Theorem 4.4 and Examples 1.19 and 1.20, the following extension of Theorem 4.5 can be proven. The equilibrium is stable if and only if for all eigenvalues of $\mathbf{A}$, $Re\ \lambda_i \leq 0$ (or $|\lambda_i| \leq 1$), and if λ_i is a repeated eigenvalue of $\mathbf{A}$ such that $Re\ \lambda_i = 0$ (or $|\lambda_i| = 1$) then the size of each block containing λ_i in the Jordan canonical form of $\mathbf{A}$ is 1×1.

COROLLARY 4.2
If for at least one eigenvalue of $\mathbf{A}$, $Re\ \lambda_i > 0$ (or $|\lambda_i| > 1$), then the system is unstable.

REMARK 4.3   The equilibria of inhomogeneous equations are stable or asymptotically stable if and only if the same holds for the equilibria of the corresponding homogeneous equations.   ∎

The conditions of Theorem 4.5 are illustrated next.

Example 4.4

Consider again the continuous system

$$\dot{\mathbf{x}} = \begin{pmatrix} 0 & \omega \\ -\omega & 0 \end{pmatrix} \mathbf{x} + \begin{pmatrix} 0 \\ 1 \end{pmatrix} ,$$

the stability of which was analyzed earlier in Examples 4.1 and 4.2 by using the Lyapunov function method and the boundedness of the fundamental matrix.

The characteristic polynomial of the coefficient matrix is

$$\varphi(\lambda) = \det \begin{pmatrix} -\lambda & \omega \\ -\omega & -\lambda \end{pmatrix} = \lambda^2 + \omega^2 ,$$

therefore, the eigenvalues are $\lambda_1 = j\omega$ and $\lambda_2 = -j\omega$. Both eigenvalues have single multiplicities, and Re λ_1 = Re $\lambda_2 = 0$. Hence, the conditions of Part (i) are satisfied, and therefore, the equilibrium is stable. The conditions of Part (ii) do not hold. Consequently, the stability is not asymptotical.

If a time-invariant system is nonlinear, then the Lyapunov method is the most popular choice for stability analysis. If the system is linear, then the direct application of Theorem 4.5 is more attractive, since the eigenvalues of the coefficient matrix $\mathbf{A}$ can be obtained by standard methods (see for example, [42]). In addition, several conditions are known from the literature that guarantee the asymptotical stability of time-invariant discrete and continuous systems even without computing the eigenvalues. In the remaining part of this section some of such conditions are presented.

For continuous systems the following result has a special importance.

THEOREM 4.6
The equilibrium of a continuous system (4.6) is asymptotically stable if and only if equation

$$\mathbf{A}^T \mathbf{Q} + \mathbf{Q} \mathbf{A} = -\mathbf{M} \tag{4.12}$$

has positive definite solution $\mathbf{Q}$ with some positive definite matrix $\mathbf{M}$.

PROOF

(a)   Assume first that Equation (4.12) has positive definite solution with arbitrary positive definite matrix $\mathbf{M}$.

Consider the Lyapunov function $V(\mathbf{x}) = \mathbf{x}^T \mathbf{Q} \mathbf{x}$ for the homogeneous equation $\dot{\mathbf{x}} = \mathbf{A}\mathbf{x}$. Select the state space $X = \mathbf{R}^n$, then obviously $V(\mathbf{x})$ is continuous and has a unique minimum point at $\bar{\mathbf{x}} = \mathbf{0}$, which is the equilibrium of the homogeneous equation. Furthermore,

$$\dot{V}(\mathbf{x}(t)) = \dot{\mathbf{x}}^T \mathbf{Q} \mathbf{x} + \mathbf{x}^T \mathbf{Q} \dot{\mathbf{x}} = (\mathbf{A}\mathbf{x})^T \mathbf{Q} \mathbf{x} + \mathbf{x}^T \mathbf{Q}(\mathbf{A}\mathbf{x})$$

$$= \mathbf{x}^T (\mathbf{A}^T \mathbf{Q} + \mathbf{Q}\mathbf{A})\mathbf{x} = -\mathbf{x}^T \mathbf{M} \mathbf{x} < 0 ,$$

unless $\mathbf{x} \neq \bar{\mathbf{x}} = \mathbf{0}$. Therefore, the equilibrium of system $\dot{\mathbf{x}} = \mathbf{A}\mathbf{x}$ is asymptotically stable, which follows from Theorem 4.2.

(b) Assume that for all eigenvalues λ_i of $\mathbf{A}$, $\mathrm{Re}\lambda_i < 0$. Let $\mathbf{M}$ be any positive definite matrix. We show next that

$$\mathbf{Q} = \int_0^\infty e^{\mathbf{A}^T t} \mathbf{M} e^{\mathbf{A} t} \, dt$$

is positive definite and satisfies Equation (4.12).
   Suppose that $\mathbf{u} \neq \mathbf{0}$, then

$$\mathbf{u}^T \mathbf{Q} \mathbf{u} = \int_0^\infty \mathbf{u}^T e^{\mathbf{A}^T t} \mathbf{M} e^{\mathbf{A} t} \mathbf{u} \, dt > 0 ,$$

since $e^{\mathbf{A} t}$ is invertible (see Theorem 2.2) and therefore, $e^{\mathbf{A} t} \mathbf{u} \neq \mathbf{0}$. Furthermore,

$$\mathbf{A}^T \mathbf{Q} + \mathbf{Q}\mathbf{A} = \int_0^\infty \mathbf{A}^T e^{\mathbf{A}^T t} \mathbf{M} e^{\mathbf{A} t} \, dt + \int_0^\infty e^{\mathbf{A}^T t} \mathbf{M} e^{\mathbf{A} t} \mathbf{A} \, dt$$

$$= \int_0^\infty \frac{d}{dt} \left(e^{\mathbf{A}^T t} \mathbf{M} e^{\mathbf{A} t} \right) \, dt = \left[e^{\mathbf{A}^T t} \mathbf{M} e^{\mathbf{A} t} \right]_0^\infty$$

$$= \mathbf{O} - \mathbf{M} = -\mathbf{M} ,$$

since $e^{\mathbf{A}^T \cdot 0} = e^{\mathbf{A} \cdot 0} = \mathbf{I}$, and both matrices $e^{\mathbf{A} t}$ and $e^{\mathbf{A}^T t}$ tend to zero as $t \to \infty$.

∎

REMARK 4.4   In practical applications $\mathbf{M}$ is usually selected as the identity matrix. ∎

   Let $\varphi(\lambda) = \lambda^n + p_{n-1}\lambda^{n-1} + \cdots + p_1\lambda + p_0$ be the characteristic polynomial of matrix $\mathbf{A}$. Let λ_l denote any real eigenvalue, then the linear factor $\lambda - \lambda_l$ has positive coefficients, if $\lambda_l < 0$. Assume next that $\alpha_l + j\beta_l$ is an eigenvalue of $\mathbf{A}$, then $\alpha_l - j\beta_l$ is also an eigenvalue with the same multiplicity. Therefore, $\varphi(\lambda)$ is a multiple of the quadratic polynomial

$$(\lambda - \alpha_l - j\beta_l)(\lambda - \alpha_l + j\beta_l) = \lambda^2 - 2\alpha_l\lambda + \alpha_l^2 + \beta_l^2 .$$

If $\alpha_l < 0$, then all coefficients of this polynomial are positive. Since $\varphi(\lambda)$ is the product of such linear and quadratic factors, we have the following result.

THEOREM 4.7
Assume that all eigenvalues of matrix $\mathbf{A}$ have negative real parts. Then $p_i > 0$ ($i = 0, 1, \ldots, n-1$).

COROLLARY 4.3
If any of the coefficients p_i is negative or zero, the equilibrium of the system with coefficient matrix $\mathbf{A}$ cannot be asymptotically stable. This result can be used as an initial stability test. However, the conditions of the theorem do not imply that the eigenvalues of $\mathbf{A}$ have negative real parts, as it will be illustrated in Example 4.6.

   Example 4.5

   In the case of matrix

$$\mathbf{A} = \begin{pmatrix} 0 & \omega \\ -\omega & 0 \end{pmatrix}$$

   the characteristic polynomial is $\varphi(\lambda) = \lambda^2 + \omega^2$. Since the coefficient of λ is zero, the system of Example 4.4 is not asymptotically stable.

   The next condition is known as the *Hurwitz criterion*. It is based on the construction of the following determinants:

$$\triangle_1 = \det(p_{n-1})$$

$$\triangle_2 = \det \begin{pmatrix} p_{n-1} & p_{n-3} \\ 1 & p_{n-2} \end{pmatrix}$$

$$\triangle_3 = \det \begin{pmatrix} p_{n-1} & p_{n-3} & p_{n-5} \\ 1 & p_{n-2} & p_{n-4} \\ 0 & p_{n-1} & p_{n-3} \end{pmatrix}$$

$$\vdots \quad \vdots \quad \vdots$$

$$\triangle_n = \det \begin{pmatrix} p_{n-1} & p_{n-3} & p_{n-5} & \cdots & 0 & 0 \\ 1 & p_{n-2} & p_{n-4} & \cdots & 0 & 0 \\ 0 & p_{n-1} & p_{n-3} & \cdots & 0 & 0 \\ 0 & 1 & p_{n-2} & \cdots & 0 & 0 \\ 0 & 0 & p_{n-1} & \cdots & 0 & 0 \\ \cdots & \cdots & \cdots & \cdots & \cdots & \cdots \\ 0 & 0 & 0 & \cdots & p_0 & 0 \\ 0 & 0 & 0 & \cdots & p_1 & 0 \\ 0 & 0 & 0 & \cdots & p_2 & p_0 \end{pmatrix}.$$

THEOREM 4.8

Assume that $p_i > 0$ ($i = 0, 1, \dots, n - 1$), then all eigenvalues of $\mathbf{A}$ have negative real parts if and only if all $\triangle_i > 0$ ($i = 1, 2, \dots, n$).

The proof of this theorem is found, for example, in [15].

Example 4.6

Consider polynomial

$$\varphi(\lambda) = \lambda^4 + 2\lambda^3 + 3\lambda^2 + 4\lambda + 5,$$

which satisfies the conditions of Theorem 4.7, since all of its coefficients are positive. However, in this case,

$$\triangle_1 = \det(2) = 2 > 0$$

and

$$\triangle_2 = \det \begin{pmatrix} 2 & 4 \\ 1 & 3 \end{pmatrix} = 2 > 0,$$

but

$$\triangle_3 = \det \begin{pmatrix} 2 & 4 & 0 \\ 1 & 3 & 5 \\ 0 & 2 & 4 \end{pmatrix} = -12 < 0.$$

Hence the conditions of Theorem 4.8 are not satisfied. That is, the system is not asymptotically stable. In Problem 4.15 the same conclusion is reached based on computing the roots of φ.

In many practical cases, a similar result, the so-called *Routh-criterion*, is applied. Its details can be found for example in [31], and we mention that it is based on determining the number of sign changes in a Routh array constructed from the polynomial coefficients.

In many applications, especially in economics, the dynamic continuous systems are described by differential equation

$$\dot{x} = KAx + Bu ,$$

where matrix K is real positive definite, in most cases diagonal with positive diagonal elements. The following theorem is usually applied in establishing the asymptotical stability of such systems.

THEOREM 4.9
Assume that K is positive definite and $A + A^T$ is negative definite. Then all eigenvalues of KA have negative real parts, that is, the system is asymptotically stable.

PROOF   From the eigenvalue equation of KA we know that

$$KAv = \lambda v ,$$

where λ is an eigenvalue of KA with associated eigenvector v. Since K is nonsingular, $v = Ku$ with some vector u. Therefore,

$$KAKu = \lambda Ku .$$

Premultiplying this equation by $\bar{u}^T$, where overbar denotes complex conjugate, yields the relation

$$\bar{u}^T KAKu = \lambda \bar{u}^T Ku .$$

The transpose conjugate of this equation is

$$\bar{u}^T KA^T Ku = \bar{\lambda} \bar{u}^T Ku ,$$

and by adding these equations,

$$\bar{u}^T K(A^T + A)Ku = (\lambda + \bar{\lambda})\bar{u}^T Ku . \qquad (4.13)$$

Since $v \neq 0$, u is nonzero. Therefore,

$$\bar{u}^T K(A^T + A)Ku = \bar{v}^T(A^T + A)v < 0$$

and

$$\bar{u}^T Ku > 0 ,$$

which imply that

$$\lambda + \bar{\lambda} = 2Re\lambda < 0 .$$

Hence, the proof is completed.   ∎

COROLLARY 4.4
Assume that $\mathbf{A} + \mathbf{A}^T$ is negative definite. Then all eigenvalues of $\mathbf{A}$ have negative real parts.

PROOF   Select $\mathbf{K} = \mathbf{I}$ in the assertion of the theorem.   ∎

REMARK 4.5   An alternative proof of the theorem can be constructed based on the selection of the special Lyapunov function $\mathbf{x}^T\mathbf{K}^{-1}\mathbf{x}$.   ∎

For discrete systems, the above stability criteria can be modified accordingly. The details are given, for example, in [14]. An easy stability checkis implied by Theorem 1.8 and can be formulated as follows. If for some matrix norm, $\|\mathbf{A}\| < 1$, then the equilibrium is asymptotically stable. Note that it is possible that all discussed matrix norms ($p = 1, 2, \infty$ and the Frobenius) are greater than one, and the system is still asymptotically stable. Such an example is presented next.

Example 4.7

In the case of matrix

$$\mathbf{A} = \begin{pmatrix} 0.5 & 0 \\ 1 & 0.5 \end{pmatrix}$$

the eigenvalues are $\lambda_1 = \lambda_2 = 0.5$. However, $\|\mathbf{A}\|_\infty = \|\mathbf{A}\|_1 = 1.5$, $\|\mathbf{A}\|_2 \approx 1.207$, and $\|\mathbf{A}\|_F = \sqrt{1.5}$. Hence, the eigenvalues are inside the unit circle, but the norms are greater than one.

In the second part of this section, stability conditions will be given based on the properties of the transfer function.

The transfer function of the continuous system

$$\dot{\mathbf{x}} = \mathbf{A}\mathbf{x} + \mathbf{B}\mathbf{u}$$

$$\mathbf{y} = \mathbf{C}\mathbf{x} \tag{4.14}$$

and that of the discrete system

$$\mathbf{x}(t + 1) = \mathbf{A}\mathbf{x}(t) + \mathbf{B}\mathbf{u}(t)$$

$$\mathbf{y}(t) = \mathbf{C}\mathbf{x}(t) \tag{4.15}$$

have the common form

$$\mathbf{H}(s) = \mathbf{C}(s\mathbf{I} - \mathbf{A})^{-1}\mathbf{B} , \tag{4.16}$$

as it was shown in Sections 3.3.2 and 3.4. Note that the poles of the system are defined as all values of s such that $s\mathbf{I} - \mathbf{A}$ is singular. In special cases some poles may be cancelled in the rational function form of $\mathbf{H}(s)$, and they might not be explicitly shown. The equilibrium of system (4.14) (or (4.15)) with constant input is stable if all poles of $\mathbf{H}(s)$ have nonpositive real parts (or absolute values less than or equal to one) and all poles with zero real part (or unit absolute value) are single. Similarly, the equilibrium is asymptotically stable if and only if all poles of $\mathbf{H}(s)$ have negative real parts (or absolute value less than one). Even in the case when some poles are canceled by zeros, we still have to consider all poles in the above criteria.

Example 4.8

Consider again the system

$$\dot{\mathbf{x}} = \begin{pmatrix} 0 & \omega \\ -\omega & 0 \end{pmatrix} \mathbf{x} + \begin{pmatrix} 0 \\ 1 \end{pmatrix} ,$$

which was discussed in earlier sections. Assume that the output equation has the form

$$y = (1,1)\mathbf{x} .$$

Then

$$\mathbf{H}(s) = \frac{s + \omega}{s^2 + \omega^2} ,$$

as it was derived in Example 3.8. The poles are $j\omega$ and $-j\omega$, which have zero real parts. Consequently, the equilibrium is stable but not asymptotically stable.

A special stability criterion concerning single-input, single-output time-invariant continuous systems will be introduced next.

Consider the continuous system

$$\dot{\mathbf{x}} = \mathbf{A}\mathbf{x} + \mathbf{b}u$$

$$y = \mathbf{c}^T\mathbf{x} , \tag{4.17}$$

where $\mathbf{A}$ is an $n \times n$ constant matrix, $\mathbf{b}$ and $\mathbf{c}$ are constant n-dimensional vectors. The transfer function of this system is

$$H_1(s) = \mathbf{c}^T(s\mathbf{I} - \mathbf{A})^{-1}\mathbf{b} ,$$

which is obviously a rational function of s. Combine next this system with the feedback-type input $u = ky$, where k is a constant. Then the feedback system can be described by the differential equation

$$\dot{\mathbf{x}} = \mathbf{A}\mathbf{x} + k\mathbf{b}\mathbf{c}^T\mathbf{x} = (\mathbf{A} + k\mathbf{b}\mathbf{c}^T)\mathbf{x} . \tag{4.18}$$

The transfer function of the feedback is the 1×1 matrix (k), since in Figure 3.6, $Y_2 = kU_2$. Then, from Equation (3.37) we conclude that the transfer function of the feedback system is

$$H(s) = \frac{H_1(s)}{1 - kH_1(s)} , \tag{4.19}$$

which is also a rational function of s.

Before presenting the stability criterion due to Nyquist, which shows the connection between the asymptotical stability of systems (4.17) and (4.18), we introduce the following definition.

DEFINITION 4.3   *Let $r(s)$ be a rational function of s. Then the locus of points*

$$L(r) = \{a + jb | a = Re(r(jv)), b = Im(r(jv)), v \in \mathbf{R}\}$$

is called the response diagram of r.

Note that $L(r)$ is the image of the imaginary line $Re(s) = 0$ under the mapping r. We shall assume that $L(r)$ is bounded, which is the case if and only if the degree of the denominator of r is not less than that of the numerator, and r has no poles on the line $Re(s) = 0$.

Example 4.9

Consider again the system

$$\dot{\mathbf{x}} = \begin{pmatrix} 0 & \omega \\ -\omega & 0 \end{pmatrix}\mathbf{x} + \begin{pmatrix} 0 \\ 1 \end{pmatrix}u$$

$$y = (1, 1)\mathbf{x} .$$

From the previous example we know that the transfer function is

$$H(s) = \frac{s + \omega}{s^2 + \omega^2} \ .$$

Now the locus $L(H)$ will be determined. If $s = jv$ ($v \in \mathbf{R}$), then

$$H(s) = \frac{jv + \omega}{-v^2 + \omega^2} = \frac{\omega}{-v^2 + \omega^2} + j \cdot \frac{v}{-v^2 + \omega^2} \ .$$

Therefore, $L(H)$ consists of the points

$$x = Rez = \frac{w}{-v^2 + \omega^2} \ , \ y = Imz = \frac{v}{-v^2 + \omega^2} \ .$$

This parameterized curve can be directed with increasing values of v as it is shown by Figure 4.2 for the special case of $\omega = 1$. A direct relation between x and y can be obtained as follows.
From the first equation we have

$$v = \pm \sqrt{\frac{x\omega^2 - \omega}{x}} \ ,$$

and the second equation implies that

$$y = \frac{\pm \sqrt{\frac{x\omega^2 - \omega}{x}}}{\frac{-x\omega^2 + \omega}{x} + \omega^2} = \pm \sqrt{x^2 - \frac{x}{\omega}} \ .$$

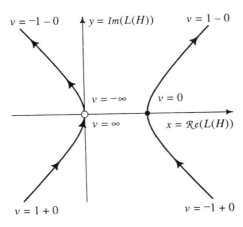

Figure 4.2   Graph of $L(H)$.

Note that this locus is not bounded, since $H(s)$ has the poles $s = \pm j\omega$ on the line $Re(s) = 0$.

THEOREM 4.10

Assume that H_1 has a bounded response diagram $L(H_1)$. If H_1 has ν poles in the half-plane $Re(s) > 0$ then H has $\rho + \nu$ poles in the half-plane $Re(s) > 0$ if the point $(1/k) + j \cdot 0$ is not on $L(H_1)$, and $L(H_1)$ encircles $(1/k) + j \cdot 0$ ρ times in the clockwise sense.

PROOF   Note first that if C is a simple closed curve, then singularities of function $H_1'(s)/H_1(s)$ (where "prime" denotes derivative) are the poles and zeros of H_1 inside C. Rewrite H_1 as

$$H_1 = (s + s_i)^{m_i} h_i(s) ,$$

and note that the residue of H_1'/H_1 at every pole of H_1 of multiplicity m_i is $-m_i$, and that the residue at every zero of multiplicity m_i is m_i; furthermore,

$$\frac{H_1'(s)}{H_1(s)} = \frac{m_i}{s + s_i} + \frac{h_i(s)}{H_1(s)} ,$$

where h_i/H_1 is analytic near $s = -s_i$. By applying the residue theorem to all such expansions with $-s_i$ being inside C, we conclude that

$$\frac{1}{2\pi j} \int_C \frac{H_1'(s)}{H_1(s)} \, ds = z - p , \tag{4.20}$$

where z and p are the numbers of zeros and poles of H_1 inside C, respectively.

By direct integration we know that

$$\frac{1}{2\pi j} \int \frac{H_1'(s)}{H_1(s)} \, ds = \frac{1}{2\pi j} ln H_1(s) ,$$

and therefore, by integrating around a closed curve C no change is obtained in the magnitude of $\ln H_1$ but the argument of H_1 is changed by $2\pi E$, where E is the number of times the image of C encircles the origin in the H_1 plane. Hence, $z - p = E$.

Consider next the transfer function $H(s)$ of the feedback system (4.18). The zeros of $H(s)$ are the same as the zeros of H_1, and the poles of $H(s)$ are the zeros of $1 - kH_1$. Select C as a contour consisting of the imaginary axis. Note that $L(H)$ encircles the origin if and only if $L(H_1)$ encircles the point $1/k$. Then the result follows from relation (4.20). ∎

COROLLARY 4.5

Assume that system (4.17) is asymptotically stable with constant input and that $L(H_1)$ is bounded and traversed in the direction of increasing v and has the point $(1/k) + j \cdot 0$ on its left. Then the feedback system (4.18) is also asymptotically stable.

This result has a lot of applications, since — as we will see in Chapter 9 — feedback systems have a crucial role in constructing stabilizers, observers, and filters for given systems. Figure 4.3 illustrates the conditions of the corollary. The application of this result is especially convenient if system (4.17) is given and only appropriate values k of the feedback are to be determined. In such cases the locus $L(H_1)$ has to be computed first, and then the region of all appropriate k values can be determined easily from the graph of $L(H_1)$.

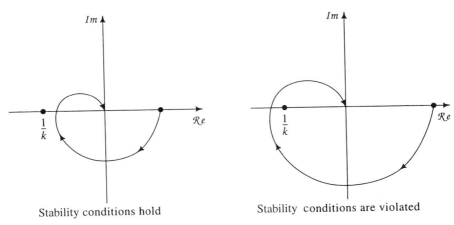

Stability conditions hold   Stability conditions are violated

Figure 4.3 Illustration of Nyquist stability criteria.

4.2 BIBO Stability

In the previous section, internal stability of dynamic time-invariant systems was examined, when some properties of the state were investigated. In this section the external stability of dynamic systems is discussed, which is usually called the BIBO (Bounded Input–Bounded Output) stability. Here we drop the simplifying assumption of the previous subsection that the system is time-invariant. We will hence examine time-variant systems.

DEFINITION 4.4   *A dynamic system is called BIBO stable if for zero initial conditions, a bounded input always evokes a bounded output.*

This kind of stability can be examined by using the direct input–output relations:

$$\mathbf{y}(t) = \mathbf{C}(t)\phi(t, t_0)\mathbf{x}_0 + \int_{t_0}^{t} \mathbf{C}(t)\phi(t, \tau)\mathbf{B}(\tau)\mathbf{u}(\tau)\, d\tau \qquad (4.21)$$

and

$$\mathbf{y}(t) = \mathbf{C}(t)\phi(t, 0)\mathbf{x}_0 + \sum_{\tau=0}^{t-1} \mathbf{C}(t)\phi(t, \tau + 1)\mathbf{B}(\tau)\mathbf{u}(\tau) \qquad (4.22)$$

for continuous and discrete time-variant linear systems, respectively.

In BIBO stability we always assume that the initial state is zero; therefore, (4.21) and (4.22) reduce to relations

$$\mathbf{y}(t) = \int_{t_0}^{t} \mathbf{T}(t, \tau)\mathbf{u}(\tau)\, d\tau \qquad (4.23)$$

with

$$\mathbf{T}(t, \tau) = \mathbf{C}(t)\phi(t, \tau)\mathbf{B}(\tau)\ ,$$

and

$$\mathbf{y}(t) = \sum_{\tau=0}^{t-1} \mathbf{T}(t, \tau)\mathbf{u}(\tau) \qquad (4.24)$$

with

$$\mathbf{T}(t, \tau) = \mathbf{C}(t)\phi(t, \tau + 1)\mathbf{B}(\tau)\ ,$$

respectively.

For continuous systems, a necessary and sufficient condition for BIBO stability can be formulated as follows.

THEOREM 4.11

Let $\mathbf{T}(t, \tau) = (t_{ij}(t, \tau))$, then the continuous time-variant linear system is BIBO stable if and only if the integral

$$\int_{t_0}^{t} |t_{ij}(t, \tau)|\, d\tau \qquad (4.25)$$

is bounded for all $t \geq t_0$, i and j.

PROOF

(a)   Assume first that for all $t \geq t_0$, i and j,

$$\int_{t_0}^{t} |t_{ij}(t,\tau)| \, d\tau \leq K_{ij} \; ;$$

furthermore,

$$|u_j(t)| \leq U_j \qquad (t \geq t_0 \text{ and all } j) \; .$$

Then

$$|y_i(t)| = \left| \int_{t_0}^{t} \sum_j t_{ij}(t,\tau) u_j(\tau) \, d\tau \right|$$

$$\leq \int_{t_0}^{t} \sum_j |t_{ij}(t,\tau)| \cdot |u_j(\tau)| \, d\tau \leq \sum_j K_{ij} U_j \; ,$$

hence the output is bounded.

(b)   Assume next that the integrals (4.25) are not all bounded. Then there exists a pair (i_0, j_0) such that for all $N > 0$,

$$\int_{t_0}^{t_N} |t_{i_0 j_0}(t_N,\tau)| \, d\tau > N$$

with some $t_N > t_0$. Select now the input function as

$$u_{j_0}(\tau) = \begin{cases} 1 \text{ if } t_{i_0 j_0}(t_N,\tau) \geq 0 \\ -1 \text{ otherwise,} \end{cases}$$

and $u_j(\tau) \equiv 0$ for $j \neq j_0$. Then

$$y_{i_0}(t_N) = \int_{t_0}^{t_N} \sum_j t_{i_0 j}(t_N,\tau) u_j(\tau) \, d\tau$$

$$= \int_{t_0}^{t_N} t_{i_0 j_0}(t_N,\tau) u_{j_0}(\tau) \, d\tau = \int_{t_0}^{t_N} |t_{i_0 j_0}(t_N,\tau)| \, d\tau > N \; .$$

Hence, the output is not bounded, which completes the proof.

∎

COROLLARY 4.6
Integrals (4.25) are all bounded if and only if

$$I(t) = \int_{t_0}^{t} \sum_i \sum_j |t_{ij}(t,\tau)| \, d\tau \qquad (4.26)$$

is bounded for $t \geq t_0$. Therefore, it is sufficient to show the boundedness of only one integral in order to establish BIBO stability.

The discrete counterpart of this theorem can be given in the following way.

THEOREM 4.12
Let $\mathbf{T}(t,\tau) = (t_{ij}(t,\tau))$, then the discrete time-variant linear system is BIBO stable if and only if the sum

$$I(t) = \sum_{\tau=0}^{t-1} |t_{ij}(t,\tau)| \qquad (4.27)$$

is bounded for all $t \geq 1$, i and j.

Since the proof of this result is analogous to the continuous case, the details are left to the reader as an exercise.

COROLLARY 4.7
The sums (4.27) are all bounded if and only if

$$\sum_{\tau=0}^{t-1} \sum_i \sum_j |t_{ij}(t,\tau)| \qquad (4.28)$$

is bounded. Therefore, it is sufficient to verify the boundness of only one sum in order to establish BIBO stability.

Consider next the time-invariant case, when $\mathbf{A}(t) \equiv \mathbf{A}$, $\mathbf{B}(t) \equiv \mathbf{B}$, and $\mathbf{C}(t) \equiv \mathbf{C}$. From the above theorems and the definition of $\mathbf{T}(t,\tau)$, we have immediately the following sufficient condition.

4.2   BIBO Stability

THEOREM 4.13
Assume that for all eigenvalues λ_i of $\mathbf{A}$, $Re\lambda_i < 0$ (or $|\lambda_i| < 1$). Then the time-invariant linear continuous (or discrete) system is BIBO stable.

Example 4.10

Consider again the continuous system

$$\dot{\mathbf{x}} = \begin{pmatrix} 0 & \omega \\ -\omega & 0 \end{pmatrix} \mathbf{x} + \begin{pmatrix} 0 \\ 1 \end{pmatrix} u$$

$$y = (1,1)\mathbf{x}\,.$$

In this case the results of Example 2.6 imply that

$$\phi(t,\tau) = \begin{pmatrix} \cos\omega(t-\tau) & \sin\omega(t-\tau) \\ -\sin\omega(t-\tau) & \cos\omega(t-\tau) \end{pmatrix};$$

therefore,

$$T(t,\tau) = (1,1) \begin{pmatrix} \cos\omega(t-\tau) & \sin\omega(t-\tau) \\ -\sin\omega(t-\tau) & \cos\omega(t-\tau) \end{pmatrix} \begin{pmatrix} 0 \\ 1 \end{pmatrix}$$

$$= \sin\omega(t-\tau) + \cos\omega(t-\tau)\,,$$

and

$$I(t) = \int_0^t |\sin\omega(t-\tau) + \cos\omega(t-\tau)|\, d\tau\,.$$

We will now show that this integral is not bounded. Note first that by introducing the new integration variable $x = \omega(t-\tau)$,

$$I(t) = \frac{1}{\omega} \int_0^{\omega t} |\sin x + \cos x|\, dx\,.$$

Observe next that

$$\int_0^{2\pi} |\sin x + \cos x|\, dx = \int_{-\frac{1}{4}\pi}^{\frac{3}{4}\pi} (\sin x + \cos x)\, dx + \int_{\frac{3}{4}\pi}^{\frac{7}{4}\pi} (-\sin x - \cos x)\, dx$$

$$= [-\cos x + \sin x]_{-\frac{1}{4}\pi}^{\frac{3}{4}\pi} + [\cos x - \sin x]_{\frac{3}{4}\pi}^{\frac{7}{4}\pi} = 4\sqrt{2}\,.$$

Hence by selecting $t = 2\pi N/\omega$,

$$I(t) = \frac{1}{\omega} 4\sqrt{2}N\,,$$

which tends to infinity as $N \to \infty$. That is, this system is not BIBO stable.

Finally we note that BIBO stability is not implied by an observation that a certain bounded input generates bounded output. All bounded inputs must generates bounded outputs in order to guarantee BIBO stability.

4.3   Applications

In this section the applications of the stability analysis of dynamic systems will be illustrated via particular systems arising in engineering and social sciences.

4.3.1   Applications in Engineering

1. Consider the simple *harmonic oscillator* introduced in Chapter 2, and given in Application 3.5.1-1, which is summarized below:

$$\dot{\mathbf{x}} = \begin{pmatrix} 0 & \omega \\ -\omega & 0 \end{pmatrix} \mathbf{x} + \begin{pmatrix} 0 \\ 1 \end{pmatrix} u \ .$$

Is it stable?

To answer this question we must find the eigenvalues of $\mathbf{A}$. Note that the characteristic equation has the form

$$\varphi(\lambda) = \det(\mathbf{A} - \lambda \mathbf{I}) = \det \begin{pmatrix} -\lambda & \omega \\ -\omega & -\lambda \end{pmatrix} = \lambda^2 + \omega^2 \ .$$

The eigenvalues are the roots of φ:

$$\lambda_{1,2} = \pm j\omega \ .$$

These values are also called the poles of the system. The poles are single and on the imaginary axis. Therefore, the system is stable, but not asymptotically stable, which means that if we leave it alone in its equilibrium state, it will remain stationary. But if we jerk on the mass it will oscillate forever. There is no damping term to remove the energy, so the energy will be transferred back and forth between potential energy in the spring and kinetic energy in the mass. A good approximation of such a harmonic oscillator is a pendulum clock. The more expensive it is (i.e., the smaller the damping), the less often we have to wind it (i.e., add energy).

2. What about the *damped linear second-order system* of Application 3.5.1-2; is it stable? From Equation (3.66) we know that the eigenvalues are

$$\lambda_{1,2} = -\zeta\omega_n \pm j\omega_n\sqrt{1 - \zeta^2} \ .$$

The locations of the poles depend on the value of ζ. Refer to Equation (3.66) and Figures 3.20 and 3.22, and note that if $\zeta > 0$ the poles are in the left half of the λ-plane and therefore, the system is asymptotically stable. If $\zeta = 0$, as in the previous problem, the poles are on the imaginary axis; therefore, the system is stable, but not asymptotically stable. If $\zeta < 0$, the poles are in the right half-plane and the system is unstable.

3. For the *electrical system* of Application 3.5.1-3 the characteristic polynomial of matrix **A** of (3.78) has the form

$$\left(-\lambda - \frac{R_1}{L}\right)\left(-\lambda - \frac{1}{CR_2}\right) + \frac{1}{LC} = 0 \ ,$$

which simplifies as

$$\lambda^2 + \lambda\left(\frac{R_1}{L} + \frac{1}{CR_2}\right) + \left(\frac{R_1}{LCR_2} + \frac{1}{LC}\right) = 0 \ .$$

Since R_1, R_2, L, and C are positive numbers, the coefficients of this equation are all positive. The constant term equals $\lambda_1\lambda_2$, and the coefficient of λ is $-(\lambda_1 + \lambda_2)$. Therefore,

$$\lambda_1 + \lambda_2 < 0 \qquad \text{and} \qquad \lambda_1\lambda_2 > 0 \ .$$

If the eigenvalues are real, then these relations hold if and only if both eigenvalues are negative. If they were positive, then $\lambda_1 + \lambda_2 > 0$. If they had different signs, then $\lambda_1\lambda_2 < 0$. Furthermore, if at least one eigenvalue is zero, then $\lambda_1\lambda_2 = 0$. Assume next that the eigenvalues are complex:

$$\lambda_{1,2} = Re\lambda \pm jIm\lambda \ .$$

Then

$$\lambda_1 + \lambda_2 = 2Re\lambda$$

and

$$\lambda_1\lambda_2 = (Re\lambda)^2 + (Im\lambda)^2 \ .$$

Hence $\lambda_1 + \lambda_2 < 0$ implies that $Re\lambda < 0$.

In summary, the system is asymptotically stable, since in both the real and complex cases the eigenvalues have negative values and negative real parts, respectively.

4. For the *transistor circuit* model (3.80) of Application 3.5.1-4, the characteristic equation is

$$\det \begin{pmatrix} -\frac{h_{ic}}{L} - \lambda & 0 \\ \frac{h_{fc}}{C} & -\lambda \end{pmatrix} = 0 \ ,$$

which can be simplified as

$$\lambda^2 + \lambda \frac{h_{ie}}{L} + 0 = 0 \ .$$

The roots are

$$\lambda_1 = 0 \text{ and } \lambda_2 = -\frac{h_{ie}}{L} \ .$$

Therefore, the system is stable, but not asymptotically stable.

5. To access the stability of the *hydraulic system* of Application 3.5.1-5, we must solve its characteristic equation

$$\det \begin{pmatrix} -a - \lambda & a \\ b & -(b+c) - \lambda \end{pmatrix} = 0 \ ,$$

where $a = 1/R_1 A_1$, $b = 1/R_1 A_2$, and $c = 1/R_2 A_2$. This equation is simplified as

$$\lambda^2 + \lambda(a + b + c) + ac = 0 \ .$$

Note that a, b, and c are all positive numbers, therefore,

$$\lambda_1 + \lambda_2 < 0 \qquad \text{and} \qquad \lambda_1 \lambda_2 > 0 \ .$$

Hence both roots are negative or they are complex conjugate numbers with negative real parts. In either case the roots are in the left half of the λ-plane, and the system is asymptotically stable.

6. In the case of our *multiple input electrical system* the stability can be easily examined by determining the characteristical polynomial of the coefficient matrix. By expanding the determinant with respect to the first row we have the following result:

$$\varphi(\lambda) = \det \begin{pmatrix} -\frac{R}{L_1} - \lambda & 0 & -\frac{1}{L_1} \\ 0 & -\lambda & -\frac{1}{L_2} \\ \frac{1}{C} & \frac{1}{C} & -\lambda \end{pmatrix}$$

$$= (-\frac{R}{L_1} - \lambda)(\lambda^2 + \frac{1}{CL_2}) - \frac{1}{L_1}(0 + \frac{\lambda}{C})$$

$$= -\lambda^3 - \lambda^2 \frac{R}{L_1} - \lambda(\frac{1}{CL_1} + \frac{1}{CL_2}) - \frac{R}{CL_1L_2} \ .$$

Therefore, the eigenvalues are the roots of the cubic equation

$$\lambda^3 + \lambda^2 \frac{R}{L_1} + \lambda(\frac{1}{CL_1} + \frac{1}{CL_2}) + \frac{R}{CL_1L_2} = 0 \ .$$

Notice that all coefficients are positive, that is, the necessary conditions of Theorem 4.7 are satisfied for the asymptotical stability of the system. In order to verify that the system is asymptotically stable, we will apply Theorem 4.8. We compute first the following determinants:

$$\Delta_1 = \det(\frac{R}{L_1}) = \frac{R}{L_1} > 0$$

$$\Delta_2 = \det \begin{pmatrix} \frac{R}{L_1} & \frac{R}{CL_1L_2} \\ 1 & \frac{1}{CL_1} + \frac{1}{CL_2} \end{pmatrix} = \frac{R}{CL_1^2} > 0$$

$$\Delta_3 = \det \begin{pmatrix} \frac{R}{L_1} & \frac{R}{CL_1L_2} & 0 \\ 1 & \frac{1}{CL_1} + \frac{1}{CL_2} & 0 \\ 0 & \frac{R}{L_1} & \frac{R}{CL_1L_2} \end{pmatrix} = \frac{1}{C^2L_1^3L_2} > 0 \ .$$

Because all the three determinants are positive, Theorem 4.8 implies that the system is asymptotically stable.

7. To find the eigenvalues for the *stick-balancing problem* of Application 3.5.1-7, find the roots of the characteristic polynomial

$$\varphi(\lambda) = \det(\mathbf{A} - \lambda\mathbf{I}) = \det \begin{pmatrix} -\lambda & 1 \\ g & -\lambda \end{pmatrix} = \lambda^2 - g \ ,$$

which are

$$\lambda_{1,2} = \pm\sqrt{g} \ .$$

One is in the right half-plane and the other is in the left half-plane, so the system is unstable. The instability is understandable, since without an input to control the system, if you are not upright with zero velocity the stick will fall over.

8. For the *cart with two sticks* model of Application 3.5.1-8 we must solve the characteristic equation

$$\det \begin{pmatrix} -\lambda & 0 & 1 & 0 \\ 0 & -\lambda & 0 & 1 \\ a_1 & a_2 & -\lambda & 0 \\ a_3 & a_4 & 0 & -\lambda \end{pmatrix} = 0$$

of matrix $\mathbf{A}$ in Equation (3.103). By expanding the determinant with respect to its last column, we have

$$\begin{pmatrix} -\lambda & 0 & 1 \\ a_1 & a_2 & -\lambda \\ a_3 & a_4 & 0 \end{pmatrix} - \lambda \begin{pmatrix} -\lambda & 0 & 1 \\ 0 & -\lambda & 0 \\ a_1 & a_2 & -\lambda \end{pmatrix} = 0 \ .$$

Simple calculation shows that it simplifies to equation

$$\lambda^4 - \lambda^2(a_4 + a_1) + (a_4 a_1 - a_2 a_3) = 0 \ .$$

It is easy to see that λ^2 is real, since the discriminant is

$$(a_4 + a_1)^2 - 4(a_4 a_1 - a_2 a_3) = (a_4 - a_1)^2 + 4a_2 a_3 > 0 \ .$$

Furthermore,

$$a_4 + a_1 > 0 \qquad \text{and} \qquad a_4 a_1 - a_2 a_3 > 0 \ ,$$

which imply that there are two distinct positive solutions for λ^2. Hence, there are two positive and two negative eigenvalues, which implies the instability of the system.

The fact that the system is not asymptotically stable follows also from Theorem 4.7, since there are a negative and two zero missing coefficients.

9. The stability of the *electrical heating system* can be also examined by computing the characteristic polynomial of the coefficient matrix:

$$\det \begin{pmatrix} -\dfrac{A_1 h_1}{C_1} - \lambda & \dfrac{A_1 h_1}{C_1} \\ \dfrac{A_1 h_1}{C_2} & -\dfrac{A_1 h_1 + A_2 h_2}{C_2} - \lambda \end{pmatrix}$$

$$= \lambda^2 + \lambda\Big(\dfrac{A_1 h_1}{C_1} + \dfrac{A_1 h_1 + A_2 h_2}{C_2}\Big) + \dfrac{A_1 A_2 h_1 h_2}{C_1 C_2} \ .$$

Because all coefficients are positive, similar to the case of the electrical system discussed in Application 3, we see that the system is asymptotically stable.

10. For the *nuclear reactor* model the characteristic polynomial of the coefficient matrix has the following form:

$$\varphi_m(\lambda) = \det \begin{pmatrix} \dfrac{\rho-\beta}{l} - \lambda & \lambda_1 & \lambda_2 & \lambda_3 & \cdots & \lambda_{m-1} & \lambda_m \\ \dfrac{\beta_1}{l} & -\lambda_1 - \lambda & 0 & 0 & \cdots & 0 & 0 \\ \vdots & \vdots & \vdots & \vdots & \ddots & \vdots & \vdots \\ \dfrac{\beta_m}{l} & 0 & 0 & 0 & \cdots & 0 & -\lambda_m - \lambda \end{pmatrix} \ .$$

4.3   Applications

We can expand this determinant with respect to its last column to obtain the recursion

$$\varphi_m(\lambda) = (-\lambda-\lambda_m)\varphi_{m-1}(\lambda)+(-1)^m\frac{\beta_m\lambda_m}{l}(\lambda+\lambda_1)(\lambda+\lambda_2)\cdots(\lambda+\lambda_{m-1})\cdot$$

If $m = 0$, then

$$\varphi_0(\lambda) = \frac{\rho - \beta}{l} - \lambda\,,$$

and if $m = 1$, then $\beta_1 = \beta$ implies that

$$\varphi_1(\lambda) = \left(\frac{\rho - \beta}{l} - \lambda\right)(-\lambda_1 - \lambda) - \frac{\beta_1\lambda_1}{l}$$

$$= \lambda^2 - \lambda\left(\frac{\rho - \beta}{l} - \lambda_1\right) + \lambda_1\left(\frac{-\rho + \beta}{l} - \frac{\beta_1}{l}\right)$$

$$= \lambda^2 - \lambda\left(\frac{\rho - \beta}{l} - \lambda_1\right) - \frac{\lambda_1\rho}{l}\,.$$

If ρ is positive, then the system is unstable and the reactor is called *supercritical*. If ρ is zero, then the two eigenvalues are

$$\lambda_1 = 0 \qquad \text{and} \qquad \lambda_2 = \frac{-\beta - l\lambda_1}{l} < 0\,.$$

Hence the system is stable, but not asymptotically stable. In this case the reactor is called *critical*. If $\rho < 0$, then the constant and the coefficient of λ are both positive. Hence the system is asymptotically stable, and the reactor is called *subcritical*.

In the more general case, if $m > 1$ a similar but more complicated derivation is needed.

4.3.2   Applications in the Social Sciences

1. Consider first the nonlinear *predator–prey* model (3.111) and (3.112), which is now repeated for convenience:

$$\dot{G}(t) = aG(t) - bG(t)W(t)$$

$$\dot{W}(t) = -cW(t) + dG(t)W(t)\,.$$

Before examining the stability of the system we introduce the new normalized variables

$$g(t) = \frac{d}{c}G(t) \quad \text{and} \quad w(t) = \frac{b}{a}W(t) ,$$

then the system reduces to

$$\dot{g}(t) = ag(t)(1 - w(t))$$

$$\dot{w}(t) = -cw(t)(1 - g(t)) \tag{4.29}$$

with the nonzero equilibrium $\bar{g} = \bar{w} = 1$.

A Lyapunov function will be now constructed that will guarantee the stability of this equilibrium.

Divide Equation (4.29) to get

$$\frac{\dot{w}}{\dot{g}} = \frac{-cw(1 - g)}{ag(1 - w)} ,$$

which implies that

$$c\dot{g} - c\frac{\dot{g}}{g} + a\dot{w} - a\frac{\dot{w}}{w} = 0 .$$

By integrating each term we obtain

$$cg - c\ln g + aw - a\ln w = C , \tag{4.30}$$

where C is a constant.

Define next the Lyapunov function

$$V(g, w) = cg - c\ln g + aw - a\ln w \tag{4.31}$$

for $g, w > 0$. It is continuous, and for every trajectory it is constant, so it is (not strictly) decreasing. The unique minimum of V is at the equilibrium (1,1), which can be proven as follows.

Note first that equations

$$\frac{\partial V}{\partial g} = c - c \cdot \frac{1}{g} = 0$$

$$\frac{\partial V}{\partial w} = a - a \cdot \frac{1}{w} = 0$$

have the unique solution (1,1); furthermore, the Hessian of function V is $diag(c/g^2, a/w^2)$, which is always positive definite. Hence function V is strictly convex and therefore, (1,1) is the only global minimizer of V. Consequently, V satisfies the conditions of Theorem 4.1; therefore, the positive equilibrium is stable. We will next show that the stability is not asymptotic. For fixed values of C, the points (g, w) satisfying Equation (4.30) form a closed curve shown in Figure 4.4. Hence $g(t)$ and $w(t)$ do not converge to the equilibrium (1,1). A computer study of the predator–prey model is reported in [44], where the closed curve solutions are determined by using numerical methods.

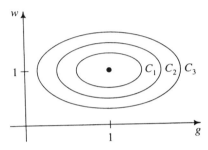

Figure 4.4 Closed curve solutions of the predator–prey model.

The linearized predator–prey model (3.114) has the coefficient matrix

$$\begin{pmatrix} 0 & -\frac{bc}{d} \\ \frac{ad}{b} & 0 \end{pmatrix}$$

with characteristic polynomial

$$\varphi(\lambda) = \lambda^2 + \frac{bc}{d} \cdot \frac{ad}{b} = \lambda^2 + ac .$$

Therefore, the eigenvalues are $j\sqrt{ac}$ and $-j\sqrt{ac}$, which satisfy the conditions of Part (i) of Theorem 4.5. Hence the equilibrium is stable and the stability is not asymptotic. Note that the same result is obtained as in the case of the original nonlinear system.

2. In the case of the *cohort population* model, the coefficient matrix

of the governing difference equation (3.115) is

$$
\mathbf{A} =
\begin{pmatrix}
b_1 & b_2 & b_3 & \cdots & b_{n-1} & b_n \\
a_1 & & & & & \\
& a_2 & & & & \\
& & & \vdots & & \\
& & & & \ddots & \\
& & & & a_{n-1} & 0
\end{pmatrix},
\tag{4.32}
$$

where $b_i \geq 0$ $(i = 1, \ldots, n)$ and $a_i \geq 0$ $(i = 1, \ldots, n - 1)$. The system is asymptotically stable if and only if all eigenvalues of $\mathbf{A}$ are inside the unit circle. An easy sufficient condition using only the matrix elements can be formulated in the following way. The system is asymptotically stable if for some matrix norm, $\|\mathbf{A}\| < 1$, because it implies that all eigenvalues of $\mathbf{A}$ are inside the unit circle (see Theorem 1.8). By selecting $\|\cdot\|_\infty$, $\|\cdot\|_1$, and $\|\cdot\|_2$, the following sufficient conditions are obtained for the asymptotical stability of the system:

(i) $b_1 + b_2 + \cdots + b_n < 1$, $a_i < 1$ $(1 \leq i \leq n - 1)$;

(ii) $b_i + a_i < 1$ $(i = 1, \ldots, n - 1)$, $b_n < 1$;

(iii) $\sum_{i=1}^{n} b_i^2 + \sum_{i=1}^{n-1} a_i^2 < 1$.

Hence, each of these conditions implies the asymptotical stability of the system.

3. In the case of the *arms-races* model (Equations (3.116) and (3.117)), the coefficient matrix has the form

$$
\mathbf{A} = \begin{pmatrix} -b & a \\ c & -d \end{pmatrix} \qquad (a, b, c, d > 0)
\tag{4.33}
$$

with characteristic polynomial

$$
\varphi(\lambda) = (-b - \lambda)(-d - \lambda) - ac = \lambda^2 + \lambda(b + d) + (-ac + bd) .
$$

If λ_1 and λ_2 are the eigenvalues, then

$$
\lambda_1 + \lambda_2 = -(b + d)
$$

$$
\lambda_1 \lambda_2 = -ac + bd ,
$$

which imply that both eigenvalues have negative real parts if and only if $b + d$ and $-ac + bd$ are both positive. Hence the system is asymptotically stable if and only if $ac < bd$. The condition has sense, since a and c show how armament levels increase, and b and d show how they decrease.

4. The *warfare model* (3.121) is based on matrix

$$\mathbf{A} = \begin{pmatrix} 0 & -h_2 \\ -h_1 & 0 \end{pmatrix} , \tag{4.34}$$

where h_1 and h_2 are positive constants. The characteristic polynomial of this matrix is

$$\varphi(\lambda) = \lambda^2 - h_1 h_2 ,$$

and the eigenvalues are $\sqrt{h_1 h_2}$ and $-\sqrt{h_1 h_2}$. Because we have a positive eigenvalue, the system is unstable.

5. In Section 3.5.2 we saw that the nonlinear *epidemics* model (3.124) is a special case of the predator–prey model (3.111) and (3.112) by selecting the special parameter values $a = 0$ and $b = d$. Therefore, the stability of nonlinear epidemics can be discussed in an analogous manner. It is easy to see that system (3.124) has infinitely many equilibrium points $(\bar{x}, 0)$, where $\bar{x} \geq 0$ is arbitrary. Simple calculation shows that the linearized model has the following form:

$$\dot{\mathbf{x}}_\delta = \begin{pmatrix} 0 & -\alpha\bar{x} \\ 0 & \alpha\bar{x} - \beta \end{pmatrix} \mathbf{x}_\delta ,$$

where we used the notation of Equation (3.13). The eigenvalues of the coefficient matrix are $\lambda_1 = 0$ and $\lambda_2 = \alpha\bar{x} - \beta$. Therefore, the system is unstable for $\bar{x} \geq \beta/\alpha$ and stable for $\bar{x} < \beta/\alpha$. The stability is not asymptotical. The case $\bar{x} \geq \beta/\alpha$ represents an expanding, very dangerous epidemic.

6. The *Harrod-type national economy* model (3.10) has the form

$$Y(t + 1) = (1 + r - rm)Y(t) - rG(t) ,$$

where $Y(t)$ is the national income and $G(t)$ is the government expenditure. Assume that $G(t)$ has the form

$$G(t) = \alpha Y(t) + \beta ,$$

that is, it is a linear function of the national income. In this case the model simplifies to the time-invariant difference equation:

$$Y(t + 1) = (1 + r - rm - r\alpha)Y(t) - r\beta .$$

The only equilibrium of this modified system is the solution $\bar{Y}$ of equation

$$\bar{Y} = (1 + r - rm - r\alpha)\bar{Y} - r\beta ,$$

which is

$$\bar{Y} = \frac{r\beta}{r - rm - r\alpha} = \frac{\beta}{1 - m - \alpha} \ .$$

Note that $\bar{Y}$ does not depend on the growth factor r, and it is asymptotically stable if and only if

$$-1 < 1 + r - rm - r\alpha < 1 \ .$$

These conditions are equivalent to relations

$$\alpha > 1 - m$$

and

$$0 < r < \frac{2}{\alpha + m - 1} \ ,$$

which can be interpreted as a relatively large share of the government in the economy, which has a small growth factor.

7. In Application 3.5.2 we saw that the nonlinear *cobweb* model (3.125) has the form

$$p(t + 1) = f(p(t)) \ ,$$

where

$$f(p) = d^{-1}(s(p)) \ .$$

It was also verified that this system has an asymptotically stable unique equilibrium if

(a) $f(p)$ exists for all $p \geq 0$;

(b) $f(p) \geq 0$ for all $p \geq 0$; and

(c) f' exists and $|f'(p)| \leq K$ for all $p \geq 0$, where $K \in [0, 1)$ is a fixed constant.

8. The linear continuous model of *interrelated market* dynamics is governed by differential equation (3.130) with the coefficient matrix

$$\mathbf{K} \cdot (\mathbf{A} - \mathbf{B}) \ , \tag{4.35}$$

where $\mathbf{K} = diag(k_1, \ldots, k_n)$ $(k_i > 0, \ i = 1, 2, \ldots, n)$ and $\mathbf{A} = (a_{ij})$ and $\mathbf{B} = (b_{ij})$ are $n \times n$ constant matrices. From assumptions (3.129) we know that

$$a_{ii} - b_{ii} < 0 \qquad \text{and} \qquad a_{ij} - b_{ij} \geq 0 \qquad (j \neq i) \ .$$

Theorem 4.9 implies that all eigenvalues of matrix (4.35) have negative real parts if $(\mathbf{A} - \mathbf{B}) + (\mathbf{A} - \mathbf{B})^T$ is negative definite, since $\mathbf{K}$ is positive definite. Observe that the diagonal and off-diagonal elements of this matrix are

$$2(a_{ii} - b_{ii}) \qquad \text{and} \qquad (a_{ij} + a_{ji}) - (b_{ij} + b_{ji}) \,,$$

respectively. Then Theorem 1.9 implies that all eigenvalues of this matrix lie in the domain

$$D = B_1 \cup B_2 \cup \cdots \cup B_n$$

where

$$B_i = \left\{ \lambda \;\middle|\; |\lambda - 2(a_{ii} - b_{ii})| \leq \sum_{j \neq i} |(a_{ij} + a_{ji}) - (b_{ij} + b_{ji})| \right\} \,.$$

Therefore, if for all i,

$$- 2(a_{ii} - b_{ii}) > \sum_{j \neq i} [(a_{ij} + a_{ji}) - (b_{ij} + b_{ji})] \,, \qquad (4.36)$$

then all eigenvalues are in the left half-plane. Hence, we proved that condition (4.36) is sufficient for the asymptotical stability of the system. In the economic theory, condition (4.36) is summarized by saying that matrix $(\mathbf{A} - \mathbf{B}) + (\mathbf{A} - \mathbf{B})^T$ is *strictly negatively diagonally dominant*.

9. Consider next the *simple discrete oligopoly problem* (3.134). Note that the matrix $\mathbf{A}_c$ of coefficients of the governing difference equation has the special form

$$\mathbf{A}_c = -\frac{1}{2}\mathbf{1} + \frac{1}{2}\mathbf{I} \,, \qquad (4.37)$$

where $\mathbf{1} = (1)$ and $\mathbf{I}$ is the identity matrix.

We first show that the eigenvalues of $\mathbf{1}$ are 0 and N. The eigenvalue equation of matrix $\mathbf{1}$ can be written as

$$v_1 + v_2 + \cdots + v_N = \lambda v_k \qquad (k = 1, \ldots, N) \,.$$

If $\lambda = 0$, then any vector $\mathbf{v} = (v_k)$ satisfying the relation $v_1 + \cdots + v_N = 0$ is an associated eigenvector. If $\lambda \neq 0$, then $v_1 = v_2 = \cdots = v_N = v^*$, and therefore,

$$Nv^* = \lambda v^* \,.$$

Hence, the nonzero eigenvalue is N.

Consequently, the eigenvalues of $\mathbf{A}_c$ are $-1/2 \cdot 0 + 1/2 \cdot 1 = 1/2$ and $-1/2 \cdot N + 1/2 \cdot 1 = (1-N)/2$, which are inside the unit circle if and only if $N = 2$. Hence, this model is asymptotically stable if and only if it is a duopoly.

The modified discrete model (3.138) with *adaptive expectations* is based on the coefficient matrix

$$\mathbf{A}_a = \begin{pmatrix} -\frac{m}{2}\mathbf{1} + \frac{m}{2}\mathbf{I} & -\frac{1-m}{2}\mathbf{I} \\ m\mathbf{1} - m\mathbf{I} & (1-m)\mathbf{I} \end{pmatrix} , \tag{4.38}$$

where for the sake of simplicity we assume that $m_1 = m_2 = \cdots = m_N = m$. The eigenvalue equation of $\mathbf{A}_a$ can be rewritten as

$$\left(-\frac{m}{2}\mathbf{1} + \frac{m}{2}\mathbf{I} \right)\mathbf{u} + \left(-\frac{1-m}{2}\mathbf{I} \right)\mathbf{v} = \lambda\mathbf{u}$$

$$(m\mathbf{1} - m\mathbf{I})\mathbf{u} + (1-m)\mathbf{I}\mathbf{v} = \lambda\mathbf{v} . \tag{4.39}$$

Add the 1/2-multiple of the second equation to the first equation to see that

$$\lambda\left(\mathbf{u} + \frac{1}{2}\mathbf{v} \right) = \mathbf{0} .$$

That is, either $\lambda = 0$ or $\mathbf{u} = -(1/2)\mathbf{v}$. The eigenvalue $\lambda = 0$ is inside the unit circle. If $\lambda \neq 0$, then substitute $\mathbf{u} = -(1/2)\mathbf{v}$ into the second equation of (4.39):

$$-\frac{m}{2}(\mathbf{1} - \mathbf{I})\mathbf{v} + (1-m)\mathbf{I}\mathbf{v} = \lambda\mathbf{v} ,$$

which is equivalent to the eigenvalue equation of matrix

$$-\frac{m}{2}\mathbf{1} + \left(1 - \frac{m}{2} \right)\mathbf{I} . \tag{4.40}$$

The eigenvalues of this matrix are

$$-\frac{m}{2}\cdot 0 + \left(1 - \frac{m}{2} \right)\cdot 1 = 1 - \frac{m}{2} \quad \text{and} \quad -\frac{m}{2}\cdot N + \left(1 - \frac{m}{2} \right)\cdot 1 = 1 - \frac{m(N+1)}{2} ,$$

which are inside the unit circle if and only if

$$0 < m < \frac{4}{N+1} .$$

Therefore, this system is asymptotically stable for all $N > 0$ if m is sufficiently small.

The optimal value m_{opt} of m can be determined by minimizing the largest eigenvalue of matrix (4.40), which ensures the fastest order of magnitude in the speed of the convergence of the solution to the equilibrium as $t \to \infty$. This optimization problem is formulated as follows:

$$\text{minimize}_m \quad \max\left\{\left|1 - \frac{m}{2}\right|; \left|1 - \frac{m(N+1)}{2}\right|\right\} .$$

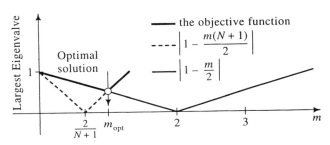

Figure 4.5   Finding the optimal value m_{opt}.

The objective function is shown in Figure 4.5, where the value of m_{opt} is found by solving equation

$$-\left(1 - \frac{m(N+1)}{2}\right) = 1 - \frac{m}{2} ,$$

which implies that

$$m_{\text{opt}} = \frac{4}{N+2} .$$

The *continuous oligopoly model* (3.139) is based on the coefficient matrix $\mathbf{M} \cdot \mathbf{A}$, where $\mathbf{M} = diag(m_1, \ldots, m_N)$ is positive definite, since $m_i > 0$ for all i; furthermore,

$$\mathbf{A} = \begin{pmatrix} 2a & a & \cdots & a \\ a & 2a & \cdots & a \\ \multicolumn{4}{c}{\dotfill} \\ a & a & \cdots & 2a \end{pmatrix} .$$

From Theorem 4.9 we know that this system is asymptotically stable, if

$\mathbf{A} + \mathbf{A}^T$ is negative definite. Note that

$$\mathbf{A} + \mathbf{A}^T = \begin{pmatrix} 4a & 2a & \cdots & 2a \\ 2a & 4a & \cdots & 2a \\ \cdots\cdots\cdots\cdots \\ 2a & 2a & \cdots & 4a \end{pmatrix} = (2\mathbf{1} + 2\mathbf{I})a$$

with eigenvalues

$$(2 \cdot 0 + 2 \cdot 1)a = 2a \qquad \text{and} \qquad (2 \cdot N + 2 \cdot 1)a = (2N + 2)a \ ,$$

which are always negative since we have assumed that $a < 0$. Hence, this system is always asymptotically stable.

Problems

1. Discuss the stability of the system

$$\dot{\mathbf{x}} = \begin{pmatrix} \frac{1}{t} & 0 \\ 0 & \frac{1}{t} \end{pmatrix} \mathbf{x} + \begin{pmatrix} 1 \\ 1 \end{pmatrix} u \ ,$$

where t is the time.

2. Is system

$$\dot{\mathbf{x}} = \begin{pmatrix} 1 & 1 \\ 2 & 2 \end{pmatrix} \mathbf{x} + \begin{pmatrix} 1 \\ 0 \end{pmatrix} u$$

stable? Is it asymptotically stable?

3. Is the following system stable? Is it asymptotically stable?

$$\dot{\mathbf{x}} = \begin{pmatrix} 2 & 1 \\ 0 & 2 \end{pmatrix} \mathbf{x} + \begin{pmatrix} 1 \\ 1 \end{pmatrix} u \ .$$

4. Discuss the stability of the mechanical system

$$\dot{\mathbf{x}} = \begin{pmatrix} 0 & 1 \\ 0 & -6 \end{pmatrix} \mathbf{x} + \begin{pmatrix} 0 \\ 2 \end{pmatrix} u$$

$$y = (1, 0)\mathbf{x}$$

introduced in Problem 3.7.

5.   Examine the stability of the discrete system

$$\mathbf{x}(t+1) = \begin{pmatrix} 1 & 1 \\ 2 & 2 \end{pmatrix} \mathbf{x}(t) + \begin{pmatrix} 1 \\ 0 \end{pmatrix} u(t) \, .$$

6.   Discuss the stability of this discrete system

$$\mathbf{x}(t+1) = \begin{pmatrix} 2 & 1 \\ 0 & 2 \end{pmatrix} \mathbf{x}(t) + \begin{pmatrix} 1 \\ 1 \end{pmatrix} u(t) \, .$$

7.   Is the electric circuit system

$$L\frac{di(t)}{dt} + (R_1 + R_2)i(t) = u(t)$$

introduced in Problem 3.13 asymptotically stable?

8.   Examine the stability of the system

$$x^{(3)} + 2\ddot{x} + 3\dot{x} + 5x = u \, .$$

9.   Examine the stability of this system

$$x^{(5)} + 3x^{(4)} + x^{(3)} + 2\ddot{x} + 3\dot{x} + 5x = u \, .$$

10.   Find the values of α so that system

$$\dot{\mathbf{x}} = \begin{pmatrix} \alpha & 1 \\ 1 & \alpha \end{pmatrix} \mathbf{x} + \begin{pmatrix} 1 \\ 1 \end{pmatrix} u$$

is stable, or asymptotically stable.

11.   Examine the stability of the system

$$\dot{\mathbf{x}} = \begin{pmatrix} -\sigma & \omega \\ -\omega & -\sigma \end{pmatrix} \mathbf{x} + \begin{pmatrix} 0 \\ 1 \end{pmatrix} u \, ,$$

which is the generalization of Example (4.1) (with $\sigma \neq 0$).

12.   What is the condition that function $V(\mathbf{x}) = \|\mathbf{x} - \bar{\mathbf{x}}\|_2^2$ is a Lyapunov function for system

$$\dot{\mathbf{x}} = \mathbf{f}(\mathbf{x}) \, ,$$

where $\mathbf{f} : \Omega \to \Omega$ is a continuously differentiable function and $\bar{\mathbf{x}}$ is the only equilibrium?

13.   Assume that $\mathbf{f} : \mathbf{R}^n \rightarrow \mathbf{R}^n$ is continuously differentiable, $\mathbf{f(0)} = 0$, and $\|\mathbf{f'(x)}\|_2 < 1$ for all $\mathbf{x} \in \mathbf{R}^n$. Prove that with arbitrary $\mathbf{x}(0) = \mathbf{x_0} \in \mathbf{R}^n$, the sequence $\mathbf{x}(t+1) = \mathbf{f(x}(t))$ converges to zero.

14.   Let
$$\mathbf{A} = \begin{pmatrix} -2 & -1 \\ 1 & -4 \end{pmatrix} .$$

(i)   Show that the eigenvalues of $\mathbf{A}$ have negative real parts.

(ii) Select $\mathbf{M} = \mathbf{I}$ in Theorem 4.6. Find matrix $\mathbf{Q}$, which satisfies Equation (4.12).

15.   Find the roots of the polynomial

$$\varphi(\lambda) = \lambda^4 + 2\lambda^3 + 3\lambda^2 + 4\lambda + 5$$

and show that a system with this characteristic polynomial is unstable.

16.   Illustrate Theorem 4.9 with matrices

$$\mathbf{K} = \begin{pmatrix} 4 & 1 \\ 1 & 4 \end{pmatrix} \quad \text{and} \quad \mathbf{A} = \begin{pmatrix} -4 & 1 \\ 2 & -6 \end{pmatrix} .$$

17.   Given
$$H(s) = \frac{s+1}{s^4 + 2s^3 + 3s^2 + 4s + 5} ,$$

is the system stable?

18.   Is system

$$\dot{\mathbf{x}} = \begin{pmatrix} -14 & -2 \\ 4 & -23 \end{pmatrix} \mathbf{x} + \begin{pmatrix} 1 \\ 1 \end{pmatrix} u$$

$$y = (1,0)\mathbf{x}$$

BIBO stable?

19.   Is the following system BIBO stable?

$$\dot{\mathbf{x}} = \begin{pmatrix} -4 & 1 \\ 2 & -6 \end{pmatrix} \mathbf{x} + \begin{pmatrix} 2 & 2 \\ 1 & 1 \end{pmatrix} u .$$

$$y = (1,1)\mathbf{x}$$

20.

(i)   Prove Theorem 4.1 for the continuous case.

(ii) Prove Theorem 4.2 for the continuous case.

(iii) Prove Theorem 4.3 for the continuous case.

(iv) Prove Theorem 4.12.

21.   Assume that all eigenvalues of an $n \times n$ real matrix $\mathbf{A}$ have negative real parts. Show that $\mathbf{A}^{-1}$ exists and

$$\mathbf{A}^{-1} = \int_{\infty}^{0} e^{\mathbf{A}t} \, dt \ .$$

22.   Prove the following generalization of Theorem 4.6. All eigenvalues of matrix $\mathbf{A}$ have real parts less than $-\alpha < 0$ if and only if for every symmetric, positive definite matrix $\mathbf{M}$ there exists a positive definite solution $\mathbf{Q}$ of equation

$$\mathbf{A}^T \mathbf{Q} + \mathbf{Q}\mathbf{A} + 2\alpha \mathbf{Q} = -\mathbf{M} \ .$$

23.   Show that if $\mathbf{A}$ is a real $n \times n$ matrix, then the continuous system $\dot{\mathbf{x}}(t) = \mathbf{A}\mathbf{x}(t)$ is asymptotically stable if and only if the discrete system $\mathbf{x}(t+1) = e^{\mathbf{A}}\mathbf{x}(t)$ is asymptotically stable.

24.   Interpret the conditions of Definition 4.2 and Theorem 4.2 for the continuous system

$$\dot{\mathbf{x}}(t) = \mathbf{f}(\mathbf{x}(t))$$

with the Lyapunov function $V(\mathbf{x}) = (\mathbf{x} - \bar{\mathbf{x}})^T \mathbf{G}(\mathbf{x} - \bar{\mathbf{x}})$, where $\mathbf{f}$ is continuously differentiable, and $\mathbf{G}$ is a real, constant, symmetric, positive definite matrix.

25.   Repeat the previous problem with Lyapunov function

$$V(\mathbf{x}) = (\mathbf{x} - \bar{\mathbf{x}})^T \mathbf{G}(\mathbf{x})(\mathbf{x} - \bar{\mathbf{x}})$$

where $\mathbf{G}(\mathbf{x})$ is a real, continuously differentiable, symmetric, positive definite matrix for all $\mathbf{x}$.

chapter five

Controllability

In previous chapters we have been concerned with the analysis of linear and nonlinear dynamics. We have developed closed formulas for predicting future states and outputs, and in addition, stability problems have been discussed. On the other hand, in control theory the control problem is to find an input that causes the state or the output to behave in a desired way. As an example, consider again the satellite problem discussed earlier in Example 3.6. Let $\mathbf{x}_1$ be a desired future state of the satellite, that is, its desired positions, radial and angular velocities. Find an input function that will drive the state to $\mathbf{x}_1$ in a finite time. A more restrictive problem is when an entire trajectory $\mathbf{x}(t)$ is given, and we wish to find an input function such that the entire state trajectory coincides with $\mathbf{x}(t)$.

In this chapter conditions will be introduced for the existence of state and output control and in addition, an input will be found that performs the desired control.

DEFINITION 5.1   *A dynamic system with initial condition $\mathbf{x}(t_0) = \mathbf{x}_0$ is said to be controllable to state $\mathbf{x}_1$ at t_1 ($> t_0$) if there exists an input $\mathbf{u}(t)$ such that $\mathbf{x}(t_1) = \mathbf{x}_1$. This concept is illustrated in Figure 5.1.*

5.1   Continuous Systems

In this section the controllability of the continuous dynamic system

$$\dot{\mathbf{x}} = \mathbf{A}(t)\mathbf{x} + \mathbf{B}(t)\mathbf{u}, \qquad \mathbf{x}(t_0) = \mathbf{x}_0 \tag{5.1}$$

$$\mathbf{y} = \mathbf{C}(t)\mathbf{x} \tag{5.2}$$

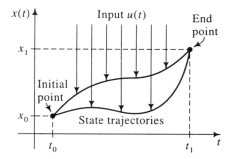

Figure 5.1   Concept of controllability.

will be analyzed. We assume that $\mathbf{A}(t)$, $\mathbf{B}(t)$, and $\mathbf{C}(t)$ are $n \times n$, $n \times m$, and $p \times n$, respectively, and they are continuous in $[t_0, \infty)$. Our discussion will start with the general case, then special results on time-invariant systems will be demonstrated.

5.1.1   General Conditions

The general solution (3.19) of continuous linear systems implies that the system is controllable to $\mathbf{x}_1$ at t_1 if and only if there exists an input $\mathbf{u}(t)$ such that

$$\mathbf{x}_1 = \phi(t_1, t_0)\mathbf{x}_0 + \int_{t_0}^{t_1} \phi(t_1, \tau)\mathbf{B}(\tau)\mathbf{u}(\tau) \, d\tau \ .$$

Since $\phi(t_0, t_1)$ is nonsingular, this relation is equivalent to equation

$$\phi(t_0, t_1)\mathbf{x}_1 - \mathbf{x}_0 = \int_{t_0}^{t_1} \phi(t_0, \tau)\mathbf{B}(\tau)\mathbf{u}(\tau) \, d\tau \ , \tag{5.3}$$

where we used Properties (i), (ii), and (iii) of Theorem 2.3.

Introduce the mapping

$$A(\mathbf{u}) = \int_{t_0}^{t_1} \phi(t_0, \tau)\mathbf{B}(\tau)\mathbf{u}(\tau) \, d\tau$$

on the set of the m-dimensional continuous functions. This mapping must not be confused with the system matrix $\mathbf{A}(t)$. Note that the range of this mapping is in $\mathbf{R}^n$. It is obvious that there exists an input $\mathbf{u}(t)$ for all $\mathbf{x}_1 \in \mathbf{R}^n$ which leads the state to $\mathbf{x}_1$ at t_1 if and only if the range $R(A)$ is the entire $\mathbf{R}^n$.

LEMMA 5.1

Vector $\mathbf{v}$ is in $R(A)$ if and only if it belongs to the range space of matrix

$$\mathbf{W}(t_0, t_1) = \int_{t_0}^{t_1} \phi(t_0, \tau)\mathbf{B}(\tau)\mathbf{B}^T(\tau)\phi^T(t_0, \tau)\, d\tau \, . \qquad (5.4)$$

PROOF

(a) Assume first that $\mathbf{v} \in R(\mathbf{W}(t_0, t_1))$, then there exists a vector $\mathbf{a}$ such that

$$\mathbf{v} = \mathbf{W}(t_0, t_1)\mathbf{a} \, . \qquad (5.5)$$

Select input

$$\mathbf{u}(t) = \mathbf{B}^T(t)\phi^T(t_0, t)\mathbf{a} \, , \qquad (5.6)$$

then

$$A(\mathbf{u}) = \int_{t_0}^{t_1} \phi(t_0, \tau)\mathbf{B}(\tau)\mathbf{B}^T(\tau)\phi^T(t_0, \tau)\mathbf{a}\, d\tau = \mathbf{W}(t_0, t_1)\mathbf{a} = \mathbf{v} \, ,$$

therefore, $\mathbf{v} \in R(A)$.

(b) Assume next that $\mathbf{v} \notin R\ (\mathbf{W}(t_0, t_1))$, then there exists a vector $\mathbf{w}$ from the null-space of $\mathbf{W}(t_0, t_1)$ which is not orthogonal to $\mathbf{v}$. This fact is the consequence of the well-known property of $n \times n$ symmetric matrices that their range and null spaces are orthogonal complementary subspaces in $\mathbf{R}^n$ (see Theorem 1.10). That is,

$$\mathbf{W}(t_0, t_1)\mathbf{w} = \mathbf{0} \qquad \text{and} \qquad \mathbf{w}^T\mathbf{v} \neq 0 \, .$$

We shall now verify that $\mathbf{v} \notin R(A)$. Contrary to this assertion assume that $\mathbf{v} \in R(A)$. Then with some function $\tilde{\mathbf{u}}(t)$,

$$\mathbf{v} = \int_{t_0}^{t_1} \phi(t_0, \tau)\mathbf{B}(\tau)\tilde{\mathbf{u}}(\tau)\, d\tau \, .$$

Therefore,

$$0 \neq \mathbf{w}^T\mathbf{v} = \int_{t_0}^{t_1} \mathbf{w}^T\phi(t_0, \tau)\mathbf{B}(\tau)\tilde{\mathbf{u}}(\tau)\, d\tau \, . \qquad (5.7)$$

The definition of $\mathbf{W}$ implies that

$$0 = \mathbf{w}^T\mathbf{W}(t_0, t_1)\mathbf{w} = \int_{t_0}^{t_1} \mathbf{w}^T\phi(t_0, \tau)\mathbf{B}(\tau)\mathbf{B}^T(\tau)\phi^T(t_0, \tau)\mathbf{w}\, d\tau$$

$$= \int_{t_0}^{t_1} \|\mathbf{B}^T(\tau)\phi^T(t_0,\tau)\mathbf{w}\|_2^2 \, d\tau \; ,$$

where $\| \cdot \|_2$ is the $p = 2$ norm of real vectors introduced in Definition 1.8. Since the integrand is continuous and nonnegative, it has to be identically zero. Consequently, for all $\tau \in [t_0, t_1]$,

$$\mathbf{B}(\tau)^T \phi^T(t_0, \tau)\mathbf{w} = \mathbf{0} \; ,$$

and by taking the transpose of both sides,

$$\mathbf{w}^T \phi(t_0, \tau)\mathbf{B}(\tau) = \mathbf{0}^T \qquad (\text{all } \tau \in [t_0, t_1])$$

which contradicts relation (5.7). Thus, the proof is completed.

∎

REMARK 5.1   There exists an input $\mathbf{u}(t)$ that drives the state of the continuous linear system from $\mathbf{x}_0$ to $\mathbf{x}_1$ at time $t_1 > t_0$ if and only if $\phi(t_0, t_1)\mathbf{x}_1 - \mathbf{x}_0 \in R(\mathbf{W}(t_0, t_1))$. This condition is equivalent to the existence of a vector a such that

$$\mathbf{W}(t_0, t_1)\mathbf{a} = \phi(t_0, t_1)\mathbf{x}_1 - \mathbf{x}_0 \; . \qquad (5.8)$$

From part (a) of the proof of the theorem, we conclude that one particular input that drives the system to $\mathbf{x}_1$ at time t_1 is given as

$$\mathbf{u}(t) = \mathbf{B}^T(t)\phi^T(t_0, t)\mathbf{a} \; . \qquad (5.9)$$

In using this equation we have to compute first matrix $\mathbf{W}(t_0, t_1)$. In most cases, numerical integration is needed. Then we solve linear equations (5.8) for a, for example, by Gauss elimination (see [42]). The method shows whether a solution exists or not. If no solution exists, then the system is not controllable to $\mathbf{x}_1$ at t_1. If there is at least one solution a, then an appropriate input can be obtained by the above formula. We note that computer programs are available to perform Gauss elimination.

In summary, an algorithm that verifies whether the system can be controlled to $\mathbf{x}_1$ at time t_1 or not, and in the case of controllability gives an appropriate input, consists of the following steps:

Step 1   Compute the fundamental matrix $\phi(t, t_0)$.

Step 2   Determine matrix $\mathbf{W}(t_0, t_1)$ by using Equation (5.4).

Step 3 Find vector $\mathbf{d} = \phi\,(t_0, t_1)\mathbf{x}_1 - \mathbf{x}_0$.

Step 4 Use Gauss elimination to determine whether linear equation

$$\mathbf{W}(t_0, t_1)\mathbf{a} = \mathbf{d}$$

has a solution. If it does, then the system can be controlled to $\mathbf{x}_1$ at t_1, otherwise not.

Step 5 If $\mathbf{a}$ is a solution of the previous step, then determine function $\mathbf{u}(t)$ by formula (5.9).

∎

This algorithm is illustrated by the following example.

Example 5.1

Assume that the state of the system

$$\dot{\mathbf{x}} = \begin{pmatrix} 2 & 1 \\ 0 & 2 \end{pmatrix} \mathbf{x} + \begin{pmatrix} 0 \\ 1 \end{pmatrix} u, \qquad \mathbf{x}(0) = \begin{pmatrix} 0 \\ 0 \end{pmatrix}$$

has to be controlled to the final state

$$\mathbf{x}(1) = \begin{pmatrix} 1 \\ 1 \end{pmatrix}.$$

Step 1. Simple calculation shows that

$$\phi(t, \tau) = \begin{pmatrix} e^{2(t-\tau)} & (t - \tau)e^{2(t-\tau)} \\ 0 & e^{2(t-\tau)} \end{pmatrix}$$

(see Problem 1.17).

Step 2. Using Equation (5.4) we have

$$\mathbf{W}(0, 1)$$

$$= \int_0^1 \begin{pmatrix} e^{-2\tau} & -\tau e^{-2\tau} \\ 0 & e^{-2\tau} \end{pmatrix} \begin{pmatrix} 0 \\ 1 \end{pmatrix} (0, 1) \begin{pmatrix} e^{-2\tau} & 0 \\ -\tau e^{-2\tau} & e^{-2\tau} \end{pmatrix} d\tau$$

$$= \int_0^1 \begin{pmatrix} \tau^2 e^{-4\tau} & -\tau e^{-4\tau} \\ -\tau e^{-4\tau} & e^{-4\tau} \end{pmatrix} d\tau.$$

The matrix elements are calculated by elementary integration:

$$\int_0^1 e^{-4\tau}d\tau = [\frac{e^{-4\tau}}{-4}]_0^1$$

$$= \frac{1 - e^{-4}}{4},$$

$$\int_0^1 -\tau e^{-4\tau}d\tau = [-\tau\frac{e^{-4\tau}}{-4}]_0^1 - \int_0^1 (-1)\frac{e^{-4\tau}}{-4}d\tau$$

$$= \frac{e^{-4}}{4} - [\frac{e^{-4\tau}}{-16}]_0^1$$

$$= \frac{-1 + 5e^{-4}}{16},$$

and

$$\int_0^1 \tau^2 e^{-4\tau}d\tau = [\tau^2\frac{e^{-4\tau}}{-4}]_0^1 - \int_0^1 2\tau\frac{e^{-4\tau}}{-4}d\tau$$

$$= \frac{e^{-4}}{-4} + \frac{1}{2}\int_0^1 \tau e^{-4\tau}d\tau$$

$$= -\frac{e^{-4}}{4} + \frac{1}{2}\frac{1 - 5e^{-4}}{16}$$

$$= \frac{1 - 13e^{-4}}{32}.$$

So,

$$\mathbf{W}(0, 1) \approx \begin{pmatrix} 0.02381 & -0.05678 \\ -0.05678 & 0.24542 \end{pmatrix}.$$

Step 3.   Simple matrix-vector algebra shows that

$$\mathbf{d} = \begin{pmatrix} e^{-2} & -e^{-2} \\ 0 & e^{-2} \end{pmatrix}\begin{pmatrix} 1 \\ 1 \end{pmatrix} - \begin{pmatrix} 0 \\ 0 \end{pmatrix} = \begin{pmatrix} 0 \\ e^{-2} \end{pmatrix} \approx \begin{pmatrix} 0 \\ 0.13534 \end{pmatrix}.$$

Step4.   We have next to solve the linear equations

$$\begin{pmatrix} 0.02381 & -0.05678 \\ -0.05678 & 0.24542 \end{pmatrix}\begin{pmatrix} a_1 \\ a_2 \end{pmatrix} = \begin{pmatrix} 0 \\ 0.13534 \end{pmatrix},$$

which can be rewritten as

$$0.02381a_1 - 0.05678a_2 = 0$$

$$-0.05678a_1 + 0.24542a_2 = 0.13534 \,.$$

From the first equation we have

$$a_2 = \frac{0.02381}{0.05678} a_1 \approx 0.41934a_1 \,,$$

and by substituting this expression into the second equation we get a single equation for a_1:

$$(-0.05678 + 0.24542 \times 0.41934)a_1 = 0.13534 \,,$$

which implies that

$$a_1 \approx 2.9336 \qquad \text{and so} \qquad a_2 \approx 1.2302 \,.$$

Step 5. And finally, the input is the following:

$$u(t) \approx (0,1) \begin{pmatrix} e^{-2t} & 0 \\ -te^{-2t} & e^{-2t} \end{pmatrix} a = e^{-2t}(-2.9336t + 1.2302) \,.$$

Lemma 5.1 implies the following important theorem.

THEOREM 5.1
The continuous system is controllable from any initial state $\mathbf{x}(t_0) = \mathbf{x}_0$ to an arbitrary state $\mathbf{x}_1$ at time $t_1 > t_0$ if and only if matrix $\mathbf{W}(t_0, t_1)$ is nonsingular.

REMARK 5.2 If a continuous linear system is controllable from an arbitrary initial state $\mathbf{x}_0$ at any t_0 to any state $\mathbf{x}_1$ at arbitrary $t_1 > t_0$ then the system is called *completely controllable*. ∎

Example 5.2

Consider again the system modeled by the differential equation

$$\dot{\mathbf{x}} = \begin{pmatrix} 0 & \omega \\ -\omega & 0 \end{pmatrix} \mathbf{x} + \begin{pmatrix} 0 \\ 1 \end{pmatrix} u, \qquad \mathbf{x}(0) = \begin{pmatrix} 1 \\ 0 \end{pmatrix} \,.$$

In Example 2.6 we have seen that

$$\phi(t,\tau) = \begin{pmatrix} \cos\omega(t-\tau) & \sin\omega(t-\tau) \\ -\sin\omega(t-\tau) & \cos\omega(t-\tau) \end{pmatrix} \,.$$

Therefore,

$\mathbf{W}(0, t_1)$

$$= \int_0^{t_1} \begin{pmatrix} \cos\omega(-\tau) & \sin\omega(-\tau) \\ -\sin\omega(-\tau) & \cos\omega(-\tau) \end{pmatrix} \begin{pmatrix} 0 \\ 1 \end{pmatrix} (0, 1) \begin{pmatrix} \cos\omega(-\tau) & -\sin\omega(-\tau) \\ \sin\omega(-\tau) & \cos\omega(-\tau) \end{pmatrix} d\tau$$

$$= \int_0^{t_1} \begin{pmatrix} -\sin\omega\tau \\ \cos\omega\tau \end{pmatrix} (-\sin\omega\tau, \cos\omega\tau) \, d\tau$$

$$= \int_0^{t_1} \begin{pmatrix} \sin^2\omega\tau & -\sin\omega\tau\cos\omega\tau \\ -\sin\omega\tau\cos\omega\tau & \cos^2\omega\tau \end{pmatrix} d\tau$$

$$= \begin{pmatrix} \frac{t_1}{2} - \frac{\sin 2\omega t_1}{4\omega} & \frac{\cos 2\omega t_1 - 1}{4\omega} \\ \frac{\cos 2\omega t_1 - 1}{4\omega} & \frac{t_1}{2} + \frac{\sin 2\omega t_1}{4\omega} \end{pmatrix}.$$

We shall next prove that this matrix is nonsingular for all $t_1 > 0$, that is, the system is controllable to all desired $\mathbf{x}_1$ at all $t_1 > 0$. The determinant of $\mathbf{W}(0, t_1)$ can be written as

$$\frac{t_1^2}{4} - \frac{\sin^2 2\omega t_1}{16\omega^2} - \frac{\cos^2 2\omega t_1 - 2\cos 2\omega t_1 + 1}{16\omega^2},$$

which equals zero if and only if

$$4\omega^2 t_1^2 + 2\cos 2\omega t_1 - 2 = 0.$$

Introduce the new variable $\alpha = 2\omega t_1 > 0$, then this equation is equivalent to relation

$$\cos\alpha = 1 - \frac{\alpha^2}{2}.$$

Consider next function

$$\varphi(\alpha) = \cos\alpha - 1 + \frac{\alpha^2}{2},$$

then easy calculation shows that $\varphi(0) = 0$ and for all $\alpha > 0$,

$$\varphi'(\alpha) = -\sin\alpha + \alpha > 0.$$

Hence $\varphi(\alpha) > 0$ for all $\alpha > 0$, and therefore, the determinant of $\mathbf{W}(0, t_1)$ is nonzero for all $t_1 > 0$.

Matrix $\mathbf{W}(t_0, t_1)$ is usually called the *controllability Gramian*. Its properties are summarized next.

THEOREM 5.2
Matrix $\mathbf{W}(t_0, t_1)$ satisfies the following properties:

(i)   *It is symmetric.*

(ii) *It is positive semidefinite.*

(iii) $(\partial/\partial t)\mathbf{W}(t, t_1) = \mathbf{A}(t)\mathbf{W}(t, t_1) + \mathbf{W}(t, t_1)\mathbf{A}^T(t) - \mathbf{B}(t)\mathbf{B}^T(t)$,
$\mathbf{W}(t_1, t_1) = \mathbf{0}$.

(iv) $\mathbf{W}(t_0, t_1) = \mathbf{W}(t_0, t) + \boldsymbol{\phi}(t_0, t)\mathbf{W}(t, t_1)\boldsymbol{\phi}^T(t_0, t)$.

PROOF

(i)   Because the integrand in (5.4) is symmetric, $\mathbf{W}(t_0, t_1)$ is also symmetric.

(ii) Let $\mathbf{v}$ be any real vector, then

$$\mathbf{v}^T \mathbf{w}(t_0, t_1)\mathbf{v} = \int_{t_0}^{t_1} \mathbf{v}^T \boldsymbol{\phi}(t_0, \tau)\mathbf{B}(\tau)\mathbf{B}^T(\tau)\boldsymbol{\phi}^T(t_0, \tau)\mathbf{v}\, d\tau$$

$$= \int_{t_0}^{t_1} \|\mathbf{B}^T(\tau)\boldsymbol{\phi}^T(t_0, \tau)\mathbf{v}\|_2^2\, d\tau \geq 0 \ .$$

(iii) We shall use the well-known fact that for smooth functions,

$$\frac{d}{dt}\int_t^{t_1} \mathbf{f}(t, \tau)\, d\tau = -\mathbf{f}(t, t) + \int_t^{t_1} \frac{\partial \mathbf{f}}{\partial t}(t, \tau)\, d\tau\ ,$$

which can easily be proven by using the definition of derivatives. In our case,

$$\frac{\partial}{\partial t}\mathbf{W}(t, t_1) = -\boldsymbol{\phi}(t, t)\mathbf{B}(t)\mathbf{B}^T(t)\boldsymbol{\phi}^T(t, t)$$

$$+ \int_t^{t_1} \frac{\partial}{\partial t}\boldsymbol{\phi}(t, \tau)\mathbf{B}(\tau)\mathbf{B}^T(\tau)\boldsymbol{\phi}^T(t, \tau)\, d\tau$$

$$+ \int_t^{t_1} \boldsymbol{\phi}(t, \tau)\mathbf{B}(\tau)\mathbf{B}^T(\tau)\left(\frac{\partial}{\partial t}\boldsymbol{\phi}(t, \tau)\right)^T d\tau$$

$$= -\mathbf{B}(t)\mathbf{B}^T(t) + \int_t^{t_1} \mathbf{A}(t)\phi(t,\tau)\mathbf{B}(\tau)\mathbf{B}^T(\tau)\phi^T(t,\tau)\,d\tau$$

$$+ \int_t^{t_1} \phi(t,\tau)\mathbf{B}(\tau)\mathbf{B}^T(\tau)(\mathbf{A}(t)\phi(t,\tau))^T\,d\tau$$

$$= -\mathbf{B}(t)\mathbf{B}^T(t) + \mathbf{A}(t)\mathbf{W}(t,t_1) + \mathbf{W}(t,t_1)\mathbf{A}^T(t)\;;$$

furthermore, the continuity of the integrand of $\mathbf{W}(t,t_1)$ implies that $\mathbf{W}(t_1,t_1) = \mathbf{0}$.

(iv)

$$\mathbf{W}(t_0,t_1) = \int_{t_0}^{t} \phi(t_0,\tau)\mathbf{B}(\tau)\mathbf{B}^T(\tau)\phi^T(t_0,\tau)\,d\tau$$

$$+ \int_t^{t_1} \phi(t_0,\tau)\mathbf{B}(\tau)\mathbf{B}^T(\tau)\phi^T(t_0,\tau)\,d\tau$$

$$= \mathbf{W}(t_0,t) + \int_t^{t_1} \phi(t_0,t)\phi(t,\tau)\mathbf{B}(\tau)\mathbf{B}^T(\tau)\phi^T(t,\tau)\phi^T(t_0,t)\,d\tau$$

$$= \mathbf{W}(t_0,t) + \phi(t_0,t)\mathbf{W}(t,t_1)\phi^T(t_0,t)\;.$$

■

In the control theory literature some authors say that a dynamic system is *completely controllable* if it is controllable from an arbitrary initial state $\mathbf{x}_0$ to $\mathbf{x}_1 = \mathbf{0}$ at every future time $t_1 > t_0$. Similarly, a dynamic system is called *completely reachable* if for all $t_1 > t_0$ and $\mathbf{x}_1$, the system is controllable from $\mathbf{x}_0 = \mathbf{0}$ to $\mathbf{x}_1$ at time t_1. In our analysis we will use our Definition 5.1 of controllability, since it contains the usual concepts of controllability and reachability as special cases by selecting $\mathbf{x}_1 = 0$ and $\mathbf{x}_0 = 0$, respectively. We mention in addition that some authors use the following modified version of the controllability Gramian:

$$\mathbf{W}_m(t_0,t_1) = \int_{t_0}^{t_1} \phi(t_1,\tau)\mathbf{B}(\tau)\mathbf{B}^T(\tau)\phi^T(t_1,\tau)\,d\tau\;,$$

obtained by replacing t_0 by t_1 in the integrand of (5.4). Since $\phi(t_0,t_1)$ is nonsingular, $\mathbf{W}_m(t_0,t_1)$ is nonsingular if and only if $\mathbf{W}(t_0,t_1)$ is nonsingular. Therefore, it makes no difference in proving controllability which version of the controllability Gramian is used. The form of $\mathbf{W}_m(t_0, t_1)$ has the obvious advantage that it is formally analogous to that of the controllability Gramian (5.25) of discrete systems, where the possible nonsingularity of matrix $\mathbf{A}(t)$ makes it impossible to define and use both versions. However, we decided to use the original form $\mathbf{W}(t_0, t_1)$ for the continuous case, since in examining duality and in deriving observability conditions later in Chapter 6, this form will have important advantages.

We conclude this section with an easy-to-check sufficient condition for complete controllability. The great advantage of this condition is the fact that it does not require the knowledge of the controllability Gramian. The major disadvantage of this approach is that it gives only sufficient condition, therefore, in many cases we cannot decide if the system is completely controllable based on only this condition.

Define first the sequence of $n \times m$ matrix functions as following:

$$\mathbf{K}_0(t) = \mathbf{B}(t)$$

$$\mathbf{K}_i(t) = -\mathbf{A}(t)\mathbf{K}_{i-1}(t) + \dot{\mathbf{K}}_{i-1}(t), \quad i = 1, 2, \ldots$$

First we show by finite induction that for all $i \geq 0$,

$$\frac{\partial^i}{\partial \tau^i}[\phi(t, \tau)\mathbf{B}(\tau)] = \phi(t, \tau)\mathbf{K}_i(\tau) .$$

This identity is obviously true for $i = 0$. Assume that it is true for an $i \geq 0$, then by using the inductive hypothesis and part (v) of Theorem 2.3 we have

$$\frac{\partial^{i+1}}{\partial \tau^{i+1}}[\phi(t, \tau)\mathbf{B}(\tau)]$$

$$= \frac{\partial}{\partial \tau}[\phi(t, \tau)\mathbf{K}_i(\tau)]$$

$$= \frac{\partial}{\partial \tau}\phi(t, \tau)\mathbf{K}_i(\tau) + \phi(t, \tau)\frac{\partial}{\partial \tau}\mathbf{K}_i(\tau)$$

$$= -\phi(t, \tau)\mathbf{A}(\tau)\mathbf{K}_i(\tau) + \phi(t, \tau)\dot{\mathbf{K}}_i(\tau)$$

$$= \phi(t, \tau)\mathbf{K}_{i+1}(\tau) .$$

THEOREM 5.3

Assume that with some positive integer q, $\mathbf{B}(t)$ is q-times continuously differentiable, and $\mathbf{A}(t)$ is $(q-1)$-times continuously differentiable on the interval $[t_0, t_1]$, furthermore for some $t^* \in [t_0, t_1]$,

$$rank(\mathbf{K}_0(t^*), \mathbf{K}_1(t^*), \dots, \mathbf{K}_q(t^*)) = n .$$

Then system (5.1) is completely controllable.

PROOF   Assume that with some t^* the rank condition holds, but the system is not completely controllable. Then $\mathbf{W}(t_0, t_1)$ is singular, and part (b) of the proof of Lemma 5.1 implies that there is a vector $\mathbf{w}$ such that

$$\mathbf{w}^T \phi(t_0, t)\mathbf{B}(t) = \mathbf{0}^T$$

for all $t \in [t_0, t_1]$. Define vector $\mathbf{z}^T = \mathbf{w}^T \phi(t_0, t^*)$, then

$$\mathbf{z}^T \phi(t^*, t)\mathbf{B}(t) = \mathbf{0}^T .$$

By substituting $t = t^*$ we see that

$$\mathbf{z}^T \mathbf{K}_0(t^*) = \mathbf{0}^T .$$

Simple differentiation shows that for $i = 1, 2, \dots, q$,

$$0 = \frac{\partial^i}{\partial t^i}[\mathbf{z}^T \phi(t^*, t)\mathbf{B}(t)] = \mathbf{z}^T \phi(t^*, t)\mathbf{K}_i(t) ,$$

and the substitution $t = t^*$ gives the equation

$$\mathbf{z}^T \mathbf{K}_i(t^*) = \mathbf{0}^T \qquad (i = 1, 2, \dots, q) .$$

Hence,

$$\mathbf{z}^T (\mathbf{K}_0(t^*), \mathbf{K}_1(t^*), \dots, \mathbf{K}_q(t^*)) = \mathbf{0}^T ,$$

which contradicts the rank condition of the theorem. Thus the proof is complete. ∎

The assertion of the theorem is illustrated by the following example.

Example 5.3

Consider again the system of the previous example. In this case,

$$\mathbf{K}_0(t) = \mathbf{B}(t) = \begin{pmatrix} 0 \\ 1 \end{pmatrix}$$

$$\mathbf{K}_1(t) = -\mathbf{A}(t)\mathbf{K}_0(t) + \dot{\mathbf{K}}_0(t) = -\begin{pmatrix} 0 & \omega \\ -\omega & 0 \end{pmatrix}\begin{pmatrix} 0 \\ 1 \end{pmatrix} = \begin{pmatrix} -\omega \\ 0 \end{pmatrix}.$$

By selecting $q = 1$, the rank of matrix

$$(\mathbf{K}_0(t), \mathbf{K}_1(t)) = \begin{pmatrix} 0 & -\omega \\ 1 & 0 \end{pmatrix}$$

is $n = 2$; therefore, the system is completely controllable.

Notice that the type of matrix $(\mathbf{K}_0(t^*), \mathbf{K}_1(t^*), ..., \mathbf{K}_q(t^*))$ is $n \times (qm)$, where n is the dimension of the state and m is the dimension of the input. If $qm < n$, then the rank condition of Theorem 5.3 cannot be satisfied even if the system is completely controllable. Therefore, the rank condition of the Theorem is only suffient but not necessary. In the special case of time invariant systems (that is, when $\mathbf{A}(t)$ and $\mathbf{B}(t)$ are constant matrices) we have

$$\mathbf{K}_0(t) = \mathbf{B}$$

$$\mathbf{K}_1(t) = -\mathbf{A}\mathbf{B}$$

$$\mathbf{K}_2(t) = \mathbf{A}^2\mathbf{B}$$

$$\vdots$$

$$\mathbf{K}_q(t) = (-1)^q\mathbf{A}^q\mathbf{B},$$

therefore,

$$(\mathbf{K}_0(t^*), \mathbf{K}_1(t^*), \ldots, \mathbf{K}_q(t^*)) = (\mathbf{B}, -\mathbf{A}\mathbf{B}, \mathbf{A}^2\mathbf{B}, \ldots, (-1)^q\mathbf{A}^q\mathbf{B}),$$

which has the same rank as matrix

$$(\mathbf{B}, \mathbf{A}\mathbf{B}, \mathbf{A}^2\mathbf{B}, \ldots, \mathbf{A}^q\mathbf{B}).$$

We will see in the next subsection that a time invariant system is completely controllable if and only if this matrix with $q = n - 1$ has full rank, showing that in this special case, the condition of Theorem 5.3 with $q = n - 1$ is sufficient and necessary.

5.1.2   Time-Invariant Systems

In this section the special case will be discussed when $\mathbf{A}(t)$ and $\mathbf{B}(t)$ are time-independent.

Introduce first the *controllability matrix*

$$\mathbf{K} = (\mathbf{B}, \mathbf{AB}, \mathbf{A}^2\mathbf{B}, \dots, \mathbf{A}^{n-1}\mathbf{B}) . \tag{5.10}$$

Note that in $\mathbf{K}$, matrices $\mathbf{B}, \mathbf{AB}, \mathbf{A}^2\mathbf{B}, \dots, \mathbf{A}^{n-1}\mathbf{B}$ are the blocks, and they are placed next to each other horizontally.

Our first result is as follows.

LEMMA 5.2
The null space and range space of $\mathbf{W}(t_0, t_1)$ for all $t_1 > t_0$ coincide with the null space and range space of matrix

$$\mathbf{W}_T = \mathbf{KK}^T . \tag{5.11}$$

PROOF   Since both $\mathbf{W}(t_0, t_1)$ and $\mathbf{W}_T$ are symmetric, and from Theorem 1.10 we know that the null and range spaces of $n \times n$ symmetric matrices are orthogonal complementary subspaces in $\mathbf{R}^n$, it is sufficient to show that the null spaces coincide, that is, $N(\mathbf{W}(t_0, t_1)) = N(\mathbf{W}_T)$.

(a) Assume first that $\mathbf{v} \in N(\mathbf{W}(t_0, t_1))$. Then $\mathbf{W}(t_0, t_1)\mathbf{v} = \mathbf{0}$, therefore,

$$0 = \mathbf{v}^T\mathbf{W}(t_0, t_1)\mathbf{v} = \int_{t_0}^{t_1} \mathbf{v}^T e^{\mathbf{A}(t_0-\tau)}\mathbf{BB}^T e^{\mathbf{A}^T(t_0-\tau)}\mathbf{v}\, d\tau$$

$$= \int_{t_0}^{t_1} \|\mathbf{B}^T e^{\mathbf{A}^T(t_0-\tau)}\mathbf{v}\|_2^2\, d\tau .$$

Since the integrand is continuous and nonnegative,

$$\mathbf{B}^T e^{\mathbf{A}^T(t_0-\tau)}\mathbf{v} = \mathbf{0} \qquad (\text{for all } \tau \in [t_0, t_1]) .$$

Use the exponential Taylor's series to see that

$$\sum_{k=0}^{\infty} \frac{1}{k!}\mathbf{B}^T(\mathbf{A}^T)^k(t_0-\tau)^k\mathbf{v} = \mathbf{0} ;$$

therefore, for all $k \geq 0$,

$$\mathbf{B}^T(\mathbf{A}^T)^k\mathbf{v} = \mathbf{0} .$$

That is,
$$\mathbf{B}^T \mathbf{v} = \mathbf{B}^T \mathbf{A}^T \mathbf{v} = \cdots = \mathbf{B}^T (\mathbf{A}^T)^{n-1} \mathbf{v} = \mathbf{0} . \tag{5.12}$$

These relations are equivalent to the property that
$$\mathbf{K}^T \mathbf{v} = \mathbf{0} ,$$

which implies that
$$\mathbf{W}_T \mathbf{v} = \mathbf{K}\mathbf{K}^T \mathbf{v} = \mathbf{0} .$$

Hence $\mathbf{v} \in N(\mathbf{W}_T)$.

(b) Assume next that $\mathbf{v} \in N(\mathbf{W}_T)$. Then $\mathbf{K}\mathbf{K}^T \mathbf{v} = \mathbf{0}$, that is, $\mathbf{v}^T \mathbf{K}\mathbf{K}^T \mathbf{v} = 0$. This equation is equivalent to relation $\|\mathbf{K}^T \mathbf{v}\|_2^2 = 0$ which implies that $\mathbf{K}^T \mathbf{v} = \mathbf{0}$, and therefore, relations (5.12) are valid. Note first that the Cayley–Hamilton theorem implies that for $l \geq n$, $\mathbf{A}^l$ is the linear combination of $\mathbf{I}, \mathbf{A}, \dots, \mathbf{A}^{n-1}$ and therefore, for all $k \geq 0$,

$$(\mathbf{A}t)^k = \sum_{l=0}^{n-1} \alpha_{kl}(t) \mathbf{A}^l$$

with some functions $\alpha_{kl}(t)$. Therefore,

$$e^{\mathbf{A}t} = \sum_{l=0}^{n-1} \beta_l(t) \mathbf{A}^l ,$$

where $\beta_l(t)$ is a function of t for $l = 0, 1, \dots, n-1$. Therefore,

$$\mathbf{v}^T \mathbf{W}(t_0, t_1) = \int_{t_0}^{t_1} \left(\sum_{l=0}^{n-1} \beta_l(t_0 - \tau) \mathbf{v}^T \mathbf{A}^l \mathbf{B} \right) \mathbf{B}^T e^{\mathbf{A}^T (t_0 - \tau)} \, d\tau = \mathbf{0}^T ,$$

since for $l = 0, 1, \dots, n-1$,

$$\mathbf{v}^T \mathbf{A}^l \mathbf{B} = \left[\mathbf{B}^T (\mathbf{A}^T)^l \mathbf{v} \right]^T = \mathbf{0}^T .$$

The symmetry of matrix $\mathbf{W}(t_0, t_1)$ implies that

$$\mathbf{W}(t_0, t_1)\mathbf{v} = \left[\mathbf{v}^T \mathbf{W}(t_0, t_1) \right]^T = \mathbf{0} ;$$

hence $\mathbf{v} \in N(\mathbf{W}(t_0, t_1))$, which completes the proof. ∎

Before formulating the main theorem of this section we remind the reader that the rank of a matrix is the maximal number of the linearly independent rows (or columns) of the matrix.

THEOREM 5.4

The time-invariant continuous linear system is completely controllable if and only if the rank of the controllability matrix **K** *equals n.*

PROOF

(a) First we prove that $N(\mathbf{K}^T) = N(\mathbf{W}_T)$. Assume that $\mathbf{v} \in N(\mathbf{K}^T)$, then

$$\mathbf{K}^T\mathbf{v} = \mathbf{0} ,$$

and multiply by matrix **K** to get

$$\mathbf{K}\mathbf{K}^T\mathbf{v} = \mathbf{0} .$$

That is, $\mathbf{W}_T\mathbf{v} = \mathbf{0}$. Hence $\mathbf{v} \in N(\mathbf{W}_T)$.

Assume next that $\mathbf{v} \in N(\mathbf{W}_T)$, then by using the same reasoning as shown at the beginning of part (b) of the proof of Lemma 5.2 we conclude that $\mathbf{K}^T\mathbf{v} = \mathbf{0}$, that is, $\mathbf{v} \in N(\mathbf{K}^T)$.

(b) From Theorem 1.10 we know that $R(\mathbf{K})$ and $N(\mathbf{K}^T)$ are orthogonal complementary subspaces in $\mathbf{R}^n$, and the same is also true for $R(\mathbf{W}_T)$ and $N(\mathbf{W}_T)$. Because we proved that $N(\mathbf{K}^T) = N(\mathbf{W}_T)$, we conclude that $R(\mathbf{K}) = R(\mathbf{W}_T)$.

(c) We know from Lemma 5.2 that

$$R(\mathbf{W}(t_0, t_1)) = R(\mathbf{W}_T) = R(\mathbf{K}) ,$$

and therefore, $\mathbf{W}(t_0, t_1)$ is nonsingular if and only if $rank(\mathbf{K}) = n$. The assertion then follows from Theorem 5.1.

∎

REMARK 5.3   Note that the conditions of the theorem hold if and only if the rows of matrix **K** are linearly independent. Because row ranks and column ranks of matrices are always equal, in many applications it is easier to find n independent columns from matrix **K** rather than prove the independence of the n (usually very long) rows. In the case of high-dimensional matrices, the rank of **K** can be obtained by using standard computer packages. ∎

Example 5.4

Consider again the system

$$\dot{\mathbf{x}} = \begin{pmatrix} 0 & \omega \\ -\omega & 0 \end{pmatrix} \mathbf{x} + \begin{pmatrix} 0 \\ 1 \end{pmatrix} u, \qquad \mathbf{x}(0) = \begin{pmatrix} 1 \\ 0 \end{pmatrix},$$

which was the subject of our earlier Example 5.2. In that example, we examined the controllability of the system by proving that the controllability Gramian $\mathbf{W}(0, t_1)$ is nonsingular. The same result now will be obtained based on the controllability matrix $\mathbf{K}$. In this case, $n = 2$, and

$$\mathbf{K} = (\mathbf{B}, \mathbf{AB}) = \begin{pmatrix} 0 & \omega \\ 1 & 0 \end{pmatrix}.$$

Obviously $rank(\mathbf{K}) = 2$ for all $\omega \neq 0$.

Note that the direct application of the controllability matrix is much more attractive than the computation of the controllability Gramian even in cases when the integral can be given in closed forms.

Example 5.5

Consider now the satellite problem, presented in Example 3.6. The controllability of this system is now examined. In this case, $n = 4$,

$$\mathbf{A} = \begin{pmatrix} 0 & 1 & 0 & 0 \\ 3\omega^2 & 0 & 0 & 2\omega \\ 0 & 0 & 0 & 1 \\ 0 & -2\omega & 0 & 0 \end{pmatrix}, \qquad \mathbf{B} = \begin{pmatrix} 0 & 0 \\ 1 & 0 \\ 0 & 0 \\ 0 & 1 \end{pmatrix};$$

therefore, the controllability matrix becomes

$$\mathbf{K} = (\mathbf{B}, \mathbf{AB}, \mathbf{A}^2\mathbf{B}, \mathbf{A}^3\mathbf{B})$$

$$= \begin{pmatrix} 0 & 0 & 1 & 0 & 0 & 2\omega & -\omega^2 & 0 \\ 1 & 0 & 0 & 2\omega & -\omega^2 & 0 & 0 & 2\omega^3 \\ 0 & 0 & 0 & 1 & -2\omega & 0 & 0 & -4\omega^2 \\ 0 & 1 & -2\omega & 0 & 0 & -4\omega^2 & 2\omega^3 & 0 \end{pmatrix}.$$

Observe that the first four columns are linearly independent; therefore, $rank(\mathbf{K}) = 4$. That is, the system is completely controllable.

Next assume that one of the inputs is inoperative. Is the system still controllable?

Set $u_2 = 0$, then **B** reduces to $\mathbf{B}_1 = (0, 1, 0, 0)^T$, and so

$$\mathbf{K}_1 = (\mathbf{B}_1, \mathbf{AB}_1, \mathbf{A}^2\mathbf{B}_1, \mathbf{A}^3\mathbf{B}_1) = \begin{pmatrix} 0 & 1 & 0 & -\omega^2 \\ 1 & 0 & -\omega^2 & 0 \\ 0 & 0 & -2\omega & 0 \\ 0 & -2\omega & 0 & 2\omega^3 \end{pmatrix}.$$

Observe that the last column is the $(-\omega^2)$-multiple of the second column. Therefore, $rank(\mathbf{K}_1) < 4$, and the system with the only input u_1 is not completely controllable.

Now if $u_1 = 0$, then **B** reduces to $\mathbf{B}_2 = (0, 0, 0, 1)^T$; therefore,

$$\mathbf{K}_2 = (\mathbf{B}_2, \mathbf{AB}_2, \mathbf{A}^2\mathbf{B}_2, \mathbf{A}^3\mathbf{B}_2) = \begin{pmatrix} 0 & 0 & 2\omega & 0 \\ 0 & 2\omega & 0 & -2\omega^3 \\ 0 & 1 & 0 & -4\omega^2 \\ 1 & 0 & -4\omega^2 & 0 \end{pmatrix}.$$

It is easy to establish that for $\omega \neq 0$, $rank(\mathbf{K}_2) = 4$. That is, the system with the only input u_2 is still completely controllable. Because u_1 is the radial thrust and u_2 is the tangential thrust, we conclude that loss of radial thrust does not destroy complete controllability, but the loss of the tangential thrust does. These two models are illustrated in Figure 5.2.

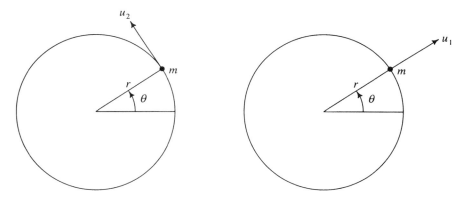

Figure 5.2   Controllable and noncontrollable satellite models.

Assume next that the system

$$\dot{\mathbf{x}} = \mathbf{Ax} + \mathbf{Bu}$$

$$y = Cx$$

is not completely controllable, then the *rank* r of the controllability matrix $\mathbf{K}$ is less than n. In this case, the controllable and uncontrollable state variables can be clearly identified by using the following result.

THEOREM 5.5
Assume that $r < n$, then there exists a nonsingular matrix $\mathbf{T}$ such that

$$\bar{\mathbf{A}} = \mathbf{TAT}^{-1} = \begin{pmatrix} \bar{\mathbf{A}}_{11} & \bar{\mathbf{A}}_{12} \\ \mathbf{O} & \bar{\mathbf{A}}_{22} \end{pmatrix},$$

$$\bar{\mathbf{B}} = \mathbf{TB} = \begin{pmatrix} \bar{\mathbf{B}}_1 \\ \mathbf{O} \end{pmatrix},$$

$$\bar{\mathbf{C}} = \mathbf{CT}^{-1} = (\bar{\mathbf{C}}_1, \bar{\mathbf{C}}_2), \tag{5.13}$$

where the sizes of the matrices $\bar{\mathbf{A}}_{11}$, $\bar{\mathbf{A}}_{12}$, and $\bar{\mathbf{A}}_{22}$ are $r \times r$, $r \times (n-r)$, and $(n-r) \times (n-r)$, respectively, and $\bar{\mathbf{B}}_1$ has r rows and $\bar{\mathbf{C}}_1$ has r columns. Furthermore,

(i)   *System $(\bar{\mathbf{A}}_{11}, \bar{\mathbf{B}}_1, \bar{\mathbf{C}}_1)$ is completely controllable.*

(ii) *The transfer function of systems $(\mathbf{A}, \mathbf{B}, \mathbf{C})$ and $(\bar{\mathbf{A}}_{11}, \bar{\mathbf{B}}_1, \bar{\mathbf{C}}_1)$ coincide.*

PROOF   Because the *rank* of $\mathbf{K}$ is r, we find r linearly independent columns, which are now denoted by $c_1, c_2, \ldots, c_r$. Select vectors $v_{r+1}, v_{r+2}, \ldots, v_n$ such that $\{c_1, c_2, \ldots, c_r, v_{r+1}, v_{r+2}, \ldots, v_n\}$ is a basis in $\mathbf{R}^n$, and define matrix

$$\mathbf{T} = (c_1, \ldots, c_r, v_{r+1}, \ldots, v_n)^{-1}.$$

Then direct calculation shows that this matrix satisfies the assertion. The details are left as an exercise.

Notice that the controllability matrix of system $(\bar{\mathbf{A}}_{11}, \bar{\mathbf{B}}_1, \bar{\mathbf{C}}_1)$ has the form

$$\begin{pmatrix} \bar{\mathbf{B}}_1 & \bar{\mathbf{A}}_{11}\bar{\mathbf{B}}_1 & \cdots & \bar{\mathbf{A}}_{11}^{r-1}\bar{\mathbf{B}}_1 & \cdots & \bar{\mathbf{A}}_{11}^{n-1}\bar{\mathbf{B}}_1 \\ \mathbf{O} & \mathbf{O} & \cdots & \mathbf{O} & \cdots & \mathbf{O} \end{pmatrix},$$

and has the same *rank* r. Therefore, the first r rows are linearly independent, and there are r linearly independent columns. Because matrix $\bar{\mathbf{A}}_{11}$ is $r \times r$, the Cayley–Hamilton theorem implies that these columns are from matrices $\bar{\mathbf{B}}_1, \bar{\mathbf{A}}_{11}\bar{\mathbf{B}}_1, \ldots, \bar{\mathbf{A}}_{11}^{r-1}\bar{\mathbf{B}}_1$, which proves that matrix

$(\bar{\mathbf{B}}_1, \bar{\mathbf{A}}_{11}\bar{\mathbf{B}}_1, \ldots, \bar{\mathbf{A}}_{11}^{r-1}\bar{\mathbf{B}}_1)$ has full rank. Since it is the controllability matrix of system $(\bar{\mathbf{A}}_{11}, \bar{\mathbf{B}}_1, \bar{\mathbf{C}}_1)$, Assertion (i) is verified.

In proving (ii), notice first that the transfer functions of systems $(\mathbf{A}, \mathbf{B}, \mathbf{C})$ and $(\bar{\mathbf{A}}, \bar{\mathbf{B}}, \bar{\mathbf{C}})$ coincide, because they differ from each other only in a state variable transformation. Observe furthermore that the transfer function of system $(\bar{\mathbf{A}}, \bar{\mathbf{B}}, \bar{\mathbf{C}})$ can be written as

$$\bar{\mathbf{C}}(s\mathbf{I} - \bar{\mathbf{A}})^{-1}\bar{\mathbf{B}} = (\bar{\mathbf{C}}_1, \bar{\mathbf{C}}_2) \begin{pmatrix} s\mathbf{I} - \bar{\mathbf{A}}_{11} & -\bar{\mathbf{A}}_{12} \\ \mathbf{O} & s\mathbf{I} - \bar{\mathbf{A}}_{22} \end{pmatrix}^{-1} \begin{pmatrix} \bar{\mathbf{B}}_1 \\ \mathbf{O} \end{pmatrix}$$

$$= (\bar{\mathbf{C}}_1, \bar{\mathbf{C}}_2) \begin{pmatrix} (s\mathbf{I} - \bar{\mathbf{A}}_{11})^{-1} & * \\ \mathbf{O} & (s\mathbf{I} - \bar{\mathbf{A}}_{22})^{-1} \end{pmatrix} \begin{pmatrix} \bar{\mathbf{B}}_1 \\ \mathbf{O} \end{pmatrix} ,$$

where $*$ denotes a block, the particular form of which is not important now. Therefore,

$$\bar{\mathbf{C}}(s\mathbf{I} - \bar{\mathbf{A}})^{-1}\bar{\mathbf{B}} = \bar{\mathbf{C}}_1(s\mathbf{I} - \bar{\mathbf{A}}_{11})^{-1}\bar{\mathbf{B}}_1 ,$$

which completes the proof of the theorem.   ∎

REMARK 5.4   Notice that system $(\bar{\mathbf{A}}, \bar{\mathbf{B}}, \bar{\mathbf{C}})$ can be rewritten as

$$\dot{\bar{\mathbf{x}}}_1 = \bar{\mathbf{A}}_{11}\bar{\mathbf{x}}_1 + \bar{\mathbf{A}}_{12}\bar{\mathbf{x}}_2 + \bar{\mathbf{B}}_1\mathbf{u},$$

$$\dot{\bar{\mathbf{x}}}_2 = \bar{\mathbf{A}}_{22}\bar{\mathbf{x}}_2,$$

$$\mathbf{y} = \bar{\mathbf{C}}_1\bar{\mathbf{x}}_1 + \bar{\mathbf{C}}_2\bar{\mathbf{x}}_2 .$$

Since $\dot{\bar{\mathbf{x}}}_2$ does not depend on the input, it is not controllable; furthermore, (i) implies that variable $\bar{\mathbf{x}}_1$ is completely controllable.   ∎

The controllability of a time-invariant linear system $(\mathbf{A}, \mathbf{B}, \mathbf{C})$ can be examined not only by determining the *rank* of the controllability matrix, but also by using the following result.

THEOREM 5.6
System $(\mathbf{A}, \mathbf{B}, \mathbf{C})$ is completely controllable if and only if matrix $\mathbf{A}^T$ has no eigenvector $\mathbf{a}$ *that is orthogonal to the columns of* $\mathbf{B}$.

PROOF   Assume first that there exists such a vector $\mathbf{q}$. Then $\mathbf{q}^T\mathbf{A} =$

$\lambda \mathbf{q}^T$ and $\mathbf{q}^T \mathbf{B} = \mathbf{0}$. Therefore,

$$\mathbf{q}^T \mathbf{K} = (\mathbf{q}^T \mathbf{B}, \mathbf{q}^T \mathbf{A} \mathbf{B}, \ldots, \mathbf{q}^T \mathbf{A}^{n-1} \mathbf{B}) = (\mathbf{q}^T \mathbf{B}, \lambda \mathbf{q}^T \mathbf{B}, \ldots, \lambda^{n-1} \mathbf{q}^T \mathbf{B})$$

$$= \mathbf{0}^T ,$$

so the rows of $\mathbf{K}$ are linearly dependent. Hence, the system is not completely controllable.

Assume next that the system is not completely controllable. Then it can be transformed to the form (5.13). Let $\mathbf{q}_2$ denote any eigenvector of $\mathbf{A}_{22}^T$ with associated eigenvalue λ, then

$$\bar{\mathbf{A}}^T \begin{pmatrix} \mathbf{0} \\ \mathbf{q}_2 \end{pmatrix} = \begin{pmatrix} \bar{\mathbf{A}}_{11}^T & \mathbf{O} \\ \bar{\mathbf{A}}_{12}^T & \bar{\mathbf{A}}_{22}^T \end{pmatrix} \begin{pmatrix} \mathbf{0} \\ \mathbf{q}_2 \end{pmatrix} = \begin{pmatrix} \mathbf{0} \\ \bar{\mathbf{A}}_{22}^T \mathbf{q}_2 \end{pmatrix} = \lambda \begin{pmatrix} \mathbf{0} \\ \mathbf{q}_2 \end{pmatrix}$$

and

$$(\mathbf{0}^T, \mathbf{q}_2^T) \cdot \mathbf{B} = (\mathbf{0}^T, \mathbf{q}_2^T)(\bar{\mathbf{B}}_1 \mathbf{O}) = \mathbf{0}^T .$$

Thus the proof is completed. ∎

COROLLARY 5.1
System $(\mathbf{A}, \mathbf{B}, \mathbf{C})$ is completely controllable if and only if the rank of matrix $(s\mathbf{I} - \mathbf{A}, \mathbf{B})$ is n for all s.

PROOF Notice that the *rank* of $(s\mathbf{I} - \mathbf{A}, \mathbf{B})$ is less than n if and only if there exists a nonzero vector $\mathbf{q}$ such that

$$\mathbf{q}^T (s\mathbf{I} - \mathbf{A}, \mathbf{B}) = \mathbf{0}^T .$$

This equation is equivalent to relations $\mathbf{A}^T \mathbf{q} = s\mathbf{q}$ and $\mathbf{q}^T \mathbf{B} = \mathbf{0}^T$. Hence, $\mathbf{q}$ is an eigenvector of $\mathbf{A}^T$ that is orthogonal to all columns of $\mathbf{B}$. ∎

Example 5.6

Consider again the system of Example 5.4, when $n = 2$,

$$\mathbf{A} = \begin{pmatrix} 0 & \omega \\ -\omega & 0 \end{pmatrix} \quad \text{and} \quad \mathbf{B} = \begin{pmatrix} 1 \\ 0 \end{pmatrix} .$$

In this case,

$$(s\mathbf{I} - \mathbf{A}, \mathbf{B}) = \begin{pmatrix} s & -\omega & 1 \\ \omega & s & 0 \end{pmatrix} .$$

Notice that the first and third columns are always independent, there-fore, the $rank$ of the matrix is n for all s. Hence, the system is com-pletely controllable.

5.1.3   Output and Trajectory Controllability

In many applications the entire state vector $\mathbf{x}$ is not of interest, and only a subset of its components or some linear combinations of its components are all that matters. Assume that instead of the state $\mathbf{x}$ an output vector

$$\mathbf{y} = \mathbf{C}(t)\mathbf{x} \tag{5.14}$$

is to be controlled. This type of control, which is called the output control, is examined next. The main existence theorem is as follows.

THEOREM 5.7

There exists an input $\mathbf{u}(t)$ that drives the output of the system

$$\dot{\mathbf{x}} = \mathbf{A}(t)\mathbf{x} + \mathbf{B}(t)\mathbf{u}, \qquad \mathbf{x}(t_0) = \mathbf{x}_0$$

$$\mathbf{y} = \mathbf{C}(t)\mathbf{x} \tag{5.15}$$

to $\mathbf{y}_1$ at $t_1 > t_0$ if and only if $\mathbf{y}_1 - \mathbf{C}(t_1)\phi(t_1, t_0)\mathbf{x}_0$ lies in the range space of $\mathbf{C}(t_1)\phi(t_1, t_0)\mathbf{W}(t_0, t_1)$.

PROOF   Equation (5.8) implies that the system can be controlled to a state $\mathbf{x}_1$ at $t_1 > t_0$ if and only if it can be represented as

$$\mathbf{x}_1 = \phi(t_1, t_0)\mathbf{x}_0 + \phi(t_1, t_0)\mathbf{W}(t_0, t_1)\mathbf{a}$$

with some vector $\mathbf{a}$. Therefore, the output vectors $\mathbf{y}_1$ to which the system can be controlled at t_1 have the form

$$\mathbf{y}_1 = \mathbf{C}(t_1)\mathbf{x}_1 = \mathbf{C}(t_1)\phi(t_1, t_0)\mathbf{x}_0 + \mathbf{C}(t_1)\phi(t_1, t_0)\mathbf{W}(t_0, t_1)\mathbf{a} \ .$$

A vector $\mathbf{y}_1$ can be written in this form if and only if

$$\mathbf{y}_1 - \mathbf{C}(t_1)\phi(t_1, t_0)\mathbf{x}_0 \in R(\mathbf{C}(t_1)\phi(t_1, t_0)\mathbf{W}(t_0, t_1)) \ ,$$

which completes the proof.   ∎

COROLLARY 5.2
The output can be controlled to any arbitrary $\mathbf{y}_1$ at $t_1 > t_0$ if and only if

$$rank(\mathbf{C}(t_1)\phi(t_1, t_0)\mathbf{W}(t_0, t_1)) = p,\qquad (5.16)$$

where p is the dimension of the output.

This rank condition is illustrated by the following example.

Example 5.7

Consider again the system of the previous example:

$$\mathbf{A} = \begin{pmatrix} 0 & \omega \\ -\omega & 0 \end{pmatrix}\mathbf{x} + \begin{pmatrix} 0 \\ 1 \end{pmatrix}\mathbf{u}, \qquad \mathbf{x}(0) = \begin{pmatrix} 1 \\ 0 \end{pmatrix}$$

with the output equation

$$\mathbf{y} = (1, 0)\mathbf{x}.$$

Notice that $\mathbf{C} = (1, 0)$, both $\phi(t_1, t_0)$ and $\mathbf{W}(t_0, t_1)$ are nonsingular matrices as it was shown in Part (iii) of Theorem 2.3 and in Example 5.2. Because the output is single, $p = 1$. Therefore, the nonsingularity of the product $\phi(t_1, t_0)\mathbf{W}(t_0, t_1)$ implies that with the nonzero row vector $\mathbf{C}$, $\mathbf{C}(t_1)\phi(t_1, t_0)\mathbf{W}(t_0, t_1)$ is nonzero, therefore, its rank equals 1. Hence, the rank condition (5.16) is satisfied. Consequently, the output of this system is controllable to any final output $\mathbf{y}_1$ at any future time $t_1 > t_0$.

The state and output controllability problems were concerned with driving the system to a given state and output at a future time t_1. In many applications, the control of the entire state function $\mathbf{x}(t)$ or output function $\mathbf{y}(t)$ is needed. That is, an input is to be determined such that the entire trajectory of the state or output coincides with a desired function. This problem is illustrated in Figure 5.3. This *trajectory control* problem is now discussed only for time-invariant continuous systems:

$$\dot{\mathbf{x}} = \mathbf{A}\mathbf{x} + \mathbf{B}\mathbf{u}, \qquad \mathbf{x}(0) = \mathbf{x}_0 \qquad (5.17)$$

$$\mathbf{y} = \mathbf{C}\mathbf{x}. \qquad (5.18)$$

If the state trajectory is given, then check the initial condition first. If it is not satisfied, then no control exists. Otherwise, by substituting the given state trajectory into Equation (5.17), a linear system of equations

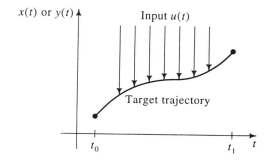

Figure 5.3   Trajectory control.

is obtained for the unknown components of the input vector. If these linear equations have no solution, then the state trajectory cannot be controlled as required. Otherwise, any solution is a suitable input.

Example 5.8

Assume that the state of the system

$$\dot{\mathbf{x}} = \begin{pmatrix} 0 & \omega \\ -\omega & 0 \end{pmatrix} \mathbf{x} + \begin{pmatrix} 0 \\ 1 \end{pmatrix} u, \qquad \mathbf{x}(0) = \begin{pmatrix} 1 \\ 0 \end{pmatrix}$$

has to be controlled to be

$$\mathbf{x}(t) = \begin{pmatrix} \omega t^2 + 1 \\ 2t \end{pmatrix} \qquad (t \geq 0).$$

Notice first that the initial condition is satisfied. Substituting this trajectory into the differential equation we have

$$\begin{pmatrix} 2\omega t \\ 2 \end{pmatrix} = \begin{pmatrix} 0 & \omega \\ -\omega & 0 \end{pmatrix} \begin{pmatrix} \omega t^2 + 1 \\ 2t \end{pmatrix} + \begin{pmatrix} 0 \\ 1 \end{pmatrix} u(t),$$

which can rewritten as

$$2\omega t = 2\omega t$$

$$2 = -\omega(\omega t^2 + 1) + u(t).$$

The first equation is always satisfied, and from the second equation we get the required input:

$$u(t) = 2 + \omega(\omega t^2 + 1) = \omega^2 t^2 + \omega + 2.$$

If the output trajectory is given, then the application of Laplace transforms is the appropriate method to find the suitable input function. From Theorem 3.2 we know that the Laplace transforms of $\mathbf{x}(t)$ and $\mathbf{y}(t)$ are given as

$$\mathbf{X}(s) = \mathbf{R}(s)\mathbf{x}_0 + \mathbf{R}(s)\mathbf{B}\mathbf{U}(s) \qquad (5.19)$$

and

$$\mathbf{Y}(s) = \mathbf{C}\mathbf{R}(s)\mathbf{x}_0 + \mathbf{H}(s)\mathbf{U}(s) , \qquad (5.20)$$

where $\mathbf{R}(s)$ is the resolvent matrix and $\mathbf{H}(s)$ is the transfer function. If the output trajectory is given, then function $\mathbf{Y}(s)$ is determined by applying the Laplace transform. Then solve the linear Equation (5.20) with parameter s to recover function $\mathbf{U}(s)$. And finally, use the inverse Laplace transform to determine the unknown input $\mathbf{u}(t)$.

Example 5.9

Consider again the system

$$\dot{\mathbf{x}} = \begin{pmatrix} 0 & \omega \\ -\omega & 0 \end{pmatrix} \mathbf{x} + \begin{pmatrix} 0 \\ 1 \end{pmatrix} u, \qquad \mathbf{x}(0) = \begin{pmatrix} 1 \\ 0 \end{pmatrix}$$

$$y = (1,1)\mathbf{x} .$$

Assume that the output of this system is to be controlled to have $y(t) \equiv 1$ for all $t \geq 0$. In Example 3.8 we have shown that

$$\mathbf{R}(s) = \frac{1}{s^2 + \omega^2} \begin{pmatrix} s & \omega \\ -\omega & s \end{pmatrix} \qquad \text{and} \qquad H(s) = \frac{s + \omega}{s^2 + \omega^2} ;$$

therefore, Equation (5.20) has now the special form

$$\frac{1}{s} = (1,1)\frac{1}{s^2 + \omega^2} \begin{pmatrix} s & \omega \\ -\omega & s \end{pmatrix} \begin{pmatrix} 1 \\ 0 \end{pmatrix} + \frac{s + \omega}{s^2 + \omega^2} U(s) ,$$

where we used item No. 2 of Table 2.1. Simplifying this equation shows that

$$\frac{1}{s} = \frac{s - \omega}{s^2 + \omega^2} + \frac{s + \omega}{s^2 + \omega^2} U(s) ,$$

from which we have

$$U(s) = \frac{\omega}{s} ,$$

that is, $u(t) = \omega$ for $t > 0$.

This result can be also verified by simple substitution of the state components

$$x_1(t) \equiv 1 \qquad \text{and} \qquad x_2(t) \equiv 0,$$

the input $u(t) = w$ and output

$$y(t) = (1,1)\mathbf{x} \equiv 1$$

into the systems equations.

Finally, we mention that this approach can be used if the state trajectory is given; however, the procedure illustrated in Example 5.8 is less complicated.

5.2   Discrete Systems

In this section the controllability of the discrete linear system

$$\mathbf{x}(t+1) = \mathbf{A}(t)\mathbf{x}(t) + \mathbf{B}(t)\mathbf{u}(t), \qquad \mathbf{x}(0) = \mathbf{x}_0 \qquad (5.21)$$

$$\mathbf{y}(t) = \mathbf{C}(t)\mathbf{x}(t) \qquad (5.22)$$

will be examined. We assume that the types of matrices $\mathbf{A}(t)$, $\mathbf{B}(t)$, and $\mathbf{C}(t)$ are $n \times n$, $n \times m$, and $p \times n$, respectively.

The general solution (3.48) of discrete linear systems implies that this system is controllable to $\mathbf{x}_1$ at t_1 if and only if there exists an input $\mathbf{u}(t)$ such that

$$\mathbf{x}_1 = \phi(t_1, 0)\mathbf{x}_0 + \sum_{\tau=0}^{t_1-1} \phi(t_1, \tau+1)\mathbf{B}(\tau)\mathbf{u}(\tau) . \qquad (5.23)$$

Introduce the mapping

$$A(\mathbf{u}) = \sum_{\tau=0}^{t_1-1} \phi(t_1, \tau+1)\mathbf{B}(\tau)\mathbf{u}(\tau) \qquad (5.24)$$

on the set of m-dimensional functions defined on $\mathbf{N} = \{0, 1, \ldots\}$. Note that the range of this mapping is in $\mathbf{R}^n$.

LEMMA 5.3
Vector $\mathbf{v}$ is in $R(A)$ if and only if it belongs to the range space of matrix

$$\mathbf{W}(0, t_1) = \sum_{\tau=0}^{t_1-1} \phi(t_1, \tau+1)\mathbf{B}(\tau)\mathbf{B}^T(\tau)\phi^T(t_1, \tau+1) . \qquad (5.25)$$

PROOF   The proof is analogous to that of Lemma 5.1.

(a)   Assume first that $\mathbf{v} \in R\left(\mathbf{W}(0, t_1)\right)$, then there exists a vector $\mathbf{a}$ such that

$$\mathbf{v} = \mathbf{W}(0, t_1)\mathbf{a} . \tag{5.26}$$

Select input

$$\mathbf{u}(t) = \mathbf{B}^T(t)\phi^T(t_1, t + 1)\mathbf{a} , \tag{5.27}$$

then

$$A(\mathbf{u}) = \sum_{\tau=0}^{t_1-1} \phi(t_1, \tau + 1)\mathbf{B}(\tau)\mathbf{B}^T(\tau)\phi^T(t_1, \tau + 1)\mathbf{a}$$

$$= \mathbf{W}(0, t_1)\mathbf{a} = \mathbf{v} ;$$

therefore, $\mathbf{v} \in R(A)$.

(b)   Assume next that $v \notin R(\mathbf{W}(0, t_1))$. Then — similarly to the proof of Lemma 5.1 — there exists a vector $\mathbf{w}$ from the null space of $\mathbf{W}(0, t_1)$ that is not orthogonal to $\mathbf{v}$. That is,

$$\mathbf{W}(0, t_1)\mathbf{w} = \mathbf{0} \qquad \text{and} \qquad \mathbf{w}^T\mathbf{v} \neq 0 .$$

We shall now verify that $\mathbf{v} \notin R(A)$. In contrary to this assertion, assume that $\mathbf{v} \in R(A)$. Then with some function $\tilde{\mathbf{u}}(t)$,

$$\mathbf{v} = \sum_{\tau=0}^{t_1-1} \phi(t_1, \tau + 1)\mathbf{B}(\tau)\tilde{\mathbf{u}}(\tau) .$$

Therefore,

$$0 \neq \mathbf{w}^T\mathbf{v} = \sum_{\tau=0}^{t_1-1} \mathbf{w}^T\phi(t_1, \tau + 1)\mathbf{B}(\tau)\tilde{\mathbf{u}}(\tau) . \tag{5.28}$$

The definition of $\mathbf{w}$ implies that

$$0 = \mathbf{w}^T\mathbf{W}(0, t_1)\mathbf{w} = \sum_{\tau=0}^{t_1-1} \mathbf{w}^T\phi(t_1, \tau + 1)\mathbf{B}(\tau)\mathbf{B}^T(\tau)\phi^T(t_1, \tau + 1)\mathbf{w}$$

$$= \sum_{\tau=0}^{t_1-1} \|\mathbf{B}^T(\tau)\phi^T(t_1, \tau + 1)\mathbf{w}\|_2^2 .$$

Since all terms are nonnegative, for all $\tau = 0, 1, \ldots, t_1 - 1$,

$$\mathbf{B}^T(\tau)\phi^T(t_1, \tau + 1)\mathbf{w} = \mathbf{0} .$$

Take transpose on both sides to get

$$\mathbf{w}^T \phi(t_1, \tau + 1)\mathbf{B}(\tau) = \mathbf{0}^T \qquad (\text{all } \tau = 0, 1, \ldots, t_1 - 1) ,$$

which contradicts to relation (5.28). Thus, the proof is completed.

∎

COROLLARY 5.3
There exists an input $\mathbf{u}(t)$ *which drives the state of the discrete linear system to* $\mathbf{x}_1$ *at time* $t_1 > 0$ *if and only if* $\mathbf{x}_1 - \phi\,(t_1, 0)\mathbf{x}_0$ *belongs to* $R(\mathbf{W}(0, t_1))$, *that is, if*

$$\mathbf{x}_1 = \phi(t_1, 0)\mathbf{x}_0 + \mathbf{W}(0, t_1)\mathbf{a}$$

with some vector $\mathbf{a}$. *The algorithm verifying controllability is similar to that shown for continuous systems.*

From the lemma we conclude the following result, which is the discrete case counterpart of Theorem 5.1.

THEOREM 5.8
The discrete system is controllable from initial state $\mathbf{x}_0$ *to arbitrary state* $\mathbf{x}_1$ *at time* $t_1 > 0$ *if and only if* $\mathbf{W}(0, t_1)$ *is nonsingular.*

REMARK 5.5   If a discrete linear system is controllable from arbitrary initial state $\mathbf{x}_0$ at 0 to arbitrary state $\mathbf{x}_1$ at any $t_1 \geq n$, then the system is called *completely controllable*.   ∎

COROLLARY 5.4
From part (a) of the proof of the lemma we conclude that a particular input that leads the system to $\mathbf{x}_1$ *at* $t_1 > 0$ *is given as*

$$\mathbf{u}(t) = \mathbf{B}^T(t)\phi^T(t_1, t + 1)\mathbf{a} ,$$

where vector $\mathbf{a}$ *is a solution of equation*

$$\mathbf{W}(0, t_1)\mathbf{a} = \mathbf{x}_1 - \phi(t_1, 0)\mathbf{x}_0 . \tag{5.29}$$

Example 5.10

Consider the discrete system modeled by difference equation

$$\mathbf{x}(t+1) = \begin{pmatrix} 1 & 1 \\ 0 & 1 \end{pmatrix} \mathbf{x}(t) + \begin{pmatrix} 0 \\ 1 \end{pmatrix} u(t), \qquad \mathbf{x}(0) = \begin{pmatrix} 1 \\ 0 \end{pmatrix}.$$

From Example 3.14 we know that

$$\phi(t,\tau) = \begin{pmatrix} 1 & t-\tau \\ 0 & 1 \end{pmatrix}.$$

Therefore,

$$\mathbf{W}(0,t_1) = \sum_{\tau=0}^{t_1-1} \begin{pmatrix} 1 & t_1-\tau-1 \\ 0 & 1 \end{pmatrix} \begin{pmatrix} 0 \\ 1 \end{pmatrix} (0,1) \begin{pmatrix} 1 & 0 \\ t_1-\tau-1 & 1 \end{pmatrix}$$

$$= \sum_{\tau=0}^{t_1-1} \begin{pmatrix} t_1-\tau-1 \\ 1 \end{pmatrix} (t_1-\tau-1, 1)$$

$$= \sum_{\tau=0}^{t_1-1} \begin{pmatrix} (t_1-\tau-1)^2 & t_1-\tau-1 \\ t_1-\tau-1 & 1 \end{pmatrix}$$

$$= \begin{pmatrix} \frac{t_1(t_1-1)(2t_1-1)}{6} & \frac{t_1(t_1-1)}{2} \\ \frac{t_1(t_1-1)}{2} & t_1 \end{pmatrix},$$

where we used the relations

$$1 + 2 + \cdots + (t_1-1) = \frac{t_1(t_1-1)}{2}$$

and

$$1^2 + 2^2 + \cdots + (t_1-1)^2 = \frac{t_1(t_1-1)(2t_1-1)}{6}.$$

Finally we show that for all $t_1 \geq 2$, this matrix is nonsingular, that is, the system is completely controllable.

The determinant of $\mathbf{W}(0,t_1)$ can be written as

$$\frac{t_1^2(t_1-1)(2t_1-1)}{6} - \frac{t_1^2(t_1-1)^2}{4} = \frac{t_1^2}{12}(4t_1^2 - 6t_1 + 2 - 3t_1^2 + 6t_1 - 3)$$

$$= \frac{t_1^2}{12}(t_1^2 - 1),$$

which is nonzero for $t_1 \geq 2$. Hence, at $t_1 \geq 2$, arbitrary state $\mathbf{x}_1$ can be obtained from arbitrary initial state $\mathbf{x}_0$, but for $t_1 = 1$ this is not true. We can demonstrate this statement by using relations (5.23) directly for $t_1 = 1$:

$$\mathbf{x}_1 = \phi(1,0)\mathbf{x}_0 + \phi(1,1)\mathbf{B}(0)u(0)$$

$$= \begin{pmatrix} 1 & 1 \\ 0 & 1 \end{pmatrix} \begin{pmatrix} 1 \\ 0 \end{pmatrix} + \begin{pmatrix} 1 & 0 \\ 0 & 1 \end{pmatrix} \begin{pmatrix} 0 \\ 1 \end{pmatrix} u(0)$$

$$= \begin{pmatrix} 1 \\ 0 \end{pmatrix} + \begin{pmatrix} 0 \\ 1 \end{pmatrix} u(0) = \begin{pmatrix} 1 \\ u(0) \end{pmatrix} .$$

Therefore, a state is feasible at $t_1 = 1$ if and only if its first component equals 1. Then its second component gives the desired input value $u(0)$.

An alternative method to find the input sequence $\{u(0), u(1), \ldots, u(t_1 - 1)\}$ that drives a discrete system to a given final state $\mathbf{x}_1 = \mathbf{x}(t_1)$ is based on solving directly Equation (5.23) for the unknowns $u(0), u(1), \ldots, u(t_1 - 1)$. This procedure is illustrated next.

Example 5.11

Assume that in the case of the system of the previous example the final state is given as

$$\mathbf{x}(2) = \begin{pmatrix} 1 \\ 1 \end{pmatrix} .$$

In this case, Equation (5.23) can be rewritten as follows:

$$\mathbf{x}(2) = \mathbf{A}^2 \mathbf{x}_0 + (\mathbf{A}\mathbf{b}u(0) + \mathbf{b}u(1))$$

with

$$\mathbf{A} = \begin{pmatrix} 1 & 1 \\ 0 & 1 \end{pmatrix}, \qquad \mathbf{b} = \begin{pmatrix} 0 \\ 1 \end{pmatrix} \qquad \text{and} \qquad \mathbf{x}_0 = \begin{pmatrix} 1 \\ 0 \end{pmatrix} .$$

That is,

$$\begin{pmatrix} 1 \\ 1 \end{pmatrix} = \begin{pmatrix} 1 & 2 \\ 0 & 1 \end{pmatrix} \begin{pmatrix} 1 \\ 0 \end{pmatrix} + \begin{pmatrix} 1 \\ 1 \end{pmatrix} u(0) + \begin{pmatrix} 0 \\ 1 \end{pmatrix} u(1) ,$$

which has the simplified form

$$\begin{pmatrix} 0 \\ 1 \end{pmatrix} = \begin{pmatrix} u(0) \\ u(0) + u(1) \end{pmatrix} .$$

The only solution is $u(0) = 0$ and $u(1) = 1$.

It is easy to modify Theorem 5.2 for discrete systems. Properties (i) and (ii) hold in the same way, and (iii) has to be modified accordingly. The details are left as an exercise to the reader.

Consider next the time-invariant case, when $\mathbf{A}(t)$, $\mathbf{B}(t)$, and $\mathbf{C}(t)$ are constant. Introduce again the *controllability matrix*

$$\mathbf{K} = (\mathbf{B}, \mathbf{AB}, \ldots, \mathbf{A}^{n-1}\mathbf{B}) \ .$$

By modifying the proof of Lemma 5.2 similarly to Lemma 5.3 the reader can easily verify that Lemma 5.2 holds also for discrete systems with $t_1 \geq n$, and Theorem 5.4 has to be modified as follows.

THEOREM 5.9
The time-invariant discrete linear system is completely controllable if and only if $rank(\mathbf{K}) = n$.

Example 5.12

In the case of the system examined in the previous example, $n = 2$ and

$$\mathbf{K} = (\mathbf{B}, \mathbf{AB}) = \begin{pmatrix} 0 & 1 \\ 1 & 1 \end{pmatrix},$$

which is nonsingular; therefore, $rank(\mathbf{K}) = 2$. Hence, arbitrary $\mathbf{x}_1$ can be obtained at arbitrary $t_1 \geq 2$. Note that the same result was obtained in Example 5.10; however, the direct use of the controllability matrix is much more attractive than the direct computation of $\mathbf{W}(0, t_1)$ and its examination.

Finally, we mention that Theorems 5.5 and 5.6 remain valid, and the output and trajectory controllability of discrete systems can be discussed analogously to the continuous case. The details are left as an exercise to the reader. We conclude this section with an example of output control.

Example 5.13

Consider again the system given in Example 5.11 with the output equation $y = (1, 1)\mathbf{x}$. Assume that the input sequence has to be determined that results in the final output $y(2) = 5$.

From Example 5.11 we see that

$$y(2) = (1, 1)\mathbf{x}(2) = (1, 1) \left[\begin{pmatrix} 1 \\ 0 \end{pmatrix} + \begin{pmatrix} 1 \\ 1 \end{pmatrix} u(0) + \begin{pmatrix} 0 \\ 1 \end{pmatrix} u(1) \right],$$

that is,

$$5 = 1 + 2u(0) + u(1) .$$

Notice that this equation has inifinitely many solutions, we may select, for example, $u(0) = 0$ and $u(1) = 4$.

5.3   Applications

In this section some systems arising in engineering and the social sciences are examined, and the controllability conditions introduced earlier in this chapter are illustrated.

5.3.1   Dynamic Systems in Engineering

1. Consider the simple *harmonic oscillator* (3.53) introduced in Chapter 2 and given in Application 3.5.1-1, which is summarized below:

$$\dot{\mathbf{x}} = \begin{pmatrix} 0 & \omega \\ -\omega & 0 \end{pmatrix} \mathbf{x} + \begin{pmatrix} 0 \\ 1 \end{pmatrix} u .$$

Is it controllable?

To answer this question let us compute the controllability matrix $\mathbf{K} = (\mathbf{b}, \mathbf{Ab})$. Since

$$\mathbf{A} = \begin{pmatrix} 0 & \omega \\ -\omega & 0 \end{pmatrix} \qquad \text{and} \qquad \mathbf{b} = \begin{pmatrix} 0 \\ 1 \end{pmatrix},$$

$$\mathbf{K} = (\mathbf{b}, \mathbf{Ab}) = \begin{pmatrix} 0 & \omega \\ 1 & 0 \end{pmatrix} .$$

Note that $rank(\mathbf{K}) = 2$. Therefore, the system is completely controllable.

2. What about the *damped linear second-order system* of Application 3.5.1-2; is it controllable? In this case,

$$\mathbf{A} = \begin{pmatrix} 0 & 1 \\ -\frac{K}{M} & -\frac{B}{M} \end{pmatrix} , \qquad \mathbf{b} = \begin{pmatrix} 0 \\ \frac{1}{M} \end{pmatrix} ,$$

and, therefore,

$$\mathbf{K} = \begin{pmatrix} 0 & \frac{1}{M} \\ \frac{1}{M} & -\frac{B}{M^2} \end{pmatrix} .$$

This matrix has a full rank; therefore, the system is completely controllable.

3. For the *electrical system* of Application 3.5.1-3,

$$\mathbf{A} = \begin{pmatrix} -\frac{R_1}{L} & -\frac{1}{L} \\ \frac{1}{C} & -\frac{1}{CR_2} \end{pmatrix} , \qquad \mathbf{b} = \begin{pmatrix} \frac{1}{L} \\ 0 \end{pmatrix} .$$

Therefore,

$$\mathbf{K} = \begin{pmatrix} \frac{1}{L} & -\frac{R_1}{L^2} \\ 0 & \frac{1}{LC} \end{pmatrix} .$$

The *rank* of $\mathbf{K}$ is 2, therefore, the system is completely controllable.

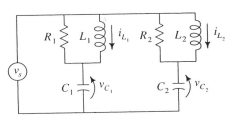

Figure 5.4   An expanded *LRC* circuit.

Investigating the controllability of this electrical circuit was not very interesting. So, let us change it as shown in Figure 5.4. Is this circuit controllable? Let us first find $\mathbf{A}$, $\mathbf{B}$, and $\mathbf{C}$. Our first question is what is the order of the system? Because there are four independent energy storage devices (two inductors and two capacitors), it should be fourth order. Next, what state variables should be used? Unless your physical intuition suggests more convenient variables, choose the energy related variables, i.e., the currents in the inductors and the voltages across the capacitors. Now we can apply Kirchhoff's law to both circuits. The equations are as follows:

$$L\frac{di_L}{dt} + v_C = u$$

$$C\frac{dv_C}{dt} = i_L + \frac{u - v_C}{R} ,$$

which can be rewritten as

$$\frac{di_L}{dt} = -\frac{1}{L}v_C + \frac{1}{L}u$$

$$\frac{dv_C}{dt} = \frac{1}{C}i_L - \frac{1}{RC}v_C + \frac{1}{RC}u .$$

Since the equations for the left and right circuits are the same, except for the subscripts, we can write four differential equations as

$$
\begin{pmatrix} \dot{i}_{L_1} \\ \dot{v}_{C_1} \\ \dot{i}_{L_2} \\ \dot{v}_{C_2} \end{pmatrix} = \begin{pmatrix} 0 & \frac{-1}{L_1} & 0 & 0 \\ \frac{1}{C_1} & \frac{-1}{C_1 R_1} & 0 & 0 \\ 0 & 0 & 0 & \frac{-1}{L_2} \\ 0 & 0 & \frac{1}{C_2} & \frac{-1}{C_2 R_2} \end{pmatrix} \begin{pmatrix} i_{L_1} \\ v_{C_1} \\ i_{L_2} \\ v_{C_2} \end{pmatrix} + \begin{pmatrix} \frac{1}{L_1} \\ \frac{1}{C_1 R_1} \\ \frac{1}{L_2} \\ \frac{1}{C_2 R_2} \end{pmatrix} u \ .
$$

Let the output be the voltage across the right capacitor:

$$
y = (0,0,0,1) \begin{pmatrix} i_{L_1} \\ v_{C_1} \\ i_{L_2} \\ v_{C_2} \end{pmatrix} \ .
$$

Now let us form the controllability matrix, $\mathbf{K}$:

$$
\mathbf{K} = \begin{pmatrix}
\frac{1}{L_1} & \frac{-1}{L_1 C_1 R_1} & \frac{-1}{L_1^2 C_1} + \frac{1}{L_1 C_1^2 R_1^2} & \frac{2}{L_1^2 C_1^2 R_1} - \frac{1}{L_1 C_1^3 R_1^3} \\
\frac{1}{C_1 R_1} & \frac{1}{L_1 C_1} - \frac{1}{C_1^2 R_1^2} & \frac{-2}{L_1 C_1^2 R_1} + \frac{1}{C_1^3 R_1^3} & \frac{-1}{L_1^2 C_1} + \frac{3}{L_1 C_1^3 R_1^2} - \frac{1}{C_1^4 R_1^4} \\
\frac{1}{L_2} & \frac{-1}{L_2 C_2 R_2} & \frac{-1}{L_2^2 C_2} + \frac{1}{L_2 C_2^2 R_2^2} & \frac{2}{L_2^2 C_2^2 R_2} - \frac{1}{L_2 C_2^3 R_2^3} \\
\frac{1}{C_2 R_2} & \frac{1}{L_2 C_2} - \frac{1}{C_2^2 R_2^2} & \frac{-2}{L_2 C_2^2 R_2} + \frac{1}{C_2^3 R_2^3} & \frac{-1}{L_2^2 C_2} + \frac{3}{L_2 C_2^3 R_2^2} - \frac{1}{C_2^4 R_2^4}
\end{pmatrix} \ .
$$

It is easy to see that if

$$
L_1 = L_2 \ , \qquad C_1 = C_2 \ , \qquad \text{and} \qquad R_1 = R_2 \ ,
$$

then rows 1 and 3 as well as rows 2 and 4 are identical. Therefore, the system is not controllable. Intuitively, this makes sense. If we have two identical circuits in parallel, there is no single input that will drive one circuit to one state and the other to a different state.

4. For the *transistor circuit* model (3.80) of Application 3.5.1-4 we know that

$$
\mathbf{A} = \begin{pmatrix} -\frac{h_{ie}}{L} & 0 \\ \frac{h_{fe}}{C} & 0 \end{pmatrix} \qquad \text{and} \qquad \mathbf{b} = \begin{pmatrix} \frac{1}{L} \\ 0 \end{pmatrix} \ .
$$

Therefore,

$$
\mathbf{K} = \begin{pmatrix} \frac{1}{L} & -\frac{h_{ie}}{L^2} \\ 0 & \frac{h_{fe}}{CL} \end{pmatrix} \ .
$$

Because $rank(\mathbf{K}) = 2$, the system is completely controllable.

5. To assess the controllability of the *hydraulic system* of Application 3.5.1-5, we can compute $\mathbf{K}$ as follows. Note first that

$$\mathbf{A} = \begin{pmatrix} -a & a \\ b & -(b+c) \end{pmatrix}, \qquad \mathbf{b} = \begin{pmatrix} d \\ 0 \end{pmatrix},$$

where

$$a = -\frac{1}{R_1 A_1}$$

$$b = \frac{1}{R_1 A_2}$$

$$c = \frac{1}{R_2 A_2}$$

$$d = \frac{1}{A_1}.$$

Therefore,

$$\mathbf{K} = \begin{pmatrix} d & -ad \\ 0 & bd \end{pmatrix}.$$

Because $rank(\mathbf{K}) = 2$, we know that the system is controllable. That is, if someone picks arbitrary h_1 and h_2, you can find a $u(t)$ that will drive the system to this state.

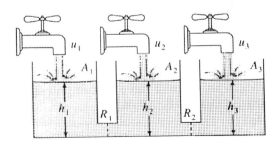

Figure 5.5   A three-tank hydraulic system.

Let us expand this problem a little as shown in Figure 5.5. Make the system symmetric with $R_1 = R_2 = 1/2$ and $A_1 = A_3 = (2/3)A_2$, and we get

$$\frac{d}{dt}\begin{pmatrix} x_1 \\ x_2 \\ x_3 \end{pmatrix} = \begin{pmatrix} -3 & 3 & 0 \\ 2 & -4 & 2 \\ 0 & 3 & -3 \end{pmatrix}\begin{pmatrix} x_1 \\ x_2 \\ x_3 \end{pmatrix} + \begin{pmatrix} b_1 & 0 & 0 \\ 0 & b_2 & 0 \\ 0 & 0 & b_3 \end{pmatrix}\begin{pmatrix} u_1 \\ u_2 \\ u_3 \end{pmatrix},$$

where b_1, b_2, and b_3 are positive constants. Now the question is, if someone requests arbitrary tank heights h_1, h_2, and h_3 can you find an input trajectory that will produce it?

We will consider three cases:

(i)   Only input u_1 is applied.

(ii) Only input u_2 is used.

(iii) Only input u_3 is applied.

We have

$$\mathbf{A} = \begin{pmatrix} -3 & 3 & 0 \\ 2 & -4 & 2 \\ 0 & 3 & -3 \end{pmatrix} \quad \text{and} \quad \mathbf{B} = \begin{pmatrix} b_1 & 0 & 0 \\ 0 & b_2 & 0 \\ 0 & 0 & b_3 \end{pmatrix}.$$

The $\mathbf{A}$ matrix is 3×3 so we will have to compute three blocks of the controllability matrix, which is $(\mathbf{B}, \mathbf{AB}, \mathbf{A}^2\mathbf{B})$:

$$\mathbf{K} = \begin{pmatrix} b_1 & 0 & 0 & \vdots & -3b_1 & 3b_2 & 0 & \vdots & 15b_1 & -21b_2 & 6b_3 \\ 0 & b_2 & 0 & \vdots & 2b_1 & -4b_2 & 2b_3 & \vdots & -14b_1 & 28b_2 & -14b_3 \\ 0 & 0 & b_3 & \vdots & 0 & 3b_2 & -3b_3 & \vdots & 6b_1 & -21b_2 & 15b_3 \end{pmatrix}.$$

Now let us try to control this system with the first input only, that is, let $b_1 = 1$ and $b_2 = b_3 = 0$. In this case,

$$\mathbf{K} = \begin{pmatrix} 1 & 0 & 0 & \vdots & -3 & 0 & 0 & \vdots & 15 & 0 & 0 \\ 0 & 0 & 0 & \vdots & 2 & 0 & 0 & \vdots & -14 & 0 & 0 \\ 0 & 0 & 0 & \vdots & 0 & 0 & 0 & \vdots & 6 & 0 & 0 \end{pmatrix},$$

we do have three linearly independent columns, which are the first, fourth, and seventh. Therefore, $rank(\mathbf{K}) = 3$, and the system is controllable with the first faucet only. However, let us now set $b_2 = 1$ and

$b_1 = b_3 = 0$. Then

$$
\mathbf{K} = \begin{pmatrix} 0\;0\;0\;\vdots\;0 & 3\;0\;\vdots\;0 & -21\;0 \\ 0\;1\;0\;\vdots\;0 & -4\;0\;\vdots\;0 & 28\;0 \\ 0\;0\;0\;\vdots\;0 & 3\;0\;\vdots\;0 & -21\;0 \end{pmatrix},
$$

and we no longer have three linearly independent rows, since the first and third rows are identical. So the system is not controllable by the second faucet alone.

If one sets $b_1 = b_2 = 0$ and $b_3 = 1$, then

$$
\mathbf{K} = \begin{pmatrix} 0\;0\;0\;\vdots\;0\;0 & 0\;\vdots\;0\;0 & 6 \\ 0\;0\;0\;\vdots\;0\;0 & 2\;\vdots\;0\;0 & -14 \\ 0\;0\;1\;\vdots\;0\;0 & -3\;\vdots\;0\;0 & 15 \end{pmatrix}.
$$

Note that the rows are linearly independent again, $rank(\mathbf{K}) = 3$. Consequently, the system is completely controllable with the third faucet only.

6. In the case of the *multiple input electrical system* we know that

$$
\mathbf{A} = \begin{pmatrix} -\frac{R}{L_1} & 0 & -\frac{1}{L_1} \\ 0 & 0 & -\frac{1}{L_2} \\ \frac{1}{C} & \frac{1}{C} & 0 \end{pmatrix} \quad \text{and} \quad \mathbf{B} = \begin{pmatrix} \frac{1}{L_1} & 0 \\ 0 & \frac{1}{L_2} \\ 0 & 0 \end{pmatrix}.
$$

The controllability matrix $\mathbf{K}$ has now the form

$$
\mathbf{K} = \begin{pmatrix} \frac{1}{L_1} & 0 & \frac{-R}{L_1^2} & 0 & \frac{R^2}{L_1^3} - \frac{1}{CL_1^2} & \frac{-1}{CL_1L_2} \\ 0 & \frac{1}{L_2} & 0 & 0 & \frac{-1}{CL_1L_2} & \frac{-1}{CL_2^2} \\ 0 & 0 & \frac{1}{CL_1} & \frac{1}{CL_2} & \frac{-R}{CL_1^2} & 0 \end{pmatrix}.
$$

Notice that the first three columns are linearly independent, therefore, $rank(\mathbf{K}) = 3$, and the system is completely controllable.

7. To check the controllability of the *stick-balancing problem* of Application 3.5.1-7, compute again the controllability matrix

$$
\mathbf{K} = \begin{pmatrix} 0 & -g \\ -g & 0 \end{pmatrix}
$$

of the system. Because the *rank* of $\mathbf{K}$ is 2, the system is completely controllable.

8. For the *cart with two sticks* of Application 3.5.1-8, we have matrices

$$\mathbf{A} = \begin{pmatrix} 0 & 0 & 1 & 0 \\ 0 & 0 & 0 & 1 \\ a_1 & a_2 & 0 & 0 \\ a_3 & a_4 & 0 & 0 \end{pmatrix} \qquad \text{and} \qquad \mathbf{b} = \begin{pmatrix} 0 \\ 0 \\ -c \\ -d \end{pmatrix} .$$

Simple calculation shows that the controllability matrix is as follows:

$$\mathbf{K} = \begin{pmatrix} 0 & -c & 0 & -a_1 c - a_2 d \\ 0 & -d & 0 & -a_3 c - a_4 d \\ -c & 0 & -a_1 c - a_2 d & 0 \\ -d & 0 & -a_3 c - a_4 d & 0 \end{pmatrix} .$$

This will have *rank* less than 4 if

$$c(a_3 c + a_4 d) = d(a_1 c + a_2 d) .$$

Substitute the definitions of a_1, a_2, a_3, a_4, c, and d to obtain equality

$$\frac{1}{ML_1} \left(\frac{M_2 g}{ML_2} \frac{1}{ML_1} + \frac{(M + M_2)g}{ML_2} \frac{1}{ML_2} \right)$$

$$= \frac{1}{ML_2} \left(\frac{(M + M_2)g}{ML_1} \frac{1}{ML_1} + \frac{M_2 g}{ML_1} \frac{1}{ML_2} \right) ,$$

which is equivalent to relation

$$L_1 = L_2 .$$

Therefore, if $L_1 \neq L_2$, the realization is completely controllable. As a consequence, an input can be found that keeps both sticks upright.

9. Our *electrical heating system* was illustrated in Figure 3.29 and is a second order system with matrices

$$\mathbf{A} = \begin{pmatrix} -\frac{A_1 h_1}{C_1} & \frac{A_1 h_1}{C_1} \\ \frac{A_1 h_1}{C_2} & -\frac{A_1 h_1 + A_2 h_2}{C_2} \end{pmatrix} \qquad \text{and} \qquad \mathbf{B} = \begin{pmatrix} 0 \\ \frac{1}{C_2} \end{pmatrix} .$$

The controllability matrix $\mathbf{K}$ is now the following:

$$\mathbf{K} = \begin{pmatrix} 0 & \frac{A_1 h_1}{C_1 C_2} \\ \frac{1}{C_2} & -\frac{A_1 h_1 + A_2 h_2}{C_2^2} \end{pmatrix}$$

which has full rank. Therefore, the system is completely controllable.

10. In the case of $m = 1$ in the *nuclear reactor model* of Application 3.5.1-10 we have

$$\mathbf{A} = \begin{pmatrix} \frac{\rho - \beta}{l} & \lambda_1 \\ \frac{\beta_1}{l} & -\lambda_1 \end{pmatrix} \qquad \text{and} \qquad \mathbf{b} = \begin{pmatrix} 1 \\ 0 \end{pmatrix} .$$

Therefore, the controllability matrix $\mathbf{K}$ has the form

$$\mathbf{K} = \begin{pmatrix} 1 & \frac{\rho - \beta}{l} \\ 0 & \frac{\beta_1}{l} \end{pmatrix} .$$

The *rank* of $\mathbf{K}$ is 2, so the system is completely controllable.

5.3.2 Applications in the Social Sciences

1. Consider first the *predator–prey* model (3.112) and (3.113) with the input "controlled by nature":

$$\dot{G}(t) = (a + u(t))G(t) - bG(t)W(t)$$

$$\dot{W}(t) = -cW(t) + dG(t)W(t) . \qquad (5.30)$$

From Section 3.5.2 we know that with zero input $\bar{u} = 0$, the only positive equilibrium is

$$\bar{G} = \frac{c}{d} , \qquad \bar{W} = \frac{a}{b} .$$

The system is first linearized. If f_1 and f_2 denote the right-hand sides of Equations (5.30), then

$$\frac{\partial f_1}{\partial G} = a + u - bW, \qquad \frac{\partial f_1}{\partial W} = -bG, \qquad \frac{\partial f_1}{\partial u} = G ,$$

$$\frac{\partial f_2}{\partial G} = dW, \qquad \frac{\partial f_2}{\partial W} = -c + dG, \qquad \frac{\partial f_2}{\partial u} = 0 .$$

Therefore, Equation (3.13) implies that the linearized equations have the form

$$\dot{G}_\delta = -\frac{bc}{d}W_\delta + \frac{c}{d}u$$

$$\dot{W}_\delta = \frac{da}{b}G_\delta , \qquad (5.31)$$

where

$$G_\delta = G - \bar{G} = G - \frac{c}{d} \quad \text{and} \quad W_\delta = W - \bar{W} = W - \frac{a}{b} .$$

It is obvious that variables G and W are controllable if and only if G_δ and W_δ are controllable; therefore, it is sufficient to examine the controllability of system (5.31).

By using the notations of Section 5.1.2,

$$\mathbf{A} = \begin{pmatrix} 0 & -\frac{bc}{d} \\ \frac{da}{b} & 0 \end{pmatrix} \quad \text{and} \quad \mathbf{B} = \begin{pmatrix} \frac{c}{d} \\ 0 \end{pmatrix} ,$$

and the controllability matrix $\mathbf{K}$ is as follows:

$$\mathbf{K} = (\mathbf{B}, \mathbf{AB}) = \begin{pmatrix} \frac{c}{d} & 0 \\ 0 & \frac{ac}{b} \end{pmatrix} .$$

From Theorem 5.3 we conclude that the system is completely controllable, since $rank(\mathbf{K}) = 2$.

This interesting result shows that both (the predator and prey) populations are completely controlled by controlling only the growth rate of the prey population.

2. Consider next the *cohort population* model (3.115). For the sake of simplicity select $n = 3$. Then

$$\mathbf{A} = \begin{pmatrix} b_1 & b_2 & b_3 \\ a_1 & 0 & 0 \\ 0 & a_2 & 0 \end{pmatrix} , \quad \mathbf{B} = \begin{pmatrix} 1 & 0 & 0 \\ 0 & 1 & 0 \\ 0 & 0 & 1 \end{pmatrix} .$$

The controllability matrix is the following:

$$\mathbf{K} = \begin{pmatrix} 1 & 0 & 0 & | & b_1 & b_2 & b_3 & | & b_1^2 + a_1 b_2 & b_1 b_2 + b_3 a_2 & b_1 b_3 \\ 0 & 1 & 0 & | & a_1 & 0 & 0 & | & a_1 b_1 & a_1 b_2 & a_1 b_3 \\ 0 & 0 & 1 & | & 0 & a_2 & 0 & | & a_1 a_2 & 0 & 0 \end{pmatrix} .$$

Because the first three columns are linearly independent, $rank(\mathbf{K}) = 3$. That is, the system is completely controllable. This result is not surprising, since the populations of the three age groups are directly controlled by the corresponding input components.

Assume next that immigration into the population is permitted only in the youngest age group, group 1. Then matrix $\mathbf{B}$ is replaced by vector

$\mathbf{b} = (1, 0, 0)^T$, and the controllability matrix is

$$\mathbf{K} = \begin{pmatrix} 1 & b_1 & b_1^2 + a_1 b_2 \\ 0 & a_1 & a_1 b_1 \\ 0 & 0 & a_1 a_2 \end{pmatrix}$$

with $rank(\mathbf{K}) = 3$. Therefore, the entire population is completely controlled by controlling the population of the youngest age group. After one time period, this control has its effect on the second age group; the third age group is then affected after one additional time period and so on. Hence, this control will indirectly control all other age groups as well after a certain time delay.

Assume next that immigration is permitted only into the second age group. Then $\mathbf{b} = (0, 1, 0)^T$, and

$$\mathbf{K} = \begin{pmatrix} 0 & b_2 & b_1 b_2 + a_2 b_3 \\ 1 & 0 & a_1 b_2 \\ 0 & a_2 & 0 \end{pmatrix} .$$

Observe that $rank(\mathbf{K}) = 3$, that is, the system is completely controllable again. Similarly, if only the third age group is controlled, the $\mathbf{b} = (0, 0, 1)^T$, and

$$\mathbf{K} = \begin{pmatrix} 0 & b_3 & b_1 b_3 \\ 0 & 0 & a_1 b_3 \\ 1 & 0 & 0 \end{pmatrix}$$

with $rank(\mathbf{K}) = 3$.

In explaining the last two results note that the control of any age group $i \geq 2$ will have its effect on the first age group via the newborn population after one time period. And then, all other age groups are affected by aging.

3. Consider the modified *arms races* model (3.119), where

$$\mathbf{A} = \begin{pmatrix} -b & a \\ c & -d \end{pmatrix} \qquad \text{and} \qquad \mathbf{B} = \begin{pmatrix} \alpha \\ \beta \end{pmatrix} .$$

The controllability matrix is

$$\mathbf{K} = \begin{pmatrix} \alpha & -b\alpha + a\beta \\ \beta & c\alpha - d\beta \end{pmatrix} .$$

This matrix has full rank if and only if

$$\alpha(c\alpha - d\beta) - \beta(-b\alpha + a\beta) \neq 0 ,$$

that is,

$$\alpha^2 c + \alpha\beta(b - d) - a\beta^2 \neq 0 .$$

This relation holds if and only if

$$\frac{\alpha}{\beta} \neq \frac{d - b \pm \sqrt{(b - d)^2 + 4ac}}{2c} .$$

Hence, the system is completely controllable except for two special values of α/β depending on the coefficients a, b, c, and d.

4. Let us now modify the *warfare* model in the following way. Assume that a guerrilla organization helps the second nation in its war. Then model (3.121) is modified as

$$\dot{X}_1 = -h_2 X_2 - h_3 u(t)$$

$$\dot{X}_2 = -h_1 X_1 , \tag{5.32}$$

where u is the force of the guerrilla organization and h_3 is its hitting power. This time-invariant system is characterized by matrices

$$\mathbf{A} = \begin{pmatrix} 0 & -h_2 \\ -h_1 & 0 \end{pmatrix} \quad \text{and} \quad \mathbf{B} = \begin{pmatrix} -h_3 \\ 0 \end{pmatrix} .$$

The controllability matrix of this system has the form

$$\mathbf{K} = \begin{pmatrix} -h_3 & 0 \\ 0 & h_1 h_3 \end{pmatrix}$$

which has full rank. Hence, the system is completely controllable, which means that the guerrilla organization is able to drive the war to any state $(X_1(t_1), (X_2(t_1))^T$ by the appropriate selection of its activity $u(t)$.

5. Consider again the *epidemics* model (3.124), and assume that the number of infected and circulating individuals can be influenced by an input u (e.g., better and more frequent screening). The corresponding mathematical model then can be written as

$$\dot{x} = -\alpha xy$$

$$\dot{y} = \alpha xy - \beta y - u . \tag{5.33}$$

It is easy to see that with zero input ($\bar{u} = 0$), ($\bar{x}, 0$) is an equilibrium state, where $\bar{x} \geq 0$ is arbitrary. The linearized model has the following form:

$$\dot{\mathbf{x}}_\delta = \begin{pmatrix} 0 & -\alpha\bar{x} \\ 0 & \alpha\bar{x} - \beta \end{pmatrix} \mathbf{x}_\delta + \begin{pmatrix} 0 \\ -1 \end{pmatrix} u \ .$$

The controllability of this linear system can be investigated on the basis of its controllability matrix

$$\mathbf{K} = \begin{pmatrix} 0 & \alpha\bar{x} \\ -1 & -\alpha\bar{x} + \beta \end{pmatrix} \ .$$

If $\bar{x} \neq 0$, then this matrix is nonsingular; that is, the system is completely controllable. If $\bar{x} = 0$, then $\mathbf{K}$ is singular; that is, the system is not controllable.

6. In the case of the *Harrod-type national economy* model (3.10) we can consider $G(t)$ as input. Then

$$\mathbf{A} = (1 + r - rm) \quad \text{and} \quad \mathbf{B} = (-r) \ ,$$

and since $n = 1$ (that is, this is a single-dimensional case), the controllability matrix $\mathbf{K}$ equals $\mathbf{B}$, which is a nonzero constant. Therefore, it has full rank. Hence, the system is completely controllable.

7. Consider now the linear *cobweb* model (3.127), and assume that the demand function can be influenced by an appropriate input u as $d(p) = ap + a_0 + u$. In this case the model is modified as

$$p(t + 1) = \frac{b}{a}p(t) + \frac{b_0 - a_0 - u}{a} \ ,$$

and by introducing a new input

$$\tilde{u} = \frac{b_0 - a_0 - u}{a} \ ,$$

the model reduces to

$$p(t + 1) = \frac{b}{a}p(t) + \tilde{u} \ . \tag{5.34}$$

This system is obviously completely controllable, since the controllability matrix is $\mathbf{K} = (1)$, which is nonzero and hence nonsingular.

8. We will now modify the model (3.128) of *interrelated markets* similar to the previous case by introducing the inputs u_i, which add to the constants a_{i0} of the demand functions. Their meaning is the same as was presented for the input of the previous model. In this case, system (3.130) is modified as

$$\dot{\mathbf{p}} = \mathbf{K}((\mathbf{A} - \mathbf{B})\mathbf{p} + \mathbf{a}_0 - \mathbf{b}_0 + \mathbf{u}) \ . \tag{5.35}$$

Introduce the transformed input

$$\hat{u}(t) = K \cdot (a_0 - b_0 + u(t)) \, ,$$

and note that any value of $\hat{u}(t)$ can be obtained by the suitable selection of $u(t)$:

$$u(t) = K^{-1}\hat{u}(t) + b_0 - a_0 \, .$$

Therefore, the controllability of systems (5.35) and

$$\dot{p} = K(A - B)p + \hat{u} \tag{5.36}$$

are equivalent. The controllability matrix

$$(I, K(A - B), \ldots, [K(A - B)]^{n-1})$$

of system (5.36) has full rank, since the first n columns are linearly independent. Hence, system (5.35) is completely controllable.

9. Consider finally the *oligopoly* problem. Assume that the government can control the market (with certain tax breaks, export subvention, etc.) by a single input $u(t)$, which shows the cost reduction of the firms per unit output. Therefore, the modified cost functions are

$$C_k(x_k) = (b_k - u)x_k + c_k \qquad (k = 1, 2, \ldots, N) \, ,$$

and as the consequence of the additional term $u x_k$, the discrete model (3.134) is modified as

$$x(t + 1) = A_c x(t) + f_c - \frac{1}{2a}1u(t) \, , \tag{5.37}$$

where $\mathbf{1}$ is now the N-dimensional vector with all components being unity. Consider the solution $z(t)$ $(t = 0, 1, \ldots)$ of the initial value problem

$$z(t + 1) = A_c z(t) + f_c, \qquad z(0) = 0 \, .$$

Then $y(t) = x(t) - z(t)$ $(t = 0, 1, \ldots)$ satisfies the difference equation

$$y(t + 1) = A_c y(t) - \frac{1}{2a}1u(t) \, ; \tag{5.38}$$

furthermore, $y(0) = x(0)$. It is obvious that system (5.37) is completely controllable if and only if the same holds for system (5.38) Note first that

$$A_c \cdot 1 = \frac{1 - N}{2}1 \, ,$$

which implies that the controllability matrix is the following:

$$\mathbf{K} = -\left(\frac{1}{2a}\cdot 1, \frac{1}{2a}\cdot\frac{1-N}{2}\cdot 1, \ldots, \frac{1}{2a}\left(\frac{1-N}{2}\right)^{N-1}1\right).$$

The columns of this matrix are dependent; therefore, systems (5.38) and (5.37) are not completely controllable.

Assume next that the cost of each firm is controlled by different input components as

$$C_k(x_k) = (b_k - u_k)x_k + c_k \qquad (k = 1, 2, \ldots, N).$$

Then model (3.134) is modified as

$$\mathbf{x}(t+1) = \mathbf{A}_c\mathbf{x}(t) + \mathbf{f}_c - \frac{1}{2a}\mathbf{u}(t).\tag{5.39}$$

Introduce again function $\mathbf{y}(t) = \mathbf{x}(t) - \mathbf{z}(t)$, where $\mathbf{z}(t)$ is the same as before, to get system

$$\mathbf{y}(t+1) = \mathbf{A}_c\mathbf{y}(t) - \frac{1}{2a}\mathbf{Iu}(t).\tag{5.40}$$

In this case the controllability matrix is as follows:

$$\mathbf{K} = \left(-\frac{1}{2a}\mathbf{I}, -\frac{1}{2a}\mathbf{A}_c, \ldots, -\frac{1}{2a}\mathbf{A}_c^{N-1}\right),$$

which has full rank, since the first N columns are linearly independent. Hence, systems (5.39) and (5.40) are completely controllable.

Finally, we note the the controllability of the alternative oligopoly models (3.138) and (3.139) can be examined in an analogous manner.

Problems

1. Examine the controllability of system

$$\dot{\mathbf{x}} = \begin{pmatrix} \frac{1}{t} & 0 \\ 0 & \frac{1}{t} \end{pmatrix}\mathbf{x} + \begin{pmatrix} 1 \\ 1 \end{pmatrix}u.$$

2. Is system

$$\dot{\mathbf{x}} = \begin{pmatrix} 1 & 1 \\ 2 & 2 \end{pmatrix}\mathbf{x} + \begin{pmatrix} 1 \\ 0 \end{pmatrix}u$$

completely controllable in $[0, 1]$? Use Theorem 5.1.

3. Compute matrix $\mathbf{W}(t_0, t_1)$ for the system described below, and illustrate Properties (i) and (ii) of Theorem 5.2. Select the $[0, 1]$ interval.

$$\dot{\mathbf{x}} = \begin{pmatrix} 1 & 1 \\ 2 & 2 \end{pmatrix} \mathbf{x} + \begin{pmatrix} 1 \\ 0 \end{pmatrix} u \ .$$

4. Determine if the system

$$\dot{\mathbf{x}} = \begin{pmatrix} 1 & 1 \\ 2 & 2 \end{pmatrix} \mathbf{x} + \begin{pmatrix} 1 \\ 0 \end{pmatrix} u$$

is completely controllable by using the controllability matrix (5.10).

5. Use Theorem 5.1 to determine if this system

$$\dot{\mathbf{x}} = \begin{pmatrix} 2 & 1 \\ 0 & 2 \end{pmatrix} \mathbf{x} + \begin{pmatrix} 1 \\ 1 \end{pmatrix} u$$

is completely controllable in $[0, 1]$.

6. Compute matrix $\mathbf{W}(t_0, t_1)$ for system

$$\dot{\mathbf{x}} = \begin{pmatrix} 2 & 1 \\ 0 & 2 \end{pmatrix} \mathbf{x} + \begin{pmatrix} 1 \\ 1 \end{pmatrix} u \ ,$$

and illustrate Properties (i) and (ii) of Theorem 5.2. Select the $[0, 1]$ interval.

7. Determine if system

$$\dot{\mathbf{x}} = \begin{pmatrix} 2 & 1 \\ 0 & 2 \end{pmatrix} \mathbf{x} + \begin{pmatrix} 1 \\ 1 \end{pmatrix} u$$

is completely controllable by using the controllability matrix (5.10).

8. Is there any input for system

$$\dot{\mathbf{x}} = \begin{pmatrix} 1 & 1 \\ 2 & 2 \end{pmatrix} \mathbf{x} + \begin{pmatrix} 1 \\ 0 \end{pmatrix} u, \qquad \mathbf{x}(0) = \begin{pmatrix} 1 \\ 1 \end{pmatrix} ,$$

which controls the trajectory to

$$\mathbf{x}(t) = \begin{pmatrix} t + 1 \\ 1 \end{pmatrix}$$

in the interval $[0, 1]$?

9. Is there any input for system

$$\dot{x} = \begin{pmatrix} 2 & 1 \\ 0 & 2 \end{pmatrix} x + \begin{pmatrix} 1 \\ 1 \end{pmatrix} u, \qquad x(0) = \begin{pmatrix} 0 \\ 1 \end{pmatrix},$$

which controls the trajectory to

$$x(t) = \begin{pmatrix} t \\ t+1 \end{pmatrix}$$

in the interval $[0, 1]$?

10. Is the electric circuit system

$$L\frac{di(t)}{dt} + (R_1 + R_2)i(t) = u(t)$$

introduced in Problem 3.13 completely controllable?

11. Discuss the controllability of the mechanical system

$$\dot{x} = \begin{pmatrix} 0 & 1 \\ 0 & -6 \end{pmatrix} x + \begin{pmatrix} 0 \\ 2 \end{pmatrix} u$$

introduced in Problem 3.7.

12. Is the system

$$x(t+1) = \begin{pmatrix} 1 & 1 \\ 2 & 2 \end{pmatrix} x(t) + \begin{pmatrix} 1 \\ 0 \end{pmatrix} u(t)$$

completely controllable? Use Theorem 5.8, and select $t_1 = 2$.

13. Use the controllability matrix K to determine if the following system is completely controllable:

$$x(t+1) = \begin{pmatrix} 1 & 1 \\ 2 & 2 \end{pmatrix} x(t) + \begin{pmatrix} 1 \\ 0 \end{pmatrix} u(t).$$

14. Use Theorem 5.8 and $t_1 = 2$ to determine if the following discrete system is completely controllable:

$$x(t+1) = \begin{pmatrix} 2 & 1 \\ 0 & 2 \end{pmatrix} x(t) + \begin{pmatrix} 1 \\ 1 \end{pmatrix} u(t).$$

15. Use the controllability matrix $\mathbf{K}$ to determine if the following system is completely controllable:

$$\mathbf{x}(t+1) = \begin{pmatrix} 2 & 1 \\ 0 & 2 \end{pmatrix} \mathbf{x}(t) + \begin{pmatrix} 1 \\ 1 \end{pmatrix} u(t).$$

16. Is the output of the system

$$\dot{\mathbf{x}} = \begin{pmatrix} 1 & 1 \\ 2 & 2 \end{pmatrix} \mathbf{x} + \begin{pmatrix} 1 \\ 0 \end{pmatrix} u$$

$$y = (1,1)\mathbf{x},$$

controllable in [0,1]?

17. Is the output of the system

$$\dot{\mathbf{x}} = \begin{pmatrix} 2 & 1 \\ 0 & 2 \end{pmatrix} \mathbf{x} + \begin{pmatrix} 1 \\ 1 \end{pmatrix} u$$

$$y = (0,1)\mathbf{x},$$

controllable in [0,1]?

18. Discuss the output controllability of system

$$\dot{\mathbf{x}} = \mathbf{A}\mathbf{x} + \mathbf{B}\mathbf{u}$$

$$\mathbf{y} = \mathbf{C}\mathbf{x} + \mathbf{D}\mathbf{u}.$$

19. Prove that for any $n \times n$ continuous matrix $\mathbf{A}(t)$ there is a continuous n-vector $\mathbf{b}(t)$ such that system

$$\dot{\mathbf{x}}(t) = \mathbf{A}(t)\mathbf{x}(t) + \mathbf{b}(t)u(t)$$

is completely controllable. That is, if time-dependent $\mathbf{b}$ is allowed, appropriate single-dimensional input always can control the system.

20. (i) Prove Lemma 5.2 for discrete systems. (ii) Prove Theorem 5.9.

21. Assume that the time invariant system $\dot{\mathbf{x}} = \mathbf{A}\mathbf{x} + \mathbf{B}\mathbf{u}$ is completely controllable, and matrix $\bar{\mathbf{A}}$ is sufficiently close to $\mathbf{A}$ and $\bar{\mathbf{B}}$ is sufficiently close to $\mathbf{B}$. Prove that system $\dot{\mathbf{x}} = \bar{\mathbf{A}}\mathbf{x} + \bar{\mathbf{B}}\mathbf{u}$ is also completely controllable.

22.   Let α be a real constant. Prove that the time invariant linear system $\dot{\mathbf{x}} = \mathbf{Ax} + \mathbf{Bu}$ is completely controllable if and only if $\dot{\mathbf{x}} = (\mathbf{A} + \alpha\mathbf{I})\mathbf{x} + \mathbf{Bu}$ is completely controllable.

23.   Prove that system $\dot{\mathbf{x}} = \mathbf{Ax} + \mathbf{Bu}$ is completely controllable if and only if system $\dot{\mathbf{x}} = \mathbf{Ax} + \mathbf{BB}^T\mathbf{v}$ is completely controllable.

24.   Let m be the degree of the minimal-polynomial of $\mathbf{A}$. Show that $rank(\mathbf{K}) = rank(\mathbf{B}, \mathbf{AB}, ..., \mathbf{A}^{m-1}\mathbf{B})$.

25.   Show that matrices $\mathbf{K}$ and $e^{\mathbf{A}}\mathbf{K}$ have identical range spaces, where $\mathbf{K}$ is the controllability matrix of system $\dot{\mathbf{x}} = \mathbf{Ax} + \mathbf{Bu}$.

chapter six

Observability

In the case of many practical systems, the state cannot be measured directly; only the input and output are known. However, complete knowledge of the state is needed for predictions and to describe the dynamic law driving the system. In certain cases we are able to determine the state on the basis of the applied input and observed output values; in other cases we cannot. In this chapter we will discuss the observability of the state of dynamic systems.

DEFINITION 6.1   *The initial state* $\mathbf{x}_0$ *of a continuous (or discrete) system is said to be observable in interval* $[t_0, t_1]$ *if the trajectories of* $\mathbf{u}(t)$ *and* $\mathbf{y}(t)$ *for* $t \in [t_0, t_1]$ *(or for* $t = 0, 1, \ldots, t_1 - 1$*) uniquely determine* $\mathbf{x}_0$*. This concept is illustrated in Figure 6.1.*

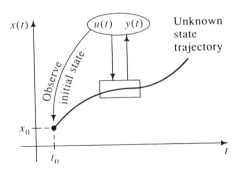

Figure 6.1   Concept of observability.

In this chapter, practical conditions will be derived for the observability of the initial state and methods will be introduced for determining it. In the first part of this chapter continuous linear systems will be discussed, and in the second part discrete linear systems will be examined. We will introduce and use an analogy between the derived observability criteria and the conditions for the complete controllability of linear systems.

6.1 Continuous Systems

In this section the observability of the initial state $\mathbf{x}_0$ of the continuous linear system

$$\dot{\mathbf{x}} = \mathbf{A}(t)\mathbf{x} + \mathbf{B}(t)\mathbf{u}, \quad \mathbf{x}(t_0) = \mathbf{x}_0 \tag{6.1}$$

$$\mathbf{y} = \mathbf{C}(t)\mathbf{x} \tag{6.2}$$

will be examined, where $\mathbf{A}(t)$, $\mathbf{B}(t)$, and $\mathbf{C}(t)$ are continuous matrices for $t \geq t_0$ and their sizes are $n \times n$, $n \times m$, and $p \times n$, respectively. By using the methodology of this chapter, the observability of the state of the system at any time period $t^*(t^* \geq t_0)$ can be examined, since we can always consider t^* as the initial time period, and the input, state and output of the system can be considered only for $t \geq t^*$.

6.1.1 General Conditions

The general solution formula (3.20) for the output of a continuous linear system can be written as

$$\mathbf{C}(t)\phi(t, t_0)\mathbf{x}_0 = \mathbf{y}(t) - \int_{t_0}^{t} \mathbf{C}(t)\phi(t, \tau)\mathbf{B}(\tau)\mathbf{u}(\tau)\, d\tau \ . \tag{6.3}$$

Assume that for an interval $[t_0, t_1]$ the values of $\mathbf{u}(t)$ and $\mathbf{y}(t)$ are known; then, for all $t \in [t_0, t_1]$, the right-hand side of Equation (6.3) is known. Our first condition for the observability of the initial state $\mathbf{x}_0$ is based on the mapping

$$B(\mathbf{x})(t) = \mathbf{C}(t)\phi(t, t_0)\mathbf{x}$$

with domain $\mathbf{R}^n$ and range in the set of the p-dimensional continuous functions. Once again we note that there must not be confusion between this mapping and the system matrix $\mathbf{B}(t)$. The null space of this mapping is defined as

$$N(B) = \{\mathbf{x} \mid B(\mathbf{x})(t) = \mathbf{0} \quad \text{for all} \quad t \in [t_0, t_1]\} \ .$$

It is obvious from the linearity of mapping B that Equation (6.3) uniquely determines $\mathbf{x}_0$ if and only if $N(B) = \{\mathbf{0}\}$.

Before particular observability conditions are derived, the relation between the observability of system (6.1) and (6.2) and the controllability of its adjoint system

$$\dot{\mathbf{z}} = -\mathbf{A}^T(t)\mathbf{z} + \mathbf{C}^T(t)\mathbf{v}$$

$$\mathbf{w} = \mathbf{B}^T(t)\mathbf{z} \tag{6.4}$$

will be analyzed.

We know from the previous chapter that the state of the adjoint system can be controlled to $\mathbf{z}_1$ in the interval $[t_0, t_1]$ if and only if vector $\phi^T(t_1, t_0)\mathbf{z}_1 - \mathbf{z}_0$ belongs to the range space of mapping

$$A_a(\mathbf{v}) = \int_{t_0}^{t_1} \phi^T(\tau, t_0)\mathbf{C}^T(\tau)\mathbf{v}(\tau)\,d\tau \ .$$

Here we use the fact that the fundamental matrix of Equation (6.4) is given as $\phi^T(t_0, t)$ (see Section 3.3.5). Our analysis is based on the following results.

LEMMA 6.1
$R(A_a)$ and $N(B)$ are orthogonal complementary subspaces in $\mathbf{R}^n$.

PROOF   Assume first that $\mathbf{r} \in R(A_a)$ and $\mathbf{s} \in N(B)$. Then

$$\mathbf{r} = \int_{t_0}^{t_1} \phi^T(\tau, t_0)\mathbf{C}^T(\tau)\mathbf{v}(\tau)\,d\tau$$

with some input function $\mathbf{v}(t)$, and

$$\mathbf{C}(t)\phi(t, t_0)\mathbf{s} = \mathbf{0}$$

for all $t \in [t_0, t_1]$. Therefore,

$$\mathbf{r}^T\mathbf{s} = \int_{t_0}^{t_1} \mathbf{v}^T(\tau)(\mathbf{C}(\tau)\phi(\tau, t_0)\mathbf{s})\,d\tau = 0 \ ,$$

because the integrand is zero for all t.

Assume next that for a vector $\mathbf{s}$, $\mathbf{r}^T\mathbf{s} = 0$ with all $\mathbf{r} \in R(A_a)$. We shall prove that $\mathbf{s} \in N(B)$. Note first that by selecting the input function as $\mathbf{v}(t) = \mathbf{C}(t)\,\phi(t, t_0)\mathbf{s}$, we conclude that vector

$$\mathbf{r} = \int_{t_0}^{t_1} \phi^T(\tau, t_0)\mathbf{C}^T(\tau)\mathbf{C}(\tau)\phi(\tau, t_0)\mathbf{s}\,d\tau$$

belongs to $R(A_a)$. Consequently,

$$0 = \mathbf{r}^T \mathbf{s} = \int_{t_0}^{t_1} \mathbf{s}^T \phi^T(\tau, t_0) \mathbf{C}^T(\tau) \mathbf{C}(\tau) \phi(\tau, t_0) \mathbf{s} \, d\tau$$

$$= \int_{t_0}^{t_1} \| \mathbf{C}(\tau) \phi(\tau, t_0) \mathbf{s} \|_2^2 \, d\tau \; ,$$

which implies that $\mathbf{C}(\tau)\phi(\tau, t_0)\mathbf{s} = \mathbf{0}$ for all $t \in [t_0, t_1]$, that is, $\mathbf{s} \in N(B)$. Thus, the proof is complete.   ∎

COROLLARY 6.1

System (6.1) and (6.2) is observable in interval $[t_0, t_1]$ if and only if the state of the adjoint system (6.4) can be controlled to arbitrary state $\mathbf{z}_1$ at time t_1 from any initial state $\mathbf{z}_0$ at t_0. Because the adjoint of the adjoint system is the original system, the state of system (6.1) and (6.2) can be controlled to any arbitrary state $\mathbf{x}_1$ at t_1 from any initial state $\mathbf{x}_0$ at t_0 if and only if its adjoint is observable in the interval $[t_0, t_1]$. Therefore, the observability of any continuous linear system can be examined by using the methodology of the previous chapter. This idea will be used in the discussions to follow.

Use Lemma 5.1 to see that $R(A_a)$ coincides with the range space of the controllability Gramian of the adjoint system, which is now denoted by

$$\mathbf{M}(t_0, t_1) = \int_{t_0}^{t_1} \phi^T(\tau, t_0) \mathbf{C}^T(\tau) \mathbf{C}(\tau) \phi(\tau, t_0) \, d\tau \; . \qquad (6.5)$$

The following result is therefore, the obvious consequence of Lemma 6.1 and Theorem 1.10.

LEMMA 6.2

Vector $\mathbf{v}$ is in $N(B)$ if and only if it belongs to the null space of matrix $\mathbf{M}(t_0, t_1)$.

Our first observability criteria are given in the following theorem.

THEOREM 6.1

It is possible to determine $\mathbf{x}_0$ with in an additive constant vector, which is in $N(\mathbf{M}(t_0, t_1))$. If $\mathbf{M}(t_0, t_1)$ is nonsingular, then $\mathbf{x}_0$ can be determined uniquely.

PROOF Let $\mathbf{x}_0$ and $\tilde{\mathbf{x}}_0$ be two solutions of Equation (6.3). Then

$$\mathbf{C}(t)\phi(t,t_0)\mathbf{x}_0 = \mathbf{C}(t)\phi(t,t_0)\tilde{\mathbf{x}}_0 \qquad \text{(for all } t \in [t_0,t_1]) ,$$

which can be written as

$$\mathbf{C}(t)\phi(t,t_0)(\mathbf{x}_0 - \tilde{\mathbf{x}}_0) = 0 \qquad \text{(for all } t \in [t_0,t_1]) .$$

This relation holds if and only if

$$\mathbf{x}_0 - \tilde{\mathbf{x}}_0 \in N(B) , \tag{6.6}$$

that is, if $\mathbf{x}_0$ and $\tilde{\mathbf{x}}_0$ differ by a vector belonging to $N(B)$.

If $\mathbf{M}(t_0,t_1)$ is nonsingular, then $N(\mathbf{M}(t_0,t_1)) = \{\mathbf{0}\}$. From Lemma 6.2 we conclude that $N(B) = \{\mathbf{0}\}$; therefore, (6.6) holds if and only if $\mathbf{x}_0 - \tilde{\mathbf{x}}_0 = \mathbf{0}$. Hence, Equation (6.3) has a unique solution. ∎

REMARK 6.1 If the initial state $\mathbf{x}_0$ of a continuous linear system can be uniquely determined on the basis of input and output values on any interval $[t_0,t_1]$ $(t_1 > t_0)$, then the system is called *completely observable*. ∎

The algorithm that decides whether a given system is completely observable or not consists of the following steps:

Step 1 Compute the fundamental matrix $\phi(t,\tau)$.

Step 2 Determine matrix $\mathbf{M}(t_0,t_1)$.

Step 3 Find $rank(\mathbf{M}(t_0,t_1))$. If it equals n, then the system is completely observable; otherwise it is not.

We note here that standard computer packages are available for determining $\phi(t,\tau)$, for computing $\mathbf{M}(t_0,t_1)$ by numerical integration, and for finding the rank of $\mathbf{M}(t_0,t_1)$.

If the system is observable, then $\mathbf{x}_0$ can be determined in the following way. Equation (6.3) implies that at $t = t_0$,

$$\mathbf{C}(t_0)\mathbf{x}_0 = \mathbf{y}(t_0) ,$$

and select sequentially different values of $t^{(1)}, t^{(2)}, \ldots, t^{(k)}$ and substitute them into Equation (6.3). Then a system of linear equations is obtained for the unknown $\mathbf{x}_0$:

$$\mathbf{C}(t_0)\mathbf{x}_0 = \mathbf{y}(t_0)$$

$$\mathbf{C}(t^{(i)})\phi(t^{(i)}, t_0)\mathbf{x}_0 = \mathbf{y}(t^{(i)}) - \int_{t_0}^{t^{(i)}} \mathbf{C}(t^{(i)})\phi(t^{(i)}, \tau)\mathbf{B}(\tau)\mathbf{u}(\tau)d\tau$$

for $i = 1, 2, \ldots, k$. If there is a unique solution, it has to be selected as
$\mathbf{x}_0$. Otherwise select a new value $t^{(k+1)}$, and check the uniqueness of the
solution of the new system of equations, which has the same equations
as the previous system and in addition the new equation with this newly
selected $t^{(k+1)}$. Repeat this process until a unique solution is obtained.

Example 6.1

Consider again the system

$$\dot{\mathbf{x}} = \begin{pmatrix} 0 & \omega \\ -\omega & 0 \end{pmatrix} \mathbf{x} + \begin{pmatrix} 0 \\ 1 \end{pmatrix} u, \qquad \mathbf{x}(0) = \mathbf{x}_0 \,,$$

$$y = (1, 1)\mathbf{x} \,,$$

which was the subject of earlier examples. In Example 2.6 we have
seen that

$$\phi(t, \tau) = \begin{pmatrix} \cos \omega(t - \tau) & \sin \omega(t - \tau) \\ -\sin \omega(t - \tau) & \cos \omega(t - \tau) \end{pmatrix} \,.$$

Therefore,

$\mathbf{M}(0, t_1)$

$$= \int_0^{t_1} \begin{pmatrix} \cos \omega\tau & -\sin \omega\tau \\ \sin \omega\tau & \cos \omega\tau \end{pmatrix} \begin{pmatrix} 1 \\ 1 \end{pmatrix} (1, 1) \begin{pmatrix} \cos \omega\tau & \sin \omega\tau \\ -\sin \omega\tau & \cos \omega\tau \end{pmatrix} d\tau$$

$$= \int_0^{t_1} \begin{pmatrix} \cos \omega\tau - \sin \omega\tau \\ \sin \tau + \cos \omega\tau \end{pmatrix} (\cos \omega\tau - \sin \omega\tau, \sin \omega\tau + \cos \omega\tau) d\tau$$

$$= \int_0^{t_1} \begin{pmatrix} 1 - 2\sin \omega\tau \cos \omega\tau & \cos^2 \omega\tau - \sin^2 \omega\tau \\ \cos^2 \omega\tau - \sin^2 \omega\tau & 1 + 2\sin \omega\tau \cos \omega\tau \end{pmatrix} d\tau$$

$$= \int_0^{t_1} \begin{pmatrix} 1 - \sin 2\omega\tau & \cos 2\omega\tau \\ \cos 2\omega\tau & 1 + \sin 2\omega\tau \end{pmatrix} d\tau$$

$$= \begin{pmatrix} t_1 + \frac{\cos 2\omega t_1 - 1}{2\omega} & \frac{\sin 2\omega t_1}{2\omega} \\ \frac{\sin 2\omega t_1}{2\omega} & t_1 - \frac{\cos 2\omega t_1 - 1}{2\omega} \end{pmatrix} = \begin{pmatrix} t_1 - \frac{\sin^2 \omega t_1}{\omega} & \frac{\sin \omega t_1 \cos \omega t_1}{\omega} \\ \frac{\sin \omega t_1 \cos \omega t_1}{\omega} & t_1 + \frac{\sin^2 \omega t_1}{\omega} \end{pmatrix} \,.$$

The nonsingularity of this matrix can be examined by computing its determinant, which is

$$t_1^2 - \frac{\sin^4 \omega t_1}{\omega^2} - \frac{\sin^2 \omega t_1 \cos^2 \omega t_1}{\omega^2}$$

$$= t_1^2 - \frac{\sin^4 \omega t_1 + \sin^2 \omega t_1 (1 - \sin^2 \omega t_1)}{\omega^2} = t_1^2 - \frac{\sin^2 \omega t_1}{\omega^2} .$$

Introduce the new variable $\alpha = \omega t_1$, then the determinant equals

$$\frac{\alpha^2}{\omega^2} - \frac{\sin^2 \alpha}{\omega^2} = \frac{1}{\omega^2}(\alpha^2 - \sin^2 \alpha) ,$$

which is positive for all $\alpha > 0$. Hence, $\mathbf{M}(0, t_1)$ is nonsingular for all $t_1 > 0$; therefore, the initial state $\mathbf{x}_0$ is observable for all $t_1 > 0$. Hence the system is completely observable.

Matrix $\mathbf{M}(t_0, t_1)$ is usually called the *observability Gramian*. Since it is the controllability Gramian of the adjoint system (6.4), Theorem 5.2 implies the following result.

THEOREM 6.2
Matrix $\mathbf{M}(t_0, t_1)$ satisfies the following properties:

(i)   *It is symmetric.*

(ii) *It is positive semidefinite.*

(iii) $(\partial/\partial t)\mathbf{M}(t, t_1) = -\mathbf{A}^T(t)\mathbf{M}(t, t_1) - \mathbf{M}(t, t_1)\mathbf{A}(t) - \mathbf{C}^T(t)\mathbf{C}(t), \mathbf{M}(t_1, t_1) = 0.$

(iv) $\mathbf{M}(t_0, t_1) = \mathbf{M}(t_0, t) + \phi^T(t, t_0)\mathbf{M}(t, t_1)\phi(t, t_0).$

6.1.2   *Time-Invariant Systems*

In this section the special case of time-invariant systems will be discussed. That is, assume that matrices $\mathbf{A}(t), \mathbf{B}(t)$, and $\mathbf{C}(t)$ are time-independent.

Introduce first the *observability matrix*

$$\mathbf{L} = \begin{pmatrix} \mathbf{C} \\ \mathbf{CA} \\ \mathbf{CA}^2 \\ \cdots \\ \mathbf{CA}^{n-1} \end{pmatrix} . \tag{6.7}$$

From the corollary of Lemma 6.1 we get the following results.

LEMMA 6.3

The null space and range space of $\mathbf{M}(t_0, t_1)$ *for all* $t_1 > t_0$ *coincide with the null space and range space of matrix*

$$\mathbf{M}_T = \mathbf{L}^T \mathbf{L} \,. \tag{6.8}$$

THEOREM 6.3

The time-invariant continuous linear system is observable for arbitrary $t_1 > t_0$ *if and only if the rank of the observability matrix* $\mathbf{L}$ *equals* n.

Note that the condition of the theorem holds if and only if the columns of matrix $\mathbf{L}$ are linearly independent.

Example 6.2

Consider again the system

$$\dot{\mathbf{x}} = \begin{pmatrix} 0 & w \\ -w & 0 \end{pmatrix} \mathbf{x} + \begin{pmatrix} 0 \\ 1 \end{pmatrix} u, \qquad \mathbf{x}(0) = \mathbf{x}_0 \,,$$

$$y = (1, 1)\mathbf{x}$$

which was the subject of our earlier Example 6.1. In that example we examined the observability of the system by verifying that the observability Gramian $\mathbf{M}(0, t_1)$ is nonsingular. The same result will be obtained now based on the observability matrix $\mathbf{L}$. Note that in this case $n = 2$, and

$$\mathbf{L} = \begin{pmatrix} \mathbf{C} \\ \mathbf{CA} \end{pmatrix} = \begin{pmatrix} 1 & 1 \\ -w & w \end{pmatrix} \,.$$

Obviously $rank(\mathbf{L}) = 2$ for all $w \neq 0$. Hence the system is completely observable. This example illustrates that the direct application of the observability matrix is much more attractive (similar to controllability) than the computation of the observability Gramian even in cases when the integrals can be given in closed form.

Example 6.3

Consider next the satellite problem, which was the subject of our earlier Example 5.5. The observability of this system is now examined. In this case $n = 4$ and

$$
A = \begin{pmatrix} 0 & 1 & 0 & 0 \\ 3\omega^2 & 0 & 0 & 2\omega \\ 0 & 0 & 0 & 1 \\ 0 & -2\omega & 0 & 0 \end{pmatrix} .
$$

Assume first that the radius r and angle θ can be measured. Then x_1 and x_3 are the components of the output; therefore,

$$
C = \begin{pmatrix} 1 & 0 & 0 & 0 \\ 0 & 0 & 1 & 0 \end{pmatrix}
$$

and the observability matrix has the form

$$
L = \begin{pmatrix} C \\ CA \\ CA^2 \\ CA^3 \end{pmatrix} = \begin{pmatrix} 1 & 0 & 0 & 0 \\ 0 & 0 & 1 & 0 \\ 0 & 1 & 0 & 0 \\ 0 & 0 & 0 & 1 \\ 3\omega^2 & 0 & 0 & 2\omega \\ 0 & -2\omega & 0 & 0 \\ 0 & -\omega^2 & 0 & 0 \\ -6\omega^3 & 0 & 0 & -4\omega^2 \end{pmatrix} .
$$

It is easy to see that the first four rows are linearly independent, so $rank(L) = 4$. That is, x_0 is completely observable.

Assume next that only the radius is measurable. In this case

$$
C_1 = (1, 0, 0, 0) ,
$$

and

$$
L_1 = \begin{pmatrix} C_1 \\ C_1 A \\ C_1 A^2 \\ C_1 A^3 \end{pmatrix} = \begin{pmatrix} 0 & 0 & 1 & 0 \\ 0 & 1 & 0 & 0 \\ 3\omega^2 & 0 & 0 & 2\omega \\ 0 & -\omega^2 & 0 & 0 \end{pmatrix} .
$$

Observe that the last row is the $(-\omega^2)$ multiple of the second row. Therefore, $rank(L_1) < 4$, that is, the system is not observable with the only output $y_1 = x_1$.

Assume now that only the angle θ is measurable. In this case

$$
C_2 = (0, 0, 1, 0) ,
$$

and

$$
L_2 = \begin{pmatrix} C_2 \\ C_2 A \\ C_2 A^2 \\ C_2 A^3 \end{pmatrix} = \begin{pmatrix} 0 & 0 & 1 & 0 \\ 0 & 0 & 0 & 1 \\ 0 & -2\omega & 0 & 0 \\ -6\omega^3 & 0 & 0 & -4\omega^2 \end{pmatrix} .
$$

It is easy to see that $rank(\mathbf{L}_2) = 4$, that is, the system is completely observable with the only output $y_2 = x_3$.

In summary, the loss of the measurements on the radius does not destroy observability, but the loss of measurements of the angle θ does. This property is illustrated in Figure 6.2.

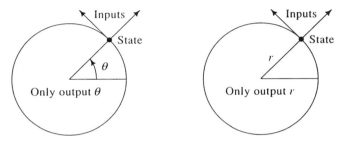

Figure 6.2 Observable and non-observable satellite models.

Note that observability does not depend on the properties of $\mathbf{B}$; hence all results of this section remain true if $\mathbf{B}$ is time-dependent.

Note that for large systems, the rank of matrix $\mathbf{L}$ can be determined by using standard program packages.

Similar to the case of controllability, one can easily verify the following results.

THEOREM 6.4

Assume that the rank r of matrix $\mathbf{L}$ is less than n. Then there exists a nonsingular matrix $\mathbf{T}$ such that

$$\bar{\mathbf{A}} = \mathbf{T}\mathbf{A}\mathbf{T}^{-1} = \begin{pmatrix} \bar{\mathbf{A}}_{11} & \mathbf{O} \\ \bar{\mathbf{A}}_{21} & \bar{\mathbf{A}}_{22} \end{pmatrix},$$

$$\bar{\mathbf{B}} = \mathbf{T}\mathbf{B} = \begin{pmatrix} \bar{\mathbf{B}}_1 \\ \bar{\mathbf{B}}_2 \end{pmatrix},$$

$$\bar{\mathbf{C}} = \mathbf{C}\mathbf{T}^{-1} = (\bar{\mathbf{C}}_1, \mathbf{O}), \tag{6.9}$$

where the sizes of matrices $\bar{\mathbf{A}}_{11}, \bar{\mathbf{A}}_{21}, \bar{\mathbf{A}}_{22}$ are $r \times r$, $(n-r) \times r$, $(n-r) \times (n-r)$, respectively, and $\bar{\mathbf{B}}_1$ has r rows and $\bar{\mathbf{C}}_1$ has r columns. Furthermore,

(i)   system $(\bar{\mathbf{A}}_{11}, \bar{\mathbf{B}}_1, \bar{\mathbf{C}}_1)$ is completely observable, and

(ii)   *the transfer function of systems* $(\mathbf{A}, \mathbf{B}, \mathbf{C})$ *and* $(\bar{\mathbf{A}}_{11}, \bar{\mathbf{B}}_1, \bar{\mathbf{C}}_1)$ *coincide.*

REMARK 6.2   Notice that system (6.9) can be rewritten as

$$\dot{\bar{\mathbf{x}}}_1 = \bar{\mathbf{A}}_{11}\bar{\mathbf{x}}_1 + \bar{\mathbf{B}}_1\mathbf{u}$$

$$\dot{\bar{\mathbf{x}}}_2 = \bar{\mathbf{A}}_{21}\bar{\mathbf{x}}_1 + \bar{\mathbf{A}}_{22}\bar{\mathbf{x}}_2 + \bar{\mathbf{B}}_2\mathbf{u}$$

$$\mathbf{y} = \bar{\mathbf{C}}_1\bar{\mathbf{x}}_1 \ .$$

Because the output does not depend on $\bar{\mathbf{x}}_2$, this variable is not observable, and part (i) implies that $\bar{\mathbf{x}}_1$ is completely observable.   ∎

THEOREM 6.5
System $(\mathbf{A}, \mathbf{B}, \mathbf{C})$ *is completely observable if and only if matrix* $\mathbf{A}$ *has no eigenvector* $\mathbf{q}$ *that is orthogonal to the rows of* $\mathbf{C}$.

COROLLARY 6.2
System $(\mathbf{A}, \mathbf{B}, \mathbf{C})$ *is completely observable if and only if the rank of matrix* $(s\mathbf{I} - \mathbf{A}^T, \mathbf{C}^T)$ *is n for all s.*

Example 6.4

Consider again the system of Example 6.2, where $n = 2$,

$$\mathbf{A} = \begin{pmatrix} 0 & w \\ -w & 0 \end{pmatrix} \quad \text{and} \quad \mathbf{C} = (1, 1) \ .$$

In this case,

$$(s\mathbf{I} - \mathbf{A}^T, \mathbf{C}^T) = \begin{pmatrix} s & w & 1 \\ -w & s & 1 \end{pmatrix} \ .$$

Notice that for $s = w$, the first and third columns are independent; otherwise the second and third columns are independent. Hence, the *rank* of the matrix is always n, that is, the system is completely observable.

6.2   *Discrete Systems*

In this section the observability of the discrete linear system

$$\mathbf{x}(t+1) = \mathbf{A}(t)\mathbf{x}(t) + \mathbf{B}(t)\mathbf{u}(t), \qquad \mathbf{x}(0) = \mathbf{x}_0 \qquad (6.10)$$

$$\mathbf{y}(t) = \mathbf{C}(t)\mathbf{x}(t) \tag{6.11}$$

will be examined. We assume that the sizes of matrices $\mathbf{A}(t)$, $\mathbf{B}(t)$, and $\mathbf{C}(t)$ are $n \times n$, $n \times m$, and $p \times n$, respectively.

The general solution (2.44) of linear difference equations implies that

$$\mathbf{y}(t) = \mathbf{C}(t)\phi(t,0)\mathbf{x}_0 + \sum_{\tau=0}^{t-1} \mathbf{C}(t)\phi(t,\tau+1)\mathbf{B}(\tau)\mathbf{u}(\tau) .$$

Assume that the values of $\mathbf{u}(t)$ and $\mathbf{y}(t)$ are known for $t = 0, 1, 2, \ldots, t_1 - 1$. Then the right-hand side of equation

$$\mathbf{C}(t)\phi(t,0)\mathbf{x}_0 = \mathbf{y}(t) - \sum_{\tau=0}^{t-1} \mathbf{C}(t)\phi(t,\tau+1)\mathbf{B}(\tau)\mathbf{u}(\tau) \tag{6.12}$$

is known. Similar to the continuous case, the observability of the initial state $\mathbf{x}_0$ is based on the mapping

$$B(\mathbf{x})(t) = \mathbf{C}(t)\phi(t,0)\mathbf{x} \tag{6.13}$$

with domain $\mathbf{R}^n$ and the range in the set of the *p-dimensional* functions defined on the set $\{0, 1, 2, \ldots, t_1 - 1\}$. The null space of this mapping is given as

$$N(B) = \{\mathbf{x} \mid B(\mathbf{x})(t) = \mathbf{0} \quad \text{for all} \quad t \in \{0, 1, 2, \ldots, t_1 - 1\}\} .$$

The linearity of mapping B implies that Equation (6.12) uniquely determines $\mathbf{x}_0$ if and only if $N(B) = \{\mathbf{0}\}$.

LEMMA 6.4
Vector $\mathbf{v}$ is in $N(B)$ if and only if it belongs to the null space of matrix

$$\mathbf{M}(0, t_1) = \sum_{\tau=0}^{t_1-1} \phi^T(\tau, 0)\mathbf{C}^T(\tau)\mathbf{C}(\tau)\phi(\tau, 0) . \tag{6.14}$$

THEOREM 6.6
It is possible to determine $\mathbf{x}_0$ within an additive constant vector, which is in $N(\mathbf{M}(0, t_1))$. If $\mathbf{M}(0, t_1)$ is nonsingular, then $\mathbf{x}_0$ can be determined uniquely.

REMARK 6.3   If the initial state $\mathbf{x}_0$ of a discrete linear system can be uniquely determined on the basis of input and output values for $t =$

$0, 1, \ldots, t_1$ with $t_1 \geq n$, then the system is called *completely observable*. The algorithm to determine whether a given discrete linear system is completely observable is the same as was given for continuous systems.

Example 6.5

Consider again the discrete system of Example 3.14:

$$\mathbf{x}(t+1) = \begin{pmatrix} 1 & 1 \\ 0 & 1 \end{pmatrix} \mathbf{x}(t) + \begin{pmatrix} 0 \\ 1 \end{pmatrix} u(t), \qquad \mathbf{x}(0) = \mathbf{x}_0$$

$$y(t) = (1, 1)\mathbf{x}(t) \, ,$$

from which we know that

$$\phi(t, \tau) = \begin{pmatrix} 1 & t - \tau \\ 0 & 1 \end{pmatrix} \, .$$

Therefore,

$$\mathbf{M}(0, t_1) = \sum_{\tau=0}^{t_1-1} \begin{pmatrix} 1 & 0 \\ \tau & 1 \end{pmatrix} \begin{pmatrix} 1 \\ 1 \end{pmatrix} (1, 1) \begin{pmatrix} 1 & \tau \\ 0 & 1 \end{pmatrix}$$

$$= \sum_{\tau=0}^{t_1-1} \begin{pmatrix} 1 \\ \tau + 1 \end{pmatrix} (1, \tau + 1) = \sum_{\tau=0}^{t_1-1} \begin{pmatrix} 1 & \tau + 1 \\ \tau + 1 & (\tau + 1)^2 \end{pmatrix} \, .$$

By using the relations

$$1 + 2 + \cdots + t_1 = \frac{t_1(t_1 + 1)}{2} \qquad \text{and} \qquad 1^2 + 2^2 + \cdots + t_1^2 = \frac{t_1(t_1 + 1)(2t_1 + 1)}{6}$$

we obtain that

$$\mathbf{M}(0, t_1) = \begin{pmatrix} t_1 & \frac{t_1(t_1+1)}{2} \\ \frac{t_1(t_1+1)}{2} & \frac{t_1(t_1+1)(2t_1+1)}{6} \end{pmatrix} \, .$$

The determinant of $\mathbf{M}(0, t_1)$ can be written as

$$\frac{t_1^2(t_1 + 1)(2t_1 + 1)}{6} - \frac{t_1^2(t_1 + 1)^2}{4} = \frac{t_1^2}{12}(4t_1^2 + 6t_1 + 2 - 3t_1^2 - 6t_1 - 3)$$

$$= \frac{t_1^2}{12}(t_1^2 - 1) > 0 \, .$$

Hence, for $t_1 \geq 2$, $\mathbf{M}(0, t_1)$ is nonsingular and the initial state is observable, but for $t_1 = 1$ it is not. If $t_1 = 1$, then we have only one observation $y(0)$ of the one-dimensional output, which is not sufficient to determine the two-dimensional initial state $\mathbf{x}_0$.

It is easy to modify Theorem 6.2 for discrete systems. Properties (i) and (ii) hold in the same way, and (iii) has to be modified accordingly. The details are left as an exercise to the reader.

Consider next the special case, when $\mathbf{A}$ and $\mathbf{C}$ are constant matrices. Introduce again the *observability matrix*

$$\mathbf{L} = \begin{pmatrix} \mathbf{C} \\ \mathbf{CA} \\ \mathbf{CA}^2 \\ \cdots \\ \mathbf{CA}^{n-1} \end{pmatrix}.$$

One may easily verify that Lemma 6.3 remains true for discrete systems with $t_1 \geq n$, and Theorem 6.3 has to be modified as follows.

THEOREM 6.7

The time-invariant discrete linear system is observable at arbitrary $t_1 \geq n$ if and only if the rank of the observability matrix $\mathbf{L}$ equals n.

Example 6.6

In the case of the discrete system being examined in the previous example, $n = 2$ and

$$\mathbf{L} = \begin{pmatrix} \mathbf{C} \\ \mathbf{CA} \end{pmatrix} = \begin{pmatrix} 1 & 1 \\ 1 & 2 \end{pmatrix},$$

which is nonsingular. That is, $rank(\mathbf{L}) = 2$. Hence, for all $t_1 \geq 2$, the initial state $\mathbf{x}_0$ is observable. Note that the same result was obtained in the previous example, but the direct use of the observability matrix is much easier than the computation of $\mathbf{M}(0, t_1)$ and its examination. Assume next that the following measurements are known:

$$u(0) = 1, \quad y(0) = 2, \quad y(1) = 0.$$

We will now find the initial state $\mathbf{x}_0$. At $t = 0$,

$$2 = y(0) = (1, 1)\mathbf{x}(0) = (1, 1)\mathbf{x}_0,$$

and at $t = 1$,

$$0 = y(1) = (1,1)\mathbf{x}(1) = (1,1)[\begin{pmatrix} 1 & 1 \\ 0 & 1 \end{pmatrix} \mathbf{x}_0 + \begin{pmatrix} 0 \\ 1 \end{pmatrix} u(0)]$$

$$= (1,2)\mathbf{x}_0 + u(0).$$

These equations simplify as

$$(1,1)\mathbf{x}_0 = 2$$

$$(1,2)\mathbf{x}_0 = -1,$$

that is,

$$\begin{pmatrix} 1 & 1 \\ 1 & 2 \end{pmatrix} \mathbf{x}_0 = \begin{pmatrix} 2 \\ -1 \end{pmatrix}.$$

The unique solution is

$$\mathbf{x}_0 = \begin{pmatrix} 5 \\ -3 \end{pmatrix}.$$

Finally, note that Theorems 6.4 and 6.5 remain valid in the case of discrete systems.

6.3   *Duality*

In Section 6.1.1 the relation between the observability of a linear system and the controllability of its adjoint system was analyzed. In the case of time-invariant systems, similar properties hold for the dual. These results are the subjects of this section.

We first remind the reader (see Section 3.3.5) that dual systems are defined as follows:

(i)   The *dual* of the time-invariant continuous system

$$P_c : \dot{\mathbf{x}} = \mathbf{A}\mathbf{x} + \mathbf{B}\mathbf{u}$$

$$\mathbf{y} = \mathbf{C}\mathbf{x}$$

is given as

$$D_c : \dot{\mathbf{z}} = \mathbf{A}^T\mathbf{z} + \mathbf{C}^T\mathbf{v}$$

$$\mathbf{w} = \mathbf{B}^T\mathbf{z},$$

where $\mathbf{z}$, $\mathbf{v}$, and $\mathbf{w}$ denote the state, input, and output, respectively.

(ii)   The *dual* of the time-invariant discrete system

$$P_d : \mathbf{x}(t+1) = \mathbf{A}\mathbf{x}(t) + \mathbf{B}\mathbf{u}(t)$$

$$\mathbf{y}(t) = \mathbf{C}\mathbf{x}(t)$$

is given as

$$D_d : \mathbf{z}(t+1) = \mathbf{A}^T\mathbf{z}(t) + \mathbf{C}^T\mathbf{v}(t)$$

$$\mathbf{w}(t) = \mathbf{B}^T\mathbf{z}(t),$$

where $\mathbf{z}$, $\mathbf{v}$, and $\mathbf{w}$ denote the state, input, and output, respectively.

The original systems are called *primal*, and they are denoted by P_c and P_d, where the subscripts refer to the types (continuous or discrete) of the systems. Similarly, D_c and D_d denote the *duals*. Primal and dual systems are illustrated in Figure 6.3.

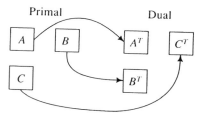

Figure 6.3   Primal–dual systems.

Note first that the construction of a dual system is very similar to that of a linear programming problem. Obviously, the dual of a dual system is the primal, which is implied by the simple property that the transpose of the transpose of a matrix equals the matrix itself. The most important relation between the primal and dual systems can be presented as follows.

THEOREM 6.8

The primal of a time-invariant continuous (or discrete) system is completely controllable if and only if its dual is completely observable.

PROOF   This assertion follows from Theorems 5.4, 5.9, 6.3, and 6.7, and from the observations that with $\mathbf{K}_P$ being the controllability matrix of the primal and $\mathbf{L}_D$ being the observability matrix of the dual,

$$
\mathbf{K}_P^T = (\mathbf{B}, \mathbf{AB}, \ldots, \mathbf{A}^{n-1}\mathbf{B})^T = \begin{pmatrix} \mathbf{B}^T \\ \mathbf{B}^T(\mathbf{A}^T) \\ \vdots \\ \mathbf{B}^T(\mathbf{A}^T)^{n-1} \end{pmatrix} = \mathbf{L}_D .
$$

■

REMARK 6.4   Because the dual of the dual is the primal, we have the following modification of the theorem:

The primal of a time-invariant continuous (or discrete) system is completely observable if and only if its dual is completely controllable.   ■

Example 6.7

For the continuous linear system

$$
\dot{\mathbf{x}} = \begin{pmatrix} 0 & \omega \\ -\omega & 0 \end{pmatrix} \mathbf{x} + \begin{pmatrix} 0 \\ 1 \end{pmatrix} u
$$

$$
y = (1, 1)\mathbf{x} ,
$$

its dual is

$$
\dot{\mathbf{z}} = \begin{pmatrix} 0 & -\omega \\ \omega & 0 \end{pmatrix} \mathbf{z} + \begin{pmatrix} 1 \\ 1 \end{pmatrix} v
$$

$$
w = (0, 1)\mathbf{z} .
$$

Similarly, for the discrete linear system

$$
\mathbf{x}(t + 1) = \begin{pmatrix} 1 & 1 \\ 0 & 1 \end{pmatrix} \mathbf{x}(t) + \begin{pmatrix} 0 \\ 1 \end{pmatrix} u(t),
$$

$$
y(t) = (1, 1)\mathbf{x}(t) ,
$$

the dual is as follows:

$$
\mathbf{z}(t + 1) = \begin{pmatrix} 1 & 0 \\ 1 & 1 \end{pmatrix} \mathbf{z}(t) + \begin{pmatrix} 1 \\ 1 \end{pmatrix} v(t)
$$

$$w(t) = (0, 1)\mathbf{z}(t) .$$

Hence, the duals are determined in both cases.

The concept of duality has many applications in linear systems theory. The observability of a time-invariant system can be examined by the controllability of its dual, and the controllability of a time-invariant system can be investigated by examining the observability of its dual. Further applications will be introduced in the next chapter, when duality will be used in deriving observability canonical forms; duality will be applied also in Chapter 8 in obtaining standard observable realizations of given transfer functions. Since the concept of duality is a consequence of the developments of this and the previous chapters, it is introduced here; however, its main applications will be discussed in later chapters of this book.

6.4 Applications

In this section we present some applications of the observability theory and duality of linear systems in engineering and in the social sciences.

6.4.1 Dynamic Systems in Engineering

1. Consider the simple *harmonic oscillator* (3.57) introduced in Chapter 2 and given in Application 3.5.1-1, which is summarized below:

$$\dot{\mathbf{x}} = \begin{pmatrix} 0 & \omega \\ -\omega & 0 \end{pmatrix} \mathbf{x} + \begin{pmatrix} 0 \\ 1 \end{pmatrix} u .$$

Also let the output be $y = x_1$; therefore,

$$\mathbf{c}^T = (1, 0) .$$

Is it observable?

To answer this question let us compute the observability matrix

$$\mathbf{L} = \begin{pmatrix} \mathbf{c}^T \\ \mathbf{c}^T \mathbf{A} \end{pmatrix} = \begin{pmatrix} 1 & 0 \\ 0 & \omega \end{pmatrix} .$$

Since $rank(\mathbf{L}) = 2$, the system is completely observable.

2. What about the *damped linear second-order system* of Application 3.5.1-2; is it observable? In this case,

$$\mathbf{A} = \begin{pmatrix} 0 & 1 \\ -\frac{K}{M} & -\frac{B}{M} \end{pmatrix} , \qquad \mathbf{b} = \begin{pmatrix} 0 \\ \frac{1}{M} \end{pmatrix} .$$

Let the output be position; therefore,

$$\mathbf{c}^T = (1,0) \ .$$

Let's compute the observability matrix $\mathbf{L}$:

$$\mathbf{L} = \begin{pmatrix} 1 & 0 \\ 0 & 1 \end{pmatrix} \ .$$

The $rank(\mathbf{L}) = 2$; therefore, the system is completely observable.
   3. For the *electrical system* of Application 3.5.1-3,

$$\mathbf{A} = \begin{pmatrix} -\frac{R_1}{L} & -\frac{1}{L} \\ \frac{1}{C} & -\frac{1}{CR_2} \end{pmatrix}, \qquad \mathbf{b} = \begin{pmatrix} \frac{1}{L} \\ 0 \end{pmatrix} \ .$$

Let the output be the voltage across the capacitor, then

$$\mathbf{c}^T = (0,1) \ .$$

Let's compute the observability matrix $\mathbf{L}$:

$$\mathbf{L} = \begin{pmatrix} 0 & 1 \\ \frac{1}{C} & \frac{-1}{CR_2} \end{pmatrix} \ .$$

Since $rank(\mathbf{L}) = 2$, the system is completely observable.
   Investigating the observability of this electrical circuit was not very interesting. So, let us investigate the modified circuit of Figure 5.4 that has the following equations:

$$\begin{pmatrix} \dot{i}_{L_1} \\ \dot{v}_{C_1} \\ \dot{i}_{L_2} \\ \dot{v}_{C_2} \end{pmatrix} = \begin{pmatrix} 0 & \frac{-1}{L_1} & 0 & 0 \\ \frac{1}{C_1} & \frac{-1}{C_1 R_1} & 0 & 0 \\ 0 & 0 & 0 & \frac{-1}{L_2} \\ 0 & 0 & \frac{1}{C_2} & \frac{-1}{C_2 R_2} \end{pmatrix} \begin{pmatrix} i_{L_1} \\ v_{C_1} \\ i_{L_2} \\ v_{C_2} \end{pmatrix} + \begin{pmatrix} \frac{1}{L_1} \\ \frac{1}{C_1 R_1} \\ \frac{1}{L_2} \\ \frac{1}{C_2 R_2} \end{pmatrix} u \ .$$

Let the output be the voltage across the right capacitor. That is,

$$y = (0,0,0,1) \begin{pmatrix} i_{L_1} \\ v_{C_1} \\ i_{L_2} \\ v_{C_2} \end{pmatrix} \ ,$$

and so

$$\mathbf{c}^T = (0,0,0,1) \ .$$

Now let us form the observability matrix, $\mathbf{L}$:

$$\mathbf{L} = \begin{pmatrix} 0 & 0 & 0 & 1 \\ 0 & 0 & \frac{1}{C_2} & \frac{-1}{C_2 R_2} \\ 0 & 0 & \frac{-1}{C_2^2 R_2} & \frac{-1}{L_2 C_2} + \frac{1}{C_2^2 R_2^2} \\ 0 & 0 & a & b \end{pmatrix} .$$

We do not have to calculate the constants a and b, because we can already see that the system is not observable because the first two columns are zero. What will happen if we change the outputs, so that

$$\mathbf{c}^T = (0, 1, 0, 1) .$$

The new observability matrix $\mathbf{L}^*$ is

$$\mathbf{L}^* = \begin{pmatrix} 0 & 1 & 0 & 1 \\ \frac{1}{C_1} & \frac{-1}{C_1 R_1} & \frac{1}{C_2} & \frac{-1}{C_2 R_2} \\ \frac{-1}{C_1^2 R_1} & \frac{-1}{L_1 C_1} + \frac{1}{C_1^2 R_1^2} & \frac{-1}{C_2^2 R_2} & \frac{-1}{L_2 C_2} + \frac{1}{C_2^2 R_2^2} \\ \frac{-1}{L_1 C_1^2} + \frac{1}{C_1^3 R_1^2} & \frac{2}{L_1 R_1 C_1^2} - \frac{1}{C_1^3 R_1^3} & \frac{-1}{L_2 C_2^2} + \frac{1}{C_2^3 R_2^2} & \frac{2}{L_2 R_2 C_2^2} - \frac{1}{C_2^3 R_2^3} \end{pmatrix} .$$

If $L_1 = L_2$, $C_1 = C_2$, and $R_1 = R_2$, then columns 1 and 3 and columns 2 and 4 are identical and the system is not observable. Otherwise, the system is observable.

4. For the *transistor circuit* model (3.80) of Application 3.5.1-4,

$$\mathbf{A} = \begin{pmatrix} -\frac{h_{ie}}{L} & 0 \\ \frac{h_{fe}}{C} & 0 \end{pmatrix}, \qquad \mathbf{b} = \begin{pmatrix} \frac{1}{L} \\ 0 \end{pmatrix} .$$

If we let the output be the voltage across the capacitor,

$$\mathbf{c}_1^T = (0, 1) .$$

Let's compute the observability matrix $\mathbf{L}_1$:

$$\mathbf{L}_1 = \begin{pmatrix} 0 & 1 \\ \frac{h_{fe}}{C} & 0 \end{pmatrix} .$$

Since $rank(\mathbf{L}_1) = 2$, the system is completely observable.

However, if we let the output be the base current

$$\mathbf{c}_2^T = (1, 0) ,$$

the new observability matrix $\mathbf{L}_2$ is

$$\mathbf{L}_2 = \begin{pmatrix} 1 & 0 \\ \frac{h_{ie}}{L} & 0 \end{pmatrix} .$$

The *rank* of $\mathbf{L}_2$ is only 1; therefore, the system is not completely observable from the base current. This makes sense, because i_b depends on e_s and h_{ie}. It is independent of the voltage across the capacitor. So we cannot devise an experiment that will allow us to determine $v_c(0)$ by observing only i_b.

5. To access the observability of the *hydraulic system* of Application 3.5.1-5, we can compute $\mathbf{L}$ as follows. Because

$$\mathbf{A}_1 = \begin{pmatrix} -a & a \\ b & -(b+c) \end{pmatrix} \quad \text{and} \quad \mathbf{b}_1 = \begin{pmatrix} d \\ 0 \end{pmatrix} ,$$

by selecting

$$\mathbf{c}_1^T = (0,1)$$

we get

$$\mathbf{L}_1 = \begin{pmatrix} 0 & 1 \\ b & -(b+c) \end{pmatrix} .$$

The *rank* of $\mathbf{L}_1$ is 2, so the system is completely observable.

Once again let us expand the system to the *three-tank system* of Figure 5.5. In this case,

$$\mathbf{c}_2^T = (1,0,0),$$

$$\mathbf{c}_2^T \mathbf{A}_2 = (-3,3,0),$$

$$\mathbf{c}_2^T \mathbf{A}_2^2 = (-3,3,0) \begin{pmatrix} -3 & 3 & 0 \\ 2 & -4 & 2 \\ 0 & 3 & -3 \end{pmatrix} = (15,-21,6) .$$

So

$$\mathbf{L}_2 = \begin{pmatrix} 1 & 0 & 0 \\ -3 & 3 & 0 \\ 15 & -21 & 6 \end{pmatrix} .$$

The *rank* is 3, so the system is observable by looking at the level of water in tank one. However, let us now change the output vector to

$$\mathbf{c}_3^T = (0,1,0) ,$$

then

$$\mathbf{c}_3^T \mathbf{A}_2 = (2, -4, 2) \ ,$$

$$\mathbf{c}_3^T \mathbf{A}_2^2 = (2, -4, 2) \begin{pmatrix} -3 & 3 & 0 \\ 2 & -4 & 2 \\ 0 & 3 & -3 \end{pmatrix} = (-14, 28, -14) \ .$$

So

$$\mathbf{L}_3 = \begin{pmatrix} 0 & 1 & 0 \\ 2 & -4 & 2 \\ -14 & 28 & -14 \end{pmatrix} \ .$$

Since the first and third columns are identical, $rank(\mathbf{L}_3) < 3$; therefore, the system is not observable with tank two only. The physical reason for this is that you could have the initial level in tank-1 high and that in tank-3 low, or vice versa, and you cannot tell the difference by looking at level of tank-2.

6. In the case of the *multiple input electrical system*

$$\mathbf{A} = \begin{pmatrix} -\frac{R}{L_1} & 0 & -\frac{1}{L_1} \\ 0 & 0 & -\frac{1}{L_2} \\ \frac{1}{C} & \frac{1}{C} & 0 \end{pmatrix} , \qquad \mathbf{C} = (0, 0, 1) \ ,$$

therefore, the observability matrix is

$$\mathbf{L} = \begin{pmatrix} 0 & 0 & 1 \\ \frac{1}{C} & \frac{1}{C} & 0 \\ -\frac{R}{CL_1} & 0 & -\frac{1}{C}(\frac{1}{L_1} + \frac{1}{L_2}) \end{pmatrix} \ .$$

Because this is a square matrix, it has full rank if and only if its determinant is nonzero. By expanding the determinant with respect to its first row we have

$$det(\mathbf{L}) = 1 \cdot det \begin{pmatrix} \frac{1}{C} & \frac{1}{C} \\ -\frac{R}{CL_1} & 0 \end{pmatrix} = \frac{1}{C^2 L_1} \neq 0.$$

Therefore, the system is observable.

7. To compute observability of the *stick-balancing problem*, let the output be the position of the end of the stick. That is,

$$\mathbf{c}_1^T = (1, 0) \ ,$$

then we can compute the observability matrix:

$$\mathbf{L}_1 = \begin{pmatrix} 1 & 0 \\ 0 & 1 \end{pmatrix} .$$

The *rank* of $\mathbf{L}_1$ is 2, so the system is completely observable.

Next let the output be velocity of the end of the stick. That is,

$$\mathbf{c}_2^T = (0, 1) ,$$

and the observability matrix is as follows:

$$\mathbf{L}_2 = \begin{pmatrix} 0 & 1 \\ g & 0 \end{pmatrix} .$$

Again, the *rank* of $\mathbf{L}_2$ is 2, so the system is completely observable.

However, if we let the output vector be

$$\mathbf{c}_3^T = (a, b)$$

and compute

$$\mathbf{L}_3 = \begin{pmatrix} a & b \\ bg & a \end{pmatrix} ,$$

we find the *rank* of $\mathbf{L}_3$ is less than 2, if

$$a^2 = b^2 g ,$$

which means this system could be unobservable with certain outputs.

8. For the *cart with two sticks* model of Application 3.5.1-8,

$$\mathbf{A} = \begin{pmatrix} 0 & 0 & 1 & 0 \\ 0 & 0 & 0 & 1 \\ a_1 & a_2 & 0 & 0 \\ a_3 & a_4 & 0 & 0 \end{pmatrix} \quad \text{and} \quad \mathbf{b} = \begin{pmatrix} 0 \\ 0 \\ -c \\ -d \end{pmatrix} .$$

If we let

$$\mathbf{c}^T = (1, 0, 0, 0) ,$$

then we can compute observability matrix $\mathbf{L}$ as

$$\mathbf{L} = \begin{pmatrix} 1 & 0 & 0 & 0 \\ 0 & 0 & 1 & 0 \\ a_1 & a_2 & 0 & 0 \\ 0 & 0 & a_1 & a_2 \end{pmatrix} ,$$

and if we switch column 2 and column 3, we have

$$\mathbf{L}^* = \begin{pmatrix} 1 & 0 & 0 & 0 \\ 0 & 1 & 0 & 0 \\ a_1 & 0 & a_2 & 0 \\ 0 & a_1 & 0 & a_2 \end{pmatrix} .$$

Since $a_2 \neq 0$, the *rank* of $\mathbf{L}^*$ is 4; therefore, the system is completely observable. Here we used the fact that a triangle (or a diagonal matrix in the further special case) with nonzero diagonal elements is always nonsingular.

9. In the case of our *electrical heating system* we have

$$\mathbf{A} = \begin{pmatrix} -\frac{A_1 h_1}{C_1} & \frac{A_1 h_1}{C_1} \\ \frac{A_1 h_1}{C_2} & -\frac{A_1 h_1 + A_2 h_2}{C_2} \end{pmatrix} \quad \text{and} \quad \mathbf{C} = (1, 0) ,$$

therefore, the observability matrix has the form

$$\mathbf{L} = \begin{pmatrix} 1 & 0 \\ -\frac{A_1 h_1}{C_1} & \frac{A_1 h_1}{C_1} \end{pmatrix} .$$

Since this matrix is lower triangular with nonzero diagonal elements, $rank(\mathbf{L}) = 2$, consequently the system is observable.

10. In the case of $m = 1$ in the *nuclear reactor model* of Application 3.5.1-10 we have

$$\mathbf{A} = \begin{pmatrix} \frac{\rho - \beta}{l} & \lambda_1 \\ \frac{\beta_1}{l} & -\lambda_1 \end{pmatrix} \quad \text{and} \quad \mathbf{b} = \begin{pmatrix} 1 \\ 0 \end{pmatrix} .$$

If we let

$$\mathbf{c}^T = (1, 0) ,$$

then

$$\mathbf{L} = \begin{pmatrix} 1 & 0 \\ \frac{\rho - \beta}{l} & \lambda_1 \end{pmatrix} .$$

The *rank* of $\mathbf{L}$ is 2, so the system is completely observable.

6.4.2   *Applications in the Social Sciences*

1. Consider first the linearized *predator–prey* model (3.114) and assume that the output y is the predator population. Then the resulting system

is given as

$$\dot{G} = -\frac{bc}{d}W + \frac{ac}{d}$$

$$\dot{W} = \frac{da}{b}G - \frac{ac}{b}$$

$$y = W \ . \tag{6.15}$$

By using the notation of Section 6.1.2 we have

$$\mathbf{A} = \begin{pmatrix} 0 & -\frac{bc}{d} \\ \frac{ad}{b} & 0 \end{pmatrix} \qquad \text{and} \qquad \mathbf{C} = (0, 1)$$

and, therefore, the observability matrix is

$$\mathbf{L} = \begin{pmatrix} 0 & 1 \\ \frac{ad}{b} & 0 \end{pmatrix} \ .$$

Since $rank(\mathbf{L}) = 2$, the system is completely observable.

Consider next the linearized model (5.31) and assume again that the predator population is the output. Then we have the systems model

$$\dot{G}_\delta = -\frac{bc}{d}W_\delta + \frac{c}{d}u$$

$$\dot{W}_\delta = \frac{da}{b}G_\delta$$

$$y = W_\delta \ . \tag{6.16}$$

This system is also completely observable, since $\mathbf{A}$ and $\mathbf{C}$ are the same as before. Without showing any application of duality, we note that the dual of this system is given as

$$\dot{z}_1 = \frac{ad}{b}z_2$$

$$\dot{z}_2 = -\frac{bc}{d}z_1 + v$$

$$w = \frac{c}{d}z_1 \ .$$

2. Consider now the three-dimensional *cohort population* model (3.115), and assume that the total population is the output and immigration is permitted to all age groups. Then the system has the form

$$
\begin{pmatrix} P_1(t+1) \\ P_2(t+1) \\ P_3(t+1) \end{pmatrix} = \begin{pmatrix} b_1 & b_2 & b_3 \\ a_1 & 0 & 0 \\ 0 & a_2 & 0 \end{pmatrix} \begin{pmatrix} P_1(t) \\ P_2(t) \\ P_3(t) \end{pmatrix} + \begin{pmatrix} u_1(t) \\ u_2(t) \\ u_3(t) \end{pmatrix}
$$

$$
y(t) = P_1(t) + P_2(t) + P_3(t) \ . \tag{6.17}
$$

Since

$$
\mathbf{A} = \begin{pmatrix} b_1 & b_2 & b_3 \\ a_1 & 0 & 0 \\ 0 & a_2 & 0 \end{pmatrix} \qquad \text{and} \qquad \mathbf{C} = (1,1,1) \ ,
$$

the observability matrix can be written as

$$
\mathbf{L} = \begin{pmatrix} 1 & 1 & 1 \\ b_1 + a_1 & b_2 + a_2 & b_3 \\ b_1^2 + b_1 a_1 + a_1 b_2 + a_1 a_2 & b_1 b_2 + b_2 a_1 + a_2 b_3 & b_3 b_1 + b_3 a_1 \end{pmatrix} \ .
$$

This matrix has full rank if $\mathbf{L}$ is nonsingular. Note that this property depends on the particular values of the model parameters a_i and b_i. For example, if

$$
a_1 = a_2 = b_1 = b_2 = \frac{1}{2} \qquad \text{and} \qquad b_3 = 1 \ ,
$$

then the first two rows are the same, which implies that $rank(\mathbf{L}) < 3$. That is, the system is not observable in this case, and the state vector cannot be determined uniquely.

Assume next that the output is the population of the oldest group. Then matrix $\mathbf{A}$ does not change, but in this case

$$
\mathbf{C} = (0,0,1) \ ,
$$

and, therefore,

$$
\mathbf{L} = \begin{pmatrix} 0 & 0 & 1 \\ 0 & a_2 & 0 \\ a_1 a_2 & 0 & 0 \end{pmatrix} \ ,
$$

which has full rank. Hence, based on the measurements on only the oldest population, the system becomes completely observable. This result can be explained by noting that after certain time delay all other age groups will enter the oldest population group.

3. In the case of the *arms races* model (3.119), assume that the output is $X(t)$, that is, nation 1 can observe its own armament level but cannot monitor the armament level of the other nation. In this case the model has the form

$$\dot{\mathbf{x}}(t) = \begin{pmatrix} -b & a \\ c & -d \end{pmatrix} \mathbf{x}(t) + \begin{pmatrix} \alpha \\ \beta \end{pmatrix} u(t) \ ,$$

$$y(t) = (1,0)\mathbf{x}(t) \ . \tag{6.18}$$

The observability matrix

$$\mathbf{L} = \begin{pmatrix} 1 & 0 \\ -b & a \end{pmatrix}$$

has full rank; therefore, the system is completely observable. That is, observations on $u(t)$ and $y(t)$ uniquely determine the state.

4. Assume that in a *warfare* (model (5.32)), each nation can monitor her own force X_1 only, that is, the output from the viewpoint of the first nation is X_1. The resulting model is

$$\dot{X}_1 = -h_2 X_2 - h_3 u(t)$$

$$\dot{X}_2 = -h_1 X_1 \ ,$$

$$y = X_1 \ . \tag{6.19}$$

Since

$$\mathbf{A} = \begin{pmatrix} 0 & -h_2 \\ -h_1 & 0 \end{pmatrix} \qquad \text{and} \qquad \mathbf{C} = (1,0) \ ,$$

the observability matrix is

$$\mathbf{L} = \begin{pmatrix} 1 & 0 \\ 0 & -h_2 \end{pmatrix} \ .$$

Note that $rank(\mathbf{L}) = 2$, which implies that the system is completely observable.

Modify the above model by assuming that the output is $X_1 + X_2$, which is the total combined force of the two nations. In this case $\mathbf{A}$ does not change, but $\mathbf{C} = (1,1)$ and, therefore,

$$\mathbf{L} = \begin{pmatrix} 1 & 1 \\ -h_1 & -h_2 \end{pmatrix} \ .$$

This matrix has full rank if and only if $h_1 \neq h_2$. Hence, the system is completely observable if and only if $h_1 \neq h_2$.

5. The linear *epidemics* model was discussed in Application 5.3.2-5, where an input u was introduced to influence the number of infected and circulating individuals. Assuming that their number is the output, then the model can be written as follows:

$$\dot{\mathbf{x}} = \begin{pmatrix} 0 & -\alpha\bar{x} \\ 0 & \alpha\bar{x} - \beta \end{pmatrix} \mathbf{x}(t) + \begin{pmatrix} 0 \\ -1 \end{pmatrix} u \ ,$$

$$y = (0,1)\mathbf{x} \ ,$$

where $\bar{x} \geq 0$ is arbitrary. The observability matrix of this system is the following:

$$\mathbf{L} = \begin{pmatrix} 0 & 1 \\ 0 & \alpha\bar{x} - \beta \end{pmatrix} \ .$$

Since the first column is zero, $rank(\mathbf{L}) = 1$. Hence, the system is not observable. This result is expected, since neither the governing differential equation nor the output equation depends on x_1. Therefore, x_1 cannot be observable.

6. Consider next a *Harrod-type national economy* and assume that consumption $C(t)$ is observed as output. Then model (3.10) is modified as

$$Y(t+1) = [1 + r - rm]Y(t) - rG(t)$$

$$C(t) = m \cdot Y(t) \ , \tag{6.20}$$

where G is the input. Any observation of $C(t)$ immediately gives the corresponding value of $Y(t) = (1/m)C(t)$; therefore, the system is completely observable.

7. We find a similar situation in the case of the *linear cobweb* model (5.34), when we assume that, for example, the supply is the observed output. This situation is modeled as

$$p(t+1) = \frac{b}{a}p(t) + u(t)$$

$$y(t) = bp(t) + b_0 \ . \tag{6.21}$$

Note that any observation on $y(t)$ implicitly implies the corresponding value of the state variable, since from the output equation,

$$p(t) = \frac{1}{b}(y(t) - b_0) \ .$$

Hence, the system is completely observable.

8. Consider next the model (5.36) of *interrelated markets* with the additional assumption that the output is the average price $(1/n)(p_1(t) + \cdots + p_n(t))$. The corresponding model is now

$$\dot{\mathbf{p}} = \mathbf{K}(\mathbf{A} - \mathbf{B})\mathbf{p} + \mathbf{u}$$

$$y = \frac{1}{n}\mathbf{1}^T\mathbf{p} \ , \tag{6.22}$$

where $\mathbf{1}^T = (1, 1, \ldots, 1)$. The observability matrix

$$\mathbf{L} = \frac{1}{n}\begin{pmatrix} \mathbf{1}^T \\ \mathbf{1}^T\mathbf{K}(\mathbf{A} - \mathbf{B}) \\ \vdots \\ \mathbf{1}^T[\mathbf{K}(\mathbf{A} - \mathbf{B})]^{n-1} \end{pmatrix} \ .$$

is $n \times n$, and, therefore, the system is completely observable if and only if $\mathbf{L}$ is nonsingular. A trivial case of a singular $\mathbf{L}$ occurs when the sum of the rows of matrix $\mathbf{K}(\mathbf{A} - \mathbf{B})$ has identical elements, that is, when the sum is the constant multiple of $\mathbf{1}^T$.

9. In the case of an *oligopoly*, assume that we are interested in only the total output of the industry. Then model (5.38) is completed by the corresponding output equation as

$$\mathbf{y}(t+1) = \mathbf{A}_c\mathbf{y}(t) - \frac{1}{2a}\mathbf{1}u(t)$$

$$y(t) = \mathbf{1}^T\mathbf{y}(t) \ , \tag{6.23}$$

where $y(t)$ is the sum of the elements of vector $\mathbf{y}(t)$,

$$\mathbf{A}_c = \begin{pmatrix} 0 & -\frac{1}{2} & -\frac{1}{2} & \cdots & -\frac{1}{2} \\ -\frac{1}{2} & 0 & -\frac{1}{2} & \cdots & -\frac{1}{2} \\ \vdots & \vdots & \vdots & \ddots & \vdots \\ -\frac{1}{2} & -\frac{1}{2} & -\frac{1}{2} & \cdots & 0 \end{pmatrix} \ ,$$

and we used the fact that measuring $\mathbf{1}^T\mathbf{x}(t)$ is equivalent to measure $\mathbf{1}^T\mathbf{y}(t)$, since

$$\mathbf{1}^T\mathbf{x}(t) = \mathbf{1}^T\mathbf{y}(t) + \mathbf{1}^T\mathbf{z}(t)$$

where $\mathbf{z}(t)$ is known. (For the details, see the derivation of system (5.38).)
Before determining the observability matrix, observe that

$$\mathbf{1}^T\mathbf{A}_c = \frac{1-N}{2}\mathbf{1}^T .$$

Therefore,

$$\mathbf{L} = \begin{pmatrix} \mathbf{1}^T \\ \frac{1-N}{2}\mathbf{1}^T \\ \dots \\ (\frac{1-N}{2})^{N-1}\mathbf{1}^T \end{pmatrix}$$

with $rank(\mathbf{L}) = 1$. Hence the system is not observable. That is, from measurements of the total output of the industry it is impossible to determine the individual outputs of the firms.

Problems

1. Discuss the observability of system

$$\dot{\mathbf{x}} = \begin{pmatrix} \frac{1}{t} & 0 \\ 0 & \frac{1}{t} \end{pmatrix}\mathbf{x} + \begin{pmatrix} 1 \\ 1 \end{pmatrix}u$$

$$y = (1,1)\mathbf{x} .$$

2. Examine the observability of system

$$\dot{\mathbf{x}} = \begin{pmatrix} 1 & 1 \\ 2 & 2 \end{pmatrix}\mathbf{x} + \begin{pmatrix} 1 \\ 0 \end{pmatrix}u$$

$$y = (1,1)\mathbf{x} .$$

Use Theorem 6.1, and select $t_0 = 0$.

3. Compute matrix $\mathbf{M}(t_0, t_1)$ for system

$$\dot{\mathbf{x}} = \begin{pmatrix} 1 & 1 \\ 2 & 2 \end{pmatrix}\mathbf{x} + \begin{pmatrix} 1 \\ 0 \end{pmatrix}u$$

$$y = (1,1)\mathbf{x} ,$$

and illustrate the Properties (i) and (ii) of Theorem 6.2. Select the $[0, 1]$
interval.

4.   Examine the observability of the following system by using the
observability matrix

$$\dot{\mathbf{x}} = \begin{pmatrix} 1 & 1 \\ 2 & 2 \end{pmatrix} \mathbf{x} + \begin{pmatrix} 1 \\ 0 \end{pmatrix} u$$

$$y = (1, 1)\mathbf{x} .$$

5.   Use Theorem 6.1 to examine the observability of this system:

$$\dot{\mathbf{x}} = \begin{pmatrix} 2 & 1 \\ 0 & 2 \end{pmatrix} \mathbf{x} + \begin{pmatrix} 1 \\ 1 \end{pmatrix} u$$

$$y = (0, 1)\mathbf{x} .$$

Select $t_0 = 0$.

6.   Compute matrix $\mathbf{M}(t_0, t_1)$ for system

$$\dot{\mathbf{x}} = \begin{pmatrix} 2 & 1 \\ 0 & 2 \end{pmatrix} \mathbf{x} + \begin{pmatrix} 1 \\ 1 \end{pmatrix} u$$

$$y = (0, 1)\mathbf{x} ,$$

and illustrate Properties (i) and (ii) of Theorem 6.2. Select $t_0 = 0$.

7.   Examine the observability of system

$$\dot{\mathbf{x}} = \begin{pmatrix} 2 & 1 \\ 0 & 2 \end{pmatrix} \mathbf{x} + \begin{pmatrix} 1 \\ 1 \end{pmatrix} u$$

$$y = (0, 1)\mathbf{x} ,$$

by using the observability matrix (6.7).

8.   Is the electric circuit

$$L\frac{di(t)}{dt} + (R_1 + R_2)i(t) = u(t)$$

$$y(t) = R_2 i(t)$$

introduced in Problem 3.13 completely observable?

9. Discuss the observability of the mechanical system

$$\dot{\mathbf{x}} = \begin{pmatrix} 0 & 1 \\ 0 & -6 \end{pmatrix} \mathbf{x} + \begin{pmatrix} 0 \\ 2 \end{pmatrix} u$$

$$y = (1,0)\mathbf{x}$$

introduced in Problem 3.7.

10. Is the discrete system

$$\mathbf{x}(t+1) = \begin{pmatrix} 1 & 1 \\ 2 & 2 \end{pmatrix} \mathbf{x}(t) + \begin{pmatrix} 1 \\ 0 \end{pmatrix} u(t)$$

$$y(t) = (1,1)\mathbf{x}(t)$$

completely observable? Use Theorem 6.6, and select $t_1 = 2$.

11. Use the observability matrix $\mathbf{L}$ to determine if the following system is completely observable:

$$\mathbf{x}(t+1) = \begin{pmatrix} 1 & 1 \\ 2 & 2 \end{pmatrix} \mathbf{x}(t) + \begin{pmatrix} 1 \\ 0 \end{pmatrix} u(t)$$

$$y(t) = (1,1)\mathbf{x}(t) .$$

12. Use Theorem 6.6 to determine if the following discrete system is completely observable:

$$\mathbf{x}(t+1) = \begin{pmatrix} 2 & 1 \\ 0 & 2 \end{pmatrix} \mathbf{x}(t) + \begin{pmatrix} 1 \\ 1 \end{pmatrix} u(t)$$

$$y(t) = (0,1)\mathbf{x}(t) .$$

13. Discuss the observability of system

$$\mathbf{x}(t+1) = \begin{pmatrix} 2 & 1 \\ 0 & 2 \end{pmatrix} \mathbf{x}(t) + \begin{pmatrix} 1 \\ 1 \end{pmatrix} u(t)$$

$$y(t) = (0,1)\mathbf{x}(t)$$

by using the observability matrix $\mathbf{L}$.

14. Discuss the observability of system

$$\dot{x} = Ax + Bu$$

$$y = Cx + Du .$$

15. Find the dual of the system

$$\dot{x} = \begin{pmatrix} 1 & 1 \\ 2 & 2 \end{pmatrix} x + \begin{pmatrix} 1 \\ 0 \end{pmatrix} u$$

$$y = (1,1)x .$$

16. Find the dual of the system

$$\dot{x} = \begin{pmatrix} 2 & 1 \\ 0 & 2 \end{pmatrix} x + \begin{pmatrix} 1 \\ 1 \end{pmatrix} u$$

$$y = (0,1)x .$$

17. Find the dual of the system

$$\dot{x} = \begin{pmatrix} 0 & 1 \\ 0 & -6 \end{pmatrix} x + \begin{pmatrix} 0 \\ 2 \end{pmatrix} u$$

$$y = (1,0)x .$$

18. Find the dual of the system

$$x(t+1) = \begin{pmatrix} 1 & 1 \\ 2 & 2 \end{pmatrix} x(t) + \begin{pmatrix} 1 \\ 0 \end{pmatrix} u(t)$$

$$y(t) = (1,1)x(t) .$$

19. Find the dual of the system

$$x(t+1) = \begin{pmatrix} 2 & 1 \\ 0 & 2 \end{pmatrix} x(t) + \begin{pmatrix} 1 \\ 1 \end{pmatrix} u(t)$$

$$y(t) = (0,1)x(t) .$$

20. (i) Prove Lemma 6.3.

(ii)   Prove Theorem 6.3.

(iii)   Prove Theorem 6.6.

(iv)   Prove Lemma 6.3 for discrete systems with $t_1 \geq n$.

(v)   Prove Theorem 6.7.

21.   Prove that for any $n \times n$ continuous matrices $\mathbf{A}(t)$ and $\mathbf{B}(t)$ there is a continuous row vector $\mathbf{c}^T(t)$ such that system

$$\dot{\mathbf{x}}(t) = \mathbf{A}(t)\mathbf{x}(t) + \mathbf{B}(t)\mathbf{u}(t)$$

$$y(t) = \mathbf{c}^T(t)\mathbf{x}(t)$$

is observable.

22.   Assume that the time invariant system $\dot{\mathbf{x}} = \mathbf{A}\mathbf{x} + \mathbf{B}\mathbf{u}$, $y = \mathbf{C}\mathbf{x}$ is observable, and $\mathbf{A}$, $\bar{\mathbf{A}}$ and $\mathbf{C}$, $\bar{\mathbf{C}}$ are sufficiently close to each other. Prove that $\dot{\mathbf{x}} = \bar{\mathbf{A}}\mathbf{x} + \mathbf{B}\mathbf{u}$, $y = \bar{\mathbf{C}}\mathbf{x}$ is also observable.

23.   Let α be a real constant. Prove that the time invariant linear system $\dot{\mathbf{x}} = \mathbf{A}\mathbf{x} + \mathbf{B}\mathbf{u}$, $y = \mathbf{C}\mathbf{x}$ is observable if and only if $\dot{\mathbf{x}} = (\mathbf{A} + \alpha\mathbf{I})\mathbf{x} + \mathbf{B}\mathbf{u}$, $y = \mathbf{C}\mathbf{x}$ is observable.

24.   Prove that system $\dot{\mathbf{x}} = \mathbf{A}\mathbf{x} + \mathbf{B}\mathbf{u}$, $y = \mathbf{C}\mathbf{x}$ is observable if and only if system $\dot{\mathbf{x}} = \mathbf{A}\mathbf{x} + \mathbf{B}\mathbf{u}$, $y = \mathbf{C}^T\mathbf{C}\mathbf{x}$ is observable.

25.   Let m be the degree of the minimal-polynomial of $\mathbf{A}$. Prove that

$$rank(\mathbf{L}) = rank \begin{pmatrix} \mathbf{C} \\ \mathbf{C}\mathbf{A} \\ \vdots \\ \mathbf{C}\mathbf{A}^{m-1} \end{pmatrix}.$$

chapter seven

Canonical Forms

In this chapter some special transformations of time-invariant linear systems will be introduced, and their properties will be discussed. These special forms make the computer solutions and the investigation of the systems properties much easier.

Let $\mathbf{A}$, $\mathbf{B}$, and $\mathbf{C}$ be constant matrices of the size $n \times n$, $n \times m$, and $p \times n$, respectively. For the sake of simplicity, the continuous system

$$\dot{\mathbf{x}} = \mathbf{A}\mathbf{x} + \mathbf{B}\mathbf{u}$$

$$\mathbf{y} = \mathbf{C}\mathbf{x} \tag{7.1}$$

or the analogous discrete system

$$\mathbf{x}(t+1) = \mathbf{A}\mathbf{x}(t) + \mathbf{B}\mathbf{u}(t)$$

$$\mathbf{y}(t) = \mathbf{C}\mathbf{x}(t) \tag{7.2}$$

will be called the $(\mathbf{A}, \mathbf{B}, \mathbf{C})$-system.

Introduce the new state variable

$$\tilde{\mathbf{x}} = \mathbf{T}\mathbf{x} , \tag{7.3}$$

where $\mathbf{T}$ is a nonsingular matrix. Use the first equations of the systems models (7.1) and (7.2) to derive

$$\dot{\tilde{\mathbf{x}}} = \mathbf{T}\dot{\mathbf{x}} = \mathbf{T}\mathbf{A}\mathbf{x} + \mathbf{T}\mathbf{B}\mathbf{u} = (\mathbf{T}\mathbf{A}\mathbf{T}^{-1})\mathbf{T}\mathbf{x} + (\mathbf{T}\mathbf{B})\mathbf{u}$$

$$= (\mathbf{T}\mathbf{A}\mathbf{T}^{-1})\tilde{\mathbf{x}} + (\mathbf{T}\mathbf{B})\mathbf{u}$$

or

$$\tilde{\mathbf{x}}(t+1) = \mathbf{T}\mathbf{x}(t+1) = \mathbf{TA}\mathbf{x}(t) + \mathbf{TB}\mathbf{u}(t) = (\mathbf{TAT}^{-1})\mathbf{T}\mathbf{x}(t) + (\mathbf{TB})\mathbf{u}(t)$$

$$= (\mathbf{TAT})^{-1}\tilde{\mathbf{x}}(t) + (\mathbf{TB})\mathbf{u}(t) \ .$$

Similarly, the output equations can be transformed as

$$\mathbf{y} = \mathbf{C}\mathbf{x} = (\mathbf{CT}^{-1})\tilde{\mathbf{x}} \ .$$

The above derivations imply that by introducing the new variable (7.3), the $(\mathbf{A}, \mathbf{B}, \mathbf{C})$-system is transformed into an $(\tilde{\mathbf{A}}, \tilde{\mathbf{B}}, \tilde{\mathbf{C}})$-system, where

$$\tilde{\mathbf{A}} = \mathbf{TAT}^{-1}, \quad \tilde{\mathbf{B}} = \mathbf{TB}, \quad \text{and} \quad \tilde{\mathbf{C}} = \mathbf{CT}^{-1} \ . \tag{7.4}$$

The original $(\mathbf{A}, \mathbf{B}, \mathbf{C})$-system and the transformed $(\tilde{\mathbf{A}}, \tilde{\mathbf{B}}, \tilde{\mathbf{C}})$-system have some common properties:

(i)   First we remind the reader that in Theorem 3.4 we proved that the two systems have the same transfer function for continuous systems, and in Section 3.4 we mentioned that this property also holds for discrete systems.

(ii) If one of the two systems is completely controllable (or completely observable) then the same holds for the other system as well. This assertion can be proven as follows. Let $\mathbf{K}$ and $\tilde{\mathbf{K}}$ denote the controllability matrices, and let $\mathbf{L}$ and $\tilde{\mathbf{L}}$ denote the observability matrices of systems $(\mathbf{A}, \mathbf{B}, \mathbf{C})$ and $(\tilde{\mathbf{A}}, \tilde{\mathbf{B}}, \tilde{\mathbf{C}})$, respectively. Note first that

$$\tilde{\mathbf{K}} = (\tilde{\mathbf{B}}, \tilde{\mathbf{A}}\tilde{\mathbf{B}}, \ldots, \tilde{\mathbf{A}}^{n-1}\tilde{\mathbf{B}}) = (\mathbf{TB}, \mathbf{TAT}^{-1}\mathbf{TB}, \ldots, \mathbf{TA}^{n-1}\mathbf{T}^{-1}\mathbf{TB}) \ ,$$

where we used the fact that for $k \geq 1$,

$$(\mathbf{TAT}^{-1})^k = \mathbf{TA}(\mathbf{T}^{-1}\mathbf{T})\mathbf{A}(\mathbf{T}^{-1} \ldots \mathbf{T})\mathbf{AT}^{-1} = \mathbf{TA}^k\mathbf{T}^{-1} \ .$$

Therefore,

$$\tilde{\mathbf{K}} = \mathbf{T} \cdot (\mathbf{B}, \mathbf{AB}, \ldots, \mathbf{A}^{n-1}\mathbf{B}) = \mathbf{T} \cdot \mathbf{K} \ ,$$

and since $\mathbf{T}$ is nonsingular, $rank(\mathbf{K}) = n$ if and only if $rank(\tilde{\mathbf{K}}) =$

n. Similarly,

$$
\tilde{\mathbf{L}} = \begin{pmatrix} \tilde{\mathbf{C}} \\ \tilde{\mathbf{C}}\tilde{\mathbf{A}} \\ \vdots \\ \tilde{\mathbf{C}}\tilde{\mathbf{A}}^{n-1} \end{pmatrix} = \begin{pmatrix} \mathbf{C}\mathbf{T}^{-1} \\ \mathbf{C}\mathbf{T}^{-1}\mathbf{T}\mathbf{A}\mathbf{T}^{-1} \\ \vdots \\ \mathbf{C}\mathbf{T}^{-1}\mathbf{T}\mathbf{A}^{n-1}\mathbf{T}^{-1} \end{pmatrix} = \begin{pmatrix} \mathbf{C} \\ \mathbf{C}\mathbf{A} \\ \vdots \\ \mathbf{C}\mathbf{A}^{n-1} \end{pmatrix} \cdot \mathbf{T}^{-1}
$$

$$
= \mathbf{L} \cdot \mathbf{T}^{-1} \, ,
$$

and, therefore, $rank(\tilde{\mathbf{L}}) = n$ if and only if $rank(\mathbf{L}) = n$.
Transformation (7.4) allows us to transform a system into special
forms without losing the essential properties of the system, which
can then be solved much easier than the original system, and certain properties (such as controllability, observability, etc.) can be
verified immediately without further calculation. These special
forms are the subject of this chapter. This transformation principle is illustrated in Figure 7.1. Finally, we note that all results of
this chapter hold for both discrete and continuous systems.

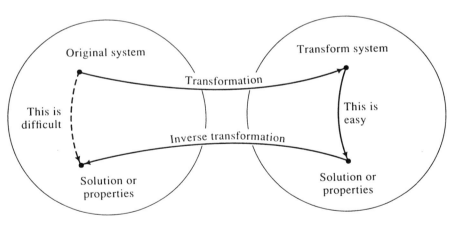

Figure 7.1   Principle of the transformation method.

7.1   *Diagonal and Jordan Forms*

Assume first that matrix $\mathbf{A}$ can be diagonalized by similarity transformation. That is,

$$
\mathbf{T}\mathbf{A}\mathbf{T}^{-1} = diag(\lambda_1, \dots, \lambda_n)
$$

with some nonsingular matrix $\mathbf{T}$. Introduce the new variable $\tilde{\mathbf{x}} = \mathbf{Tx}$, then the transformed system has the property that $\tilde{\mathbf{A}}$ is diagonal. Therefore, the $(\tilde{\mathbf{A}}, \tilde{\mathbf{B}}, \tilde{\mathbf{C}})$-system has the very special form

$$\dot{\tilde{x}}_1 = \lambda_1 \tilde{x}_1 + \tilde{u}_1$$

$$\dot{\tilde{x}}_2 = \lambda_2 \tilde{x}_2 + \tilde{u}_2$$

$$\vdots$$

$$\dot{\tilde{x}}_n = \lambda_n \tilde{x}_n + \tilde{u}_n \ , \tag{7.5}$$

where for $k = 1, 2, \ldots, n$,

$$\tilde{u}_k = \tilde{b}_{k1} u_1 + \tilde{b}_{k2} u_2 + \cdots + \tilde{b}_{km} u_m \ . \tag{7.6}$$

Here we use the notation

$$\tilde{\mathbf{B}} = (\tilde{b}_{kl}) \qquad \text{and} \qquad \mathbf{u} = (u_l) \ .$$

If the input function is known, then all functions $\tilde{u}_k$ are also known. Therefore, the solution of the original system is reduced to solving n independent single-dimensional linear differential equations, which is a much easier task than the solution of the original n-dimensional linear differential equation. After the $\tilde{x}_k$s are determined, the original state variable is obtained as $\mathbf{x} = \mathbf{T}^{-1}\tilde{\mathbf{x}}$. If the new coefficient matrix $\tilde{\mathbf{A}}$ is diagonal, then the transfer function of the system can be written as

$$\mathbf{H}(s) = \tilde{\mathbf{C}}(s\mathbf{I} - \tilde{\mathbf{A}})^{-1}\tilde{\mathbf{B}}$$

$$= \tilde{\mathbf{C}}\{diag(s - \lambda_1, \ldots, s - \lambda_n)\}^{-1} \cdot \tilde{\mathbf{B}}$$

$$= \tilde{\mathbf{C}} \, diag\left(\frac{1}{s - \lambda_1}, \ldots, \frac{1}{s - \lambda_n}\right) \cdot \tilde{\mathbf{B}} \ ;$$

therefore, the (i, j) element of $\mathbf{H}(s)$ is as follows:

$$\sum_l \tilde{c}_{il} \tilde{b}_{lj} \cdot \frac{1}{s - \lambda_l} \ .$$

This observation implies that $\mathbf{H}(s)$ is the sum of functions of the form $1/(s - \lambda_l)\tilde{\mathbf{D}}_l$, where $\tilde{\mathbf{D}}_l$ is a matrix with (i, j) element $\tilde{c}_{il} \tilde{b}_{lj}$. Hence, $\mathbf{H}(s)$ is the parallel combination of these special transfer functions.

In the general case, matrix $\mathbf{A}$ can be transformed into Jordan canonical form:

$$\mathbf{TAT}^{-1} = \begin{pmatrix} \mathbf{J}_1 & & & \mathbf{O} \\ & \mathbf{J}_2 & & \\ & & \ddots & \\ \mathbf{O} & & & \mathbf{J}_s \end{pmatrix} ,$$

where for $j = 1, 2, \ldots, s,$

$$\mathbf{J}_j = \begin{pmatrix} \lambda_i & 1 & & & \mathbf{O} \\ & \lambda_i & 1 & & \\ & & \ddots & \ddots & \\ & & & \lambda_i & 1 \\ \mathbf{O} & & & & \lambda_i \end{pmatrix}$$

is a νth-order square matrix, λ_i being an eigenvalue.

Note that the details of diagonal and Jordan form transformations were discussed in Section 1.3.2.

By introducing the new variable $\tilde{\mathbf{x}} = \mathbf{Tx}$, the transformed system has the property that $\tilde{\mathbf{A}}$ is block-diagonal. Therefore, the $(\tilde{\mathbf{A}}, \tilde{\mathbf{B}}, \tilde{\mathbf{C}})$-system has the special form

$$\dot{\tilde{\mathbf{x}}}_1 = \mathbf{J}_1 \tilde{\mathbf{x}}_1 + \tilde{\mathbf{u}}_1$$

$$\dot{\tilde{\mathbf{x}}}_2 = \mathbf{J}_2 \tilde{\mathbf{x}}_2 + \tilde{\mathbf{u}}_2$$

$$\vdots$$

$$\dot{\tilde{\mathbf{x}}}_s = \mathbf{J}_s \tilde{\mathbf{x}}_s + \tilde{\mathbf{u}}_s , \tag{7.7}$$

where for $j = 1, 2, \ldots, s$, vectors $\tilde{\mathbf{x}}_j$ and $\tilde{\mathbf{u}}_j$ are ν_j-dimensional, and if the input $\mathbf{u}$ is known, then all functions $\tilde{\mathbf{u}}_j$ are also known. Therefore, the solution of the original system is reduced to the solution of s ν_j-dimensional problems. Similar to the diagonal case, this reduction saves a lot of computations. In addition, in solving each block, the special structure of matrix $\mathbf{J}_j$ makes the solution very simple, as is shown next.

Consider the jth block of Equation (7.7):

$$\dot{\tilde{x}}_{j1} = \lambda_i \tilde{x}_{j1} + \tilde{x}_{j2} + \tilde{u}_{j1}$$

$$\dot{\tilde{x}}_{j2} = \lambda_i \tilde{x}_{j2} + \tilde{x}_{j3} + \tilde{u}_{j2}$$

$$\vdots$$

$$\dot{\tilde{x}}_{j,\nu_j-1} = \lambda_i \tilde{x}_{j,\nu_j-1} + \tilde{x}_{j,\nu_j} + \tilde{u}_{j,\nu_j-1}$$

$$\dot{\tilde{x}}_{j\nu_j} = \lambda_i \tilde{x}_{j\nu_j} + \tilde{u}_{j\nu_j} . \tag{7.8}$$

From the last equation $\tilde{x}_{j\nu_j}$ can be obtained easily, because it is only a single-dimensional equation. After $\tilde{x}_{j\nu_j}$ is determined, $\tilde{x}_{j,\nu_j-1}$ can be obtained from the (ν_j-1)st equation, which is again single-dimensional. Continuing this process until the first equation, all components of $\tilde{\mathbf{x}}_j$ are determined recursively in the backward order $\tilde{x}_{j\nu_j}, \tilde{x}_{j\nu_j-1}, \ldots, \tilde{x}_{j2},$ $\tilde{x}_{j1}$. Note that at each step only a single-dimensional linear equation is solved, which makes this process very attractive.

Example 7.1

We first solve the diagonal system

$$\dot{\tilde{x}}_1 = 2 \cdot \tilde{x}_1 + e^t , \ \tilde{x}_1(0) = 0$$
$$\dot{\tilde{x}}_2 = \tilde{x}_2 + 2e^t , \quad \tilde{x}_2(0) = 0$$
$$\dot{\tilde{x}}_3 = \tilde{x}_3 + 3e^t , \quad \tilde{x}_3(0) = 0 .$$

Note that this system consists of three independent single-dimensional equations. By applying standard techniques from the theory of linear differential equations, we have the solutions

$$\tilde{x}_1(t) = e^{2t} - e^t$$

$$\tilde{x}_2(t) = 2te^t$$

$$\tilde{x}_3(t) = 3te^t .$$

Hence, the solution is obtained very easily.

Example 7.2

Next the system

$$\dot{\tilde{x}}_1 = \tilde{x}_1 + \tilde{x}_2 + e^t , \ \tilde{x}_1(0) = 0$$
$$\dot{\tilde{x}}_2 = \tilde{x}_2 + \tilde{x}_3 + 2e^t , \ \tilde{x}_2(0) = 0$$
$$\dot{\tilde{x}}_3 = \tilde{x}_3 + 3e^t , \quad\quad \tilde{x}_3(0) = 0$$

with a Jordan block coefficient matrix, is solved.

The last equation is a single-dimensional problem with solution

$$\tilde{x}_3(t) = 3te^t \ .$$

By substituting this function into the second equation, a single-dimensional problem is obtained for $\tilde{x}_2$:

$$\dot{\tilde{x}}_2 = \tilde{x}_2 + (3t+2)e^t, \qquad \tilde{x}_2(0) = 0 \ .$$

The solution of this problem is

$$\tilde{x}_2(t) = \left(\frac{3t^2}{2} + 2t \right) e^t \ .$$

And finally, substitute this function into the first equation to get the single-variable equation for $\tilde{x}_1$:

$$\dot{\tilde{x}}_1 = \tilde{x}_1 + \left(\frac{3t^2}{2} + 2t + 1 \right) e^t, \qquad \tilde{x}_1(0) = 0 \ ,$$

which has the solution

$$\tilde{x}_1(t) = \left(\frac{t^3}{2} + t^2 + t \right) e^t \ .$$

Hence, the solution is obtained again very easily.

Note that standard computer packages are available to transform matrices into diagonal or Jordan canonical form.

7.2   *Controllability Canonical Forms*

In this and also in the next section, single-input and single-output systems of the forms

$$\dot{\mathbf{x}} = \mathbf{A}\mathbf{x} + \mathbf{b}u$$

$$y = \mathbf{c}^T\mathbf{x} \tag{7.9}$$

and

$$\mathbf{x}(t+1) = \mathbf{A}\mathbf{x}(t) + \mathbf{b}u(t)$$

$$y(t) = \mathbf{c}^T\mathbf{x}(t) \tag{7.10}$$

are discussed, where $\mathbf{A}$ is an $n \times n$ constant matrix and vectors $\mathbf{b}$ and $\mathbf{c}$ are n-dimensional.

First we verify the *first type* of controllability canonical forms.

THEOREM 7.1

Assume that system $(\mathbf{A}, \mathbf{b}, \mathbf{c}^T)$ is completely controllable, then it can be transformed into an $(\tilde{\mathbf{A}}, \tilde{\mathbf{b}}, \tilde{\mathbf{c}}^T)$-system, where

$$\tilde{\mathbf{A}} = \begin{pmatrix} 0 & 0 & 0 & \cdots & 0 & a_0 \\ 1 & 0 & 0 & \cdots & 0 & a_1 \\ 0 & 1 & 0 & \cdots & 0 & a_2 \\ \vdots & \vdots & \vdots & \ddots & \vdots & \vdots \\ 0 & 0 & 0 & \cdots & 1 & a_{n-1} \end{pmatrix} \quad and \quad \tilde{\mathbf{b}} = \begin{pmatrix} 1 \\ 0 \\ 0 \\ \vdots \\ 0 \end{pmatrix}. \tag{7.11}$$

PROOF   Because system $(\mathbf{A}, \mathbf{b}, \mathbf{c}^T)$ is completely controllable, the *rank* of the controllability matrix is n. In our case,

$$\mathbf{K} = (\mathbf{b}, \mathbf{Ab}, \mathbf{A}^2\mathbf{b}, \ldots, \mathbf{A}^{n-1}\mathbf{b}),$$

which is $n \times n$; therefore, it is nonsingular.

Select the transformation matrix $\mathbf{T} = \mathbf{K}^{-1}$. Note first that its rows $\mathbf{t}_1^T, \ldots, \mathbf{t}_n^T$ satisfy relation

$$\begin{pmatrix} \mathbf{t}_1^T \\ \mathbf{t}_2^T \\ \vdots \\ \mathbf{t}_n^T \end{pmatrix} (\mathbf{b}, \mathbf{Ab}, \mathbf{A}^2\mathbf{b}, \ldots, \mathbf{A}^{n-1}\mathbf{b}) = \mathbf{I},$$

which holds if and only if

$$\mathbf{t}_k^T \mathbf{A}^{k-1}\mathbf{b} = 1 \quad and \quad \mathbf{t}_k^T \mathbf{A}^{l-1}\mathbf{b} = 0 \quad (l \neq k). \tag{7.12}$$

Therefore, relations (7.12) imply that

$$\tilde{\mathbf{b}} = \mathbf{Tb} = \begin{pmatrix} \mathbf{t}_1^T \\ \mathbf{t}_2^T \\ \vdots \\ \mathbf{t}_n^T \end{pmatrix} \mathbf{b} = \begin{pmatrix} 1 \\ 0 \\ \vdots \\ 0 \end{pmatrix},$$

and

$$\tilde{\mathbf{A}} = \mathbf{T}\mathbf{A}\mathbf{T}^{-1} = \begin{pmatrix} \mathbf{t}_1^T \\ \mathbf{t}_2^T \\ \vdots \\ \mathbf{t}_n^T \end{pmatrix} \mathbf{A}(\mathbf{b}, \mathbf{A}\mathbf{b}, \ldots, \mathbf{A}^{n-1}\mathbf{b})$$

$$= \begin{pmatrix} \mathbf{t}_1^T \\ \mathbf{t}_2^T \\ \vdots \\ \mathbf{t}_n^T \end{pmatrix} (\mathbf{A}\mathbf{b}, \mathbf{A}^2\mathbf{b}, \ldots, \mathbf{A}^n\mathbf{b})$$

$$= \begin{pmatrix} 0 & 0 & 0 & \cdots & 0 & a_0 \\ 1 & 0 & 0 & \cdots & 0 & a_1 \\ 0 & 1 & 0 & \cdots & 0 & a_2 \\ \vdots & \vdots & \vdots & \ddots & \vdots & \vdots \\ 0 & 0 & 0 & \cdots & 1 & a_{n-1} \end{pmatrix},$$

where

$$a_0 = \mathbf{t}_1^T \mathbf{A}^n \mathbf{b}, \ldots, a_{n-1} = \mathbf{t}_n^T \mathbf{A}^n \mathbf{b}.$$

Thus, the proof is completed. ∎

REMARK 7.1   There is nothing special about vector $\tilde{\mathbf{c}}$. ∎

An algorithm to find canonical form (7.11) consists of the following steps:

Step 1   Find matrix $\mathbf{K}$.

Step 2   Compute $\mathbf{T} = \mathbf{K}^{-1}$.

Step 3   Determine $\tilde{\mathbf{A}}$, $\tilde{\mathbf{b}}$, and $\tilde{\mathbf{c}}^T$ by using relations (7.12).

Example 7.3

This algorithm is now illustrated in the case of system

$$\dot{\mathbf{x}} = \begin{pmatrix} 0 & \omega \\ -\omega & 0 \end{pmatrix} \mathbf{x} + \begin{pmatrix} 0 \\ 1 \end{pmatrix} u$$

$$y = (1, 1)\mathbf{x},$$

which was examined in earlier chapters.

Step 1: The definition of the controllability matrix implies that

$$\mathbf{K} = \begin{pmatrix} 0 & \omega \\ 1 & 0 \end{pmatrix} .$$

Step 2: By inverting $\mathbf{K}$,

$$\mathbf{T} = \mathbf{K}^{-1} = \frac{1}{\omega} \begin{pmatrix} 0 & \omega \\ 1 & 0 \end{pmatrix} .$$

Step 3: Use relations (7.12) to get

$$\tilde{\mathbf{A}} = \mathbf{TAT}^{-1} = \frac{1}{\omega} \begin{pmatrix} 0 & \omega \\ 1 & 0 \end{pmatrix} \begin{pmatrix} 0 & \omega \\ -\omega & 0 \end{pmatrix} \begin{pmatrix} 0 & \omega \\ 1 & 0 \end{pmatrix} = \begin{pmatrix} 0 & -\omega^2 \\ 1 & 0 \end{pmatrix} ,$$

$$\tilde{\mathbf{b}} = \mathbf{Tb} = \frac{1}{\omega} \begin{pmatrix} 0 & \omega \\ 1 & 0 \end{pmatrix} \begin{pmatrix} 0 \\ 1 \end{pmatrix} = \begin{pmatrix} 1 \\ 0 \end{pmatrix} ,$$

and

$$\tilde{\mathbf{c}}^T = \mathbf{c}^T \cdot \mathbf{T}^{-1} = (1,1) \begin{pmatrix} 0 & \omega \\ 1 & 0 \end{pmatrix} = (1, \omega) .$$

COROLLARY 7.1

Expanding the characteristic polynomial of $\tilde{\mathbf{A}}$ with respect to the last column, it is easy to verify that it equals

$$\varphi(\lambda) = \lambda^n - a_{n-1}\lambda^{n-1} - \cdots - a_1\lambda - a_0 . \qquad (7.13)$$

Since $\mathbf{A}$ and $\tilde{\mathbf{A}}$ are similar matrices, this is the characteristic polynomial of $\mathbf{A}$ as well. Therefore, the method presented in the proof of the theorem can also be considered as a numerical method for constructing the characteristic polynomial of real matrices. Note that this method can be used if there exists a real vector $\mathbf{b}$ such that $\mathbf{b}, \mathbf{Ab}, \ldots, \mathbf{A}^{n-1}\mathbf{b}$ are linearly independent.

Example 7.4

Consider matrix $\mathbf{A}$ of the system of the previous example. Its characteristic polynomial is

$$\varphi(\lambda) = \det \begin{pmatrix} -\lambda & \omega \\ -\omega & -\lambda \end{pmatrix} = \lambda^2 + \omega^2 .$$

From the previous example we know that

$$\tilde{\mathbf{A}} = \begin{pmatrix} 0 & -\omega^2 \\ 1 & 0 \end{pmatrix} ;$$

therefore, $a_0 = -\omega^2$, $a_1 = 0$, and from Equation (7.13),

$$\varphi(\lambda) = \lambda^2 - 0 \cdot \lambda + \omega^2 ,$$

which coincides with the result obtained by the direct computation of $\varphi(\lambda)$. For small n, the method has no practical importance, but for large values of n it certainly does.

A *second type* of controllability canonical form is presented next.

THEOREM 7.2
Assume that system $(\mathbf{A}, \mathbf{b}, \mathbf{c}^T)$ is completely controllable, then it can be transformed into an $(\tilde{\mathbf{A}}, \tilde{\mathbf{b}}, \tilde{\mathbf{c}}^T)$-system, where

$$\tilde{\mathbf{A}} = \begin{pmatrix} 0 & 1 & 0 & \cdots & 0 \\ 0 & 0 & 1 & \cdots & 0 \\ \vdots & \vdots & \vdots & \ddots & \vdots \\ 0 & 0 & 0 & \cdots & 1 \\ a_0 & a_1 & a_2 & \cdots & a_{n-1} \end{pmatrix} \quad and \quad \tilde{\mathbf{b}} = \begin{pmatrix} 0 \\ 0 \\ \vdots \\ 0 \\ 1 \end{pmatrix} . \tag{7.14}$$

PROOF First we prove that vectors

$$\mathbf{t}_n^T, \mathbf{t}_n^T \mathbf{A}, \ldots, \mathbf{t}_n^T \mathbf{A}^{n-1}$$

are linearly independent, where $\mathbf{t}_n^T$ denotes the last row of $\mathbf{K}^{-1}$, as in the proof of the previous theorem. Assume not, then there exist constants $\alpha_0, \alpha_1, \ldots, \alpha_{n-1}$ such that at least one α_k is nonzero and

$$\alpha_0 \mathbf{t}_n^T + \alpha_1 \mathbf{t}_n^T \mathbf{A} + \cdots + \alpha_{n-1} \mathbf{t}_n^T \mathbf{A}^{n-1} = \mathbf{0}^T . \tag{7.15}$$

Multiply this equation by $\mathbf{b}$ to get

$$\alpha_0 \mathbf{t}_n^T \mathbf{b} + \alpha_1 \mathbf{t}_n^T \mathbf{A} \mathbf{b} + \cdots + \alpha_{n-1} \mathbf{t}_n^T \mathbf{A}^{n-1} \mathbf{b} = 0 .$$

Observe that relation (7.12) implies that the first $n - 1$ terms of the lefthand side are equal to zero, and the last term equals $\alpha_{n-1} \cdot 1 = \alpha_{n-1}$. Therefore, $\alpha_{n-1} = 0$, and (7.15) reduces to equation

$$\alpha_0 \mathbf{t}_n^T + \alpha_1 \mathbf{t}_n^T \mathbf{A} + \cdots + \alpha_{n-2} \mathbf{t}_n^T \mathbf{A}^{n-2} = \mathbf{0}^T . \tag{7.16}$$

Multiply this equation now by $\mathbf{Ab}$ to get

$$\alpha_0 \mathbf{t}_n^T \mathbf{Ab} + \alpha_1 \mathbf{t}_n^T \mathbf{A}^2 \mathbf{b} + \cdots + \alpha_{n-2} \mathbf{t}_n^T \mathbf{A}^{n-1} \mathbf{b} = 0 \,.$$

Similar to the previous case, from relations (7.12) we conclude that $\alpha_{n-2} = 0$. Therefore, the last term of the left-hand side of Equation (7.16) equals zero. Continuing the same process by multiplying the resulting equation by $\mathbf{A}^2 \mathbf{b}$, and so on, one can easily verify that all coefficients $\alpha_0, \alpha_1, \ldots, \alpha_{n-1}$ are equal to zero.

Select now the transformation matrix

$$\mathbf{T} = \begin{pmatrix} \mathbf{t}_n^T \\ \mathbf{t}_n^T \mathbf{A} \\ \vdots \\ \mathbf{t}_n^T \mathbf{A}^{n-1} \end{pmatrix} \,,$$

and denote the columns of $\mathbf{T}^{-1}$ by $\mathbf{c}_1, \ldots, \mathbf{c}_n$. Then

$$\begin{pmatrix} \mathbf{t}_n^T \\ \mathbf{t}_n^T \mathbf{A} \\ \vdots \\ \mathbf{t}_n^T \mathbf{A}^{n-1} \end{pmatrix} (\mathbf{c}_1, \ldots, \mathbf{c}_n) = \mathbf{I} \,,$$

which holds if and only if

$$\mathbf{t}_n^T \mathbf{A}^{k-1} \mathbf{c}_k = 1 \qquad \text{and} \qquad \mathbf{t}_n^T \mathbf{A}^{k-1} \mathbf{c}_l = 0 \qquad (l \neq k) \,.$$

Therefore, relations (7.12) imply that

$$\tilde{\mathbf{b}} = \mathbf{Tb} = \begin{pmatrix} \mathbf{t}_n^T \\ \mathbf{t}_n^T \mathbf{A} \\ \vdots \\ \mathbf{t}_n^T \mathbf{A}^{n-2} \\ \mathbf{t}_n^T \mathbf{A}^{n-1} \end{pmatrix} \mathbf{b} = \begin{pmatrix} 0 \\ 0 \\ \vdots \\ 0 \\ 1 \end{pmatrix} \,,$$

and

$$\tilde{\mathbf{A}} = \mathbf{TAT}^{-1} = \begin{pmatrix} \mathbf{t}_n^T \\ \mathbf{t}_n^T \mathbf{A} \\ \vdots \\ \mathbf{t}_n^T \mathbf{A}^{n-1} \end{pmatrix} \mathbf{A}(\mathbf{c}_1, \ldots, \mathbf{c}_n)$$

$$= \begin{pmatrix} \mathbf{t}_n^T \mathbf{A} \\ \mathbf{t}_n^T \mathbf{A}^2 \\ \vdots \\ \mathbf{t}_n^T \mathbf{A}^n \end{pmatrix} (\mathbf{c}_1, \ldots, \mathbf{c}_n)$$

$$= \begin{pmatrix} 0 & 1 & 0 & \cdots & 0 \\ 0 & 0 & 1 & \cdots & 0 \\ \vdots & \vdots & \vdots & \ddots & \vdots \\ 0 & 0 & 0 & \cdots & 1 \\ a_0 & a_1 & a_2 & \cdots & a_{n-1} \end{pmatrix},$$

where

$$a_0 = \mathbf{t}_n^T \mathbf{A}^n \mathbf{c}_1, \ldots, a_{n-1} = \mathbf{t}_n^T \mathbf{A}^n \mathbf{c}_n .$$

Thus, the proof is completed.   ∎

REMARK 7.2   No special property holds for vector $\tilde{\mathbf{c}}$. Note that this canonical form is the same as the model (3.29) and (3.30) for systems given in input–output form.   ∎

COROLLARY 7.2
Similar to the case of the canonical form (7.11), the common characteristic polynomial of $\tilde{\mathbf{A}}$ and $\mathbf{A}$ is the one given by Equation (7.13).

The algorithm for determining the canonical form (7.14) consists of the following steps:

Step 1   Compute matrix $\mathbf{K}$.

Step 2   Find matrix $\mathbf{T}$.

Step 3   Compute $\tilde{\mathbf{A}}$, $\tilde{\mathbf{b}}$, and $\tilde{\mathbf{c}}^T$ by using Equation (7.4).

Example 7.5

Consider again the system

$$\dot{\mathbf{x}} = \begin{pmatrix} 0 & \omega \\ -\omega & 0 \end{pmatrix} \mathbf{x} + \begin{pmatrix} 0 \\ 1 \end{pmatrix} u$$

$$y = (1, 1)\mathbf{x} .$$

The canonical form (7.6) will now be determined.

Step 1: In Example 7.3 we derived that

$$\mathbf{K} = \begin{pmatrix} 0 & \omega \\ 1 & 0 \end{pmatrix} ,$$

and, therefore,

$$\mathbf{t}_1^T = \frac{1}{\omega}(0, \omega), \qquad \mathbf{t}_2^T = \frac{1}{\omega}(1, 0) .$$

Step 2: Simple substitution shows that

$$\mathbf{T} = \frac{1}{\omega} \begin{pmatrix} 1 & 0 \\ 0 & \omega \end{pmatrix} .$$

Step 3: Invert $\mathbf{T}$:

$$\mathbf{T}^{-1} = \begin{pmatrix} \omega & 0 \\ 0 & 1 \end{pmatrix} ,$$

and use relations (7.4) to get

$$\tilde{\mathbf{A}} = \mathbf{T}\mathbf{A}\mathbf{T}^{-1} = \frac{1}{\omega} \begin{pmatrix} 1 & 0 \\ 0 & \omega \end{pmatrix} \begin{pmatrix} 0 & \omega \\ -\omega & 0 \end{pmatrix} \begin{pmatrix} \omega & 0 \\ 0 & 1 \end{pmatrix} = \begin{pmatrix} 0 & 1 \\ -\omega^2 & 0 \end{pmatrix} ,$$

$$\tilde{\mathbf{b}} = \mathbf{T}\mathbf{b} = \frac{1}{\omega} \begin{pmatrix} 1 & 0 \\ 0 & \omega \end{pmatrix} \begin{pmatrix} 0 \\ 1 \end{pmatrix} = \begin{pmatrix} 0 \\ 1 \end{pmatrix} ,$$

and

$$\tilde{\mathbf{c}}^T = \mathbf{c}\mathbf{T}^{-1} = (1, 1) \begin{pmatrix} \omega & 0 \\ 0 & 1 \end{pmatrix} = (\omega, 1) .$$

From the last row of $\tilde{\mathbf{A}}$ we see that the common characteristic polynomial of $\tilde{\mathbf{A}}$ and $\mathbf{A}$ is

$$\varphi(\lambda) = \lambda^2 - 0 \cdot \lambda + \omega^2 = \lambda^2 + \omega^2 ,$$

which coincides with the result obtained in the previous example.

The above canonical forms remain valid for multiple-output systems, when the input is still one-dimensional, since only $\mathbf{A}$ and $\mathbf{b}$ are transformed into special forms. In the general case of multiple inputs, the transformation to a canonical form becomes more complicated. As Example 7.6 will show, the proofs of Theorems 7.1 and 7.2 cannot be applied in the general case. However, a multi-input canonical form is given as

$$\tilde{\mathbf{A}} = \begin{pmatrix} \tilde{\mathbf{A}}_{11} & \tilde{\mathbf{A}}_{12} & \cdots & \tilde{\mathbf{A}}_{1s} \\ \tilde{\mathbf{A}}_{21} & \tilde{\mathbf{A}}_{22} & \cdots & \tilde{\mathbf{A}}_{2s} \\ \vdots & \vdots & \ddots & \vdots \\ \tilde{\mathbf{A}}_{s1} & \tilde{\mathbf{A}}_{s2} & \cdots & \tilde{\mathbf{A}}_{ss} \end{pmatrix} \quad \text{and} \quad \tilde{\mathbf{B}} = \begin{pmatrix} \tilde{\mathbf{B}}_1 \\ \tilde{\mathbf{B}}_2 \\ \vdots \\ \tilde{\mathbf{B}}_s \end{pmatrix}, \quad (7.17)$$

where the types of matrices

$$\tilde{\mathbf{A}}_{kk} = \begin{pmatrix} 0 & 1 & 0 & \cdots & 0 \\ 0 & 0 & 1 & \cdots & 0 \\ \vdots & \vdots & \vdots & \ddots & \vdots \\ 0 & 0 & 0 & \cdots & 1 \\ a_0^{(k,k)} & a_1^{(k,k)} & a_2^{(k,k)} & \cdots & a_{\nu_k-1}^{(k,k)} \end{pmatrix}$$

and

$$\tilde{\mathbf{A}}_{kl} = \begin{pmatrix} 0 & 0 & \cdots & 0 \\ 0 & 0 & \cdots & 0 \\ 0 & 0 & \cdots & 0 \\ \vdots & \vdots & & \vdots \\ 0 & 0 & \cdots & 0 \\ a_0^{(k,l)} & a_1^{(k,l)} & \cdots & a_{\nu_l-1}^{(k,l)} \end{pmatrix} \quad (l \neq k)$$

are $\nu_k \times \nu_k$ and $\nu_k \times \nu_l$, respectively. Furthermore, the type of matrix

$$\tilde{\mathbf{B}}_k = \begin{pmatrix} 0 & \cdots & 0 & 0 & 0 & \cdots & 0 \\ 0 & \cdots & 0 & 0 & 0 & \cdots & 0 \\ \vdots & & \vdots & \vdots & \vdots & & \vdots \\ 0 & \cdots & 0 & 0 & 0 & \cdots & 0 \\ 0 & \cdots & 0 & 1 & 0 & \cdots & 0 \end{pmatrix},$$

is $\nu_k \times m$, where unity shows up only at the kth element of the last row. Note that $\nu_1 + \nu_2 + \cdots + \nu_s = n$. Other canonical form variants are also known from the literature.

Example 7.6

Consider the system

$$\dot{\mathbf{x}} = \begin{pmatrix} 1 & 0 \\ 0 & 1 \end{pmatrix} \mathbf{x} + \begin{pmatrix} 1 & 0 \\ 0 & 1 \end{pmatrix} \mathbf{u}$$

$$y = (1,1)\mathbf{x}\,,$$

which is completely controllable, since the *rank* of the controllability matrix

$$\mathbf{K} = (\mathbf{B}, \mathbf{AB}) = \begin{pmatrix} 1\ 0\ 1\ 0 \\ 0\ 1\ 0\ 1 \end{pmatrix}$$

is obviously 2. We can show, however, that there is no vector $\mathbf{b}$, such that matrix $(\mathbf{b}, \mathbf{Ab})$ is nonsingular. In our case $\mathbf{Ab} = \mathbf{b}$; therefore, the two columns of this matrix always equal. Hence, this matrix is always singular.

7.3   Observability Canonical Forms

Assume now that system (7.9) or (7.10) is completely observable. Then its dual is completely controllable; therefore, the dual can be transformed into controllability canonical forms. Take the duals of these canonical forms to get the *observability canonical forms* of the original system. This principle is illustrated in Figure 7.2 and is summarized by the following theorems.

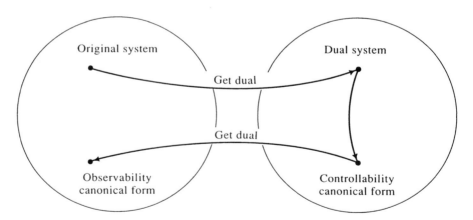

Figure 7.2   Computation of observability canonical form.

THEOREM 7.3
Assume that system $(\mathbf{A}, \mathbf{b}, \mathbf{c}^T)$ is completely observable; it can then be trans-

formed into an $(\tilde{\mathbf{A}}, \tilde{\mathbf{b}}, \tilde{\mathbf{c}}^T)$*-system, where*

$$\tilde{\mathbf{A}} = \begin{pmatrix} 0 & 1 & 0 & \cdots & 0 \\ 0 & 0 & 1 & \cdots & 0 \\ \vdots & \vdots & \vdots & \ddots & \vdots \\ 0 & 0 & 0 & \cdots & 1 \\ a_0 & a_1 & a_2 & \cdots & a_{n-1} \end{pmatrix} \qquad and \qquad \tilde{\mathbf{c}}^T = (1, 0, \ldots, 0, 0) . \quad (7.18)$$

THEOREM 7.4
Assume that system $(\mathbf{A}, \mathbf{b}, \mathbf{c}^T)$ *is completely observable; it can then be transformed into an* $(\tilde{\mathbf{A}}, \tilde{\mathbf{b}}, \tilde{\mathbf{c}}^T)$*-system, where*

$$\tilde{\mathbf{A}} = \begin{pmatrix} 0 & 0 & \cdots & 0 & a_0 \\ 1 & 0 & \cdots & 0 & a_1 \\ 0 & 1 & \cdots & 0 & a_2 \\ \vdots & \vdots & \ddots & \vdots & \vdots \\ 0 & 0 & \cdots & 1 & a_{n-1} \end{pmatrix} \qquad and \qquad \tilde{\mathbf{c}}^T = (0, 0, \ldots, 0, 1) . \quad (7.19)$$

A general algorithm to find any of the observability canonical forms consists of the following steps:

Step 1   Determine the dual system.

Step 2   Find the corresponding controllability canonical form of the dual.

Step 3   Compute the dual of the resulting canonical form.

Example 7.7

Consider again the system

$$\dot{\mathbf{x}} = \begin{pmatrix} 0 & \omega \\ -\omega & 0 \end{pmatrix} \mathbf{x} + \begin{pmatrix} 0 \\ 1 \end{pmatrix} u ,$$

$$y = (1, 1)\mathbf{x} ,$$

which was the subject of the examples of the previous section. The observability canonical forms given in the above two theorems will be determined.
    Step 1: The dual system can be written as follows:

$$\dot{\mathbf{z}} = \begin{pmatrix} 0 & -\omega \\ \omega & 0 \end{pmatrix} \mathbf{z} + \begin{pmatrix} 1 \\ 1 \end{pmatrix} v ,$$

$$\omega = (0,1)\mathbf{z} \ .$$

(a) **Step 2:** The canonical form (7.11) of the dual is based on the transformation matrix

$$\mathbf{T} = \mathbf{K}_D^{-1} = \begin{pmatrix} 1 & -\omega \\ 1 & \omega \end{pmatrix}^{-1} = \frac{1}{2\omega} \begin{pmatrix} \omega & \omega \\ -1 & 1 \end{pmatrix} \ .$$

Therefore,

$$\tilde{\mathbf{A}}_D = \mathbf{T}\mathbf{A}_D\mathbf{T}^{-1} = \frac{1}{2\omega} \begin{pmatrix} \omega & \omega \\ -1 & 1 \end{pmatrix} \begin{pmatrix} 0 & -\omega \\ \omega & 0 \end{pmatrix} \begin{pmatrix} 1 & -\omega \\ 1 & \omega \end{pmatrix}$$

$$= \begin{pmatrix} 0 & -\omega^2 \\ 1 & 0 \end{pmatrix} \ ,$$

$$\tilde{\mathbf{b}}_D = \mathbf{T}\mathbf{b}_D = \frac{1}{2\omega} \begin{pmatrix} \omega & \omega \\ -1 & 1 \end{pmatrix} \begin{pmatrix} 1 \\ 1 \end{pmatrix} = \begin{pmatrix} 1 \\ 0 \end{pmatrix}$$

and

$$\tilde{\mathbf{c}}_D^T = \mathbf{c}_D^T\mathbf{T}^{-1} = (0,1) \begin{pmatrix} 1 & -\omega \\ 1 & \omega \end{pmatrix} = (1,\omega) \ .$$

Step 3: Therefore, the corresponding observability canonical form $(\tilde{\mathbf{A}}, \tilde{\mathbf{b}}, \tilde{\mathbf{c}}^T)$ is the following:

$$\tilde{\mathbf{A}} = \tilde{\mathbf{A}}_D^T = \begin{pmatrix} 0 & 1 \\ -\omega^2 & 0 \end{pmatrix} \ , \quad \tilde{\mathbf{b}} = \tilde{\mathbf{c}}_D = \begin{pmatrix} 1 \\ \omega \end{pmatrix} \ , \text{ and } \quad \tilde{\mathbf{c}}^T = \tilde{\mathbf{b}}_D^T = (1,0) \ .$$

(b) **Step 2:** The canonical form (7.14) of the dual is based on the transformation matrix

$$\mathbf{T} = \frac{1}{2\omega} \begin{pmatrix} -1 & 1 \\ \omega & \omega \end{pmatrix} \ ,$$

since from Part (i) we have that

$$\mathbf{t}_n^T = \frac{1}{2\omega}(-1,1) \ ,$$

and simple calculation shows that

$$\mathbf{t}_n^T\mathbf{A}_D = \frac{1}{2\omega}(-1,1) \begin{pmatrix} 0 & -\omega \\ \omega & 0 \end{pmatrix} = \left(\frac{1}{2}, \frac{1}{2}\right) \ .$$

It is easy to verify that

$$\mathbf{T}^{-1} = \begin{pmatrix} -\omega & 1 \\ \omega & 1 \end{pmatrix} ;$$

therefore,

$$\tilde{\mathbf{A}}_D = \mathbf{T}\mathbf{A}_D\mathbf{T}^{-1} = \frac{1}{2\omega} \begin{pmatrix} -1 & 1 \\ \omega & \omega \end{pmatrix} \begin{pmatrix} 0 & -\omega \\ \omega & 0 \end{pmatrix} \begin{pmatrix} -\omega & 1 \\ \omega & 1 \end{pmatrix}$$

$$= \begin{pmatrix} 0 & 1 \\ -\omega^2 & 0 \end{pmatrix} ,$$

$$\tilde{\mathbf{b}}_D = \mathbf{T}\mathbf{b}_D = \frac{1}{2\omega} \begin{pmatrix} -1 & 1 \\ \omega & \omega \end{pmatrix} \begin{pmatrix} 1 \\ 1 \end{pmatrix} = \begin{pmatrix} 0 \\ 1 \end{pmatrix}$$

and

$$\tilde{\mathbf{c}}_D^T = \mathbf{c}_D^T\mathbf{T}^{-1} = (0, 1) \begin{pmatrix} -\omega & 1 \\ \omega & 1 \end{pmatrix} = (\omega, 1) .$$

Step 3: The corresponding observability canonical form $(\tilde{\mathbf{A}}, \tilde{\mathbf{b}}, \tilde{\mathbf{c}}^T)$ is the following:

$$\tilde{\mathbf{A}} = \tilde{\mathbf{A}}_D^T = \begin{pmatrix} 0 & -\omega^2 \\ 1 & 0 \end{pmatrix} , \quad \tilde{\mathbf{b}} = \tilde{\mathbf{c}}_D = \begin{pmatrix} \omega \\ 1 \end{pmatrix} , \quad \text{and} \quad \tilde{\mathbf{c}}^T = \tilde{\mathbf{b}}_D^T = (0, 1) .$$

In the case of multiple outputs, the observability canonical forms are more complicated. As an example, we mention a variant of general observability canonical forms, which is associated to the general controllability canonical form (7.17):

$$\tilde{\mathbf{A}} = \begin{pmatrix} \tilde{\mathbf{A}}_{11} & \tilde{\mathbf{A}}_{12} & \cdots & \tilde{\mathbf{A}}_{1s} \\ \tilde{\mathbf{A}}_{21} & \tilde{\mathbf{A}}_{22} & \cdots & \tilde{\mathbf{A}}_{2s} \\ \vdots & \vdots & \ddots & \vdots \\ \tilde{\mathbf{A}}_{s1} & \tilde{\mathbf{A}}_{s2} & \cdots & \tilde{\mathbf{A}}_{ss} \end{pmatrix} \quad \text{and} \quad \tilde{\mathbf{C}} = (\tilde{\mathbf{C}}_1, \tilde{\mathbf{C}}_2, \ldots, \tilde{\mathbf{C}}_s) \qquad (7.20)$$

where the types of matrices

$$\tilde{\mathbf{A}}_{kk} = \begin{pmatrix} 0 & 0 & \cdots & 0 & a_0^{(k,k)} \\ 1 & 0 & \cdots & 0 & a_1^{(k,k)} \\ 0 & 1 & \cdots & 0 & a_2^{(k,k)} \\ \vdots & \vdots & \ddots & \vdots & \vdots \\ 0 & 0 & \cdots & 1 & a_{\nu_k-1}^{(k,k)} \end{pmatrix}$$

and

$$
\tilde{\mathbf{A}}_{kl} = \begin{pmatrix} 0\,0 \cdots 0\ a_0^{(k,l)} \\ 0\,0 \cdots 0\ a_1^{(k,l)} \\ 0\,0 \cdots 0\ a_2^{(k,l)} \\ \vdots\ \vdots\ \ddots\ \vdots\quad \vdots \\ 0\,0 \cdots 0\ a_{\nu_k-1}^{(k,l)} \end{pmatrix} \qquad (l \neq k)
$$

are $\nu_k \times \nu_k$ and $\nu_k \times \nu_l$, respectively. Furthermore, the type of matrix

$$
\tilde{\mathbf{C}}_k = \begin{pmatrix} 0\,0 \cdots 0\,0 \\ \vdots\ \vdots\quad \vdots\ \vdots \\ 0\,0 \cdots 0\,0 \\ 0\,0 \cdots 0\,1 \\ 0\,0 \cdots 0\,0 \\ \vdots\ \vdots\quad \vdots\ \vdots \\ 0\,0 \cdots 0\,0 \end{pmatrix}
$$

is $p \times \nu_k$, where only the kth element of the last column equals unity. Note, that similarly to the general controllability canonical form (7.17), $\nu_1 + \nu_2 + \cdots + \nu_s = n$.

7.4 Applications

In this section we further develop the application examples introduced and discussed earlier.

7.4.1 Dynamic Systems in Engineering

1. The simple *harmonic oscillator* was described with

$$
\mathbf{A} = \begin{pmatrix} 0 & \omega \\ -\omega & 0 \end{pmatrix}, \qquad \mathbf{b} = \begin{pmatrix} 0 \\ 1 \end{pmatrix}, \qquad \text{and} \qquad \mathbf{c}^T = (1,0) \ .
$$

We showed in Example 1.13 that

$$
\mathbf{A} = \begin{pmatrix} 0 & \omega \\ -\omega & 0 \end{pmatrix} = \begin{pmatrix} 1 & 1 \\ j & -j \end{pmatrix} \begin{pmatrix} j\omega & 0 \\ 0 & -j\omega \end{pmatrix} \begin{pmatrix} \frac{1}{2} & -\frac{j}{2} \\ \frac{1}{2} & \frac{j}{2} \end{pmatrix} \ .
$$

Noting that

$$
\mathbf{A} = \mathbf{T}^{-1}\tilde{\mathbf{A}}\mathbf{T} \ ,
$$

we find that the diagonalized representation is

$$\tilde{\mathbf{A}} = \begin{pmatrix} j\omega & 0 \\ 0 & -j\omega \end{pmatrix} ,$$

$$\tilde{\mathbf{b}} = \begin{pmatrix} \frac{1}{2} & -\frac{j}{2} \\ \frac{1}{2} & \frac{j}{2} \end{pmatrix} \begin{pmatrix} 0 \\ 1 \end{pmatrix} = \begin{pmatrix} -\frac{j}{2} \\ \frac{j}{2} \end{pmatrix} ,$$

$$\tilde{\mathbf{c}}^T = \mathbf{c}^T \mathbf{T}^{-1} = (1, 0) \begin{pmatrix} 1 & 1 \\ j & -j \end{pmatrix} = (1, 1) .$$

Finally, we note that the controllability and observability canonical forms were derived in Sections 7.2 and 7.3.

2. Our *damped linear second-order system* was described with

$$\mathbf{A} = \begin{pmatrix} 0 & 1 \\ -\frac{K}{M} & -\frac{B}{M} \end{pmatrix} , \qquad \mathbf{b} = \begin{pmatrix} 0 \\ \frac{1}{M} \end{pmatrix} ,$$

and

$$\mathbf{c}^T = (1, 0) .$$

This *is* an observability canonical form (7.18); therefore, we know that the system is completely observable without computing the matrix $\mathbf{L}$ as we did in Chapter 6.

3. The *electrical system* of Application 3.5.1-3 was described with matrices

$$\mathbf{A} = \begin{pmatrix} -\frac{R_1}{L} & -\frac{1}{L} \\ \frac{1}{C} & -\frac{1}{CR_2} \end{pmatrix} , \qquad \mathbf{b} = \begin{pmatrix} \frac{1}{L} \\ 0 \end{pmatrix} , \qquad \text{and} \qquad \mathbf{c}^T = (0, 1) .$$

Let us generalize this problem as follows:

$$\mathbf{A} = \begin{pmatrix} a_{11} & a_{12} \\ a_{21} & a_{22} \end{pmatrix} , \qquad \mathbf{b} = \begin{pmatrix} \frac{1}{L} \\ 0 \end{pmatrix} , \qquad \text{and} \qquad \mathbf{c}^T = (0, 1) .$$

The canonical form (7.11) of this generalized system will now be determined. From Chapter 5 we have

$$\mathbf{K} = \begin{pmatrix} \frac{1}{L} & \frac{a_{11}}{L} \\ 0 & \frac{a_{21}}{L} \end{pmatrix} ,$$

and we know that $\mathbf{K} = \mathbf{T}^{-1}$, so we can compute

$$\mathbf{T} = \mathbf{K}^{-1} = \frac{L^2}{a_{21}} \begin{pmatrix} \frac{a_{21}}{L} & -\frac{a_{11}}{L} \\ 0 & \frac{1}{L} \end{pmatrix} = \frac{L}{a_{21}} \begin{pmatrix} a_{21} & -a_{11} \\ 0 & 1 \end{pmatrix} .$$

Use Equation (7.4) to get

$$\tilde{\mathbf{A}} = \frac{1}{a_{21}} \begin{pmatrix} a_{21} & -a_{11} \\ 0 & 1 \end{pmatrix} \begin{pmatrix} a_{11} & a_{12} \\ a_{21} & a_{22} \end{pmatrix} \begin{pmatrix} 1 & a_{11} \\ 0 & a_{21} \end{pmatrix}$$

$$= \frac{1}{a_{21}} \begin{pmatrix} 0 & a_{21}(a_{11}^2 + a_{12}a_{21}) - a_{11}(a_{21}a_{11} + a_{22}a_{21}) \\ a_{21} & a_{21}a_{11} + a_{22}a_{21} \end{pmatrix} .$$

Hence,

$$\tilde{\mathbf{A}} = \begin{pmatrix} 0 & a_{12}a_{21} - a_{11}a_{22} \\ 1 & a_{11} + a_{22} \end{pmatrix} .$$

From (7.4) we also conclude that

$$\tilde{\mathbf{b}} = \frac{L}{a_{21}} \begin{pmatrix} a_{21} & -a_{11} \\ 0 & 1 \end{pmatrix} \begin{pmatrix} \frac{1}{L} \\ 0 \end{pmatrix} = \begin{pmatrix} 1 \\ 0 \end{pmatrix} ,$$

$$\tilde{\mathbf{c}}^T = \mathbf{c}^T \mathbf{T}^{-1} = (0, 1) \begin{pmatrix} \frac{1}{L} & \frac{a_{11}}{L} \\ 0 & \frac{a_{21}}{L} \end{pmatrix} = \left(0, \frac{a_{21}}{L} \right) .$$

We can check this result by computing the characteristic polynomial of the original $\mathbf{A}$ matrix and noting that its coefficients are the same as the negatives of the coefficients in the last column of $\tilde{\mathbf{A}}$. Since

$$\mathbf{A} = \begin{pmatrix} a_{11} & a_{12} \\ a_{21} & a_{22} \end{pmatrix} ,$$

$$\varphi(\lambda) = \det \begin{pmatrix} a_{11} - \lambda & a_{12} \\ a_{21} & a_{22} - \lambda \end{pmatrix}$$

$$= (a_{11} - \lambda)(a_{22} - \lambda) - a_{12}a_{21}$$

$$= \lambda^2 - \lambda(a_{11} + a_{22}) + a_{11}a_{22} - a_{12}a_{21} .$$

4. For the *transistor circuit* model of Application 6.4.2-4, we have

$$\mathbf{A} = \begin{pmatrix} -\frac{h_{ie}}{L} & 0 \\ \frac{h_{fe}}{C} & 0 \end{pmatrix} , \qquad \mathbf{b} = \begin{pmatrix} \frac{1}{L} \\ 0 \end{pmatrix} , \qquad \text{and} \qquad \mathbf{c}_1^T = (0, 1) .$$

Let us find an observability canonical form by the technique of Section 7.3. In our case, with suitable choices for α and β,

$$\mathbf{A} = \begin{pmatrix} -\alpha & 0 \\ \beta & 0 \end{pmatrix} , \quad \mathbf{b} = \begin{pmatrix} \frac{1}{L} \\ 0 \end{pmatrix} , \quad \text{and} \quad \mathbf{c}_1^T = (0,1) .$$

The algorithm is as follows:

Step 1: Get the dual from Section 6.3 to find

$$\mathbf{A}_D = \begin{pmatrix} -\alpha & \beta \\ 0 & 0 \end{pmatrix} , \quad \mathbf{b}_D = \begin{pmatrix} 0 \\ 1 \end{pmatrix} , \quad \text{and} \quad \mathbf{c}_{1D}^T = \begin{pmatrix} \frac{1}{L}, 0 \end{pmatrix} .$$

Step 2: Get the controllability form for the dual:

$$\mathbf{K}_D = \begin{pmatrix} 0 & \beta \\ 1 & 0 \end{pmatrix} = \mathbf{T}_D^{-1} ;$$

therefore,

$$\mathbf{T}_D = \frac{-1}{\beta} \begin{pmatrix} 0 & -\beta \\ -1 & 0 \end{pmatrix} = \begin{pmatrix} 0 & 1 \\ \frac{1}{\beta} & 0 \end{pmatrix} .$$

From (7.4) we know that

$$\tilde{\mathbf{A}}_D = \mathbf{T}_D \mathbf{A}_D \mathbf{T}_D^{-1} = \begin{pmatrix} 0 & 1 \\ \frac{1}{\beta} & 0 \end{pmatrix} \begin{pmatrix} -\alpha & \beta \\ 0 & 0 \end{pmatrix} \begin{pmatrix} 0 & \beta \\ 1 & 0 \end{pmatrix}$$

$$= \begin{pmatrix} 0 & 0 \\ 1 & -\alpha \end{pmatrix} ,$$

$$\tilde{\mathbf{b}}_D = \mathbf{T}_D \mathbf{b}_D = \begin{pmatrix} 0 & 1 \\ \frac{1}{\beta} & 0 \end{pmatrix} \begin{pmatrix} 0 \\ 1 \end{pmatrix} = \begin{pmatrix} 1 \\ 0 \end{pmatrix} ,$$

and

$$\tilde{\mathbf{c}}_{1D}^T = \mathbf{c}_{1D}^T \mathbf{T}_D^{-1} = \begin{pmatrix} \frac{1}{L}, 0 \end{pmatrix} \begin{pmatrix} 0 & \beta \\ 1 & 0 \end{pmatrix} = \begin{pmatrix} 0, \frac{\beta}{L} \end{pmatrix} .$$

Step 3: Get dual of result:

$$\tilde{\mathbf{A}} = \begin{pmatrix} 0 & 1 \\ 0 & -\alpha \end{pmatrix} , \quad \tilde{\mathbf{b}} = \begin{pmatrix} 0 \\ \frac{\beta}{L} \end{pmatrix} , \quad \tilde{\mathbf{c}}_1^T = (1,0) .$$

This is the observability canonical form (7.18). We can check this result by computing the characteristic polynomial of the original $\mathbf{A}$ matrix:

$$\varphi(\lambda) = \det \begin{pmatrix} -\alpha - \lambda & 0 \\ \beta & -\lambda \end{pmatrix}$$

$$= \lambda^2 + \alpha\lambda \ .$$

The negatives of the coefficients of this polynomial, $a_0 = 0$ and $a_1 = -\alpha$, are indeed the terms in the bottom row of $\tilde{\mathbf{A}}$, so our work is correct.

With $\mathbf{c}_2^T = (1,0)$, which we introduced in Chapter 6, no observability canonical form is possible, because the system is not completely observable. The method cannot be used, since $\mathbf{L}_2^{-1}$ has to be computed. But $\mathbf{L}_2$ is singular (second column is zero); therefore, $\mathbf{L}_2^{-1}$ does not exist.

5. There are an infinite number of ways of selecting the state variables for a given system. Some of them have physical significance and some have mathematical convenience. For the *hydraulic system* of Application 3.5.1-5 we will show five different representations. We will also illustrate these five representations with block diagrams. But first let us look at the *two-tank hydraulic system*:

$$\mathbf{A} = \begin{pmatrix} -\frac{1}{R_1 A_1} & \frac{1}{R_1 A_1} \\ \frac{1}{R_1 A_2} & -\left(\frac{1}{R_1 A_2} + \frac{1}{R_2 A_2}\right) \end{pmatrix} , \qquad \mathbf{b} = \begin{pmatrix} \frac{1}{A_1} \\ 0 \end{pmatrix} ,$$

and

$$\mathbf{c}^T = \left(\frac{1}{R_1} , -\frac{1}{R_1} \right) .$$

Let us note that the $\mathbf{A}$ matrix is full and, therefore, this system is the same as that of Application 7.4.1-3. Therefore, we will not repeat that analysis.

However, let us now look at one particular instance of the three-tank system presented in Application 5.3.1-5. We will present five different forms to represent this system.

A. The first is called *physical variables* because the state variables have physical significance. They are the heights of the water levels in the tanks. This representation is

$$\mathbf{A} = \begin{pmatrix} -3 & 3 & 0 \\ 2 & -4 & 2 \\ 0 & 3 & -3 \end{pmatrix} , \qquad \mathbf{b} = \begin{pmatrix} 1 \\ 0 \\ 0 \end{pmatrix} , \qquad \mathbf{c}^T = (1,0,0) .$$

This representation is illustrated in Figure 7.3.

B. Let us find first a *controllability canonical form*. Since

$$\mathbf{K} = \begin{pmatrix} 1 & -3 & 15 \\ 0 & 2 & -14 \\ 0 & 0 & 6 \end{pmatrix} = \mathbf{T}^{-1} ,$$

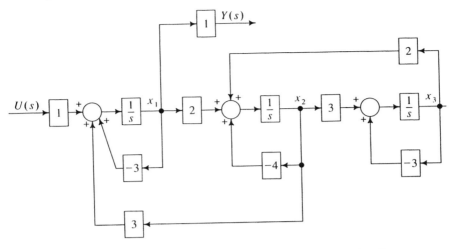

Figure 7.3   Physical variable representation for the three-tank hydraulic example.

$$\mathbf{T} = \begin{pmatrix} 1 & \frac{3}{2} & 1 \\ 0 & \frac{1}{2} & \frac{7}{6} \\ 0 & 0 & \frac{1}{6} \end{pmatrix} .$$

Therefore, from Equation (7.4) we have

$$\tilde{\mathbf{A}} = \begin{pmatrix} 1 & \frac{3}{2} & 1 \\ 0 & \frac{1}{2} & \frac{7}{6} \\ 0 & 0 & \frac{1}{6} \end{pmatrix} \begin{pmatrix} -3 & 3 & 0 \\ 2 & -4 & 2 \\ 0 & 3 & -3 \end{pmatrix} \begin{pmatrix} 1 & -3 & 15 \\ 0 & 2 & -14 \\ 0 & 0 & 6 \end{pmatrix}$$

$$= \begin{pmatrix} 0 & 0 & 0 \\ 1 & 0 & -21 \\ 0 & 1 & -10 \end{pmatrix} ,$$

$$\tilde{\mathbf{b}} = \begin{pmatrix} 1 & \frac{3}{2} & 1 \\ 0 & \frac{1}{2} & \frac{7}{6} \\ 0 & 0 & \frac{1}{6} \end{pmatrix} \begin{pmatrix} 1 \\ 0 \\ 0 \end{pmatrix} = \begin{pmatrix} 1 \\ 0 \\ 0 \end{pmatrix} ,$$

and

$$\tilde{\mathbf{c}}^T = \mathbf{c}^T \mathbf{T}^{-1} = (1,0,0) \begin{pmatrix} 1 & -3 & 15 \\ 0 & 2 & -14 \\ 0 & 0 & 6 \end{pmatrix} = (1,-3,15) .$$

This representation is illustrated in Figure 7.4.

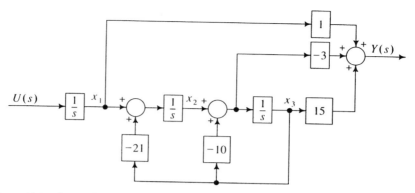

Figure 7.4   Controllability canonical form for hydraulic example.

C. Next we would like to find an *observability canonical form*. Let us do this with the method of Section 7.3. First using the techniques of Section 6.3, find the dual:

$$\mathbf{A}_D = \mathbf{A}^T = \begin{pmatrix} -3 & 2 & 0 \\ 3 & -4 & 3 \\ 0 & 2 & -3 \end{pmatrix}, \qquad \mathbf{b}_D = \begin{pmatrix} 1 \\ 0 \\ 0 \end{pmatrix}, \qquad \mathbf{c}_D^T = (1,0,0).$$

Next find the controllability canonical form for this dual. Since

$$\mathbf{K}_D = \begin{pmatrix} 1 & -3 & 15 \\ 0 & 3 & -21 \\ 0 & 0 & 6 \end{pmatrix} = \mathbf{T}_D^{-1},$$

we know that

$$\mathbf{T}_D = \begin{pmatrix} 1 & 1 & 1 \\ 0 & \frac{1}{3} & \frac{7}{6} \\ 0 & 0 & \frac{1}{6} \end{pmatrix}.$$

Then from (7.4),

$$\tilde{\mathbf{A}}_D = \mathbf{T}_D \mathbf{A}_D \mathbf{T}_D^{-1} = \begin{pmatrix} 1 & 1 & 1 \\ 0 & \frac{1}{3} & \frac{7}{6} \\ 0 & 0 & \frac{1}{6} \end{pmatrix} \begin{pmatrix} -3 & 2 & 0 \\ 3 & -4 & 3 \\ 0 & 2 & -3 \end{pmatrix} \begin{pmatrix} 1 & -3 & 15 \\ 0 & 3 & -21 \\ 0 & 0 & 6 \end{pmatrix}$$

$$= \begin{pmatrix} 0 & 0 & 0 \\ 1 & 0 & -21 \\ 0 & 1 & -10 \end{pmatrix},$$

$$\tilde{\mathbf{b}}_D = \mathbf{T}_D \mathbf{b}_D = \begin{pmatrix} 1 & 1 & 1 \\ 0 & \frac{1}{3} & \frac{7}{6} \\ 0 & 0 & \frac{1}{6} \end{pmatrix} \begin{pmatrix} 1 \\ 0 \\ 0 \end{pmatrix} = \begin{pmatrix} 1 \\ 0 \\ 0 \end{pmatrix} \, ,$$

and

$$\tilde{\mathbf{c}}_D^T = \mathbf{c}_D^T \mathbf{T}_D^{-1} = (1,0,0) \begin{pmatrix} 1 & -3 & 15 \\ 0 & 3 & -21 \\ 0 & 0 & 6 \end{pmatrix} = (1, -3, 15) \, .$$

Finally take the dual of result:

$$\tilde{\mathbf{A}} = \begin{pmatrix} 0 & 1 & 0 \\ 0 & 0 & 1 \\ 0 & -21 & -10 \end{pmatrix} \, , \qquad \tilde{\mathbf{b}} = \begin{pmatrix} 1 \\ -3 \\ 15 \end{pmatrix} \, , \qquad \tilde{\mathbf{c}}^T = (1,0,0) \, .$$

This representation is illustrated in Figure 7.5. Note that x_1 is on the right in this drawing.

The controllability canonical form is sometimes called the method of phase variables. In this method, the state variables consist of one variable and its $n - 1$ derivatives, as in Theorem 3.5.

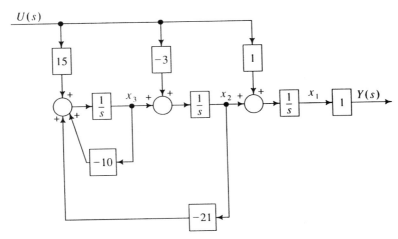

Figure 7.5   Observability canonical form for the hydrology system.

D. Next let us find a representation for this system with a *diagonalized matrix*.

Step 1: From the original $\mathbf{A}$ matrix, find the characteristic equation

$$\det \begin{pmatrix} -3-\lambda & 3 & 0 \\ 2 & -4-\lambda & 2 \\ 0 & 3 & -3-\lambda \end{pmatrix} = 0 \,,$$

that is,

$$\lambda^3 + 10\lambda^2 + 21\lambda = 0$$

or

$$\lambda_1 = 0, \quad \lambda_2 = -3, \quad \text{and} \quad \lambda_3 = -7 \,.$$

Because there are no repeated roots, we know that our $\tilde{\mathbf{A}}$ matrix is

$$\tilde{\mathbf{A}} = \begin{pmatrix} 0 & 0 & 0 \\ 0 & -3 & 0 \\ 0 & 0 & -7 \end{pmatrix} \,.$$

However, we still need to compute $\tilde{\mathbf{b}}$ and $\tilde{\mathbf{c}}^T$. To get these we need to compute the eigenvectors in order to find the transformation matrix.

Step 2: The eigenvectors are determined as follows.

If $\lambda = 0$, then the homogeneous equation is

$$\begin{pmatrix} -3 & 3 & 0 \\ 2 & -4 & 2 \\ 0 & 3 & -3 \end{pmatrix} \begin{pmatrix} x_1 \\ x_2 \\ x_3 \end{pmatrix} = \begin{pmatrix} 0 \\ 0 \\ 0 \end{pmatrix} \,;$$

therefore,

$$\mathbf{x} = \begin{pmatrix} 1 \\ 1 \\ 1 \end{pmatrix}$$

is a solution.

If $\lambda = -3$, then we have equations

$$\begin{pmatrix} 0 & 3 & 0 \\ 2 & -1 & 2 \\ 0 & 3 & 0 \end{pmatrix} \begin{pmatrix} x_1 \\ x_2 \\ x_3 \end{pmatrix} = \begin{pmatrix} 0 \\ 0 \\ 0 \end{pmatrix}$$

with solution

$$\mathbf{x} = \begin{pmatrix} 1 \\ 0 \\ -1 \end{pmatrix} \,.$$

If $\lambda = -7$, then

$$\begin{pmatrix} 4 & 3 & 0 \\ 2 & 3 & 2 \\ 0 & 3 & 4 \end{pmatrix} \begin{pmatrix} x_1 \\ x_2 \\ x_3 \end{pmatrix} = \begin{pmatrix} 0 \\ 0 \\ 0 \end{pmatrix} ,$$

which has a solution

$$\mathbf{x} = \begin{pmatrix} 3 \\ -4 \\ 3 \end{pmatrix} .$$

Step 3: Assemble the eigenvectors to get $\mathbf{T}^{-1}$:

$$\mathbf{T}^{-1} = \begin{pmatrix} 1 & 1 & 3 \\ 1 & 0 & -4 \\ 1 & -1 & 3 \end{pmatrix} ,$$

and by inversion,

$$\mathbf{T} = \frac{1}{14} \begin{pmatrix} 4 & 6 & 4 \\ 7 & 0 & -7 \\ 1 & -2 & 1 \end{pmatrix} .$$

Step 4: Find the resulting matrices by (7.4),

$$\tilde{\mathbf{A}} = \mathbf{T}\mathbf{A}\mathbf{T}^{-1} = \frac{1}{14} \begin{pmatrix} 4 & 6 & 4 \\ 7 & 0 & -7 \\ 1 & -2 & 1 \end{pmatrix} \begin{pmatrix} -3 & 3 & 0 \\ 2 & -4 & 2 \\ 0 & 3 & -3 \end{pmatrix} \begin{pmatrix} 1 & 1 & 3 \\ 1 & 0 & -4 \\ 1 & -1 & 3 \end{pmatrix}$$

$$= \frac{1}{14} \begin{pmatrix} 0 & 0 & 0 \\ 0 & -42 & 0 \\ 0 & 0 & -98 \end{pmatrix} = \begin{pmatrix} 0 & 0 & 0 \\ 0 & -3 & 0 \\ 0 & 0 & -7 \end{pmatrix} ,$$

$$\tilde{\mathbf{b}} = \mathbf{T}\mathbf{b} = \frac{1}{14} \begin{pmatrix} 4 & 6 & 4 \\ 7 & 0 & -7 \\ 1 & -2 & 1 \end{pmatrix} \begin{pmatrix} 1 \\ 0 \\ 0 \end{pmatrix} = \begin{pmatrix} \frac{2}{7} \\ \frac{1}{2} \\ \frac{1}{14} \end{pmatrix} ,$$

and

$$\tilde{\mathbf{c}}^T = \mathbf{c}^T \mathbf{T}^{-1} = (1, 0, 0) \begin{pmatrix} 1 & 1 & 3 \\ 1 & 0 & -4 \\ 1 & -1 & 3 \end{pmatrix} = (1, 1, 3) .$$

This method of representation is popular because you can immediately see what the eigenvalues of the system are. This representation is illustrated in Figure 7.6.

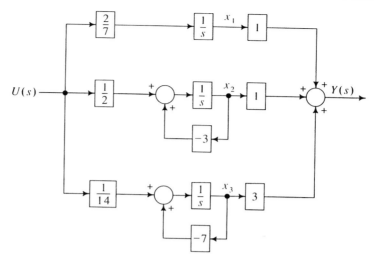

Figure 7.6   Diagonalized matrix form for the hydrology system.

E. We can also represent this system with a *transfer function*. Simple calculation shows that

$$H(s) = \mathbf{c}^T (s\mathbf{I} - \mathbf{A})^{-1}\mathbf{b}$$

$$= (1,0,0) \begin{pmatrix} s+3 & -3 & 0 \\ -2 & s+4 & -2 \\ 0 & -3 & s+3 \end{pmatrix}^{-1} \begin{pmatrix} 1 \\ 0 \\ 0 \end{pmatrix} .$$

First, find the inverse matrix $(s\mathbf{I} - \mathbf{A})^{-1}$. There are many techniques for finding the inverse of a matrix. Here we will use the cofactor matrix technique. The cofactor of a_{23}, for example, is formed by eliminating the second row and the third column from the original matrix and multiplying the determinant of the resulting 2×2 matrix by $(-1)^{2+3}$. Thus, cofactor of a_{23} is formed as

$$\text{cof}(a_{23}) = (-1)^{2+3} \det \begin{pmatrix} s+3 & -3 & 0 \\ -2 & s+4 & -2 \\ 0 & -3 & s+3 \end{pmatrix}$$

$$= (-1)^5 \det \begin{pmatrix} s+3 & -3 \\ 0 & -3 \end{pmatrix} = 3s + 9 .$$

The other cofactors can be found by a similar process to produce

$$\text{cof}(\mathbf{A}) = \begin{pmatrix} s^2 + 7s + 6 & 2s + 6 & 6 \\ 3s + 9 & s^2 + 6s + 9 & 3s + 9 \\ 6 & 2s + 6 & s^2 + 7s + 6 \end{pmatrix}.$$

Now the adjoint matrix can be formed by transposing the cofactor matrix:

$$\text{adj}(s\mathbf{I} - \mathbf{A}) = [\text{cof}(\mathbf{A})]^T = \begin{pmatrix} s^2 + 7s + 6 & 3s + 9 & 6 \\ 2s + 6 & s^2 + 6s + 9 & 2s + 6 \\ 6 & 3s + 9 & s^2 + 7s + 6 \end{pmatrix}.$$

The inverse can be now computed as

$$(s\mathbf{I} - \mathbf{A})^{-1} = \frac{\text{adj}(s\mathbf{I} - \mathbf{A})}{\det(s\mathbf{I} - \mathbf{A})},$$

where $\det(s\mathbf{I} - \mathbf{A}) = s^3 + 10s^2 + 21s$.

Finally the transfer function becomes

$$H(s) = \frac{s^2 + 7s + 6}{s^3 + 10s^2 + 21s} = \frac{(s + 1)(s + 6)}{s(s + 3)(s + 7)}.$$

This representation is illustrated in Figure 7.7.

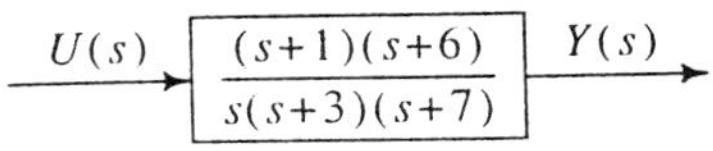

Figure 7.7 Transfer function for the hydrology system.

We could make the calculation easier by observing that $(s\mathbf{I} - \mathbf{A})^{-1}\mathbf{b}$ can be obtained by solving linear equations with coefficient matrix $s\mathbf{I} - \mathbf{A}$ and right-hand side vector $\mathbf{b}$. And then, the solution vector has to be multiplied by $\mathbf{c}^T$.

Note that a partial fraction expansion of this transfer function produces

$$\frac{(s + 1)(s + 6)}{s(s + 3)(s + 7)} = \frac{2/7}{s} + \frac{1/2}{s + 3} + \frac{3/14}{s + 7},$$

which is the same result that can be observed from the diagonalized matrix of Figure 7.6.

Many more representations for systems are possible, and indeed are used in the Systems and Control literature. But with these five examples we hope to convince the reader that you can easily change from any one form to any other. Therefore, the particular representation technique does not limit the generality of the methods presented in this book.

6. Since the *multiple-input electronic system* has two input variables, the usual canonical forms cannot be obtained. However the more general forms (7.17) and (7.20) can be used. The details will not be given here.

7. The *stick-balancing problem* is modeled by matrices

$$\mathbf{A} = \begin{pmatrix} 0 & 1 \\ g & 0 \end{pmatrix}, \qquad \mathbf{b} = \begin{pmatrix} 0 \\ -g \end{pmatrix}, \qquad \text{and} \qquad \mathbf{c}^T = (1, 0) .$$

Note that this is already in an observability canonical form (7.18), which has an immediate consequence, we know that the system is completely observable without computing the observability matrix $\mathbf{L}$.

8. In Application 3.5.1-8 we introduced the problem of trying to control *two sticks mounted on a cart*. The description of the problem was based on matrices

$$\mathbf{A} = \begin{pmatrix} 0 & 0 & 1 & 0 \\ 0 & 0 & 0 & 1 \\ a_1 & a_2 & 0 & 0 \\ a_3 & a_4 & 0 & 0 \end{pmatrix}, \qquad \mathbf{b} = \begin{pmatrix} 0 \\ 0 \\ -c \\ -d \end{pmatrix},$$

and

$$\mathbf{c}^T = (1, 0, 0, 0) .$$

We will now find the diagonal form of this system.

From Application 4.4.1-8 we know that the characteristic equation of $\mathbf{A}$ is

$$\varphi(\lambda) = \lambda^4 - \lambda^2(a_1 + a_4) + (a_1 a_4 - a_2 a_3) = 0 ,$$

therefore,

$$\lambda^2 = \frac{a_1 + a_4 \pm \sqrt{(a_1 - a_4)^2 + 4 a_2 a_3}}{2} .$$

Since $a_1 a_4 - a_2 a_3 > 0$ and $a_1 + a_4 > 0$, there are two distinct positive roots for λ^2. Consequently, the eigenvalues are

$$\lambda_1 = \alpha_1, \quad \lambda_2 = -\alpha_1, \quad \lambda_3 = \alpha_2, \quad \lambda_4 = -\alpha_2$$

with

$$\alpha_1 = \left\{ \frac{a_1 + a_4 + \sqrt{(a_1 - a_4)^2 + 4 a_2 a_3}}{2} \right\}^{\frac{1}{2}}$$

and

$$\alpha_2 = \left\{ \frac{a_1 + a_4 - \sqrt{(a_1 - a_4)^2 + 4a_2a_3}}{2} \right\}^{\frac{1}{2}} .$$

The eigenvector equations are

$$\begin{pmatrix} -\lambda_i & 0 & 1 & 0 \\ 0 & -\lambda_i & 0 & 1 \\ a_1 & a_2 & -\lambda_i & 0 \\ a_3 & a_4 & 0 & -\lambda_i \end{pmatrix} \begin{pmatrix} x_1 \\ x_2 \\ x_3 \\ x_4 \end{pmatrix} = \begin{pmatrix} 0 \\ 0 \\ 0 \\ 0 \end{pmatrix} , \quad (i = 1, 2, 3, 4) ,$$

that is,

$$\begin{aligned} -\lambda_i x_1 \quad\quad\quad\quad +x_3 \quad\quad\quad &= 0 \\ -\lambda_i x_2 \quad\quad\quad +x_4 &= 0 \\ a_1 x_1 \;+a_2 x_2 \;-\lambda_i x_3 \quad\quad &= 0 \\ a_3 x_1 \;+a_4 x_2 \quad\quad\quad -\lambda_i x_4 &= 0 . \end{aligned}$$

Select $x_1 = 1$, then

$$x_2 = \frac{-a_1 + \lambda_i^2}{a_2} ,$$

$$x_3 = \lambda_i ,$$

and

$$x_4 = \frac{-a_1 \lambda_i + \lambda_i^3}{a_2} .$$

Therefore, the transformation matrix is as follows:

$$\mathbf{T} = \begin{pmatrix} 1 & 1 & 1 & 1 \\ \frac{-a_1 + \alpha_1^2}{a_2} & \frac{-a_1 + \alpha_1^2}{a_2} & \frac{-a_1 + \alpha_2^2}{a_2} & \frac{-a_1 + \alpha_2^2}{a_2} \\ \alpha_1 & -\alpha_1 & \alpha_2 & -\alpha_2 \\ \frac{-a_1\alpha_1 + \alpha_1^3}{a_2} & \frac{a_1\alpha_1 - \alpha_1^3}{a_2} & \frac{-a_1\alpha_2 + \alpha_2^3}{a_2} & \frac{a_1\alpha_2 - \alpha_2^3}{a_2} \end{pmatrix}^{-1} .$$

Hence, simple calculation shows that

$$\tilde{\mathbf{A}} = \begin{pmatrix} \alpha_1 & & & \mathbf{O} \\ & -\alpha_1 & & \\ & & \alpha_2 & \\ \mathbf{O} & & & -\alpha_2 \end{pmatrix} ,$$

$$\tilde{\mathbf{b}} = \begin{pmatrix} \frac{-cA_2+d}{2\alpha_1(A_2-A_1)} \\ \frac{cA_2-d}{2\alpha_1(A_2-A_1)} \\ \frac{-cA_1+d}{2\alpha_2(A_1-A_2)} \\ \frac{cA_1-d}{2\alpha_2(A_1-A_2)} \end{pmatrix} ,$$

where $A_1 = -a_1 + \alpha_1^2/a_2$ and $A_2 = -a_1 + \alpha_2^2/a_2$; furthermore

$$\tilde{\mathbf{c}}^T = (1,1,1,1) .$$

9. The *electrical heating system* of Application 3.5.1-9 has a general 2×2 coefficient matrix and $\mathbf{B}$ is the first basis vector. Therefore, it is the special case of the one being discussed earlier in this section (Application 3 of the *electrical system*) with the selection of $L = 1$. Therefore, the calculations are not repeated here.

10. The *nuclear reactor* was described with

$$\mathbf{A} = \begin{pmatrix} \frac{\rho-\beta}{l} & \lambda_1 \\ \frac{\beta_1}{l} & -\lambda_1 \end{pmatrix} , \qquad \mathbf{b} = \begin{pmatrix} 1 \\ 0 \end{pmatrix} , \qquad \text{and} \qquad \mathbf{c}^T = (1,0) .$$

This is almost the same as Application 7.4.1-3, so we will not repeat the calculation.

7.4.2   *Applications in the Social Sciences and Economics*

1. Consider first the *linearized predator–prey model* (6.16), where the coefficient matrix has the form

$$\mathbf{A} = \begin{pmatrix} 0 & -\frac{bc}{d} \\ \frac{da}{b} & 0 \end{pmatrix} ;$$

furthermore

$$\mathbf{b} = \left(\frac{c}{d}, 0 \right)^T \qquad \text{and} \qquad \mathbf{c}^T = (0,1) .$$

We illustrate first how this system can be diagonalized. The characteristic polynomial of $\mathbf{A}$ is

$$\varphi(\lambda) = \lambda^2 + \frac{bc}{d} \cdot \frac{da}{b} = \lambda^2 + ac ;$$

therefore, the eigenvalues are $\lambda_1 = j\sqrt{ac}$ and $\lambda_2 = -j\sqrt{ac}$. In order to use Theorem 1.11 for finding the diagonalizing transformation matrix

T, we first have to find the eigenvectors of matrix **A**. The eigenvector associated to λ_1 is the solution of the homogeneous linear equation

$$\begin{pmatrix} -j\sqrt{ac} & -\frac{bc}{d} \\ \frac{da}{b} & -j\sqrt{ac} \end{pmatrix} \begin{pmatrix} x_{11} \\ x_{12} \end{pmatrix} = \begin{pmatrix} 0 \\ 0 \end{pmatrix} .$$

By selecting $x_{12} = d\sqrt{ac}$, the first component implies that

$$-j\sqrt{ac}x_{11} - bc\sqrt{ac} = 0 ,$$

that is,

$$x_{11} = \frac{bc}{-j} = jbc .$$

Therefore,

$$\mathbf{x}_1 = (jbc, d\sqrt{ac})^T .$$

Similarly, the eigenvector associated to λ_2 is the solution of equation

$$\begin{pmatrix} j\sqrt{ac} & -\frac{bc}{d} \\ \frac{da}{b} & j\sqrt{ac} \end{pmatrix} \begin{pmatrix} x_{21} \\ x_{22} \end{pmatrix} = \begin{pmatrix} 0 \\ 0 \end{pmatrix} .$$

Select again $x_{22} = d\sqrt{ac}$ to obtain $x_{21} = -jbc$. That is, $\mathbf{x}_2 = (-jbc, d\sqrt{ac})^T$. The transformation matrix is then determined by using Equation (1.29) and (7.4):

$$\mathbf{T} = (\mathbf{x}_1, \mathbf{x}_2)^{-1} = \begin{pmatrix} jbc & -jbc \\ d\sqrt{ac} & d\sqrt{ac} \end{pmatrix}^{-1}$$

$$= \frac{1}{2jbcd\sqrt{ac}} \begin{pmatrix} d\sqrt{ac} & jbc \\ -d\sqrt{ac} & jbc \end{pmatrix} .$$

Since

$$\mathbf{T}^{-1} = (\mathbf{x}_1, \mathbf{x}_2) = \begin{pmatrix} jbc & -jbc \\ d\sqrt{ac} & d\sqrt{ac} \end{pmatrix} ,$$

relations (7.4) imply that

$$\tilde{\mathbf{A}} = \begin{pmatrix} j\sqrt{ac} & 0 \\ 0 & -j\sqrt{ac} \end{pmatrix} ,$$

$$\tilde{\mathbf{b}} = \frac{-j}{2bd} \begin{pmatrix} 1 \\ -1 \end{pmatrix} \tag{7.21}$$

and

$$\tilde{c}^T = d\sqrt{ac}(1,1) \ .$$

Hence the system is diagonalized.

2. The general *cohort population model* (6.17) is based on matrices

$$A = \begin{pmatrix} b_1 & b_2 & \cdots & b_{n-1} & b_n \\ a_1 & 0 & \cdots & 0 & 0 \\ 0 & a_2 & \cdots & 0 & 0 \\ \vdots & \vdots & \ddots & \vdots & \vdots \\ 0 & 0 & \cdots & a_{n-1} & 0 \end{pmatrix} , \qquad B = I$$

and

$$C = (1, 1, \ldots, 1) \ .$$

As an example, we will transform matrix A into the form $\tilde{A}$ given by the canonical form (7.14).

Note first that matrix A has a structure similar to the coefficient matrix of the canonical form (7.14). Based on this similarity it will be easier to construct the transformation matrix directly rather than to use the general method given in Section 7.2.

Permute first the components $P_1, P_2, \ldots, P_n$ of the state vector as $(P_n, P_{n-1}, \ldots, P_2, P_1)$, and also, introduce the new input vector $(u_n, u_{n-1}, \ldots, u_2, u_1)$. This permutation of the state and input components result in the corresponding permutations of the rows and columns, and, therefore, the resulting matrix has the form

$$A_1 = \begin{pmatrix} 0 & a_{n-1} & 0 & \cdots & 0 \\ 0 & 0 & a_{n-2} & \cdots & 0 \\ \vdots & \vdots & \vdots & \ddots & \vdots \\ 0 & 0 & 0 & \cdots & a_1 \\ b_n & b_{n-1} & b_{n-2} & \cdots & b_1 \end{pmatrix} .$$

This form is now closer to (7.14), since the nonzero elements have the same locations. Next, all coefficients $a_{n-1}, \ldots, a_2, a_1$ will be transformed to be equal to one. Select a diagonal transformation matrix $T = diag(x_1, \ldots, x_n)$, where the x_i's are unknown. Since $T^{-1} = diag(1/x_1, \ldots, 1/x_n)$, from (7.4) we have

$$\tilde{A} = TA_1T^{-1} = diag(x_1, \ldots, x_n)A_1 diag\left(\frac{1}{x_1}, \ldots, \frac{1}{x_n}\right)$$

$$
= \begin{pmatrix}
0 & a_{n-1}\frac{x_1}{x_2} & 0 & \cdots & 0 \\
0 & 0 & a_{n-2}\frac{x_2}{x_3} & \cdots & 0 \\
\vdots & \vdots & \vdots & \ddots & \vdots \\
0 & 0 & 0 & \cdots & a_1\frac{x_{n-1}}{x_n} \\
b_n\frac{x_n}{x_1} & b_{n-1}\frac{x_n}{x_2} & b_{n-2}\frac{x_n}{x_3} & \cdots & b_1\frac{x_n}{x_n}
\end{pmatrix} .
$$

We now select the $x_i's$ to satisfy relations

$$
a_{n-1}\frac{x_1}{x_2} = a_{n-2}\frac{x_2}{x_3} = \cdots = a_1\frac{x_{n-1}}{x_n} = 1 .
$$

If $x_1 = 1$, then we have

$$
x_2 = a_{n-1}, x_3 = a_{n-2}a_{n-1}, \cdots, x_n = a_1 a_2 \cdots a_{n-2}a_{n-1} . \qquad (7.22)
$$

Hence, the transformed matrix is as follows:

$$
\tilde{\mathbf{A}} = \begin{pmatrix}
0 & 1 & 0 & \cdots & 0 \\
0 & 0 & 1 & \cdots & 0 \\
\vdots & \vdots & \vdots & \ddots & \vdots \\
0 & 0 & 0 & \cdots & 1 \\
\tilde{b}_n & \tilde{b}_{n-1} & \tilde{b}_{n-2} & \cdots & \tilde{b}_1
\end{pmatrix} , \qquad (7.23)
$$

where

$$
\tilde{b}_1 = b_1, \qquad \tilde{b}_2 = a_1 b_2, \ldots, \tilde{b}_{n-1} = a_1 a_2 \ldots a_{n-2} b_{n-1} ,
$$

and

$$
\tilde{b}_n = a_1 a_2 \ldots a_{n-2} a_{n-1} b_n .
$$

Thus, the required canonical form is determined.

3. In the case of the *arms races* model (6.18), the computation of the observability canonical form (7.18) will be illustrated.

Note first that the dual of the problem has the form

$$
\dot{\mathbf{z}} = \begin{pmatrix} -b & c \\ a & -d \end{pmatrix} \mathbf{z} + \begin{pmatrix} 1 \\ 0 \end{pmatrix} v ,
$$

$$
w = (\alpha, \beta)\mathbf{z} . \qquad (7.24)
$$

The controllability canonical form (7.11) of the dual will be first determined, and the desired observability canonical form is its dual. From the proof of Theorem 7.1 we know that the transformation matrix is as follows:

$$\mathbf{T} = \begin{pmatrix} 1 & -b \\ 0 & a \end{pmatrix}^{-1} = \frac{1}{a} \begin{pmatrix} a & b \\ 0 & 1 \end{pmatrix} \ .$$

Therefore, the controllability canonical form of the dual is given by matrix

$$\tilde{\mathbf{A}} = \mathbf{T}\mathbf{A}\mathbf{T}^{-1} = \frac{1}{a} \begin{pmatrix} a & b \\ 0 & 1 \end{pmatrix} \begin{pmatrix} -b & c \\ a & -d \end{pmatrix} \begin{pmatrix} 1 & -b \\ 0 & a \end{pmatrix} = \begin{pmatrix} 0 & ac-bd \\ 1 & -b-d \end{pmatrix} \ ,$$

and vectors

$$\tilde{\mathbf{B}} = \mathbf{T}\mathbf{B} = \frac{1}{a} \begin{pmatrix} a & b \\ 0 & 1 \end{pmatrix} \begin{pmatrix} 1 \\ 0 \end{pmatrix} = \begin{pmatrix} 1 \\ 0 \end{pmatrix}$$

and

$$\tilde{\mathbf{C}} = \mathbf{C}\mathbf{T}^{-1} = (\alpha, \beta) \begin{pmatrix} 1 & -b \\ 0 & a \end{pmatrix} = (\alpha, -\alpha b + \beta a) \ .$$

Therefore, the dual system is

$$\dot{\tilde{\mathbf{x}}} = \begin{pmatrix} 0 & 1 \\ ac-bd & -b-d \end{pmatrix} \tilde{\mathbf{x}} + \begin{pmatrix} \alpha \\ -\alpha b + \beta a \end{pmatrix} \tilde{u} \ ,$$

$$\tilde{y} = (1,0)\tilde{\mathbf{x}} \ . \tag{7.25}$$

By using the general formulation (7.18) we have

$$n = 2, \qquad a_0 = ac - bd, \qquad \text{and} \qquad a_1 = -b - d \ .$$

Hence, the required canonical form is obtained.

4. The observability canonical form of the *warfare* model (6.19) is a special case of the above results, since by selecting

$$a = -h_2, \quad b = 0, \quad c = -h_1, \quad d = 0, \quad \alpha = -h_3, \quad \text{and} \quad \beta = 0$$

the arms-races model reduces to the warfare model. Hence, the above result implies that the observability canonical form of this model is as follows:

$$\dot{\tilde{\mathbf{x}}} = \begin{pmatrix} 0 & 1 \\ h_1 h_2 & 0 \end{pmatrix} \tilde{\mathbf{x}} + \begin{pmatrix} -h_3 \\ 0 \end{pmatrix} \tilde{u} \ ,$$

$$\tilde{y} = (1,0)\tilde{\mathbf{x}} \ . \tag{7.26}$$

5. In Application 6.4.2-5 we saw that the linear *epidemics* model is based on matrices

$$\mathbf{A} = \begin{pmatrix} 0 & -\alpha\bar{x} \\ 0 & \alpha\bar{x} - \beta \end{pmatrix}, \quad \mathbf{b} = \begin{pmatrix} 0 \\ -1 \end{pmatrix}, \quad \text{and} \quad \mathbf{c}^T = (0,1),$$

where $\bar{x} \geq 0$ is arbitrary. By selecting $\bar{x} = \beta/\alpha$, matrix $\mathbf{A}$ reduces to

$$\mathbf{A} = \begin{pmatrix} 0 & -\beta \\ 0 & 0 \end{pmatrix}.$$

If x_1 and x_2 denote the state components, then the system reduces to the following:

$$\dot{x}_1 = -\beta x_2$$

$$\dot{x}_2 = -u$$

$$y = x_2 .$$

Introduce the new variable $\tilde{x}_2 = -\beta x_2$, then these equations are transformed as

$$\dot{x}_1 = \tilde{x}_2$$

$$\dot{\tilde{x}}_2 = \beta u$$

$$y = -\frac{1}{\beta}\tilde{x}_2 .$$

This form is a Jordan canonical form, since the modified coefficient matrix

$$\tilde{\mathbf{A}} = \begin{pmatrix} 0 & 1 \\ 0 & 0 \end{pmatrix}$$

is a special 2×2 Jordan block with eigenvalue $\lambda = 0$.

6. The *Harrod-type national economy* model (6.20) is based on the 1×1 matrices

$$\mathbf{A} = (1 + r - rm), \quad \mathbf{B} = (-r), \quad \text{and} \quad \mathbf{C} = (m) .$$

Here $\mathbf{A}$ can be considered diagonal; therefore, this system can be considered also as given in diagonal form.

7. Similar to the previous application, the *linear cobweb* model (6.21) is based on the 1×1 matrices

$$\mathbf{A} = \begin{pmatrix} b \\ a \end{pmatrix}, \qquad \mathbf{B} = (1), \quad \text{and} \quad \mathbf{C} = (b).$$

Therefore, this system also can be considered as given in diagonal form.

8. The model (6.22) of *interrelated markets* is completely controllable, as was verified in Application 5.3.2-8; however, it is not always observable. In order to illustrate a special canonical form of the system, consider the following numerical example:

$$\mathbf{A} = \begin{pmatrix} -2 & 1 \\ 0 & -3 \end{pmatrix}, \qquad \mathbf{B} = \begin{pmatrix} 1 & -1 \\ 0 & 1 \end{pmatrix}, \quad \text{and} \quad \mathbf{K} = \begin{pmatrix} 1 & 0 \\ 0 & 2 \end{pmatrix}.$$

Then the system has the particular form

$$\dot{\mathbf{p}} = \begin{pmatrix} 1 & 0 \\ 0 & 2 \end{pmatrix} \left[\begin{pmatrix} -2 & 1 \\ 0 & -3 \end{pmatrix} - \begin{pmatrix} 1 & -1 \\ 0 & 1 \end{pmatrix} \right] \mathbf{p} + \mathbf{u}$$

$$= \begin{pmatrix} 1 & 0 \\ 0 & 2 \end{pmatrix} \begin{pmatrix} -3 & 2 \\ 0 & -4 \end{pmatrix} \mathbf{p} + \mathbf{u}$$

$$= \begin{pmatrix} -3 & 2 \\ 0 & -8 \end{pmatrix} \mathbf{p} + \begin{pmatrix} 1 & 0 \\ 0 & 1 \end{pmatrix} \mathbf{u} \qquad\qquad (7.27)$$

with output equation

$$y = \begin{pmatrix} \dfrac{1}{2}, \dfrac{1}{2} \end{pmatrix} \mathbf{p}.$$

The diagonal form of this system can be obtained as follows. First the eigenvalues of the coefficient matrix are determined. Since this matrix is triangular, the eigenvalues are the diagonal elements $\lambda_1 = -3$ and $\lambda_2 = -8$. Simple calculation shows that the associated eigenvectors are

$$\begin{pmatrix} 1 \\ 0 \end{pmatrix} \qquad \text{and} \qquad \begin{pmatrix} 2 \\ -5 \end{pmatrix};$$

therefore, the transformation matrix is

$$\mathbf{T} = \begin{pmatrix} 1 & 2 \\ 0 & -5 \end{pmatrix}^{-1} = \frac{1}{5} \begin{pmatrix} 5 & 2 \\ 0 & -1 \end{pmatrix}.$$

Use finally relations (7.4) to get the transformed system:

$$\dot{\tilde{\mathbf{p}}} = \begin{pmatrix} -3 & 0 \\ 0 & -8 \end{pmatrix} \tilde{\mathbf{p}} + \frac{1}{5} \begin{pmatrix} 5 & 2 \\ 0 & -1 \end{pmatrix} \mathbf{u}$$

$$y = \begin{pmatrix} \frac{1}{2}, -\frac{3}{2} \end{pmatrix} \tilde{\mathbf{p}} . \tag{7.28}$$

We can check our results by verifying that the transfer functions of the original and this diagonal representation coincide. Simple calculation shows that the transfer function of system (7.27) is the following:

$$\mathbf{H}(s) = \begin{pmatrix} \frac{1}{2}, \frac{1}{2} \end{pmatrix} \begin{pmatrix} s+3 & -2 \\ 0 & s+8 \end{pmatrix}^{-1} \begin{pmatrix} 1 & 0 \\ 0 & 1 \end{pmatrix}$$

$$= \begin{pmatrix} \frac{1}{2}, \frac{1}{2} \end{pmatrix} \frac{1}{(s+3)(s+8)} \begin{pmatrix} s+8 & 2 \\ 0 & s+3 \end{pmatrix}$$

$$= \begin{pmatrix} \frac{1}{2(s+3)}, \frac{s+5}{2(s+3)(s+8)} \end{pmatrix} .$$

The transfer function of the transformed system (7.28) is given as

$$\tilde{\mathbf{H}}(s) = \begin{pmatrix} \frac{1}{2}, -\frac{3}{2} \end{pmatrix} \begin{pmatrix} s+3 & 0 \\ 0 & s+8 \end{pmatrix}^{-1} \cdot \frac{1}{5} \begin{pmatrix} 5 & 2 \\ 0 & -1 \end{pmatrix}$$

$$= \begin{pmatrix} \frac{1}{2}, -\frac{3}{2} \end{pmatrix} \frac{1}{5(s+3)(s+8)} \begin{pmatrix} s+8 & 0 \\ 0 & s+3 \end{pmatrix} \begin{pmatrix} 5 & 2 \\ 0 & -1 \end{pmatrix}$$

$$= \begin{pmatrix} \frac{1}{2}, -\frac{3}{2} \end{pmatrix} \frac{1}{5(s+3)(s+8)} \begin{pmatrix} 5s+40 & 2s+16 \\ 0 & -s-3 \end{pmatrix}$$

$$= \begin{pmatrix} \frac{1}{2(s+3)}, \frac{s+5}{2(s+3)(s+8)} \end{pmatrix} .$$

Hence, $\mathbf{H}(s) = \tilde{\mathbf{H}}(s)$.

9. In our last example we shall diagonalize the coefficient matrix $\mathbf{A}_C$ of the *oligopoly* model (6.23). Note first that the coefficient matrix can be written as

$$\mathbf{A}_C = -\frac{1}{2}\mathbf{1} + \frac{1}{2}\mathbf{I} ,$$

and in order to find the nonsingular matrix $\mathbf{T}$ which transforms this matrix into the diagonal form, we need to determine first the eigenvalues of $\mathbf{A}_C$ and the associated eigenvectors. In Section 3.5.2 we have seen that the eigenvalues of matrix $\mathbf{1}$ are $\lambda_1 = 0$ and $\lambda_2 = N$; furthermore any vector $\mathbf{u} = (u_i)$ with the property $u_1 + u_2 + \cdots + u_N = 0$ is associated eigenvector to $\lambda_1 = 0$, and any vector having identical components is an eigenvector associated to $\lambda_2 = N$. Therefore, a complete system of eigenvectors is given as

$$
\begin{pmatrix} -N+1 \\ 1 \\ 1 \\ \vdots \\ 1 \end{pmatrix}, \begin{pmatrix} 1 \\ -N+1 \\ 1 \\ \vdots \\ 1 \end{pmatrix}, \cdots, \begin{pmatrix} 1 \\ 1 \\ \vdots \\ -N+1 \\ 1 \end{pmatrix}, \text{ and } \begin{pmatrix} 1 \\ 1 \\ \vdots \\ 1 \\ 1 \end{pmatrix},
$$

where the first $N-1$ vectors are associated to $\lambda_1 = 0$ and the last vector is associated to $\lambda_2 = N$. From Equations (1.29) and (7.4), we conclude that the transformation matrix is as follows:

$$
\mathbf{T} = \begin{pmatrix} -N+1 & 1 & \cdots & 1 & 1 \\ 1 & -N+1 & \cdots & 1 & 1 \\ 1 & 1 & \cdots & 1 & 1 \\ \vdots & \vdots & \ddots & \vdots & \vdots \\ 1 & 1 & \cdots & -N+1 & 1 \\ 1 & 1 & \cdots & 1 & 1 \end{pmatrix}^{-1}.
$$

Look for this inverse matrix in the special form

$$
\mathbf{T} = \begin{pmatrix} a & b & \cdots & b & c \\ b & a & \cdots & b & c \\ b & b & \cdots & b & c \\ \vdots & \vdots & \ddots & \vdots & \vdots \\ b & b & \cdots & a & c \\ c & c & \cdots & c & d \end{pmatrix},
$$

then equation $\mathbf{T}^{-1}\mathbf{T} = \mathbf{I}$ implies that

$$
(-N+1)a + (N-2)b + c = 1
$$

$$
a + (-N+1)b + (N-3)b + c = 0
$$

$$
a + (N-2)b + c = 0
$$

$$(N - 1)c + d = 1 \ .$$

Simple calculation shows that the solution of these equations is the following:

$$a = -\frac{1}{N}, \quad b = 0, \quad \text{and} \quad c = d = \frac{1}{N} \ .$$

Hence

$$\mathbf{T} = \frac{1}{N}\begin{pmatrix} -1 & & & \mathbf{O} & 1 \\ & -1 & & & 1 \\ & & \ddots & & 1 \\ \mathbf{O} & & & -1 & 1 \\ 1 & 1 & \cdots & 1 & 1 \end{pmatrix} ,$$

and, therefore, from relation (7.4) we conclude that

$$\tilde{\mathbf{A}}_C = \mathbf{T}\mathbf{A}_C\mathbf{T}^{-1} = \begin{pmatrix} \frac{1}{2} & & & & \mathbf{O} \\ & \frac{1}{2} & & & \\ & & \ddots & & \\ & & & \frac{1}{2} & \\ \mathbf{O} & & & & \frac{1-N}{2} \end{pmatrix} .$$

Note that the diagonal elements are the eigenvalues of $\mathbf{A}_C$ as they should be.

Problems

1. Diagonalize system

$$\dot{\mathbf{x}} = \begin{pmatrix} 1 & 1 \\ 2 & 2 \end{pmatrix}\mathbf{x} + \begin{pmatrix} 1 \\ 0 \end{pmatrix} u$$

$$y = (1, 1)\mathbf{x} \ .$$

2. Diagonalize system

$$\dot{\mathbf{x}} = \begin{pmatrix} 0 & 1 & 2 \\ 0 & 1 & 1 \\ 0 & 0 & 2 \end{pmatrix}\mathbf{x} + \begin{pmatrix} 1 \\ 1 \\ 1 \end{pmatrix} u$$

$$y = (1, 0, 0)\mathbf{x} \ .$$

3.   Repeat Example 7.1 for system

$$\dot{\mathbf{x}} = \begin{pmatrix} 0\,0\,0 \\ 0\,1\,0 \\ 0\,0\,2 \end{pmatrix} \mathbf{x} + \begin{pmatrix} 1 \\ 1 \\ 1 \end{pmatrix} u\,, \qquad \mathbf{x}(0) = \begin{pmatrix} 1 \\ 0 \\ 0 \end{pmatrix}$$

with the constant input $u(t) \equiv 1$.

4.   Repeat Example 7.2 for system

$$\dot{\mathbf{x}} = \begin{pmatrix} 2\,1 \\ 0\,2 \end{pmatrix} \mathbf{x} + \begin{pmatrix} 1 \\ 1 \end{pmatrix} u\,, \qquad \mathbf{x}(0) = \begin{pmatrix} 1 \\ 0 \end{pmatrix}$$

with the input function $u(t) \equiv t$.

5.   Solve the system

$$\dot{\mathbf{x}} = \begin{pmatrix} 0\,1\,2 \\ 0\,1\,1 \\ 0\,0\,2 \end{pmatrix} \mathbf{x} + \begin{pmatrix} 1 \\ 1 \\ 1 \end{pmatrix} u$$

$$y = (1, 0, 0)\mathbf{x}$$

with the constant input $u(t) \equiv 1$ by the transformation method. Solve first the diagonal form obtained in Problem 7.2, and then compute the solution from Equation (7.3).

6.   Find the controllability canonical form (7.11) for system

$$\dot{\mathbf{x}} = \begin{pmatrix} 1\,0\,1 \\ 0\,1\,1 \\ 1\,1\,1 \end{pmatrix} \mathbf{x} + \begin{pmatrix} 0 \\ 1 \\ 0 \end{pmatrix} u$$

$$y = (0, 1, 0)\mathbf{x}\,.$$

7.   Find the controllability canonical form (7.11) for system

$$\dot{\mathbf{x}} = \begin{pmatrix} 1\,1\,1 \\ 1\,2\,1 \\ 1\,1\,0 \end{pmatrix} \mathbf{x} + \begin{pmatrix} 1 \\ 0 \\ 0 \end{pmatrix} u$$

$$y = (1, 0, 0)\mathbf{x}\,.$$

8. Find the controllability canonical form (7.11) for system

$$\dot{x} = \begin{pmatrix} 1 & 0 & 1 \\ 0 & 3 & 0 \\ 1 & 0 & 1 \end{pmatrix} x + \begin{pmatrix} 1 \\ 1 \\ 0 \end{pmatrix} u$$

$$y = (1, 1, 0)x .$$

9. Find the controllability canonical form (7.14) for system

$$\dot{x} = \begin{pmatrix} 1 & 0 & 1 \\ 0 & 1 & 1 \\ 1 & 1 & 1 \end{pmatrix} x + \begin{pmatrix} 0 \\ 1 \\ 0 \end{pmatrix} u$$

$$y = (0, 1, 0)x .$$

10. Find the controllability canonical form (7.14) for system

$$\dot{x} = \begin{pmatrix} 1 & 1 & 1 \\ 1 & 2 & 1 \\ 1 & 1 & 0 \end{pmatrix} x + \begin{pmatrix} 1 \\ 0 \\ 0 \end{pmatrix} u$$

$$y = (1, 0, 0)x .$$

11. Find the controllability canonical form (7.14) for the system

$$\dot{x} = \begin{pmatrix} 1 & 0 & 1 \\ 0 & 3 & 0 \\ 1 & 0 & 1 \end{pmatrix} x + \begin{pmatrix} 1 \\ 1 \\ 0 \end{pmatrix} u$$

$$y = (1, 1, 0)x .$$

12. Find the observability canonical form (7.18) for system

$$\dot{x} = \begin{pmatrix} 1 & 0 & 1 \\ 0 & 1 & 1 \\ 1 & 1 & 1 \end{pmatrix} x + \begin{pmatrix} 0 \\ 1 \\ 0 \end{pmatrix} u$$

$$y = (0, 1, 0)x .$$

13.   Find the observability canonical form (7.18) for system

$$\dot{x} = \begin{pmatrix} 1 & 1 & 1 \\ 1 & 2 & 1 \\ 1 & 1 & 0 \end{pmatrix} x + \begin{pmatrix} 1 \\ 0 \\ 0 \end{pmatrix} u$$

$$y = (1, 0, 0)x .$$

14.   Find the observability canonical form (7.18) for the system

$$\dot{x} = \begin{pmatrix} 1 & 0 & 1 \\ 0 & 3 & 0 \\ 1 & 0 & 1 \end{pmatrix} x + \begin{pmatrix} 1 \\ 1 \\ 0 \end{pmatrix} u$$

$$y = (1, 1, 0)x .$$

15.   Find the observability canonical form (7.19) for system

$$\dot{x} = \begin{pmatrix} 1 & 0 & 1 \\ 0 & 1 & 1 \\ 1 & 1 & 1 \end{pmatrix} x + \begin{pmatrix} 0 \\ 1 \\ 0 \end{pmatrix} u$$

$$y = (0, 1, 0)x .$$

16.   Find the observability canonical form (7.19) for system

$$\dot{x} = \begin{pmatrix} 1 & 1 & 1 \\ 1 & 2 & 1 \\ 1 & 1 & 0 \end{pmatrix} x + \begin{pmatrix} 1 \\ 0 \\ 0 \end{pmatrix} u$$

$$y = (1, 0, 0)x .$$

17.   Find the observability canonical form (7.19) for the system

$$\dot{x} = \begin{pmatrix} 1 & 0 & 1 \\ 0 & 3 & 0 \\ 1 & 0 & 1 \end{pmatrix} x + \begin{pmatrix} 1 \\ 1 \\ 0 \end{pmatrix} u$$

$$y = (1, 1, 0)x .$$

18.   Verify directly that the canonical forms (7.11) and (7.14) are completely controllable.

19.   Verify directly that the canonical forms (7.18) and (7.19) are completely observable.

20.   Verify the corollary of Theorem 7.1 by showing that the characteristic polynomial of matrix

$$
\mathbf{A} = \begin{pmatrix}
0 & 0 & 0 & \cdots & 0 & a_0 \\
1 & 0 & 0 & \cdots & 0 & a_1 \\
0 & 1 & 0 & \cdots & 0 & a_2 \\
\vdots & \vdots & \vdots & \ddots & \vdots & \vdots \\
0 & 0 & 0 & \cdots & 1 & a_{n-1}
\end{pmatrix}
$$

is

$$
\varphi(\lambda) = \lambda^n - a_{n-1}\lambda^{n-1} - \cdots - a_1\lambda - a_0 .
$$

21.   Systems $\dot{\mathbf{x}} = \mathbf{A}\mathbf{x} + \mathbf{b}u$ and $\dot{\mathbf{z}} = \tilde{\mathbf{A}}\mathbf{z} + \tilde{\mathbf{b}}u$ are related by the state transformation $\mathbf{z} = \mathbf{T}\mathbf{x}$. Prove that matrix $\mathbf{T}$ is unique if and only if the systems are completely controllable.

22.   Is the statement of the previous problem true for multiple input system?

23.   Reformulate and verify the statement of Problem 7/21 for observable systems.

24.   Explain how the method introduced in the proof of Theorem 7.1 can be used to compute the eigenvalues of a real $n \times n$ matrix $\mathbf{A}$.

25.   Assume that $\mathbf{A}$ is a real $n \times n$ matrix with distinct eigenvalues. Using the diagonal form of $\mathbf{A}$ find an n-element real vector $\mathbf{b}$ such that system

$$
\dot{\mathbf{x}} = \mathbf{A}\mathbf{x} + \mathbf{b}u
$$

is completely controllable.

chapter eight

Realization

In Chapter 6 we were concerned with the problem of estimating the initial state of a system on the basis of input and output measurements in an interval $[t_0, t_1]$. In this chapter, an even more difficult problem is addressed, namely, how to recover the system's model itself based on the known relation between the input and output of a system.

We know from Chapter 3 that the input and output of a continuous linear system are interrelated by equation

$$\mathbf{y}(t) = \mathbf{C}(t)\phi(t, t_0)\mathbf{x}_0 + \int_{t_0}^{t} \mathbf{C}(t)\phi(t, \tau)\mathbf{B}(\tau)\mathbf{u}(\tau)\, d\tau , \qquad (8.1)$$

and for discrete linear systems this equation is modified as

$$\mathbf{y}(t) = \mathbf{C}(t)\phi(t, 0)\mathbf{x}_0 + \sum_{\tau=0}^{t-1} \mathbf{C}(t)\phi(t, \tau+1)\mathbf{B}(\tau)\mathbf{u}(\tau) . \qquad (8.2)$$

Matrix
$$\mathbf{T}(t, \tau) = \mathbf{C}(t)\phi(t, \tau)\mathbf{B}(\tau) \qquad (8.3)$$

for continuous systems and matrix

$$\mathbf{T}(t, \tau) = \mathbf{C}(t)\phi(t, \tau+1)\mathbf{B}(\tau) \qquad (8.4)$$

for discrete systems are called the *weighting patterns*.

DEFINITION 8.1   *A weighting pattern* $\mathbf{T}(t, \tau)$ *is said to be realizable in* $[t_0, t_1]$ *if there exist matrices* $\mathbf{A}(t)$, $\mathbf{B}(t)$, *and* $\mathbf{C}(t)$ *such that (8.3) (or (8.4)) is satisfied for all* $t \in [t_0, t_1]$ *with* ϕ *being the fundamental matrix of the* $(\mathbf{A}(t), \mathbf{B}(t), \mathbf{C}(t))$ *continuous (or discrete) system.*

In the first section of this chapter, the realizability of weighting patterns is discussed. Necessary and sufficient conditions will be given and the uniqueness of the realization will be analyzed.

In the case of time-invariant systems, the problem of realizability can also be addressed in a different way. If $\mathbf{U}$ and $\mathbf{Y}$ denote the Laplace (or Z) transforms of the input and output, then from Chapter 3 we know that for continuous systems,

$$\mathbf{Y}(s) = \mathbf{CR}(s)\mathbf{x}_0 + \mathbf{H}(s)\mathbf{U}(s) \; , \tag{8.5}$$

and for discrete systems,

$$\mathbf{Y}(z) = \mathbf{CR}(z)z\mathbf{x}_0 + \mathbf{H}(z)\mathbf{U}(z) \; , \tag{8.6}$$

where $\mathbf{R}(s)$ (or $\mathbf{R}(z)$) is the resolvent matrix, and $\mathbf{H}(s)$ (or $\mathbf{H}(z)$) is the transfer function. The form of the transfer function is the same for continuous and discrete systems:

$$\mathbf{H}(s) = \mathbf{C}(s\mathbf{I} - \mathbf{A})^{-1}\mathbf{B} \; ; \tag{8.7}$$

only the variable s is renamed as z in the discrete case.

DEFINITION 8.2   *A transfer function $\mathbf{H}(s)$ is said to be realizable if there exist constant matrices $\mathbf{A}, \mathbf{B},$ and $\mathbf{C}$ such that (8.7) holds for all s.*

Similar to the realizability problem of weighting patterns, necessary and sufficient conditions will be developed for the realizability of given transfer functions.

After a realization of a weighting pattern or transfer function is found, a new question arises. Is it possible to reduce the dimension of the state variable in order to simplify the system and to reduce construction and computation costs? In this chapter we also give methods to find a realization with minimal dimensional state variable, and, in addition, necessary and sufficient conditions will be presented to determine whether a given realization is minimal or not.

8.1   Realizability of Weighting Patterns

In this section, necessary and sufficient conditions are given for the realizability of a given weighting pattern. Only continuous systems are discussed, since the realizability of weighting patterns of discrete systems is analogous. The details are left as an exercise.

8.1.1 Realizability Conditions

Let $\mathbf{T}(t,\tau)$ be a bivariable function of $t, \tau \geq t_0$, where $t_0 \geq 0$ is given.

THEOREM 8.1
$\mathbf{T}(t,\tau)$ is realizable if and only if it is separable as

$$\mathbf{T}(t,\tau) = \mathbf{D}(t)\mathbf{G}(\tau) \tag{8.8}$$

with continuous functions $\mathbf{D}$ and $\mathbf{G}$.

PROOF   (a) If $\mathbf{T}(t,\tau)$ is realizable, then there exist continuous functions $\mathbf{B}$, $\mathbf{C}$ and a fundamental matrix ϕ such that

$$\mathbf{T}(t,\tau) = \mathbf{C}(t)\phi(t,\tau)\mathbf{B}(\tau) \ .$$

By using Property (ii) of Theorem 2.3 we conclude that with some t^*,

$$\mathbf{T}(t,\tau) = \mathbf{C}(t)\phi(t,t^*)\phi(t^*,\tau)\mathbf{B}(\tau) \ .$$

Then select

$$\mathbf{D}(t) = \mathbf{C}(t)\phi(t,t^*) \qquad \text{and} \qquad \mathbf{G}(\tau) = \phi(t^*,\tau)\mathbf{B}(\tau)$$

to get the form (8.3).
   (b) Assume next that relation (8.8) holds. Then we verify that $(\mathbf{O}, \mathbf{G}(t), \mathbf{D}(t))$ is a realization of $\mathbf{T}(t,\tau)$. Observe first that in the case of $\mathbf{A}(t) \equiv \mathbf{O}$,

$$\phi(t,\tau) = e^{\mathbf{O} \cdot (t-\tau)} = \mathbf{I} \ ,$$

and, therefore, with selecting $\mathbf{B} = \mathbf{G}$ and $\mathbf{C} = \mathbf{D}$,

$$\mathbf{C}(t)\phi(t,\tau)\mathbf{B}(\tau) = \mathbf{D}(t)\mathbf{I}\mathbf{G}(\tau) = \mathbf{D}(t)\mathbf{G}(\tau) = \mathbf{T}(t,\tau) \ ,$$

which completes the proof.   ∎

COROLLARY 8.1
If $\mathbf{T}(t,\tau)$ is realizable with a system having n-dimensional state, then part (a) of the proof implies that there are matrices $\mathbf{D}(t)$ with n columns and $\mathbf{G}(\tau)$ with n rows such that (8.8) holds.

A realization procedure can be formulated as follows:

Step 1   Factor the weighting pattern as $\mathbf{T}(t, \tau) = \mathbf{D}(t)\mathbf{G}(\tau)$.

Step 2   Select $\mathbf{A} = \mathbf{O}$, $\mathbf{B} = \mathbf{G}$, and $\mathbf{C} = \mathbf{D}$.

Example 8.1

Consider the real-valued function

$$\mathbf{T}(t, \tau) = t + \tau \ .$$

Since

$$\mathbf{T}(t, \tau) = (t, 1) \cdot \begin{pmatrix} 1 \\ \tau \end{pmatrix} \ ,$$

$\mathbf{T}(t, \tau)$ is realizable in $[0, \infty)$ with a system having two-dimensional state, where

$$\mathbf{A} = \mathbf{O}, \qquad \mathbf{B}(t) = \begin{pmatrix} 1 \\ t \end{pmatrix} \ , \ \text{and} \qquad \mathbf{C}(t) = (t, 1) \ .$$

Next we show that no realization exists with single-dimensional state variables. Contrary to this assertion, assume that there exist continuous real functions D and G such that

$$t + \tau = D(t) \cdot G(\tau) \quad \text{for all} \quad t, \tau \geq 0 \ . \tag{8.9}$$

Select $\tau = 0$; then for all $t \geq 0$,

$$t = D(t) \cdot G(0) \ ,$$

that is,

$$D(t) = \alpha \cdot t$$

with some constant α. Similarly, select $t = 0$ to show that

$$G(\tau) = \beta \cdot \tau$$

with some constant β. Then (8.9) can be rewritten as

$$t + \tau = \alpha \beta t \tau \quad \text{for all} \quad t, \tau \geq 0 \ ,$$

which must not hold as the selection $\tau = 0, t \neq 0$ shows.

In the case of realizable weighting patterns, the realization is not necessarily unique:

THEOREM 8.2
Assume that $(\mathbf{A}(t), \mathbf{B}(t), \mathbf{C}(t))$ is a realization of $\mathbf{T}(t, \tau)$ in $[t_0, t_1]$, and $\mathbf{P}(t)$ is invertible and differentiable for all $t \in [t_0, t_1]$. Then $(\tilde{\mathbf{A}}(t), \tilde{\mathbf{B}}(t), \tilde{\mathbf{C}}(t))$ is also a realization of $\mathbf{T}(t, \tau)$ with

$$\tilde{\mathbf{A}}(t) = \mathbf{P}(t)\mathbf{A}(t)\mathbf{P}^{-1}(t) + \dot{\mathbf{P}}(t)\mathbf{P}^{-1}(t), \qquad \tilde{\mathbf{B}}(t) = \mathbf{P}(t)\mathbf{B}(t)$$

and

$$\tilde{\mathbf{C}}(t) = \mathbf{C}(t)\mathbf{P}^{-1}(t) .$$

PROOF Introduce the new variable $\mathbf{z}(t) = \mathbf{P}(t)\mathbf{x}(t)$ in system $(\mathbf{A}(t), \mathbf{B}(t), \mathbf{C}(t))$. Then

$$\dot{\mathbf{z}}(t) = \mathbf{P}(t)\dot{\mathbf{x}}(t) + \dot{\mathbf{P}}(t)\mathbf{x}(t) = \mathbf{P}(t)[\mathbf{A}(t)\mathbf{x}(t) + \mathbf{B}(t)\mathbf{u}(t)] + \dot{\mathbf{P}}(t)\mathbf{x}(t)$$

$$= \mathbf{P}(t)[\mathbf{A}(t)\mathbf{P}^{-1}(t)\mathbf{z}(t) + \mathbf{B}(t)\mathbf{u}(t)] + \dot{\mathbf{P}}(t)\mathbf{P}^{-1}(t)\mathbf{z}(t)$$

$$= \tilde{\mathbf{A}}(t)\mathbf{z}(t) + \tilde{\mathbf{B}}(t)\mathbf{u}(t) ,$$

and

$$\mathbf{y}(t) = \mathbf{C}(t)\mathbf{x}(t) = \mathbf{C}(t)\mathbf{P}^{-1}(t)\mathbf{z}(t) = \tilde{\mathbf{C}}(t)\mathbf{z}(t) .$$

Since introducing the new state variable $\mathbf{z}$ does not change the input–output relation, the weighting pattern $\mathbf{T}(t, \tau)$ is also realized by the system $(\tilde{\mathbf{A}}(t), \tilde{\mathbf{B}}(t), \tilde{\mathbf{C}}(t))$. ∎

REMARK 8.1 Realizations $(\tilde{\mathbf{A}}(t), \tilde{\mathbf{B}}(t), \tilde{\mathbf{C}}(t))$ and $(\mathbf{A}(t), \mathbf{B}(t), \mathbf{C}(t))$ have the same dimensions in the state variables. The next example shows that even the dimension of the state variable in a realization of a given weighting pattern is not necessarily unique. ∎

Example 8.2

Consider system

$$\dot{\mathbf{x}} = \begin{pmatrix} 1 & 0 \\ 0 & 1 \end{pmatrix}\mathbf{x} + \begin{pmatrix} 1 \\ 0 \end{pmatrix}u$$

$$y = (1, 0)\mathbf{x} .$$

Since the input has no effect on the second component x_2 of $\mathbf{x}$ and the output does not depend on x_2, the input–output relation is determined only by the first equation

$$\dot{x}_1 = x_1 + u$$

and output relation

$$y = x_1 \,,$$

which is a single-dimensional realization of the same input–output relation.

In many applications we look for periodic realizations of a given weighting pattern, when $\mathbf{A}(t), \mathbf{B}(t), \mathbf{C}(t)$ are periodic with the same period. It can be proved that $\mathbf{T}(t, \tau)$ is realizable by a periodic system $(\mathbf{A}(t), \mathbf{B}(t), \mathbf{C}(t))$ if and only if it is realizable and there is a $T > 0$ such that

$$\mathbf{T}(t, \tau) = \mathbf{T}(t + T, \tau + T)$$

for all t and τ.

8.1.2   *Minimal Realizations*

In most cases, especially in system design, it is important to reduce the dimension of the system. In this section we introduce methods that can be used to determine whether a realization of a given weighting pattern has minimal state dimension or not, and in addition, if a realization is not minimal how to find a minimal realization.

DEFINITION 8.3   *A realization* $(\mathbf{A}(t), \mathbf{B}(t), \mathbf{C}(t))$ *of a given weighting pattern is called* minimal *in an interval* $[t_0, t_1]$ *if there is no other realization with lower dimensional state variable.*

First a necessary and sufficient condition is presented for the minimality of a given realization.

THEOREM 8.3
A realization $(\mathbf{A}(t), \mathbf{B}(t), \mathbf{C}(t))$ *is minimal in an interval* $[t_0, t_1]$ *if and only if the controllability Gramian* $\mathbf{W}(t_0, t_1)$ *and the observability Gramian* $\mathbf{M}(t_0, t_1)$ *are both nonsingular.*

PROOF   (a) Assume first that realization $(\mathbf{A}(t), \mathbf{B}(t), \mathbf{C}(t))$ is not minimal. Then there exists a realization

$$\dot{\mathbf{z}} = \tilde{\mathbf{A}}(t)\mathbf{z} + \tilde{\mathbf{B}}(t)\mathbf{u}$$

$$\mathbf{y} = \tilde{\mathbf{C}}(t)\mathbf{z} \tag{8.10}$$

with dim $\mathbf{z}$ < dim $\mathbf{x}$. Let ν and n denote the dimensions of $\mathbf{z}$ and $\mathbf{x}$, respectively. Since $(\mathbf{A}(t), \mathbf{B}(t), \mathbf{C}(t))$ is a realization,

$$\mathbf{T}(t, \tau) = \mathbf{C}(t)\phi(t, \tau)\mathbf{B}(\tau) = \mathbf{C}(t)\phi(t, t_0)\phi(t_0, \tau)\mathbf{B}(\tau)$$

$$= \mathbf{D}(t)\mathbf{G}(\tau)$$

with

$$\mathbf{D}(t) = \mathbf{C}(t)\phi(t, t_0) \qquad \text{and} \qquad \mathbf{G}(\tau) = \phi(t_0, \tau)\mathbf{B}(\tau) .$$

The corollary of Theorem 8.1 implies that

$$\mathbf{T}(t, \tau) = \tilde{\mathbf{D}}(t)\tilde{\mathbf{G}}(\tau) ,$$

where $\tilde{\mathbf{D}}$ has ν columns and $\tilde{\mathbf{G}}$ has ν rows, since $(\tilde{\mathbf{A}}(t), \tilde{\mathbf{B}}(t), \tilde{\mathbf{C}}(t))$ is a realization of $\mathbf{T}(t, \tau)$ with ν-dimensional state.

The definitions of $\mathbf{W}(t_0, t_1)$ and $\mathbf{M}(t_0, t_1)$ imply that

$$\mathbf{M}(t_0, t_1)\mathbf{W}(t_0, t_1) = \int_{t_0}^{t_1} \mathbf{D}^T(t)\mathbf{D}(t) \, dt \cdot \int_{t_0}^{t_1} \mathbf{G}(\tau)\mathbf{G}^T(\tau) \, d\tau$$

$$= \int_{t_0}^{t_1} \int_{t_0}^{t_1} \mathbf{D}^T(t)\mathbf{D}(t)\mathbf{G}(\tau)\mathbf{G}^T(\tau) \, d\tau \, dt$$

$$= \int_{t_0}^{t_1} \int_{t_0}^{t_1} \mathbf{D}^T(t)\tilde{\mathbf{D}}(t)\tilde{\mathbf{G}}(\tau)\mathbf{G}^T(\tau) \, d\tau \, dt$$

$$= \int_{t_0}^{t_1} \mathbf{D}^T(t)\tilde{\mathbf{D}}(t) \, dt \cdot \int_{t_0}^{t_1} \tilde{\mathbf{G}}(\tau)\mathbf{G}^T(\tau) \, d\tau .$$

Note that $\mathbf{D}^T(t)\tilde{\mathbf{D}}(t)$ has ν columns and $\tilde{\mathbf{G}}(\tau)\mathbf{G}^T(\tau)$ has ν rows; therefore, the ranks of both integrals are not larger than ν. This implies that

$$rank(\mathbf{M}(t_0, t_1)\mathbf{W}(t_0, t_1)) \leq \nu < n .$$

Hence at least one of matrices $\mathbf{M}(t_0, t_1)$ and $\mathbf{W}(t_0, t_1)$ must be singular.

(b) Assume next that at least one of matrices $\mathbf{M}(t_0, t_1)$ and $\mathbf{W}(t_0, t_1)$ is singular. From Property (ii) of Theorems 5.2 and 6.2 we know that both matrices are positive semidefinite. The corollary of Theorem 1.14 implies that there exist nonsingular matrices $\mathbf{P}$ and $\mathbf{Q}$ such that

$$\mathbf{W}(t_0, t_1) = \mathbf{PS}_1\mathbf{P}^T \qquad \text{and} \qquad \mathbf{M}(t_0, t_1) = \mathbf{Q}^T\mathbf{S}_2\mathbf{Q}\,, \qquad (8.11)$$

where $\mathbf{S}_1$ and $\mathbf{S}_2$ are diagonal matrices of the form $diag(1,\ldots,1,0,\ldots,0)$ with the additional properties that $rank(\mathbf{S}_1) = rank(\mathbf{W}(t_0, t_1))$ and $rank(\mathbf{S}_2) = rank(\mathbf{M}(t_0, t_1))$.

Next we prove that for all t,

$$\mathbf{PS}_1\mathbf{P}^{-1}\mathbf{G}(t) = \mathbf{G}(t) \qquad \text{and} \qquad \mathbf{D}(t)\mathbf{Q}^{-1}\mathbf{S}_2\mathbf{Q} = \mathbf{D}(t)\,. \qquad (8.12)$$

Simple calculation shows that matrix

$$\mathbf{Z}(t) = \mathbf{PS}_1\mathbf{P}^{-1}\mathbf{G}(t) - \mathbf{G}(t)$$

satisfies equation

$$\int_{t_0}^{t_1} \mathbf{Z}(t)\mathbf{Z}^T(t)\,dt = \int_{t_0}^{t_1} [\mathbf{PS}_1\mathbf{P}^{-1}\mathbf{G}(t) - \mathbf{G}(t)]$$

$$[\mathbf{G}^T(t)(\mathbf{P}^T)^{-1}\mathbf{S}_1^T\mathbf{P}^T - \mathbf{G}^T(t)]\,dt$$

$$= \mathbf{PS}_1\mathbf{P}^{-1}\mathbf{W}(t_0, t_1)(\mathbf{P}^T)^{-1}\mathbf{S}_1\mathbf{P}^T - \mathbf{PS}_1\mathbf{P}^{-1}\mathbf{W}(t_0, t_1)$$

$$-\,\mathbf{W}(t_0, t_1)(\mathbf{P}^T)^{-1}\mathbf{S}_1\mathbf{P}^T + \mathbf{W}(t_0, t_1)$$

$$= \mathbf{PS}_1\mathbf{P}^{-1}\mathbf{PS}_1\mathbf{P}^T(\mathbf{P}^T)^{-1}\mathbf{S}_1\mathbf{P}^T - \mathbf{PS}_1\mathbf{P}^{-1}\mathbf{PS}_1\mathbf{P}^T$$

$$-\,\mathbf{PS}_1\mathbf{P}^T(\mathbf{P}^T)^{-1}\mathbf{S}_1\mathbf{P}^T + \mathbf{PS}_1\mathbf{P}^T$$

$$= \mathbf{PS}_1^3\mathbf{P}^T - \mathbf{PS}_1^2\mathbf{P}^T - \mathbf{PS}_1^2\mathbf{P}^T + \mathbf{PS}_1\mathbf{P}^T = \mathbf{O}\,,$$

since
$$\mathbf{S}_1 = \mathbf{S}_1^2 = \mathbf{S}_1^3, \qquad \text{and} \qquad \mathbf{S}_1^T = \mathbf{S}_1\,.$$

Note that for all vectors $\mathbf{z}$,

$$0 = \mathbf{z}^T \int_{t_0}^{t_1} \mathbf{Z}(t)\mathbf{Z}^T(t)\,dt\mathbf{z} = \int_{t_0}^{t_1} \mathbf{z}^T\mathbf{Z}(t)\mathbf{Z}^T(t)\mathbf{z}\,dt$$

$$= \int_{t_0}^{t_1} \|\mathbf{Z}^T(t)\mathbf{z}\|_2^2\,dt\,,$$

and, therefore, $\mathbf{Z}^T(t)\mathbf{z} \equiv \mathbf{0}$. Consequently $\mathbf{Z}^T(t) \equiv \mathbf{O}$, which implies that $\mathbf{Z}(t) = \mathbf{O}$ for all t.

The other identity of (8.12) can be proven in the same way; therefore, the details are omitted.

From (8.12) we have

$$D(t)G(\tau) = D(t)(Q^{-1}S_2QPS_1P^{-1})G(\tau) . \tag{8.13}$$

Since at least one of $\mathbf{W}(t_0, t_1)$ and $\mathbf{M}(t_0, t_1)$ is singular, either $\mathbf{S}_1$ or $\mathbf{S}_2$ has lower $rank$ than n. Let r denote the $rank$ of matrix $\mathbf{Q}^{-1}\mathbf{S}_2\mathbf{QPS}_1\mathbf{P}^{-1}$, then it is less than n, and Theorem 1.15 implies that

$$Q^{-1}S_2QPS_1P^{-1} = D_1G_1 , \tag{8.14}$$

where $\mathbf{D}_1$ has r columns and $\mathbf{G}_1$ has r rows.

Observe next that (8.13) implies that

$$T(t,\tau) = D(t)G(\tau) = (D(t)D_1)(G_1G(\tau)) = \tilde{D}(t)\tilde{G}(\tau)$$

with

$$\tilde{D}(t) = D(t)D_1 \quad \text{and} \quad \tilde{G}(\tau) = G_1G(\tau) . \tag{8.15}$$

Note that $\tilde{\mathbf{D}}(t)$ has r columns and $\tilde{\mathbf{G}}(\tau)$ has r rows.

Apply finally part (b) of the proof of Theorem 8.1 to see that

$$\tilde{A} = O, \quad \tilde{B}(t) = \tilde{G}(t), \quad \text{and} \quad \tilde{C}(t) = \tilde{D}(t) \tag{8.16}$$

is a realization of $\mathbf{T}(t, \tau)$ with r-dimensional state variable. Hence realization $(\mathbf{A}(t), \mathbf{B}(t), \mathbf{C}(t))$ is not minimal, which completes the proof. ∎

COROLLARY 8.2
Theorems 5.1 and 6.1 imply that a realization is minimal if and only if it is completely controllable and completely observable.

Example 8.3

Consider the system

$$\dot{\mathbf{x}} = \begin{pmatrix} 0 & \omega \\ -\omega & 0 \end{pmatrix} \mathbf{x} + \begin{pmatrix} 0 \\ 1 \end{pmatrix} u ,$$

$$y = (1, 1)\mathbf{x} .$$

In Examples 5.4 and 6.2 we verified that the system is completely controllable and completely observable. Therefore, it is a minimal realization of its weighting pattern. In other words, the same input–output relation cannot be represented by a single-dimensional linear system.

Example 8.4

Consider again the satellite problem, which was discussed earlier in Examples 5.5 and 6.3 with the inputs being the radial and tangential thrusts and the outputs being the radius r and angle θ. Then the system is completely controllable and completely observable; therefore, it is a minimal realization.

COROLLARY 8.3

Part (b) of the proof of Theorem 8.3 provides an algorithm to reduce the dimension of the state variable in a given nonminimal realization. The steps of the algorithm are as follows:

Step 1   *Compute* $\mathbf{W}(t_0, t_1)$, $\mathbf{M}(t_0, t_1)$, $\mathbf{D}(t)$, *and* $\mathbf{G}(\tau)$.

Step 2   *Find decompositions (8.11).*

Step 3   *Compute matrix*

$$\mathbf{E} = \mathbf{Q}^{-1}\mathbf{S}_2\mathbf{Q}\mathbf{P}\mathbf{S}_1\mathbf{P}^{-1} .$$

Step 4   *Find factorization (8.14).*

Step 5   *Compute matrices* $\tilde{\mathbf{D}}(t)$ *and* $\tilde{\mathbf{G}}(\tau)$ *by using relations (8.15) and find the lower dimensional realization as given in Equations (8.16).*

We mention that each matrix manipulation of this algorithm can be performed with standard computer packages.

   This algorithm is illustrated next.

Example 8.5

Consider now the system

$$\dot{\mathbf{x}} = \begin{pmatrix} 1 & 0 \\ 0 & 1 \end{pmatrix}\mathbf{x} + \begin{pmatrix} 1 \\ 0 \end{pmatrix}u ,$$

$$y = (1, 0)\mathbf{x} ,$$

which was the subject of our earlier Example 8.2. Since $n = 2$ and

$$\mathbf{K} = \begin{pmatrix} 1 & 1 \\ 0 & 0 \end{pmatrix}$$

with $rank(\mathbf{K}) = 1 < n$, the system is not controllable, therefore, this is not a minimal realization. We will now reduce the dimension of this realization by using the above algorithm. Assume for the sake of simplicity that $t_0 = 0$ and $t_1 = 1$. Since the system is time-invariant, we do not lose generality by this assumption.

Step 1: Since the fundamental matrix is

$$\phi(t, t_0) = \begin{pmatrix} e^{t-t_0} & 0 \\ 0 & e^{t-t_0} \end{pmatrix},$$

$$\mathbf{W}(0, 1) = \int_0^1 \begin{pmatrix} e^{-t} & 0 \\ 0 & e^{-t} \end{pmatrix} \begin{pmatrix} 1 \\ 0 \end{pmatrix} (1, 0) \begin{pmatrix} e^{-t} & 0 \\ 0 & e^{-t} \end{pmatrix} dt$$

$$= \int_0^1 \begin{pmatrix} e^{-2t} & 0 \\ 0 & 0 \end{pmatrix} dt = \begin{pmatrix} \frac{1-e^{-2}}{2} & 0 \\ 0 & 0 \end{pmatrix}$$

and

$$\mathbf{M}(0, 1) = \int_0^1 \begin{pmatrix} e^t & 0 \\ 0 & e^t \end{pmatrix} \begin{pmatrix} 1 \\ 0 \end{pmatrix} (1, 0) \begin{pmatrix} e^t & 0 \\ 0 & e^t \end{pmatrix} dt$$

$$= \int_0^1 \begin{pmatrix} e^{2t} & 0 \\ 0 & 0 \end{pmatrix} dt = \begin{pmatrix} \frac{e^2-1}{2} & 0 \\ 0 & 0 \end{pmatrix}.$$

Furthermore,

$$\mathbf{D}(t) = (1, 0) \begin{pmatrix} e^t & 0 \\ 0 & e^t \end{pmatrix} = (e^t, 0)$$

and

$$\mathbf{G}(\tau) = \begin{pmatrix} e^{-\tau} & 0 \\ 0 & e^{-\tau} \end{pmatrix} \begin{pmatrix} 1 \\ 0 \end{pmatrix} = \begin{pmatrix} e^{-\tau} \\ 0 \end{pmatrix}.$$

Step 2: It is easy to see that decompositions (8.11) hold with

$$\mathbf{P} = \mathbf{P}^T = \begin{pmatrix} \sqrt{\frac{1-e^{-2}}{2}} & 0 \\ 0 & 1 \end{pmatrix}, \qquad \mathbf{S}_1 = \begin{pmatrix} 1 & 0 \\ 0 & 0 \end{pmatrix}$$

and

$$\mathbf{Q} = \mathbf{Q}^T = \begin{pmatrix} \sqrt{\frac{e^2-1}{2}} & 0 \\ 0 & 1 \end{pmatrix}, \qquad \mathbf{S}_2 = \begin{pmatrix} 1 & 0 \\ 0 & 0 \end{pmatrix}.$$

Step 3: In this case,

$$\mathbf{E} = \begin{pmatrix} \sqrt{\frac{2}{e^2-1}} & 0 \\ 0 & 1 \end{pmatrix} \begin{pmatrix} 1 & 0 \\ 0 & 0 \end{pmatrix} \begin{pmatrix} \sqrt{\frac{e^2-1}{2}} & 0 \\ 0 & 1 \end{pmatrix} \begin{pmatrix} \sqrt{\frac{1-e^{-2}}{2}} & 0 \\ 0 & 1 \end{pmatrix} \begin{pmatrix} 1 & 0 \\ 0 & 0 \end{pmatrix}$$

$$\begin{pmatrix} \sqrt{\frac{2}{1-e^{-2}}} & 0 \\ 0 & 1 \end{pmatrix}$$

$$= \begin{pmatrix} 1 & 0 \\ 0 & 0 \end{pmatrix} \begin{pmatrix} 1 & 0 \\ 0 & 0 \end{pmatrix} = \begin{pmatrix} 1 & 0 \\ 0 & 0 \end{pmatrix} .$$

Step 4: Obviously $rank(\mathbf{E}) = 1$ and

$$\mathbf{E} = \begin{pmatrix} 1 \\ 0 \end{pmatrix} (1,0) \,,$$

and, therefore, we may select

$$\mathbf{D}_1 = \begin{pmatrix} 1 \\ 0 \end{pmatrix} \quad \text{and} \quad \mathbf{G}_1 = (1,0) \,.$$

Step 5: And finally, a lower dimensional realization is given by the 1×1 matrices

$$\tilde{\mathbf{A}} = (0), \quad \tilde{\mathbf{B}}(t) = \mathbf{G}_1 \mathbf{G}(t) = (e^{-t}), \quad \text{and} \quad \tilde{\mathbf{C}}(t) = \mathbf{D}(t)\mathbf{D}_1 = (e^t) \,.$$

Hence, the reduced dimensional realization is given as

$$\dot{z} = e^{-t} \cdot u$$

$$y = e^t \cdot z \,.$$

Since the state is single-dimensional, we obtained a minimal realization.

REMARK 8.2   We know from Example 8.2 that system

$$\dot{x}_1 = x_1 + u$$

$$y = x_1$$

is also a realization of the same input–output relation. Therefore, minimal realizations need not to be unique. Note furthermore that the original system was time-invariant; however, the application of the algorithm resulted in a time-variant system. That is, the dimension reduction procedure resulted in a more complicated system structure. We will see, however, in the next section that by using a slight modification of the above algorithm time-invariant nonminimal systems can be reduced to lower dimensional time-invariant realizations. That is, time invariance can be preserved by dimension reductions. ∎

8.1.3   Time-Invariant Realizations

Among the realizations of a given weighting pattern time-invariant realizations have a special role, since the solution and the verification of the properties (e.g., controllability, observability) of such systems are much easier tasks than those in the general case.

THEOREM 8.4
$\mathbf{T}(t, \tau)$ *has a time-invariant realization if and only if*

(i) $\mathbf{T}(t, \tau) = \mathbf{D}(t)\mathbf{G}(\tau)$ *with differentiable* $\mathbf{D}$ *and* $\mathbf{G}$.

(ii) $\mathbf{T}(t, \tau) = \mathbf{T}(t - \tau, 0)$.

PROOF   (a) Assume first that $\mathbf{T}(t, \tau)$ has a time-invariant realization $(\mathbf{A}, \mathbf{B}, \mathbf{C})$. Then

$$\mathbf{T}(t, \tau) = \mathbf{C}e^{\mathbf{A}(t-\tau)}\mathbf{B} = \mathbf{C}e^{\mathbf{A}t} \cdot e^{-\mathbf{A}\tau}\mathbf{B} ;$$

therefore, we may select

$$\mathbf{D}(t) = \mathbf{C}e^{\mathbf{A}t} \quad \text{and} \quad \mathbf{G}(\tau) = e^{-\mathbf{A}\tau}\mathbf{B} ,$$

which are differentiable. Furthermore,

$$\mathbf{T}(t, \tau) = \mathbf{C}e^{\mathbf{A}(t-\tau)}\mathbf{B} = \mathbf{C}e^{\mathbf{A}[(t-\tau)-0]}\mathbf{B} = \mathbf{T}(t - \tau, 0) .$$

(b) Assume next that $\mathbf{T}(t, \tau)$ satisfies Conditions (i) and (ii). Note first that all conditions of Theorem 8.1 are satisfied; therefore, $\mathbf{T}(t, \tau)$ is realizable, and there is a minimal realization. Let the corresponding factorization of $\mathbf{T}(t, \tau)$ be given as

$$\mathbf{T}(t, \tau) = \tilde{\mathbf{D}}(t)\tilde{\mathbf{G}}(\tau) , \tag{8.17}$$

and let $\tilde{\mathbf{M}}(t_0, t_1)$ and $\tilde{\mathbf{W}}(t_0, t_1)$ denote the corresponding controllability and observability Gramians, respectively. From Theorem 8.3 we know that $\tilde{\mathbf{M}}(t_0, t_1)$ and $\tilde{\mathbf{W}}(t_0, t_1)$ are both nonsingular.

Note first that Condition (i) implies that $\mathbf{T}(t, \tau)$ is differentiable with respect to both variables. Therefore, $\tilde{\mathbf{D}}$ and $\tilde{\mathbf{G}}$ are differentiable functions.

Use Condition (ii) to get

$$\mathbf{O} = \frac{\partial}{\partial t}\mathbf{T}(t, \tau) + \frac{\partial}{\partial \tau}\mathbf{T}(t, \tau) = \dot{\tilde{\mathbf{D}}}(t)\tilde{\mathbf{G}}(\tau) + \tilde{\mathbf{D}}(t)\dot{\tilde{\mathbf{G}}}(\tau) .$$

Post-multiply both sides by $\tilde{\mathbf{G}}^T(\tau)$ and integrate the resulting equality on $[t_0, t_1]$ with respect to τ to see that

$$\dot{\tilde{\mathbf{D}}}(t)\tilde{\mathbf{W}}(t_0, t_1) + \tilde{\mathbf{D}}(t)\tilde{\mathbf{W}}_1(t_0, t_1) = \mathbf{O} ,$$

where $\tilde{\mathbf{W}}(t_0, t_1)$ is the controllability Gramian, and

$$\tilde{\mathbf{W}}_1(t_0, t_1) = \int_{t_0}^{t_1} \dot{\tilde{\mathbf{G}}}(\tau)\tilde{\mathbf{G}}^T(\tau)\, d\tau . \tag{8.18}$$

Solve this equation for $\dot{\tilde{\mathbf{D}}}(t)$:

$$\dot{\tilde{\mathbf{D}}}(t) = \tilde{\mathbf{D}}(t)\tilde{\mathbf{A}} \tag{8.19}$$

with

$$\tilde{\mathbf{A}} = -\tilde{\mathbf{W}}_1(t_0, t_1)\tilde{\mathbf{W}}^{-1}(t_0, t_1) . \tag{8.20}$$

By transposing (8.19),

$$\dot{\tilde{\mathbf{D}}}^T(t) = \tilde{\mathbf{A}}^T\tilde{\mathbf{D}}^T(t) ,$$

which implies that

$$\tilde{\mathbf{D}}^T(t) = e^{\tilde{\mathbf{A}}^T t}\tilde{\mathbf{D}}^T(0) ,$$

that is

$$\tilde{\mathbf{D}}(t) = \tilde{\mathbf{D}}(0)e^{\tilde{\mathbf{A}}t} .$$

And finally, use Condition (ii) to show that

$$\mathbf{T}(t, \tau) = \mathbf{T}(t - \tau, 0) = \tilde{\mathbf{D}}(t - \tau)\tilde{\mathbf{G}}(0)$$

$$= \tilde{\mathbf{D}}(0)e^{\tilde{\mathbf{A}}(t - \tau)}\tilde{\mathbf{G}}(0) .$$

Hence, $(\tilde{\mathbf{A}}, \tilde{\mathbf{G}}(0), \tilde{\mathbf{D}}(0))$ is a time-invariant realization of $\mathbf{T}(t, \tau)$.   ∎

REMARK 8.3   Assume that the minimal realization (which was intro-duced in part (b) of the proof) has ν-dimensional state variable. Then $\tilde{\mathbf{D}}$ has ν columns and $\tilde{\mathbf{G}}$ has ν rows; therefore, both matrices $\tilde{\mathbf{W}}$ and $\tilde{\mathbf{W}}_1$ are $\nu \times \nu$, and the same holds for matrix $\tilde{\mathbf{A}}$. Consequently, $(\tilde{\mathbf{A}}, \tilde{\mathbf{G}}(0), \tilde{\mathbf{D}}(0))$ is a time-invariant minimal realization. Hence, part (b) of the proof provides a method for constructing a time-invariant minimal realiza-tion, assuming that a time-variant minimal realization is known. This algorithm can be summarized as follows:

Step 1   Compute $\tilde{\mathbf{D}}(t)$ and $\tilde{\mathbf{G}}(\tau)$ from the given minimal realization.

Step 2   Determine matrices $\tilde{\mathbf{W}}_1(t_0, t_1)$ and $\tilde{\mathbf{W}}(t_0, t_1)$.

Step 3   Find matrix $\tilde{\mathbf{A}}$ by using Equation (8.20), and compute $\tilde{\mathbf{G}}(0)$ and $\tilde{\mathbf{D}}(0)$. Then the time-invariant minimal realization is given by $(\tilde{\mathbf{A}}, \tilde{\mathbf{G}}(0), \tilde{\mathbf{D}}(0))$.   ∎

Example 8.6

In Example 8.5 we saw that

$$\dot{z} = e^{-t} u$$

$$y = e^t z \tag{8.21}$$

is a minimal realization of a two-dimensional time-invariant system. We now illustrate the above algorithm to obtain a time-invariant min-imal realization.

Step 1: For the sake of simplicity, select $t_0 = 0$ and $t_1 = 1$, then

$$\tilde{D}(t) = \tilde{C}(t)\tilde{\phi}(t, t_0) = (e^t) \cdot (1) = (e^t)$$

$$\tilde{G}(\tau) = \tilde{\phi}(t_0, \tau)\tilde{B}(\tau) = (1) \cdot (e^{-\tau}) = (e^{-\tau}).$$

Step 2: Simple calculation shows that

$$\tilde{W}_1(t_0, t_1) = \int_0^1 (-e^{-\tau})(e^{-\tau})\, d\tau = \left(\frac{e^{-2} - 1}{2}\right)$$

and

$$\tilde{W}(t_0, t_1) = \int_0^1 (e^{-\tau})(e^{-\tau})\, d\tau = \left(\frac{1 - e^{-2}}{2}\right).$$

Step 3: Therefore, from (8.20),

$$\tilde{A} = -\left(\frac{e^{-2}-1}{2}\right) \cdot \left(\frac{2}{1-e^{-2}}\right) = (1) \, .$$

Since

$$\tilde{G}(0) = \tilde{D}(0) = 1 \, ,$$

the time-invariant minimal realization is as follows:

$$\dot{x} = x + u$$

$$y = x \, .$$

Note that this result is also known from Example 8.2.

On the basis of Theorems 8.3 and 8.4 the following two-stage process can be proposed to find a time-invariant minimal realization of a given weighting pattern:

Step 1 Find a minimal realization by the repeated application of the algorithm suggested by the proof of Theorem 8.3.

Step 2 Starting from this minimal realization, apply the algorithm of Theorem 8.4 to obtain a time-invariant minimal realization.

Assume next that a nonminimal time-invariant realization is known. The above algorithm can obviously be used for finding a minimal time-invariant realization; however, more simple procedures are available in this special case. In this section two such algorithms are discussed. The first method is a slight modification of the algorithm of Theorem 8.3, and the second one is based on separating the controllable and non-controllable, the observable and nonobservable states as was shown in Chapters 5 and 6.

The first method modifies the algorithm of Theorem 8.3 as follows:

Replace matrices $\mathbf{W}(t_0, t_1)$ and $\mathbf{M}(t_0, t_1)$ by $\mathbf{W}_T = \mathbf{K}\mathbf{K}^T$ and $\mathbf{M}_T = \mathbf{L}^T\mathbf{L}$, respectively, where $\mathbf{K}$ is the controllability matrix and $\mathbf{L}$ is the observability matrix of the given realization. Apply the algorithm of Theorem 8.3 with $\mathbf{W}_T$ and $\mathbf{M}_T$, then it results in a minimal realization of the same form $(\mathbf{O}, \mathbf{G}_1 \cdot \mathbf{G}(t), \mathbf{D}(t) \cdot \mathbf{D}_1)$ as the original algorithm, where in this case

$$\mathbf{G}(t) = e^{-\mathbf{A}t}\mathbf{B} \qquad \text{and} \qquad \mathbf{D}(t) = \mathbf{C}e^{-\mathbf{A}t} \, .$$

In addition, similar to the proof of Theorem 8.4, one can easily show that after the above minimal realization is determined, the time-invariant realization

$$(\mathbf{G}_1\mathbf{A}\mathbf{W}_T\mathbf{G}_1^T(\mathbf{G}_1\mathbf{W}_T\mathbf{G}_1^T)^{-1}, \mathbf{G}_1\mathbf{B}, \mathbf{C}\mathbf{D}_1) \tag{8.22}$$

is also minimal.

Example 8.7

Consider again the system

$$\dot{\mathbf{x}} = \begin{pmatrix} 1 & 0 \\ 0 & 1 \end{pmatrix} \mathbf{x} + \begin{pmatrix} 1 \\ 0 \end{pmatrix} u$$

$$y = (1,0)\mathbf{x}\,.$$

We first apply the modified algorithm to determine a minimal time-variant realization.

Step 1: Since

$$\mathbf{K} = (\mathbf{b}, \mathbf{Ab}) = \begin{pmatrix} 1 & 1 \\ 0 & 0 \end{pmatrix}$$

and

$$\mathbf{L} = \begin{pmatrix} \mathbf{c}^T \\ \mathbf{c}^T\mathbf{A} \end{pmatrix} = \begin{pmatrix} 1 & 0 \\ 1 & 0 \end{pmatrix},$$

we have

$$\mathbf{W}_T = \mathbf{M}_T = \begin{pmatrix} 1 & 1 \\ 0 & 0 \end{pmatrix}\begin{pmatrix} 1 & 0 \\ 1 & 0 \end{pmatrix} = \begin{pmatrix} 2 & 0 \\ 0 & 0 \end{pmatrix}.$$

Step 2: Since

$$\mathbf{W}_T = \mathbf{M}_T = \begin{pmatrix} \sqrt{2} & 0 \\ 0 & 1 \end{pmatrix}\begin{pmatrix} 1 & 0 \\ 0 & 0 \end{pmatrix}\begin{pmatrix} \sqrt{2} & 0 \\ 0 & 1 \end{pmatrix},$$

in decomposition (8.11) we may select

$$\mathbf{P} = \mathbf{Q} = \begin{pmatrix} \sqrt{2} & 0 \\ 0 & 1 \end{pmatrix} \quad \text{and} \quad \mathbf{S}_1 = \mathbf{S}_2 = \begin{pmatrix} 1 & 0 \\ 0 & 0 \end{pmatrix}.$$

Step 3: Therefore,

$$\mathbf{E} = \begin{pmatrix} \frac{1}{\sqrt{2}} & 0 \\ 0 & 1 \end{pmatrix}\begin{pmatrix} 1 & 0 \\ 0 & 0 \end{pmatrix}\begin{pmatrix} \sqrt{2} & 0 \\ 0 & 1 \end{pmatrix}\begin{pmatrix} \sqrt{2} & 0 \\ 0 & 1 \end{pmatrix}\begin{pmatrix} 1 & 0 \\ 0 & 0 \end{pmatrix}\begin{pmatrix} \frac{1}{\sqrt{2}} & 0 \\ 0 & 1 \end{pmatrix}$$

$$= \begin{pmatrix} 1 & 0 \\ 0 & 0 \end{pmatrix}.$$

Step 4: Obviously $rank(\mathbf{E}) = 1$ and

$$\mathbf{E} = \begin{pmatrix} 1 \\ 0 \end{pmatrix} (1,0),$$

and, therefore, we may select

$$\mathbf{D}_1 = \begin{pmatrix} 1 \\ 0 \end{pmatrix} \quad \text{and} \quad \mathbf{G}_1 = (1,0).$$

Step 5: And finally,

$$\mathbf{G}_1 \mathbf{G}(t) = \mathbf{G}_1 e^{-\mathbf{A}t}\mathbf{B} = (1,0)e^{-\mathbf{I}t} \begin{pmatrix} 1 \\ 0 \end{pmatrix} = (e^{-t})$$

and

$$\mathbf{D}(t)\mathbf{D}_1 = \mathbf{C}e^{\mathbf{A}t}\mathbf{D}_1 = (1,0)e^{\mathbf{I}t} \begin{pmatrix} 1 \\ 0 \end{pmatrix} = (e^t).$$

Note that the resulting realization $((0), (e^{-t}), (e^t))$ is the same one that was obtained earlier in Example 8.5.

Next, realization (8.22) is determined. Simple calculation shows that

$$\mathbf{G}_1 \mathbf{A} \mathbf{W}_T \mathbf{G}_1^T (\mathbf{G}_1 \mathbf{W}_T \mathbf{G}_1^T)^{-1}$$

$$= (1,0) \begin{pmatrix} 1 & 0 \\ 0 & 1 \end{pmatrix} \begin{pmatrix} 2 & 0 \\ 0 & 0 \end{pmatrix} \begin{pmatrix} 1 \\ 0 \end{pmatrix} \left[(1,0) \begin{pmatrix} 2 & 0 \\ 0 & 0 \end{pmatrix} \begin{pmatrix} 1 \\ 0 \end{pmatrix} \right]^{-1}$$

$$= (2) \cdot (2)^{-1} = 1,$$

$$\mathbf{G}_1 \mathbf{B} = (1,0) \begin{pmatrix} 1 \\ 0 \end{pmatrix} = (1),$$

and

$$\mathbf{C}\mathbf{D}_1 = (1,0) \begin{pmatrix} 1 \\ 0 \end{pmatrix} = (1).$$

Therefore, realization (8.22) has the form

$$\dot{z} = z + u$$

$$y = z,$$

which is the same result that was obtained in Example 8.2.

Note that the above modified algorithm is much more attractive than the original algorithm where the computation of matrices $\mathbf{W}(t_0, t_1)$, $\mathbf{W}_1(t_0, t_1)$, and $\mathbf{M}(t_0, t_1)$ usually requires the application of numerical integration. That is a difficult task when the elements of the integrands are complicated functions. However, the computation of matrices $\mathbf{W}_T$ and $\mathbf{M}_T$ in the modified algorithm involves only elementary matrix operations.

An alternative approach is based on separating the controllable and noncontrollable, and observable and nonobservable, states. First apply Theorem 5.5 to transform the system to the special form

$$\dot{z} = \bar{\mathbf{A}}\mathbf{z} + \bar{\mathbf{B}}\mathbf{u}$$

$$\mathbf{y} = \bar{\mathbf{C}}\mathbf{z} \,,$$

where

$$\bar{\mathbf{A}} = \begin{pmatrix} \bar{\mathbf{A}}_{11} & \bar{\mathbf{A}}_{12} \\ \mathbf{O} & \bar{\mathbf{A}}_{22} \end{pmatrix}, \qquad \bar{\mathbf{B}} = \begin{pmatrix} \bar{\mathbf{B}}_1 \\ \mathbf{O} \end{pmatrix}, \qquad \bar{\mathbf{C}} = (\bar{\mathbf{C}}_1, \bar{\mathbf{C}}_2) \,,$$

and system $(\bar{\mathbf{A}}_{11}, \bar{\mathbf{B}}_1, \bar{\mathbf{C}}_1)$ is completely controllable and has the same input–output relation as the original system. Apply next Theorem 6.4 to the three nonzero blocks of $\bar{\mathbf{A}}$, then the following decomposition is obtained:

$$\tilde{\mathbf{A}} = \begin{pmatrix} \tilde{\mathbf{A}}_{11} & \mathbf{O} & \tilde{\mathbf{A}}_{13} & \mathbf{O} \\ \tilde{\mathbf{A}}_{21} & \tilde{\mathbf{A}}_{22} & \tilde{\mathbf{A}}_{23} & \tilde{\mathbf{A}}_{24} \\ \mathbf{O} & \mathbf{O} & \tilde{\mathbf{A}}_{33} & \mathbf{O} \\ \mathbf{O} & \mathbf{O} & \tilde{\mathbf{A}}_{43} & \tilde{\mathbf{A}}_{44} \end{pmatrix}, \qquad \tilde{\mathbf{B}} = \begin{pmatrix} \tilde{\mathbf{B}}_1 \\ \tilde{\mathbf{B}}_2 \\ \mathbf{O} \\ \mathbf{O} \end{pmatrix},$$

$$\tilde{\mathbf{C}} = (\tilde{\mathbf{C}}_1, \mathbf{O}, \tilde{\mathbf{C}}_3, \mathbf{O}) \,,$$

where

(i)   System $(\tilde{\mathbf{A}}_{11}, \tilde{\mathbf{B}}_1, \tilde{\mathbf{C}}_1)$ is completely controllable and observable, hence minimal realization.

(ii) The input–output relation of systems $(\mathbf{A}, \mathbf{B}, \mathbf{C})$ and $(\tilde{\mathbf{A}}_{11}, \tilde{\mathbf{B}}_1, \tilde{\mathbf{C}}_1)$ coincide.

(iii) System

$$\left(\begin{pmatrix} \tilde{\mathbf{A}}_{11} & \mathbf{O} \\ \tilde{\mathbf{A}}_{21} & \tilde{\mathbf{A}}_{22} \end{pmatrix}, \begin{pmatrix} \tilde{\mathbf{B}}_1 \\ \tilde{\mathbf{B}}_2 \end{pmatrix}, (\tilde{\mathbf{C}}_1, \mathbf{O}) \right)$$

is completely controllable.

(iv)   System

$$\left(\begin{pmatrix} \tilde{\mathbf{A}}_{11} & \tilde{\mathbf{A}}_{13} \\ \mathbf{O} & \tilde{\mathbf{A}}_{33} \end{pmatrix} , \begin{pmatrix} \tilde{\mathbf{B}}_1 \\ \mathbf{O} \end{pmatrix} , (\tilde{\mathbf{C}}_1, \tilde{\mathbf{C}}_3) \right)$$

is completely observable.

(v)   System $(\tilde{\mathbf{A}}_{44}, \mathbf{O}, \mathbf{O})$ is neither completely controllable nor observable.

We note that the actual values of the possible nonzero blocks are not all unique, but the dimensions of the various blocks are unique. The resulting system $(\tilde{\mathbf{A}}_{11}, \tilde{\mathbf{B}}_1, \tilde{\mathbf{C}}_1)$ is a time-invariant minimal realization.

From Examples 8.2 and 8.5 we know that the minimal realization of a given weighting pattern in not unique. However, time-invariant minimal realizations are equivalent in the sense that they are related by state transformations.

THEOREM 8.5

If systems $(\mathbf{A}, \mathbf{B}, \mathbf{C})$ *and* $(\tilde{\mathbf{A}}, \tilde{\mathbf{B}}, \tilde{\mathbf{C}})$ *are both time-invariant minimal realizations of the same weighting pattern, then there exists nonsingular matrix* $\mathbf{T}$ *such that*

$$\tilde{\mathbf{A}} = \mathbf{T}\mathbf{A}\mathbf{T}^{-1}, \qquad \tilde{\mathbf{B}} = \mathbf{T}\mathbf{B}, \quad and \quad \tilde{\mathbf{C}} = \mathbf{C}\mathbf{T}^{-1} .$$

PROOF   Since both systems are realizations of $\mathbf{T}(t, \tau)$,

$$\mathbf{T}(t, \tau) = \mathbf{C}e^{\mathbf{A}(t-\tau)}\mathbf{B} = \tilde{\mathbf{C}}e^{\tilde{\mathbf{A}}(t-\tau)}\tilde{\mathbf{B}} .$$

Replace t by $s + t$, then equation

$$\mathbf{C}e^{\mathbf{A}(s+t-\tau)}\mathbf{B} = \tilde{\mathbf{C}}e^{\tilde{\mathbf{A}}(s+t-\tau)}\tilde{\mathbf{B}} \tag{8.23}$$

is obtained.

(a) Premultiply the above equation by $e^{\mathbf{A}^T s}\mathbf{C}^T$ and postmultiply it by $\mathbf{B}^T e^{-\mathbf{A}^T \tau}$, and integrate the resulting equation on $[0, t_1]$ with respect to s and τ:

$$\int_0^{t_1} e^{\mathbf{A}^T s}\mathbf{C}^T\mathbf{C}e^{\mathbf{A}s} \, ds \cdot e^{\mathbf{A}t} \cdot \int_0^{t_1} e^{-\mathbf{A}^T \tau}\mathbf{B}\mathbf{B}^T e^{-\mathbf{A}^T \tau} \, d\tau$$

$$= \int_0^{t_1} e^{\mathbf{A}^T s}\mathbf{C}^T\tilde{\mathbf{C}}e^{\tilde{\mathbf{A}}s} \, ds \cdot e^{\tilde{\mathbf{A}}t} \cdot \int_0^{t_1} e^{-\tilde{\mathbf{A}}\tau}\tilde{\mathbf{B}}\mathbf{B}^T e^{-\mathbf{A}^T \tau} \, d\tau .$$

That is,

$$\mathbf{M}(0, t_1) e^{\mathbf{A}t} \mathbf{W}(0, t_1) = \mathbf{M}_0 e^{\tilde{\mathbf{A}}t} \mathbf{W}_0 \,, \tag{8.24}$$

where $\mathbf{M}_0$ and $\mathbf{W}_0$ are the two integrals on the right-hand side.

Since $(\mathbf{A}, \mathbf{B}, \mathbf{C})$ is a minimal realization, both $\mathbf{M}(0, t_1)$ and $\mathbf{W}(0, t_1)$ are nonsingular, and, therefore, Equation (8.24) implies that

$$e^{\mathbf{A}t} = \mathbf{T}_1 e^{\tilde{\mathbf{A}}t} \mathbf{T} \tag{8.25}$$

with

$$\mathbf{T}_1 = \mathbf{M}(0, t_1)^{-1} \cdot \mathbf{M}_0 \qquad \text{and} \qquad \mathbf{T} = \mathbf{W}_0 \cdot \mathbf{W}(0, t_1)^{-1} \,.$$

Substitute $t = 0$ into Equation (8.25) to get

$$\mathbf{I} = \mathbf{T}_1 \cdot \mathbf{I} \cdot \mathbf{T} \,,$$

that is, $\mathbf{T}_1 = \mathbf{T}^{-1}$. Note that from (8.25),

$$e^{\tilde{\mathbf{A}}t} = \mathbf{T} e^{\mathbf{A}t} \mathbf{T}^{-1} = e^{(\mathbf{TAT}^{-1})t} \,.$$

Differentiate both sides and substitute $t = 0$:

$$\tilde{\mathbf{A}} e^{\tilde{\mathbf{A}} \cdot 0} = (\mathbf{TAT}^{-1}) e^{(\mathbf{TAT}^{-1}) \cdot 0} \,,$$

that is,

$$\tilde{\mathbf{A}} = \mathbf{TAT}^{-1} \,.$$

(b) Premultiply Equation (8.23) by $e^{\mathbf{A}^T s} \mathbf{C}^T$ and integrate the resulting equation on $[0, t_1]$ with respect to s; furthermore select $t = \tau = 0$:

$$\int_0^{t_1} e^{\mathbf{A}^T s} \mathbf{C}^T \mathbf{C} e^{\mathbf{A}s} \, ds \mathbf{B} = \int_0^{t_1} e^{\mathbf{A}^T s} \mathbf{C}^T \tilde{\mathbf{C}} e^{\tilde{\mathbf{A}}s} \, ds \tilde{\mathbf{B}} \,,$$

that is,

$$\mathbf{M}(0, t_1) \mathbf{B} = \mathbf{M}_0 \tilde{\mathbf{B}} \,.$$

Since $\mathbf{M}(0, t_1)$ is invertible,

$$\mathbf{B} = \mathbf{M}(0, t_1)^{-1} \mathbf{M}_0 \tilde{\mathbf{B}} = \mathbf{T}_1 \tilde{\mathbf{B}} = \mathbf{T}^{-1} \tilde{\mathbf{B}} \,,$$

that is, $\tilde{\mathbf{B}} = \mathbf{TB}$.

(c) Postmultiply Equation (8.23) by $\mathbf{B}^T e^{-\mathbf{A}^T \tau}$ and integrate the resulting equation on $[0, t_1]$ with respect to τ; furthermore select $t = s = 0$:

$$\mathbf{C} \int_0^{t_1} e^{-\mathbf{A}\tau} \mathbf{B} \mathbf{B}^T e^{-\mathbf{A}^T \tau} \, d\tau = \tilde{\mathbf{C}} \int_0^{t_1} e^{-\tilde{\mathbf{A}}\tau} \tilde{\mathbf{B}} \mathbf{B}^T e^{-\mathbf{A}^T \tau} \, d\tau \, ,$$

that is,

$$\mathbf{C}\mathbf{W}(0, t_1) = \tilde{\mathbf{C}}\mathbf{W}_0 \, .$$

Since $\mathbf{W}(0, t_1)$ is invertible,

$$\mathbf{C} = \tilde{\mathbf{C}}\mathbf{W}_0 \mathbf{W}(0, t_1)^{-1} = \tilde{\mathbf{C}}\mathbf{T} \, ,$$

that is, $\tilde{\mathbf{C}} = \mathbf{C}\mathbf{T}^{-1}$.

Thus, the proof is completed.   ∎

Example 8.8

In the previous example we saw that

$$\dot{z} = z + u$$

$$y = z$$

is a minimal realization. Therefore, all minimal realizations have the form

$$\dot{z} = z + Tu$$

$$y = \frac{1}{T}z \, ,$$

where T is a nonzero constant. Here we use the fact that a 1×1 transformation matrix $\mathbf{T}$ is nonsingular if and only if it is nonzero, and in this case its inverse is the reciprocal.

8.2   *Realizability of Transfer Functions*

In this section, necessary and sufficient conditions will be given for the realizability of a given transfer function. Then, minimal realizations will be discussed. Since transfer functions have identical forms for continuous and discrete systems, all the results of this section apply to both cases.

8.2.1 *Realizability Conditions*

First we note that the transfer function

$$\mathbf{H}(s) = \mathbf{C}(s\mathbf{I} - \mathbf{A})^{-1}\mathbf{B}$$

is a matrix with all elements being strictly proper rational functions of s. This observation follows immediately from the facts that $\det(s\mathbf{I} - \mathbf{A})$ is the $(-1)^n$-multiple of the nth degree characteristic polynomial of $\mathbf{A}$, all of its subdeterminants are polynomials of degree less than n, and the elements of $(s\mathbf{I} - \mathbf{A})^{-1}$ are the ratios of these subdeterminants and $\det(s\mathbf{I} - \mathbf{A})$. We will first verify that this property is also sufficient for a rational matrix function to be the transfer function of a continuous (or discrete) linear system.

THEOREM 8.6
Let $\mathbf{H}(s)$ be a matrix with each element being a rational function. Then there exists a linear system with transfer function $\mathbf{H}(s)$ if and only if all elements of $\mathbf{H}(s)$ are strictly proper.

PROOF   (a) The necessary part has been shown above before formulating this theorem.
   (b) The sufficiency part will be proven by constructing a particular realization of the transfer function. In the systems theory literature, two particular constructions have special importance. They are presented below.

   Method 1. Assume that

$$p(s) = s^r + p_{r-1}s^{r-1} + \cdots + p_1 s + p_0$$

is the least common multiple of the denominators of the elements of $\mathbf{H}(s)$. Then all elements of $p(s)\mathbf{H}(s)$ are polynomials:

$$p(s)\mathbf{H}(s) = \mathbf{H}_0 + \mathbf{H}_1 s + \cdots + \mathbf{H}_{r-1}s^{r-1} , \qquad (8.26)$$

where $\mathbf{H}_0, \mathbf{H}_1, \ldots, \mathbf{H}_{r-1}$ are constant matrices.
   Define

$$
\mathbf{A}_C =
\begin{pmatrix}
\mathbf{O} & \mathbf{I} & \mathbf{O} & \cdots & \mathbf{O} \\
\mathbf{O} & \mathbf{O} & \mathbf{I} & \cdots & \mathbf{O} \\
\vdots & \vdots & \vdots & \ddots & \vdots \\
\mathbf{O} & \mathbf{O} & \mathbf{O} & \cdots & \mathbf{I} \\
-p_0\mathbf{I} & -p_1\mathbf{I} & -p_2\mathbf{I} & \cdots & -p_{r-1}\mathbf{I}
\end{pmatrix}
, \qquad
\mathbf{B}_C =
\begin{pmatrix}
\mathbf{O} \\
\mathbf{O} \\
\vdots \\
\mathbf{O} \\
\mathbf{I}
\end{pmatrix}
,
$$

and
$$C_C = (\mathbf{H}_0, \mathbf{H}_1, \ldots, \mathbf{H}_{r-1}) ,$$

where each block of $\mathbf{A}_C$ and $\mathbf{B}_C$ is $m \times m$ (m being the dimension of the input). We will prove that $(\mathbf{A}_C, \mathbf{B}_C, \mathbf{C}_C)$ is a realization of $\mathbf{H}(s)$. That is, we will verify that

$$\mathbf{C}_C(\mathbf{I}s - \mathbf{A}_C)^{-1}\mathbf{B}_C = \mathbf{H}(s) .$$

Note first that $(\mathbf{I}s - \mathbf{A}_C)^{-1}\mathbf{B}_C$ is the solution of the equation

$$(\mathbf{I}s - \mathbf{A}_C)\mathbf{X} = \mathbf{B}_C ,$$

which can be written as

$$
\begin{pmatrix} s\mathbf{X}_1 \\ s\mathbf{X}_2 \\ \vdots \\ s\mathbf{X}_{r-1} \\ s\mathbf{X}_r \end{pmatrix} - \begin{pmatrix} \mathbf{O} & \mathbf{I} & \mathbf{O} & \cdots & \mathbf{O} \\ \mathbf{O} & \mathbf{O} & \mathbf{I} & \cdots & \mathbf{O} \\ \vdots & \vdots & \vdots & \ddots & \vdots \\ \mathbf{O} & \mathbf{O} & \mathbf{O} & \cdots & \mathbf{I} \\ -p_0\mathbf{I} & -p_1\mathbf{I} & -p_2\mathbf{I} & \cdots & -p_{r-1}\mathbf{I} \end{pmatrix} \begin{pmatrix} \mathbf{X}_1 \\ \mathbf{X}_2 \\ \vdots \\ \mathbf{X}_{r-1} \\ \mathbf{X}_r \end{pmatrix} = \begin{pmatrix} \mathbf{O} \\ \mathbf{O} \\ \vdots \\ \mathbf{O} \\ \mathbf{I} \end{pmatrix} .
$$

From this equality we conclude that

$$s\mathbf{X}_i - \mathbf{X}_{i+1} = \mathbf{O} \qquad (i = 1, 2, \ldots, r-1) \tag{8.27}$$

and

$$s\mathbf{X}_r + p_0\mathbf{X}_1 + p_1\mathbf{X}_2 + \cdots + p_{r-1}\mathbf{X}_r = \mathbf{I} . \tag{8.28}$$

From (8.27),

$$\mathbf{X}_2 = s\mathbf{X}_1, \mathbf{X}_3 = s\mathbf{X}_2 = s^2\mathbf{X}_1, \ldots, \mathbf{X}_r = s\mathbf{X}_{r-1} = s^{r-1}\mathbf{X}_1 .$$

Now substitute these relations into (8.28) to see that

$$p(s)\mathbf{X}_1 = \mathbf{I} .$$

Therefore,

$$\mathbf{X}_1 = \frac{1}{p(s)}\mathbf{I}, \mathbf{X}_2 = \frac{s}{p(s)}\mathbf{I}, \ldots, \mathbf{X}_r = \frac{s^{r-1}}{p(s)}\mathbf{I} .$$

Hence,

$$\mathbf{C}_C(s\mathbf{I} - \mathbf{A}_C)^{-1}\mathbf{B}_C = \mathbf{C}_C\mathbf{X} = (\mathbf{H}_0, \mathbf{H}_1, \ldots, \mathbf{H}_{r-1})\frac{1}{p(s)}\begin{pmatrix} \mathbf{I} \\ s\mathbf{I} \\ \vdots \\ s^{r-1}\mathbf{I} \end{pmatrix}$$

$$= \frac{1}{p(s)}(\mathbf{H}_0 + \mathbf{H}_1 s + \cdots + \mathbf{H}_{r-1}s^{r-1})$$

$$= \frac{1}{p(s)}p(s)\mathbf{H}(s) = \mathbf{H}(s),$$

which completes the proof.

Method 2. Expand $\mathbf{H}(s)$ about $|s| = \infty$ to get

$$\mathbf{H}(s) = \mathbf{L}_0 s^{-1} + \mathbf{L}_1 s^{-2} + \mathbf{L}_2 s^{-3} + \cdots .$$

Define $p(s)$ as before; furthermore, let

$$\mathbf{A}_O = \begin{pmatrix} \mathbf{O} & \mathbf{I} & \mathbf{O} & \cdots & \mathbf{O} \\ \mathbf{O} & \mathbf{O} & \mathbf{I} & \cdots & \mathbf{O} \\ \vdots & \vdots & \vdots & \ddots & \vdots \\ \mathbf{O} & \mathbf{O} & \mathbf{O} & \cdots & \mathbf{I} \\ -p_0\mathbf{I} & -p_1\mathbf{I} & -p_2\mathbf{I} & \cdots & -p_{r-1}\mathbf{I} \end{pmatrix}, \qquad \mathbf{B}_O = \begin{pmatrix} \mathbf{L}_0 \\ \mathbf{L}_1 \\ \vdots \\ \mathbf{L}_{r-2} \\ \mathbf{L}_{r-1} \end{pmatrix},$$

and

$$\mathbf{C}_O = (\mathbf{I}, \mathbf{O}, \ldots, \mathbf{O}, \mathbf{O}),$$

where each block of $\mathbf{A}_O$ and $\mathbf{C}_O$ is $p \times p$ (p being the dimension of the output). We will now prove that $(\mathbf{A}_O, \mathbf{B}_O, \mathbf{C}_O)$ is also a realization of $\mathbf{H}(s)$, that is,

$$\mathbf{H}(s) = \mathbf{C}_O(s\mathbf{I} - \mathbf{A}_O)^{-1}\mathbf{B}_O.$$

Note that Example 1.25 implies that

$$\mathbf{C}_O(s\mathbf{I} - \mathbf{A}_O)^{-1}\mathbf{B}_O = \frac{1}{s}\mathbf{C}_O\left(\mathbf{I} - \frac{1}{s}\mathbf{A}_O\right)^{-1}\mathbf{B}_O$$

$$= \frac{1}{s}\mathbf{C}_O\left(\mathbf{I} + \frac{1}{s}\mathbf{A}_O + \frac{1}{s^2}\mathbf{A}_O^2 + \cdots\right)\mathbf{B}_O$$

$$= \mathbf{C}_O \mathbf{B}_O s^{-1} + \mathbf{C}_O \mathbf{A}_O \mathbf{B}_O s^{-2} + \mathbf{C}_O \mathbf{A}_O^2 \mathbf{B}_O s^{-3} + \cdots \; ;$$

therefore, for proving the identity of the infinite series, it is sufficient to verify that for $k = 0, 1, 2, \ldots,$

$$\mathbf{L}_k = \mathbf{C}_O \mathbf{A}_O^k \mathbf{B}_O \; . \tag{8.29}$$

As a first step, observe that

$$\mathbf{C}_O = (\mathbf{I}, \mathbf{O}, \ldots, \mathbf{O}, \mathbf{O}) \; ,$$

$$\mathbf{C}_O \mathbf{A}_O = (\mathbf{O}, \mathbf{I}, \ldots, \mathbf{O}, \mathbf{O}),$$

$$\vdots$$

$$\mathbf{C}_O \mathbf{A}_O^{r-1} = (\mathbf{O}, \mathbf{O}, \ldots, \mathbf{O}, \mathbf{I}) \; ,$$

which imply the relations

$$\mathbf{C}_O \mathbf{B}_O = \mathbf{L}_0, \mathbf{C}_O \mathbf{A}_O \mathbf{B}_O = \mathbf{L}_1, \ldots, \mathbf{C}_O \mathbf{A}_O^{r-1} \mathbf{B}_O = \mathbf{L}_{r-1} \; .$$

That is, (8.29) holds for $k = 0, 1, 2, \ldots, r - 1$. Hence we have to prove that (8.29) also holds for larger values of k. It is sufficient to show that matrices $\mathbf{L}_k$ and $\mathbf{C}_O \mathbf{A}_O^k \mathbf{B}_O$ satisfy the same recursive relations, since they are equal for $k \leq r - 1$. Introduce next the notation $\mathbf{M}_k = \mathbf{C}_O \mathbf{A}_O^k \mathbf{B}_O$. Consider first the polynomial

$$p(s)\mathbf{H}(s) = (s^r + p_{r-1}s^{r-1} + \cdots + p_1 s + p_0)(\mathbf{L}_0 s^{-1} + \mathbf{L}_1 s^{-2} + \mathbf{L}_2 s^{-3} + \cdots).$$

The coefficient of $s^{-(k+1)}$ is zero for all $k \geq 0$. That is,

$$p_0 \mathbf{L}_k + p_1 \mathbf{L}_{k+1} + \cdots + p_{r-1} \mathbf{L}_{k+r-1} + \mathbf{L}_{k+r} = \mathbf{O} \; ,$$

which gives the recursion

$$\mathbf{L}_{k+r} = -p_0 \mathbf{L}_k - p_1 \mathbf{L}_{k+1} - \cdots - p_{r-1} \mathbf{L}_{k+r-1} \; . \tag{8.30}$$

Note next that from Section 7.2 we know that $p(s)$ is the characteristic polynomial of matrix

$$
\begin{pmatrix}
0 & 1 & 0 & \cdots & 0 \\
0 & 0 & 1 & \cdots & 0 \\
\vdots & \vdots & \vdots & \ddots & \vdots \\
0 & 0 & 0 & \cdots & 1 \\
-p_0 & -p_1 & -p_2 & \cdots & -p_{r-1}
\end{pmatrix}.
$$

It is also known from matrix calculus that the elementary operations of block matrices with commutative blocks are performed with the blocks in exactly the same way as they are performed with matrices having scalar elements. Therefore, $p(\mathbf{A}_O) = \mathbf{O}$, that is,

$$
\mathbf{A}_O^r = -p_0 \mathbf{I} - p_1 \mathbf{A}_O - \cdots - p_{r-1} \mathbf{A}_O^{r-1} .
$$

Premultiply this equality by $\mathbf{C}_O \mathbf{A}_O^k$ and postmultiply the resulting equation by $\mathbf{B}_O$ to get

$$
\mathbf{C}_O \mathbf{A}_O^{k+r} \mathbf{B}_O = -p_0 \mathbf{C}_O \mathbf{A}_O^k \mathbf{B}_O - p_1 \mathbf{C}_O \mathbf{A}_O^{k+1} \mathbf{B}_O - \cdots - p_{r-1} \mathbf{C}_O \mathbf{A}_O^{k+r-1} \mathbf{B}_O ,
$$

that is,

$$
\mathbf{M}_{k+r} = -p_0 \mathbf{M}_k - p_1 \mathbf{M}_{k+1} - \cdots - p_{r-1} \mathbf{M}_{k+r-1} .
$$

Since this recursion coincides with (8.30), the proof is completed.   ∎

The algorithm suggested by Method 1 can be summarized as follows:

Step 1   Find the least common multiple $p(s)$ of the denominators of the elements of $\mathbf{H}(s)$.

Step 2   Compute matrix polynomial $p(s)\mathbf{H}(s)$ to get matrices $\mathbf{H}_0$, $\mathbf{H}_1$, ..., $\mathbf{H}_{r-1}$.

Step 3   Determine $\mathbf{A}_C$, $\mathbf{B}_C$, and $\mathbf{C}_C$.

The algorithm suggested by Method 2 is summarized next:

Step 1   Find the least common multiple $p(s)$ of the denominators of the elements of $\mathbf{H}(s)$.

Step 2   Expand $\mathbf{H}(s)$ about $|s| = \infty$ to get matrices $\mathbf{L}_0$, $\mathbf{L}_1$, ..., $\mathbf{L}_{r-1}$.

Step 3   Determine $\mathbf{A}_O$, $\mathbf{B}_O$, and $\mathbf{C}_O$.

COROLLARY 8.4

By using duality, two more representations of $\mathbf{H}(s)$ *can be obtained. Use Methods 1 and 2 for the dual-transfer function*

$$\mathbf{H}_D(s) = \mathbf{B}^T(s\mathbf{I} - \mathbf{A}^T)^{-1}\mathbf{C}^T \, ,$$

and determine the duals of the resulting representations. This idea leads to representations $(\tilde{\mathbf{A}}_O, \tilde{\mathbf{B}}_O, \tilde{\mathbf{C}}_O)$ *and* $(\tilde{\mathbf{A}}_C, \tilde{\mathbf{B}}_C, \tilde{\mathbf{C}}_C)$, *where*

$$\tilde{\mathbf{A}}_O = \begin{pmatrix} \mathbf{O} & \mathbf{O} & \cdots & \mathbf{O} & -p_0\mathbf{I} \\ \mathbf{I} & \mathbf{O} & \cdots & \mathbf{O} & -p_1\mathbf{I} \\ \mathbf{O} & \mathbf{I} & \cdots & \mathbf{O} & -p_2\mathbf{I} \\ \vdots & \vdots & \ddots & \vdots & \vdots \\ \mathbf{O} & \mathbf{O} & \cdots & \mathbf{I} & -p_{r-1}\mathbf{I} \end{pmatrix} \, , \qquad \tilde{\mathbf{B}}_O = \begin{pmatrix} \mathbf{H}_0^T \\ \mathbf{H}_1^T \\ \mathbf{H}_2^T \\ \vdots \\ \mathbf{H}_{r-1}^T \end{pmatrix} \, ,$$

and

$$\tilde{\mathbf{C}}_O = (\mathbf{O}, \mathbf{O}, \ldots, \mathbf{O}, \mathbf{I})$$

with $p \times p$ *blocks in* $\tilde{\mathbf{A}}_O$ *and* $\tilde{\mathbf{C}}_O$; *furthermore,*

$$\tilde{\mathbf{A}}_C = \begin{pmatrix} \mathbf{O} & \mathbf{O} & \cdots & \mathbf{O} & -p_0\mathbf{I} \\ \mathbf{I} & \mathbf{O} & \cdots & \mathbf{O} & -p_1\mathbf{I} \\ \mathbf{O} & \mathbf{I} & \cdots & \mathbf{O} & -p_2\mathbf{I} \\ \vdots & \vdots & \ddots & \vdots & \vdots \\ \mathbf{O} & \mathbf{O} & \cdots & \mathbf{I} & -p_{r-1}\mathbf{I} \end{pmatrix} \, , \qquad \tilde{\mathbf{B}}_C = \begin{pmatrix} \mathbf{I} \\ \mathbf{O} \\ \mathbf{O} \\ \vdots \\ \mathbf{O} \end{pmatrix} \, ,$$

and

$$\tilde{\mathbf{C}}_C = (\mathbf{L}_0^T, \mathbf{L}_1^T, \ldots, \mathbf{L}_{r-2}^T, \mathbf{L}_{r-1}^T)$$

with $m \times m$ *blocks in* $\tilde{\mathbf{A}}_C$ *and* $\tilde{\mathbf{B}}_C$.

Note that representations $(\mathbf{A}_C, \mathbf{B}_C, \mathbf{C}_C)$ and $(\tilde{\mathbf{A}}_C, \tilde{\mathbf{B}}_C, \tilde{\mathbf{C}}_C)$ are completely controllable and $(\mathbf{A}_O, \mathbf{B}_O, \mathbf{C}_O)$ and $(\tilde{\mathbf{A}}_O, \tilde{\mathbf{B}}_O, \tilde{\mathbf{C}}_O)$ are completely observable. Therefore, they are called the *standard controllable* and *standard observable* realizations, respectively. Observe furthermore that these realizations are analogous to the controllability and observability canonical forms discussed earlier in Sections 7.2 and 7.3.

The above methods are illustrated next.

Example 8.9

Consider the transfer function

$$\mathbf{H}(s) = \begin{pmatrix} \frac{1}{s} & \frac{1}{s-1} \\ \frac{1}{s-1} & \frac{1}{s} \end{pmatrix}$$

where the input and output are two-dimensional. That is, $m = p = 2$.
   Step 1: Observe that

$$p(s) = s(s-1) = s^2 - s \,.$$

That is, $r = 2$, $p_1 = -1$ and $p_0 = 0$.
   Step 2, Method 1: Simple calculation shows that

$$p(s)\mathbf{H}(s) = \begin{pmatrix} s-1 & s \\ s & s-1 \end{pmatrix} = s \begin{pmatrix} 1 & 1 \\ 1 & 1 \end{pmatrix} + \begin{pmatrix} -1 & 0 \\ 0 & -1 \end{pmatrix} \,,$$

that is,

$$\mathbf{H}_1 = \begin{pmatrix} 1 & 1 \\ 1 & 1 \end{pmatrix} \quad \text{and} \quad \mathbf{H}_0 = \begin{pmatrix} -1 & 0 \\ 0 & -1 \end{pmatrix} \,.$$

   Step 3: Hence,

$$\mathbf{A}_C = \left(\begin{array}{cccc} 0 & 0 & \vdots & 1 & 0 \\ 0 & 0 & \vdots & 0 & 1 \\ \cdots\cdots & \cdots\cdots \\ 0 & 0 & \vdots & 1 & 0 \\ 0 & 0 & \vdots & 0 & 1 \end{array} \right) , \quad \mathbf{B}_C = \left(\begin{array}{cc} 0 & 0 \\ 0 & 0 \\ \cdots\cdots \\ 1 & 0 \\ 0 & 1 \end{array} \right) ,$$

and

$$\mathbf{C}_C = \left(\begin{array}{cc:cc} -1 & 0 & 1 & 1 \\ 0 & -1 & 1 & 1 \end{array} \right) \,.$$

   Step 2, Method 2: Note first that

$$\mathbf{H}(s) = \frac{1}{s} \begin{pmatrix} 1 & 0 \\ 0 & 1 \end{pmatrix} + \frac{1}{s-1} \begin{pmatrix} 0 & 1 \\ 1 & 0 \end{pmatrix}$$

$$= \frac{1}{s} \begin{pmatrix} 1 & 0 \\ 0 & 1 \end{pmatrix} + \begin{pmatrix} 0 & 1 \\ 1 & 0 \end{pmatrix} \cdot \frac{\frac{1}{s}}{1 - \frac{1}{s}}$$

$$= \frac{1}{s} \begin{pmatrix} 1 & 0 \\ 0 & 1 \end{pmatrix} + \begin{pmatrix} 0 & 1 \\ 1 & 0 \end{pmatrix} \left(\frac{1}{s} + \frac{1}{s^2} + \frac{1}{s^3} + \cdots \right)$$

$$= \frac{1}{s} \begin{pmatrix} 1 & 1 \\ 1 & 1 \end{pmatrix} + \begin{pmatrix} 0 & 1 \\ 1 & 0 \end{pmatrix} \left(\frac{1}{s^2} + \frac{1}{s^3} + \cdots \right) .$$

Therefore,

$$\mathbf{L}_0 = \begin{pmatrix} 1 & 1 \\ 1 & 1 \end{pmatrix} , \qquad \mathbf{L}_1 = \mathbf{L}_2 = \cdots = \begin{pmatrix} 0 & 1 \\ 1 & 0 \end{pmatrix} .$$

Step 3: Hence,

$$\mathbf{A}_O = \left(\begin{array}{cccccc} 0 & 0 & \vdots & 1 & 0 \\ 0 & 0 & \vdots & 0 & 1 \\ \cdots & \cdots & \cdots & \cdots & \cdots \\ 0 & 0 & \vdots & 1 & 0 \\ 0 & 0 & \vdots & 0 & 1 \end{array} \right) , \qquad \mathbf{B}_O = \left(\begin{array}{cc} 1 & 1 \\ 1 & 1 \\ \cdots & \cdots \\ 0 & 1 \\ 1 & 0 \end{array} \right)$$

and

$$\mathbf{C}_O = \begin{pmatrix} 1 & 0 & \vdots & 0 & 0 \\ 0 & 1 & \vdots & 0 & 0 \end{pmatrix} .$$

Hence, the standard controllable and observable realizations are determined.

8.2.2 *Minimal Realizations*

Assume that the $(\mathbf{A}, \mathbf{B}, \mathbf{C})$-system is a realization of a given transfer function $\mathbf{H}(s)$. Then from the previous section we know that

$$\mathbf{H}(s) = \mathbf{C}(s\mathbf{I} - \mathbf{A})^{-1}\mathbf{B} = \frac{1}{\varphi(s)}\mathbf{P}(s) , \qquad (8.31)$$

where $\varphi(s)$ is the characteristic polynomial of $\mathbf{A}$, and each element of $\mathbf{P}(s)$ is a polynomial of degree less than n, where matrix $\mathbf{A}$ is assumed to be $n \times n$.

First a sufficient condition is presented for the minimality of a given realization $(\mathbf{A}, \mathbf{B}, \mathbf{C})$.

THEOREM 8.7
*Assume that there is no polynomial of degree at least one which is a common
factor of $\varphi(s)$ and all elements of $\mathbf{P}(s)$. Then realization $(\mathbf{A}, \mathbf{B}, \mathbf{C})$ is minimal.*

PROOF   Assume that realization $(\mathbf{A}, \mathbf{B}, \mathbf{C})$ is not minimal. Then there
is a realization $(\tilde{\mathbf{A}}, \tilde{\mathbf{B}}, \tilde{\mathbf{C}})$ with smaller dimension in the state variable.
Then

$$\frac{1}{\varphi(s)}\mathbf{P}(s) = \frac{1}{\tilde{\varphi}(s)}\tilde{\mathbf{P}}(s) \, ,$$

where $\tilde{\varphi}(s)$ is the characteristic polynomial of $\tilde{\mathbf{A}}$, and, therefore, the
degree of $\tilde{\varphi}$ is less than that of φ. Hence, there must be a cancellation in
the numerator and denominator of the left-hand side.   ∎

REMARK 8.4   The conditions of the theorem are not necessary in
general for the minimality of a given realization as is illustrated in the
following example.   ∎

Example 8.10

Consider realization $(\mathbf{A}, \mathbf{B}, \mathbf{C})$ with

$$\mathbf{A} = \mathbf{B} = \mathbf{C} = \begin{pmatrix} 1 & 0 \\ 0 & 1 \end{pmatrix} .$$

Since the controllability and observability matrices are

$$\mathbf{K} = (\mathbf{B}, \mathbf{A}\mathbf{B}) = \begin{pmatrix} 1 & 0 \vdots 1 & 0 \\ 0 & 1 \vdots 0 & 1 \end{pmatrix}$$

and

$$\mathbf{L} = \begin{pmatrix} \mathbf{C} \\ \mathbf{C}\mathbf{A} \end{pmatrix} = \mathbf{K}^T$$

with $rank(\mathbf{K}) = rank(\mathbf{L}) = 2$, the realization is completely controllable and completely observable. Therefore, it is minimal. However,

$$\varphi(s) = (s - 1)^2 \, ,$$

and

$$\mathbf{C}(s\mathbf{I} - \mathbf{A})^{-1}\mathbf{B} = \mathbf{I}[(s - 1)\mathbf{I}]^{-1}\mathbf{I} = \frac{1}{s - 1}\mathbf{I} \, ,$$

which implies that there must be cancellation in the numerator and
denominator of fraction (8.31).

We will next prove that if there is either a single input or a single output, then the conditions of Theorem 8.7 are necessary.

THEOREM 8.8

Assume that realization $(\mathbf{A}, \mathbf{B}, \mathbf{C})$ *is minimal and has either a single input or a single output or both. Then there is no common factor with degree at least one of* $\varphi(s)$ *and all elements of* $\mathbf{P}(s)$.

PROOF   Assume that there is a cancellation, then by using standard controllable (if $\dim(\mathbf{u}) = 1$) or standard observable (if $\dim(\mathbf{y}) = 1$) realization we can obtain a smaller dimensional realization than $(\mathbf{A}, \mathbf{B}, \mathbf{C})$.
∎

Assume that either the input, the output, or both are single, and $(\mathbf{A}, \mathbf{B}, \mathbf{C})$ is a realization of a given transfer function. An algorithm to find a minimal realization consists of the following steps:

Step 1   Cancel (if necessary) all common factors of $\varphi(s)$ and all elements of $\mathbf{P}(s)$.

Step 2   Find the standard controllable realization (if $\dim(\mathbf{u}) = 1$) or the standard observable realization (if $\dim(\mathbf{y}) = 1$).

Example 8.11

Consider again the system

$$\dot{\mathbf{x}} = \begin{pmatrix} 1 & 0 \\ 0 & 1 \end{pmatrix} \mathbf{x} + \begin{pmatrix} 1 \\ 0 \end{pmatrix} u\,,$$

$$y = (1,0)\mathbf{x}\,,$$

which was earlier examined in Examples 8.5 and 8.7. The transfer function has the form

$$\mathbf{H}(s) = (1,0) \begin{pmatrix} s-1 & 0 \\ 0 & s-1 \end{pmatrix}^{-1} \begin{pmatrix} 1 \\ 0 \end{pmatrix} = \left(\frac{1}{s-1} \right).$$

Since there is no common factor in the denominator and numerator, Step 1 of the algorithm is omitted.
     Step 2: Since $p(s) = s - 1$, $r = 1$ and $p_0 = -1$. Therefore, $p(s)\mathbf{H}(s) = (1)$, and $\mathbf{H}_0 = (1)$. The fact that $r = m = 1$ implies that each of matrices $\mathbf{A}_C$, $\mathbf{B}_C$, and $\mathbf{C}_C$ has only one 1×1 block:

$$\mathbf{A}_C = (1), \qquad \mathbf{B}_C = (1), \quad and \quad \mathbf{C}_C = (1)\,.$$

Hence, the resulting minimal realization

$$\dot{z} = z + u$$

$$y = z$$

coincides with our earlier results.

8.3 *Applications*

This section presents some real-life applications of the realization theory. As in the previous chapters, engineering examples are first introduced. Case studies from the social sciences and economics are presented in the second subsection.

8.3.1 *Dynamic Systems in Engineering*

1. In Chapters 5 and 6 we found that *the harmonic motion system* is completely controllable and observable; therefore, it is minimal.

2. The *second-order mechanical system* is also completely controllable and completely observable, so it is also minimal.

3(a). The simple second-order *electrical system* is completely controllable and observable, so it is minimal.

3(b). In Chapter 5 we found that if $L_1 = L_2$, $C_1 = C_2$, and $R_1 = R_2$, then the fourth-order system was not observable; therefore, it is not minimal. The input–output behavior of this special system is described by the last two equations:

$$\begin{pmatrix} \dot{i}_{L_2} \\ \dot{v}_{C_2} \end{pmatrix} = \begin{pmatrix} 0 & -\frac{1}{L_2} \\ \frac{1}{C_2} & -\frac{1}{C_2 R_2} \end{pmatrix} \begin{pmatrix} i_{L_2} \\ v_{C_2} \end{pmatrix} + \begin{pmatrix} \frac{1}{L_2} \\ \frac{1}{C_2 R_2} \end{pmatrix} u$$

and

$$y = (0, 1) \begin{pmatrix} i_{L_2} \\ v_{C_2} \end{pmatrix} .$$

We can see that this input–output behavior is not affected by L_1, C_1, and R_1. Therefore, these elements can be removed from the circuit without affecting the input–output behavior.

Now let us ask if our reduced second-order system given above is minimal. We can compute the controllability matrix:

$$\mathbf{K} = \begin{pmatrix} \frac{1}{L_2} & -\frac{1}{L_2 C_2 R_2} \\ \frac{1}{C_2 R_2} & \frac{1}{L_2 C_2} - \frac{1}{C_2^2 R_2^2} \end{pmatrix} .$$

This matrix is of full rank:

$$\det(\mathbf{K}) = \frac{1}{C_2 L_2^2} \neq 0 \ .$$

Therefore, it is completely controllable. Now let us compute the observability matrix

$$\mathbf{L} = \begin{pmatrix} 0 & 1 \\ \frac{1}{C_2} & -\frac{1}{C_2 R_2} \end{pmatrix} \ .$$

This matrix is also of full rank, therefore, this reduced second-order system is minimal.

4(a). With $\mathbf{c}_1^T$ the *transistor circuit* is controllable and observable, therefore, it is minimal.

4(b). With $\mathbf{c}_2^T$, we have

$$\mathbf{A} = \begin{pmatrix} \alpha & 0 \\ \beta & 0 \end{pmatrix} \ , \qquad \mathbf{b} = \begin{pmatrix} \frac{1}{L} \\ 0 \end{pmatrix} \ , \qquad \mathbf{c}_2^T = (1, 0) \ ,$$

with $\alpha = -h_{ie}/L$ and $\beta = h_{fe}/C$.

The input–output behavior is completely described by the first equation and output relation:

$$\dot{x}_1 = \alpha x_1 + \frac{1}{L} u$$

$$y = x_1 \ .$$

Therefore, the right half of the circuit is irrelevant, and the system is not minimal.

5(a). The simple two-tank *hydraulic system* is observable and controllable, so it is minimal.

5(b). The three-tank system with only input u_2 and output x_2 was described by equations

$$\dot{\mathbf{x}} = \begin{pmatrix} -3 & 3 & 0 \\ 2 & -4 & 2 \\ 0 & 3 & -3 \end{pmatrix} \mathbf{x} + \begin{pmatrix} 0 \\ 1 \\ 0 \end{pmatrix} u$$

$$y = (0, 1, 0)\mathbf{x} \ .$$

This system is neither controllable or observable. Therefore, it is not minimal and we must be able to find a reduced order system that has the same input–output behavior. Let us write

$$\begin{aligned}
\dot{x}_1 &= -3x_1 +3x_2 \\
\dot{x}_2 &= 2x_1 \quad -4x_2 +2x_3 +u \\
\dot{x}_3 &= 3x_2 \qquad\quad -3x_3
\end{aligned}$$

and

$$y = x_2 .$$

Now define

$$x_4 = x_1 + x_3$$

and then adding the first and third equations above we will get

$$\begin{aligned}
\dot{x}_4 &= -3x_4 +6x_2 \\
\dot{x}_2 &= 2x_4 \ -4x_2 +u
\end{aligned}$$

and

$$y = x_2 .$$

This is a second-order system with

$$\mathbf{A} = \begin{pmatrix} -3 & 6 \\ 2 & -4 \end{pmatrix}, \qquad \mathbf{b} = \begin{pmatrix} 0 \\ 1 \end{pmatrix}, \qquad \mathbf{c}^T = (0,1) .$$

This system is minimal, which is proven as follows. Note first that

$$\mathbf{K} = \begin{pmatrix} 0 & 6 \\ 1 & -4 \end{pmatrix},$$

which has full rank. Therefore, the system is completely controllable. Furthermore,

$$\mathbf{L} = \begin{pmatrix} 0 & 1 \\ 2 & -4 \end{pmatrix},$$

which has also full rank. Therefore, the system is completely observable. Because it is completely controllable and completely observable, it is minimal.

   Now what is the physical significance of this? It means that if the only input is into the middle tank and the only output is from the middle tank, then the two side tanks can be combined into one tank without affecting the input–output behavior of the system.

   6. The *multiple input electronic system* is controllable and observable (see Applications 5.3.1-6 and 6.4.1-6), therefore, the system is minimal.

   7. The *single stick-balancing problem* is controllable and observable, so it is minimal.

8. The *cart with two inverted pendulums* is controllable if $L_1 \neq L_2$ and is always observable. Therefore, our realization is minimal. So let us now find out what happens if the lengths of the two pendulums are equal. Assume $L_1 = L_2$, and let

$$a_1 = a_4 = \alpha$$

$$a_2 = a_3 = \beta, \text{ with } \alpha \neq \beta,$$

and let $\gamma = -1/ML_1$. Then

$$\mathbf{A} = \begin{pmatrix} 0 & 0 & 1 & 0 \\ 0 & 0 & 0 & 1 \\ \alpha & \beta & 0 & 0 \\ \beta & \alpha & 0 & 0 \end{pmatrix}, \qquad \mathbf{b} = \begin{pmatrix} 0 \\ 0 \\ \gamma \\ \gamma \end{pmatrix},$$

and

$$\mathbf{c}^T = (1, 0, 0, 0) .$$

Find a minimal realization:

Step 1: Using $\mathbf{H}(s) = \mathbf{c}^T(s\mathbf{I} - \mathbf{A})^{-1}\mathbf{b}$, find the transfer function:

$$\mathbf{H}(s) = \begin{pmatrix} 1 & 0 & 0 & 0 \end{pmatrix} \begin{pmatrix} s & 0 & -1 & 0 \\ 0 & s & 0 & -1 \\ -\alpha & -\beta & s & 0 \\ -\beta & -\alpha & 0 & s \end{pmatrix}^{-1} \begin{pmatrix} 0 \\ 0 \\ \gamma \\ \gamma \end{pmatrix} .$$

The product of the second and third factor is the solution of equations

$$\begin{pmatrix} s & 0 & -1 & 0 \\ 0 & s & 0 & -1 \\ -\alpha & -\beta & s & 0 \\ -\beta & -\alpha & 0 & s \end{pmatrix} \begin{pmatrix} v_1 \\ v_2 \\ v_3 \\ v_4 \end{pmatrix} = \begin{pmatrix} 0 \\ 0 \\ \gamma \\ \gamma \end{pmatrix},$$

that is,

$$\begin{aligned}
sv_1 \qquad\qquad -v_3 \qquad\qquad &= 0 \\
sv_2 \qquad\qquad -v_4 &= 0 \\
-\alpha v_1 - \beta v_2 + sv_3 \qquad\quad &= \gamma \\
-\beta v_1 - \alpha v_1 \qquad\quad + sv_4 &= \gamma .
\end{aligned}$$

Since $\alpha \neq \beta$, the symmetry in v_1 and v_2 and also in v_3 and v_4 implies that $v_1 = v_2 = V$ and $v_3 = v_4 = V^*$. Hence

$$sV - V^* = 0$$

$$-(\alpha + \beta)V + sV^* = \gamma .$$

From the first equation, $V^* = sV$; hence, the second equation implies that

$$V(s^2 - (\alpha + \beta)) = \gamma .$$

That is,

$$V = \frac{\gamma}{s^2 - (\alpha + \beta)} \quad \text{and} \quad V^* = \frac{s\gamma}{s^2 - (\alpha + \beta)} .$$

In summary,

$$v_1 = v_2 = \frac{\gamma}{s^2 - (\alpha + \beta)} \quad \text{and} \quad v_3 = v_4 = \frac{s\gamma}{s^2 - (\alpha + \beta)} ,$$

and, therefore,

$$\mathbf{H}(s) = \begin{pmatrix} 1 & 0 & 0 & 0 \end{pmatrix} \begin{pmatrix} v_1 \\ v_2 \\ v_3 \\ v_4 \end{pmatrix} = v_1 = \frac{\gamma}{s^2 - (\alpha + \beta)} .$$

Step 2: Find the standard controllable realization: With the notation of Method 1 of the proof of Theorem 8.6,

$$p(s) = s^2 - (\alpha + \beta) ;$$

therefore,

$$r = 2, \quad p_1 = 0, \quad \text{and} \quad p_0 = -(\alpha + \beta) ,$$

and since

$$p(s)\mathbf{H}(s) = \gamma, \quad H_0 = \gamma, \quad \text{and} \quad H_1 = 0 .$$

Therefore,

$$\mathbf{A}_C = \begin{pmatrix} 0 & 1 \\ \alpha + \beta & 0 \end{pmatrix} , \quad \mathbf{B}_C = \begin{pmatrix} 0 \\ 1 \end{pmatrix} , \quad \mathbf{C}_C = \begin{pmatrix} \gamma, 0 \end{pmatrix} ,$$

that is,

$$\begin{aligned} \dot{z}_1 &= & z_2 & \\ \dot{z}_2 &= (\alpha + \beta)z_1 & +u \\ y &= & \gamma z_1 & \end{aligned}$$

is the minimal realization.
  We can easily check the results as follows:

1. Minimality:

$$\mathbf{K}_C = \begin{pmatrix} 0 & 1 \\ 1 & 0 \end{pmatrix} , \quad rank(\mathbf{K}_C) = 2 ,$$

and

$$\mathbf{L}_C = \begin{pmatrix} \gamma & 0 \\ 0 & \gamma \end{pmatrix} , \quad rank(\mathbf{L}_C) = 2 .$$

2. Transfer function:

$$\mathbf{H}_C(s) = (\gamma, 0) \begin{pmatrix} s & -1 \\ -(\alpha + \beta) & s \end{pmatrix}^{-1} \begin{pmatrix} 0 \\ 1 \end{pmatrix}$$

$$= (\gamma, 0) \frac{1}{s^2 - (\alpha + \beta)} \begin{pmatrix} s & 1 \\ \alpha + \beta & s \end{pmatrix} \begin{pmatrix} 0 \\ 1 \end{pmatrix}$$

$$= (\gamma, 0) \frac{1}{s^2 - (\alpha + \beta)} \begin{pmatrix} 1 \\ s \end{pmatrix} = \frac{\gamma}{s^2 - (\alpha + \beta)} .$$

This means that if the lengths of the pendulums are the same then the above second-order system is the best model for the cart with two inverted pendulums in the sense that the state dimension cannot be reduced further. However, as with the original system this minimal realization is unstable.

9. Our *electrical heating system* had been shown earlier to be controllable and observable, therefore, it is minimal.

10. The *nuclear reactor system* was completely controllable and observable; therefore, it is minimal.

8.3.2   Applications in the Social Sciences and Economics

1. In Sections 5.3.2 and 6.4.2 we derived that the linearized *predator–prey* model

$$\dot{G}_\delta = -\frac{bc}{d} W_\delta + \frac{c}{d} u$$

$$\dot{W}_\delta = \frac{da}{b} G_\delta$$

$$y = W_\delta \tag{8.32}$$

is completely controllable and completely observable. Therefore, Theorem 8.3 implies that there is no lower dimensional linear system with the same input–output relation.

Next we show that the same conclusion can be reached by using the transfer function approach suggested in Section 8.2. Since in this case

$$\mathbf{A} = \begin{pmatrix} 0 & -\frac{bc}{d} \\ \frac{da}{b} & 0 \end{pmatrix}, \quad \mathbf{B} = \begin{pmatrix} \frac{c}{d} \\ 0 \end{pmatrix}, \quad \text{and} \quad \mathbf{C} = (0, 1),$$

the transfer function has the form

$$\mathbf{H}(s) = (0, 1) \begin{pmatrix} s & \frac{bc}{d} \\ -\frac{da}{b} & s \end{pmatrix}^{-1} \begin{pmatrix} \frac{c}{d} \\ 0 \end{pmatrix}$$

$$= \frac{1}{s^2 + ac} (0, 1) \begin{pmatrix} s & -\frac{bc}{d} \\ \frac{da}{b} & s \end{pmatrix} \begin{pmatrix} \frac{c}{d} \\ 0 \end{pmatrix} = \frac{\frac{ac}{b}}{s^2 + ac}.$$

Since the numerator is a constant, this fraction cannot be simplified by a polynomial of degree at least one. Hence, Theorem 8.7 implies that system (8.32) is a minimal realization of its transfer function.

2. The *cohort population* model (6.17) is always completely controllable, as established in Section 5.3.2. However, in Section 6.4.2 we derived that it is not always completely observable. Therefore, it is not a minimal realization. As an illustration, consider the numerical example

$$\mathbf{p}(t+1) = \begin{pmatrix} \alpha & \alpha & 2\alpha \\ \alpha & 0 & 0 \\ 0 & \alpha & 0 \end{pmatrix} \mathbf{p}(t) + \mathbf{u}(t)$$

$$y = (1, 1, 1)\mathbf{p}, \tag{8.33}$$

where the population is assumed to be divided into three groups, the input is three-dimensional, and there is a single output, the total population. Here $\alpha > 0$ is a given parameter.

The transfer function is

$$\mathbf{H}(s) = (1, 1, 1) \begin{pmatrix} s - \alpha & -\alpha & -2\alpha \\ -\alpha & s & 0 \\ 0 & -\alpha & s \end{pmatrix}^{-1},$$

since $\mathbf{B} = \mathbf{I}$. Note that $\mathbf{H}(s)$ is a row vector, which can be determined directly without matrix inversion. If h_1, h_2, and h_3 denote the components of $\mathbf{H}(s)$, then they satisfy equation

$$(1,1,1) = (h_1, h_2, h_3) \begin{pmatrix} s - \alpha & -\alpha & -2\alpha \\ -\alpha & s & 0 \\ 0 & -\alpha & s \end{pmatrix} .$$

That is,

$$\begin{array}{rl}
h_1(s - \alpha) \quad -h_2\alpha & = 1 \\
-h_1\alpha \quad +h_2 s \quad -h_3\alpha & = 1 \\
-2h_1\alpha \qquad\qquad +h_3 s & = 1 .
\end{array}$$

Simple calculation shows that the solution is

$$h_1 = h_2 = h_3 = \frac{1}{s - 2\alpha} .$$

That is,

$$\mathbf{H}(s) = \frac{(1,1,1)}{s - 2\alpha} . \tag{8.34}$$

Since system (8.33) has a three-dimensional state variable, and the degree of the denominator of $\mathbf{H}(s)$ is only one, the system is not a minimal realization of this transfer function. From the proof of Theorem 8.8 we know that the standard observable realization of this transfer function gives a lower dimensional realization. In this case, Method 2 of the proof of Theorem 8.6 is illustrated.

Step 1: Obviously $p(s) = s - 2\alpha$, $r = 1$, and $p_0 = -2\alpha$.

Step 2: Next we expand $\mathbf{H}(s)$ about $|s| = \infty$:

$$\mathbf{H}(s) = (1,1,1)\frac{\frac{1}{s}}{1 - \frac{2\alpha}{s}} = (1,1,1)\frac{1}{s}\left(1 + \frac{2\alpha}{s} + \frac{4\alpha^2}{s^2} + \cdots\right) ,$$

which implies that $\mathbf{L}_O = (1,1,1)$.

Step 3: Therefore,

$$\mathbf{A}_O = (2\alpha) , \qquad \mathbf{B}_O = (1,1,1) , \qquad \text{and} \qquad \mathbf{C}_O = (1) ,$$

that is, the following minimal realization is obtained:

$$z(t + 1) = 2\alpha z(t) + (1,1,1)\mathbf{u}(t)$$

$$y(t) = z(t) . \tag{8.35}$$

We can easily check that the transfer function of this system is (8.34), since

$$\mathbf{C}_O(s\mathbf{I} - \mathbf{A}_O)^{-1}\mathbf{B}_O = 1 \cdot (s - 2\alpha)^{-1}(1,1,1) = \frac{(1,1,1)}{s - 2\alpha} = \mathbf{H}(s) .$$

The same minimal realization can be obtained by using a simple elementary approach, which can be used only in certain special cases. Premultiply the difference Equation (8.33) by the row vector $(1,1,1)$ to get

$$(1,1,1)\mathbf{p}(t+1) = (2\alpha, 2\alpha, 2\alpha)\mathbf{p}(t) + (1,1,1)\mathbf{u}(t) ,$$

and by introducing the new state variable

$$z = (1,1,1)\mathbf{p}$$

we get relations (8.35).

3. The *arms races* model (6.18) was investigated earlier in Sections 5.3.2. and 6.4.2, and we verified that except for very special cases the system is completely controllable and is always completely observable. Therefore, it is almost always a minimal realization of its input–output relation. Since

$$\mathbf{A} = \begin{pmatrix} -b & a \\ c & -d \end{pmatrix} , \qquad \mathbf{B} = \begin{pmatrix} \alpha \\ \beta \end{pmatrix} , \quad \text{and} \quad \mathbf{C} = (1,0) ,$$

the transfer function has the form

$$\mathbf{H}(s) = (1,0) \begin{pmatrix} s+b & -a \\ -c & s+d \end{pmatrix}^{-1} \begin{pmatrix} \alpha \\ \beta \end{pmatrix}$$

$$= (1,0)\frac{1}{(s+b)(s+d) - ac} \begin{pmatrix} s+d & a \\ c & s+b \end{pmatrix} \begin{pmatrix} \alpha \\ \beta \end{pmatrix}$$

$$= \frac{s\alpha + (\alpha d + a\beta)}{s^2 + s(b+d) + (bd - ac)} .$$

Since the input and output are both single-dimensional, the system is a minimal realization if and only if this transfer function cannot be simplified by a linear polynomial, that is, when the numerator and denominator have no common root. The only root of the numerator is

$$s = -\frac{\alpha d + a\beta}{\alpha} = -d - a\frac{\beta}{\alpha} ,$$

and the roots of the denominator are

$$s_{1,2} = \frac{-(b+d) \pm \sqrt{(b-d)^2 + 4ac}}{2}.$$

Hence, the system is minimal if and only if

$$-d - a\frac{\beta}{\alpha} \neq \frac{-(b+d) \pm \sqrt{(b-d)^2 + 4ac}}{2}.$$

Simple calculation shows that this relation is equivalent to the following:

$$\frac{\alpha}{\beta} \neq \frac{d - b \pm \sqrt{(b-d)^2 + 4ac}}{2c}.$$

Note that the same condition was found in Section 5.3.2 to be necessary and sufficient for the complete controllability of the system.

4. For the *warfare* model (6.19) we saw in Sections 5.3.2 and 6.4.2 that the system is completely controllable and completely observable, and, therefore, it is a minimal realization. The same conclusion can be obtained by examining the transfer function of the system. In this case,

$$\mathbf{H}(s) = (1,0) \begin{pmatrix} s & h_2 \\ h_1 & s \end{pmatrix}^{-1} \begin{pmatrix} -h_3 \\ 0 \end{pmatrix}$$

$$= (1,0) \frac{1}{s^2 - h_1 h_2} \begin{pmatrix} s & -h_2 \\ -h_1 & s \end{pmatrix} \begin{pmatrix} -h_3 \\ 0 \end{pmatrix} = \frac{-sh_3}{s^2 - h_1 h_2}.$$

Since the only root of the numerator is zero and the roots of the denominator are $\pm\sqrt{h_1 h_2} \neq 0$ for positive values of h_1 and h_2, this fraction cannot be simplified by a polynomial of degree at least one. Therefore, Theorem 8.7 implies the minimality of this system.

5. The linear *epidemic* model of Application 6.4.2-5 was shown to be completely controllable for $\bar{x} \neq 0$, but not observable. Therefore, it is not minimal. A minimal realization can be determined as follows. Note first that the systems equations can be rewritten as

$$\dot{x} = (-\alpha\bar{x})y$$

$$\dot{y} = (\alpha\bar{x} - \beta)y - u$$

with y being the output. Since x does not depend on the input u and the output is independent of x, the same input–output relation can be obtained by only the second equation, which has a single-dimensional state variable.

6. The *Harrod-type national economy* model (6.20) is single-dimensional; therefore, the dimension of the state variable cannot be further decreased.

7. The same holds for the *linear cobweb* model (6.21), which also has a single-dimensional state.

8. The dynamic model (6.22) of *interrelated markets* is completely controllable, as shown in Section 5.3.2. However, it is not always completely observable; therefore, the system is not minimal. For the sake of convenience we repeat the system here:

$$\dot{\mathbf{p}} = \mathbf{K}(\mathbf{A} - \mathbf{B})\mathbf{p} + \mathbf{u}$$

$$y = \frac{1}{n}\mathbf{1}^T\mathbf{p} \ . \tag{8.36}$$

In this case,
$$\mathbf{H}(s) = \frac{1}{n}\mathbf{1}^T[s\mathbf{I} - \mathbf{K}(\mathbf{A} - \mathbf{B})]^{-1} \ ,$$

since the coefficient of $\mathbf{u}$ is the identity matrix.

Consider the special case, when the markets are independent, that is, when
$$\mathbf{A} = diag(a_{11}, \ldots, a_{nn}), \qquad \mathbf{B} = diag(b_{11}, \ldots, b_{nn})$$

with $a_{ii} < 0$ and $b_{ii} > 0$ for $i = 1, 2, \ldots, n$. It is also assumed that

$$\mathbf{K} = diag(k_1, \ldots, k_n) \ ,$$

where $k_i > 0$ for $i = 1, 2, \ldots, n$. Therefore,

$$\mathbf{H}(s) = \frac{1}{n}\mathbf{1}^T diag(s - k_1(a_{11} - b_{11}), \ldots, s - k_n(a_{nn} - b_{nn}))^{-1}$$

$$= \frac{1}{n}\left(\frac{1}{s - k_1(a_{11} - b_{11})}, \ldots, \frac{1}{s - k_n(a_{nn} - b_{nn})}\right)$$

$$= \frac{1}{n}\left(\frac{Q_1(s)}{p(s)}, \ldots, \frac{Q_n(s)}{p(s)}\right) \ ,$$

where
$$p(s) = \prod_{i=1}^{n}(s - k_i(a_{ii} - b_{ii}))$$

and for all k,

$$Q_k(s) = \prod_{i=1, i \neq k}^{n} (s - k_i(a_{ii} - b_{ii})) \,.$$

The necessary and sufficient condition that $p(s)$ and all $Q_k(s)$ ($k = 1, 2, \ldots, n$) have no common divisor of degree at least one is that the numbers $k_i(a_{ii} - b_{ii})$ are all different. Hence, Theorem 8.7 implies that this is the necessary and sufficient condition for the minimality of the system.

9. Consider finally the *oligopoly* model (6.23), which is not controllable and is not observable. Therefore, the dimension of the state variable can be reduced. In this case,

$$\mathbf{A} = \begin{pmatrix} 0 & -\frac{1}{2} & -\frac{1}{2} & \cdots & -\frac{1}{2} \\ -\frac{1}{2} & 0 & -\frac{1}{2} & \cdots & -\frac{1}{2} \\ -\frac{1}{2} & -\frac{1}{2} & 0 & \cdots & -\frac{1}{2} \\ \vdots & \vdots & \vdots & \ddots & \vdots \\ -\frac{1}{2} & -\frac{1}{2} & -\frac{1}{2} & \cdots & 0 \end{pmatrix}, \qquad \mathbf{B} = -\frac{1}{2a}\mathbf{1} \,,$$

and

$$\mathbf{C} = \mathbf{1}^T \,;$$

therefore,

$$\mathbf{H}(s) = -\frac{1}{2a} \cdot \mathbf{1}^T \begin{pmatrix} s & \frac{1}{2} & \frac{1}{2} & \cdots & \frac{1}{2} \\ \frac{1}{2} & s & \frac{1}{2} & \cdots & \frac{1}{2} \\ \frac{1}{2} & \frac{1}{2} & s & \cdots & \frac{1}{2} \\ \vdots & \vdots & \vdots & \ddots & \vdots \\ \frac{1}{2} & \frac{1}{2} & \frac{1}{2} & \cdots & s \end{pmatrix}^{-1} \mathbf{1} \,.$$

We can avoid the matrix inversion by observing that

$$\mathbf{H}(s) = -\frac{1}{2a}\mathbf{1}^T \begin{pmatrix} h_1 \\ h_2 \\ \vdots \\ h_n \end{pmatrix} \,,$$

where the $h'_i s$ satisfy the linear equations

$$sh_1 + \tfrac{1}{2}h_2 + \tfrac{1}{2}h_3 + \cdots + \tfrac{1}{2}h_n = 1$$
$$\tfrac{1}{2}h_1 + sh_2 + \tfrac{1}{2}h_3 + \cdots + \tfrac{1}{2}h_n = 1$$
$$\tfrac{1}{2}h_1 + \tfrac{1}{2}h_2 + sh_3 + \cdots + \tfrac{1}{2}h_n = 1$$
$$\vdots \quad \vdots \quad \vdots \quad \vdots \quad \vdots \quad \vdots \quad \vdots$$
$$\tfrac{1}{2}h_1 + \tfrac{1}{2}h_2 + \tfrac{1}{2}h_3 + \cdots + sh_n = 1 \ .$$

Observe that these equations are symmetric in the unknowns; therefore, $h_1 = h_2 = \cdots = h_n = h$. The first equation implies that

$$\left(s + \frac{n-1}{2} \right) h = 1 \ ,$$

that is,

$$h = \frac{1}{s + \frac{n-1}{2}} \ .$$

Hence

$$\mathbf{H}(s) = -\frac{n}{2a} \cdot \frac{1}{s + \frac{n-1}{2}} = \frac{-\frac{n}{2a}}{s + \frac{n-1}{2}} \ ,$$

and by using the notation of the proof of Theorem 8.6, $p(s) = s + ((n-1)/2)$ and $\mathbf{H}_0 = (-n/2a)$. Therefore, the standard controllable realization is minimal and has the form

$$\dot{z} = -\frac{n-1}{2} z + u$$

$$y = \frac{-n}{2a} z \ .$$

Problems

1. Is the weighting pattern

$$T(t, \tau) = t + \tau + t\tau + 1$$

realizable? Find a realization of $T(t, \tau)$.

2. Find a realization of the weighting pattern

$$\mathbf{T}(t, \tau) = \begin{pmatrix} t\tau & t \\ \tau & 1 \end{pmatrix} \ .$$

3. Find a realization of the weighting pattern

$$\mathbf{T}(t, \tau) = \begin{pmatrix} t + \tau & t\tau + 1 \\ 1 + \tau & 1 + \tau \end{pmatrix}.$$

4. Find the weighting pattern for system

$$\dot{\mathbf{x}} = \begin{pmatrix} 1 & 1 \\ 2 & 2 \end{pmatrix} \mathbf{x} + \begin{pmatrix} 1 \\ 0 \end{pmatrix} u$$

$$y = (1, 1)\mathbf{x}.$$

5. Find the weighting pattern for system

$$\dot{\mathbf{x}} = \begin{pmatrix} 0 & 0 \\ 1 & 0 \end{pmatrix} \mathbf{x} + \begin{pmatrix} 1 \\ 0 \end{pmatrix} u$$

$$y = (0, 1)\mathbf{x}.$$

6. Is the following system minimal? If not, give a minimal realization. Use the state–space approach.

$$\dot{\mathbf{x}} = \begin{pmatrix} 1 & 1 \\ 2 & 2 \end{pmatrix} \mathbf{x} + \begin{pmatrix} 1 \\ 0 \end{pmatrix} u$$

$$y = (1, 1)\mathbf{x}.$$

7. Is system

$$\dot{\mathbf{x}} = \begin{pmatrix} 0 & 0 \\ 1 & 0 \end{pmatrix} \mathbf{x} + \begin{pmatrix} 1 \\ 0 \end{pmatrix} u$$

$$y = (0, 1)\mathbf{x}$$

minimal? Use the state–space approach.

8. Is system

$$\dot{\mathbf{x}} = \begin{pmatrix} \frac{1}{t} & 0 \\ 0 & \frac{1}{t} \end{pmatrix} \mathbf{x} + \begin{pmatrix} 1 \\ 1 \end{pmatrix} u$$

$$y = (1, 1)\mathbf{x}$$

minimal in $[1, 2]$?

9.   Is system

$$\dot{\mathbf{x}} = \begin{pmatrix} 1 & 0 \\ 0 & 1 \end{pmatrix} \mathbf{x} + \begin{pmatrix} 1 & 1 \\ 1 & 1 \end{pmatrix} u$$

$$y = (1, 1)\mathbf{x}$$

minimal? Use the state–space approach. Use relation (8.22) to obtain a time-invariant minimal realization.

10.   Illustrate Theorem 8.2 for system

$$\dot{\mathbf{x}} = \begin{pmatrix} 1 & 1 \\ 2 & 2 \end{pmatrix} \mathbf{x} + \begin{pmatrix} 1 \\ 0 \end{pmatrix} u$$

$$y = (1, 1)\mathbf{x}$$

with

$$\mathbf{P}(t) = \begin{pmatrix} 1 & 1 \\ 0 & 2 \end{pmatrix} .$$

11.   Find a time-invariant realization of the weighting pattern

$$\mathbf{T}(t, \tau) = t - \tau .$$

Use the algorithm suggested by Theorem 8.4 and select $[t_0, t_1] = [0, 1]$.

12.   The system

$$\dot{\mathbf{x}} = \begin{pmatrix} 1 & 1 \\ 2 & 2 \end{pmatrix} \mathbf{x} + \begin{pmatrix} 1 \\ 0 \end{pmatrix} u$$
$$y = (1, 1)\mathbf{x}$$

has a time-variant minimal realization

$$\dot{z} = 0 \cdot z + e^{-3t} u$$

$$y = e^{3t} \cdot z .$$

Repeat Example 8.6 to find a time-invariant minimal realization. Select the unit interval $[0, 1]$.

13.   Find the standard controllable realization of the transfer function

$$\mathbf{H}(s) = \frac{(2, 2)}{s - 1} .$$

14. Find the standard observable realization of the transfer function

$$\mathbf{H}(s) = \frac{(2,2)}{s-1}.$$

15. Find a time-invariant minimal realization for the system

$$\dot{\mathbf{x}} = \begin{pmatrix} 1 & 1 \\ 2 & 2 \end{pmatrix} \mathbf{x} + \begin{pmatrix} 1 \\ 0 \end{pmatrix} u$$

$$y = (1,1)\mathbf{x},$$

using the transfer function approach.

16. By using the transfer function approach, show that the system

$$\dot{\mathbf{x}} = \begin{pmatrix} 0 & 0 \\ 1 & 0 \end{pmatrix} \mathbf{x} + \begin{pmatrix} 1 \\ 0 \end{pmatrix} u$$

$$y = (0,1)\mathbf{x}$$

is minimal.

17. By using the transfer function approach show that the system

$$\dot{\mathbf{x}} = \begin{pmatrix} 1 & 0 \\ 0 & 1 \end{pmatrix} \mathbf{x} + \begin{pmatrix} 1 & 1 \\ 1 & 1 \end{pmatrix} \mathbf{u}$$

$$y = (1,1)\mathbf{x}$$

is not minimal. Find a minimal realization.

18. Discuss the minimality of the mechanical system

$$\dot{\mathbf{x}} = \begin{pmatrix} 0 & 1 \\ 0 & -6 \end{pmatrix} \mathbf{x} + \begin{pmatrix} 0 \\ 2 \end{pmatrix} u$$

$$y = (1,0)\mathbf{x}$$

that was introduced in Problem 3.7.

19. Let $\mathbf{A}(t)$ be an $n \times n$ continuous matrix. Prove that there exist continuous n-dimensional vectors $\mathbf{b}(t)$ and $\mathbf{c}(t)$ such that system

$$\dot{\mathbf{x}}(t) = \mathbf{A}(t)\mathbf{x}(t) + \mathbf{b}(t)u(t)$$

$$y(t) = \mathbf{c}^T(t)\mathbf{x}(t)$$

is minimal.

20. (i) Prove that the standard controllable realization is completely controllable.

(ii) Prove that the standard observable realization is completely observable.

21. Can you find values of parameter α such that system

$$\dot{\mathbf{x}} = \begin{pmatrix} \alpha & -1 \\ 1 & \alpha \end{pmatrix} \mathbf{x} + \begin{pmatrix} 1 \\ 0 \end{pmatrix} u$$

$$y = (1,0)\mathbf{x}$$

is not minimal.

22. Assume that realization $\dot{\mathbf{x}} = \mathbf{A}\mathbf{x} + \mathbf{B}u$, $\mathbf{y} = \mathbf{C}\mathbf{x}$ is minimal, and matrices $\bar{\mathbf{A}}, \bar{\mathbf{B}}, \bar{\mathbf{C}}$ are sufficiently good approximations of $\mathbf{A}, \mathbf{B}, \mathbf{C}$. Prove that system $\dot{\mathbf{z}} = \bar{\mathbf{A}}\mathbf{z} + \bar{\mathbf{B}}\mathbf{v}$, $\mathbf{w} = \bar{\mathbf{C}}\mathbf{z}$ is also minimal.

23. Prove Theorem 8.6 by using uncoupled representation of each element of $\mathbf{H}(s)$ and the fact that

$$\mathbf{A} = \begin{pmatrix} 0 & 1 & 0 & \cdots & 0 \\ 0 & 0 & 1 & \cdots & 0 \\ \vdots & \vdots & \vdots & \ddots & \vdots \\ 0 & 0 & 0 & \cdots & 1 \\ -a_0 & -a_1 & -a_2 & \cdots & -a_{n-1} \end{pmatrix}, \quad \mathbf{b} = \begin{pmatrix} 0 \\ 0 \\ \vdots \\ 0 \\ 1 \end{pmatrix},$$

$$\mathbf{c}^T = (d_0, d_1, \ldots, d_{n-1})$$

is a realization of the 1×1 transfer function

$$H(s) = \frac{d_{n-1}s^{n-1} + \cdots + d_1 s + d_0}{s^n + a_{n-1}s^{n-1} + \cdots + a_1 s + a_0}.$$

24. Can the Fibonacci sequence (see Example 2.15) be generated by a first order linear difference equation?

25. Assume that in an interval $[t_0, t^*]$,

$$\left\| \mathbf{T}(t,\tau) - \sum_{k=1}^{p} \mathbf{D}_k(t)\mathbf{G}_k(\tau) \right\| < \varepsilon.$$

Show that

$$\mathbf{A} = \mathbf{0}, \qquad \mathbf{B} = \begin{pmatrix} \mathbf{G}_1(t) \\ \vdots \\ \mathbf{G}_p(t) \end{pmatrix}. \qquad \mathbf{C}(t) = (\mathbf{D}_1(t), \ldots, \mathbf{D}_p(t))$$

is a realization of the approximating weighting pattern

$$\bar{\mathbf{T}}(t, \tau) = \sum_{k=1}^{p} \mathbf{D}_k(t) \mathbf{G}_k(\tau).$$

Assume that the initial states are zero, bound the discrepancy between the outputs of the true and approximating systems.

chapter nine

Estimation and Design

In Chapter 5 we investigated the controllability of linear systems by manipulating the inputs to cause the system to behave in a desirable way. In that approach, we assumed that the input function was generated by some process external to the system itself, and that this input was applied to the system. This kind of control is called *open-loop* control. However, it is usually more effective to determine the input on a continuing basis as a function of the behavior of the system. This kind of control is called *closed-loop* control, and usually the system is called a *feedback system*, since the states or outputs are fed back (in perhaps modified form) to the input.

For the sake of simplicity, only continuous time invariant systems will be considered in this chapter; discrete systems can be analyzed in an analogous manner. Consider, therefore, the system

$$\dot{\mathbf{x}} = \mathbf{A}\mathbf{x} + \mathbf{B}\mathbf{u} \tag{9.1}$$

$$\mathbf{y} = \mathbf{C}\mathbf{x} \ . \tag{9.2}$$

In the case of *state feedback* we assume that

$$\mathbf{u}(t) = \tilde{\mathbf{u}}(t) + \mathbf{K}\mathbf{x}(t) \ , \tag{9.3}$$

where $\tilde{\mathbf{u}}(t)$ is the external input, and $\mathbf{K}$ is a given constant matrix. That is, the $\mathbf{K}$-multiple of the state is fed back to the input. In the case of *output feedback*, Equation (9.3) is modified as

$$\mathbf{u}(t) = \tilde{\mathbf{u}}(t) + \mathbf{K}\mathbf{y}(t) = \tilde{\mathbf{u}}(t) + \mathbf{K}\mathbf{C}\mathbf{x}(t) \ . \tag{9.4}$$

That is, in the case of output feedback, matrix $\mathbf{K}$ is replaced by $\mathbf{K}\mathbf{C}$. We can substitute Equations (9.3) and (9.4) into Equation (9.1) to get the modified systems equations:

$$\dot{\mathbf{x}} = (\mathbf{A} + \mathbf{B}\mathbf{K})\mathbf{x} + \mathbf{B}\bar{\mathbf{u}} \tag{9.5}$$

and

$$\dot{\mathbf{x}} = (\mathbf{A} + \mathbf{B}\mathbf{K}\mathbf{C})\mathbf{x} + \mathbf{B}\bar{\mathbf{u}} \ . \tag{9.6}$$

That is, the new coefficient matrices are $\mathbf{A} + \mathbf{B}\mathbf{K}$ and $\mathbf{A} + \mathbf{B}\mathbf{K}\mathbf{C}$, respectively. Note that output feedback has many applications if only some state variables (or their linear combinations) but not all of them are available for feedback.

Closed-loop control systems have several advantages over open-loop control systems. First, in many cases the implementation of the open-loop control requires a very sophisticated (and, therefore, expensive) computing device to determine the inputs required to lead the system to a desired behavior. Second, a well-designed feedback system is inherently less sensitive to the accuracy of the mathematical model of the system. Third, a feedback system can automatically adjust to unforeseen system changes or to unanticipated disturbance inputs. Fourth, feedback can be used to alter the dynamics of the system, e.g., to decrease the response time or broaden the bandwidth.

This chapter is devoted to analyzing feedback systems and introducing some applications of feedback systems to construct observers.

9.1   The Eigenvalue Placement Theorem

As stated before, one important feature of feedback is that even unstable systems can be made stable, and stable systems can be made faster. We know from Chapter 4 that the speed of a time-invariant linear system depends on the locations of the eigenvalues of the coefficient matrix; therefore, it is natural to ask how much influence feedback can have on the eigenvalues of a system. An answer for this question is presented in the next result, which is known as the *eigenvalue placement theorem*.

THEOREM 9.1

Let $\mathbf{A}$ *be an* $n \times n$ *constant real matrix and* $\mathbf{b}$ *a real* n-*vector such that* $\mathbf{b}$, $\mathbf{A}\mathbf{b}$, $\mathbf{A}^2\mathbf{b}, \ldots, \mathbf{A}^{n-1}\mathbf{b}$ *are linearly independent. Then, given any* nth *degree polynomial* $p(\lambda) = \lambda^n + p_{n-1}\lambda^{n-1} + \cdots + p_1\lambda + p_0$, *there is an* n-*dimensional real row vector* $\mathbf{k}^T$ *such that the characteristic polynomial of matrix* $\mathbf{A} + \mathbf{b}\mathbf{k}^T$ *is the given polynomial* $p(\lambda)$.

PROOF   Consider the system described by differential equation

$$\dot{\mathbf{x}} = \mathbf{A}\mathbf{x} + \mathbf{b}u .\tag{9.7}$$

The assumption of the theorem implies that this system is completely controllable, and, therefore, Theorem 7.2 implies that there exists a non-singular matrix $\mathbf{T}$ such that

$$\tilde{\mathbf{A}} = \mathbf{TAT}^{-1} = \begin{pmatrix} 0 & 1 & 0 & \cdots & 0 \\ 0 & 0 & 1 & \cdots & 0 \\ \vdots & \vdots & \vdots & \ddots & \vdots \\ 0 & 0 & 0 & \cdots & 1 \\ a_0 & a_1 & a_2 & \cdots & a_{n-1} \end{pmatrix}, \text{ and } \tilde{\mathbf{b}} = \mathbf{Tb} = \begin{pmatrix} 0 \\ 0 \\ \vdots \\ 0 \\ 1 \end{pmatrix}.$$

If $\tilde{\mathbf{k}}^T = (k_1, \ldots, k_n)$ is any vector, then

$$\tilde{\mathbf{b}}\tilde{\mathbf{k}}^T = \begin{pmatrix} 0 & 0 & 0 & \cdots & 0 \\ 0 & 0 & 0 & \cdots & 0 \\ \vdots & \vdots & \vdots & \ddots & \vdots \\ 0 & 0 & 0 & \cdots & 0 \\ k_1 & k_2 & k_3 & \cdots & k_n \end{pmatrix};$$

therefore,

$$\tilde{\mathbf{A}} + \tilde{\mathbf{b}}\tilde{\mathbf{k}}^T = \begin{pmatrix} 0 & 1 & 0 & \cdots & 0 \\ 0 & 0 & 1 & \cdots & 0 \\ \vdots & \vdots & \vdots & \ddots & \vdots \\ 0 & 0 & 0 & \cdots & 1 \\ a_0 + k_1 & a_1 + k_2 & a_2 + k_3 & \cdots & a_{n-1} + k_n \end{pmatrix}.$$

From the corollary of Theorem 7.2 we know that the characteristic polynomial of $\tilde{\mathbf{A}} + \tilde{\mathbf{b}}\tilde{\mathbf{k}}^T$ is the polynomial

$$\lambda^n - (a_{n-1} + k_n)\lambda^{n-1} - \cdots - (a_1 + k_2)\lambda - (a_0 + k_1) ,$$

so, by selecting

$$k_1 = -a_0 - p_0, k_2 = -a_1 - p_1, \ldots, k_n = -a_{n-1} - p_{n-1}\tag{9.8}$$

the characteristic polynomial of $\tilde{\mathbf{A}} + \tilde{\mathbf{b}}\tilde{\mathbf{k}}^T$ becomes $p(\lambda)$.

Finally, we show that vector $\mathbf{k}^T = \tilde{\mathbf{k}}^T \mathbf{T}$ satisfies the assertion. Simple calculation shows that

$$\mathbf{A} + \mathbf{b}\mathbf{k}^T = \mathbf{T}^{-1}\tilde{\mathbf{A}}\mathbf{T} + \mathbf{T}^{-1}\tilde{\mathbf{b}}\tilde{\mathbf{k}}^T\mathbf{T} = \mathbf{T}^{-1}(\tilde{\mathbf{A}} + \tilde{\mathbf{b}}\tilde{\mathbf{k}}^T)\mathbf{T} .$$

Therefore, matrix $\mathbf{A} + \mathbf{b}\mathbf{k}^T$ is similar to $\tilde{\mathbf{A}} + \tilde{\mathbf{b}}\tilde{\mathbf{k}}^T$, which implies that they have the same characteristic polynomial.   ∎

The construction of vector $\mathbf{k}^T$ consists of the following steps:

Step 1   Transform system (9.7) to controllability canonical form (7.14) by applying the algorithm of Theorem 7.2.

Step 2   Compute vector $\tilde{\mathbf{k}}^T$ by using Equation (9.8).

Step 3   Determine vector $\mathbf{k}^T = \tilde{\mathbf{k}}^T \mathbf{T}$.

COROLLARY 9.1

Assume that for an $n \times n$ real matrix $\mathbf{A}$ and an n-vector $\mathbf{c}$, vectors $\mathbf{c}^T$, $\mathbf{c}^T\mathbf{A}$, $\mathbf{c}^T\mathbf{A}^2, \ldots, \mathbf{c}^T\mathbf{A}^{n-1}$ are linearly independent. Then, given any nth degree polynomial $p(\lambda) = \lambda^n + p_{n-1}\lambda^{n-1} + \cdots + p_1\lambda + p_0$, there is a real n-vector $\mathbf{k}$ such that the characteristic polynomial of matrix $\mathbf{A} + \mathbf{k}\mathbf{c}^T$ is the given polynomial $p(\lambda)$.

This assertion is a simple consequence of the theorem and the duality principle discussed earlier in Section 6.3.

Example 9.1

Consider matrix

$$\mathbf{A} = \begin{pmatrix} 0 & \omega \\ -\omega & 0 \end{pmatrix} \quad \text{and vector} \quad \mathbf{b} = \begin{pmatrix} 0 \\ 1 \end{pmatrix} ,$$

and define polynomial $p(\lambda) = \lambda^2 + 2\lambda + 1$. We now illustrate the above algorithm.

Step 1: From Example 7.5 we know that canonical form (7.14) of this system is given by

$$\tilde{\mathbf{A}} = \begin{pmatrix} 0 & 1 \\ -\omega^2 & 0 \end{pmatrix}, \quad \tilde{\mathbf{b}} = \begin{pmatrix} 0 \\ 1 \end{pmatrix}, \quad \text{and} \quad \mathbf{T} = \frac{1}{\omega}\begin{pmatrix} 1 & 0 \\ 0 & \omega \end{pmatrix} .$$

Step 2: Since $a_0 = -\omega^2$, $a_1 = 0$, $p_0 = 1$, and $p_1 = 2$, from relations (9.8) we have

$$k_1 = -a_0 - p_0 = \omega^2 - 1 \quad \text{and} \quad k_2 = -a_1 - p_1 = 0 - 2 = -2 .$$

That is,

$$\tilde{\mathbf{k}}^T = (\omega^2 - 1, -2).$$

Step 3: Finally,

$$\mathbf{k}^T = \tilde{\mathbf{k}}^T \mathbf{T} = (\omega^2 - 1, -2)\frac{1}{\omega}\begin{pmatrix} 1 & 0 \\ 0 & \omega \end{pmatrix} = \left(\omega - \frac{1}{\omega}, -2\right).$$

We can check this result by simply computing the characteristic polynomial of matrix $\mathbf{A} + \mathbf{bk}^T$. In this case,

$$\mathbf{A} + \mathbf{bk}^T = \begin{pmatrix} 0 & \omega \\ -\omega & 0 \end{pmatrix} + \begin{pmatrix} 0 \\ 1 \end{pmatrix}\left(\omega - \frac{1}{\omega}, -2\right)$$

$$= \begin{pmatrix} 0 & \omega \\ -\omega & 0 \end{pmatrix} + \begin{pmatrix} 0 & 0 \\ \omega - \frac{1}{\omega} & -2 \end{pmatrix} = \begin{pmatrix} 0 & \omega \\ -\frac{1}{\omega} & -2 \end{pmatrix}$$

with characteristic polynomial

$$\varphi(\lambda) = \det\begin{pmatrix} -\lambda & \omega \\ -\frac{1}{\omega} & -2-\lambda \end{pmatrix} = -\lambda(-2-\lambda) - \omega\left(-\frac{1}{\omega}\right) = \lambda^2 + 2\lambda + 1.$$

Note that the same result is obtained by the following direct method,
which is very useful in the case of low-dimensional systems. Assume
that $\mathbf{k}^T = (k_1, k_2)$, then

$$\mathbf{A} + \mathbf{bk}^T = \begin{pmatrix} 0 & \omega \\ -\omega & 0 \end{pmatrix} + \begin{pmatrix} 0 \\ 1 \end{pmatrix}(k_1, k_2) = \begin{pmatrix} 0 & \omega \\ -\omega + k_1 & k_2 \end{pmatrix}.$$

The characteristic polynomial of this matrix is as follows:

$$\varphi(\lambda) = \det\begin{pmatrix} -\lambda & \omega \\ -\omega + k_1 & k_2 - \lambda \end{pmatrix} = \lambda^2 - \lambda k_2 + (\omega^2 - \omega k_1).$$

Equating the like coefficients of this polynomial and those of polynomial $p(\lambda) = \lambda^2 + 2\lambda + 1$, we obtain the equations

$$-k_2 = 2$$

$$\omega^2 - \omega k_1 = 1,$$

which imply that $k_1 = (\omega^2 - 1)/\omega = \omega - 1/\omega$ and $k_2 = -2$.

Finally, we remark that Theorem 9.1 remains true in the more general
case of multiple inputs and/or multiple outputs. We present the following theorem without proof.

THEOREM 9.2
Assume that system $(\mathbf{A}, \mathbf{B}, \mathbf{C})$ *with n-dimensional state variables is completely controllable (or observable), and let p be a given nth degree polynomial. Then there exists matrix* $\mathbf{K}$ *such that the characteristic polynomial of matrix* $\mathbf{A} + \mathbf{B}\mathbf{K}$ *(or* $\mathbf{A} + \mathbf{K}\mathbf{C}$) *is the given polynomial p.*

This theorem is illustrated in the following example.

Example 9.2

Consider the same coefficient matrix

$$\mathbf{A} = \begin{pmatrix} 0 & \omega \\ -\omega & 0 \end{pmatrix}$$

which was investigated in the previous example. Assume furthermore that $p(\lambda) = \lambda^2 + 2\lambda + 1$, but assume now that

$$\mathbf{B} = \begin{pmatrix} 0 & 1 \\ 1 & 1 \end{pmatrix}.$$

That is, the system now has a two-dimensional input. If k_{ij} $(i, j = 1, 2)$ denote the elements of matrix $\mathbf{K}$, then

$$\mathbf{A} + \mathbf{B}\mathbf{K} = \begin{pmatrix} 0 & \omega \\ -\omega & 0 \end{pmatrix} + \begin{pmatrix} 0 & 1 \\ 1 & 1 \end{pmatrix} \begin{pmatrix} k_{11} & k_{12} \\ k_{21} & k_{22} \end{pmatrix}$$

$$= \begin{pmatrix} k_{21} & \omega + k_{22} \\ -\omega + k_{11} + k_{21} & k_{12} + k_{22} \end{pmatrix}.$$

The characteristic polynomial of this matrix is given as

$$\varphi(\lambda) = \det \begin{pmatrix} k_{21} - \lambda & \omega + k_{22} \\ -\omega + k_{11} + k_{21} & k_{12} + k_{22} - \lambda \end{pmatrix}$$

$$= \lambda^2 - \lambda(k_{21} + k_{12} + k_{22}) + k_{21}(k_{12} + k_{22})$$

$$- (\omega + k_{22})(-\omega + k_{11} + k_{21}).$$

By equating the like coefficients of this matrix and those of polynomial $p(\lambda) = \lambda^2 + 2\lambda + 1$ we get the following equations:

$$k_{21} + k_{12} + k_{22} = -2$$

$$-k_{21}(k_{12} + k_{22}) + (\omega + k_{22})(-\omega + k_{11} + k_{21}) = -1.$$

We have two equations for four unknowns; therefore, we can choose arbitrary values of two of the unknowns. For example, select $k_{21} = k_{12} = 0$ in order to obtain a diagonal matrix $\mathbf{K}$. Then the above equations reduce to

$$k_{22} = -2$$

$$(\omega - 2)(-\omega + k_{11}) = -1 ,$$

that is,

$$k_{11} = \frac{-1}{\omega - 2} + \omega = \frac{\omega^2 - 2\omega - 1}{\omega - 2} \quad \text{and} \quad k_{22} = -2 .$$

Hence,

$$\mathbf{K} = \begin{pmatrix} \frac{\omega^2 - 2\omega - 1}{\omega - 2} & 0 \\ 0 & -2 \end{pmatrix} .$$

We can check this result by simply calculating matrix $\mathbf{A} + \mathbf{BK}$ and determining its characteristic polynomial. In our case,

$$\mathbf{A} + \mathbf{BK} = \begin{pmatrix} 0 & \omega \\ -\omega & 0 \end{pmatrix} + \begin{pmatrix} 0 & 1 \\ 1 & 1 \end{pmatrix} \begin{pmatrix} \frac{\omega^2 - 2\omega - 1}{\omega - 2} & 0 \\ 0 & -2 \end{pmatrix}$$

$$= \begin{pmatrix} 0 & \omega - 2 \\ -\omega + \frac{\omega^2 - 2\omega - 1}{\omega - 2} & -2 \end{pmatrix} = \begin{pmatrix} 0 & \omega - 2 \\ \frac{-1}{\omega - 2} & -2 \end{pmatrix}$$

with characteristic polynomial

$$\varphi(\lambda) = \det \begin{pmatrix} -\lambda & \omega - 2 \\ \frac{-1}{\omega - 2} & -2 - \lambda \end{pmatrix} = \lambda^2 + 2\lambda + 1 ,$$

which coincides with $p(\lambda)$.

9.2 *Observers*

In many practical systems, the entire state vector may not be available. In physical systems, some components of the state are inaccessible internal variables, which either cannot be measured or the measurements require the use of very costly measurement devices. Therefore, it is not feasible, or it is very expensive, to measure all state components. We may face similar situations in large social or economic systems, when the measurements of all state variables are very expensive due to the extensive surveys and the complex record keeping procedures.

A common way to estimate the state of a system is to build a model of the original system, and then measure the state of this model. This trivial solution is shown in Figure 9.1 for continuous systems, and it is known as the *open-loop observer.*

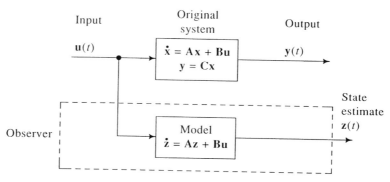

Figure 9.1 Open-loop observer.

The estimate $\mathbf{z}(t)$ provided by the measurements from the model does not utilize the available information on the output $\mathbf{y}(t)$ of the original system. If the initial state $\mathbf{z}(0)$ equals $\mathbf{x}(0)$, and the model is accurate, then it will follow the original system exactly. However, if $\mathbf{x}(0)$ is not available, and the model is started with an initial state, that differs from $\mathbf{x}(0)$, then $\mathbf{z}(t)$ may differ from $\mathbf{x}(t)$ for all future times. If we denote the error $\mathbf{x}(t) - \mathbf{z}(t)$ in the state variables by $\mathbf{x}_e(t)$, then $\mathbf{x}_e$ satisfies the differential equation

$$\dot{\mathbf{x}}_e = \mathbf{A}\mathbf{x}_e \ , \tag{9.9}$$

since

$$\dot{\mathbf{x}}_e = \dot{\mathbf{x}} - \dot{\mathbf{z}} = (\mathbf{A}\mathbf{x} + \mathbf{B}\mathbf{u}) - (\mathbf{A}\mathbf{z} + \mathbf{B}\mathbf{u}) = \mathbf{A}(\mathbf{x} - \mathbf{z}) = \mathbf{A}\mathbf{x}_e \ .$$

There is no guarantee, in general, that with increasing t, this error $\mathbf{x}_e$ will die out. We know from Chapter 4 that $\mathbf{x}_e(t) \rightarrow \mathbf{0}$ as $t \rightarrow \infty$ if and only if all eigenvalues of $\mathbf{A}$ have negative real parts.

In the case of discrete systems, the error $\mathbf{x}_e$ satisfies the homogeneous difference equation

$$\mathbf{x}_e(t + 1) = \mathbf{A}\mathbf{x}_e(t) \ , \tag{9.10}$$

and the error tends to zero as $t \rightarrow \infty$ if and only if all eigenvalues of $\mathbf{A}$ are inside the unit circle. In both cases, this kind of stabilization depends on the locations of the eigenvalues of $\mathbf{A}$.

In many cases the model is useless, since if the system is not asymptotically stable, the error does not tend to zero, even $\|\mathbf{x}_e(t)\|$ may tend to infinity as $t \to \infty$. For these reasons, *closed-loop observers* have been developed, where the output $\mathbf{y}$ of the original system is compared to the computed output $\mathbf{Cz}$ of the model, and the error $\mathbf{y}_e = \mathbf{y} - \mathbf{Cz}$ is fed back to this system, as shown in Figure 9.2.

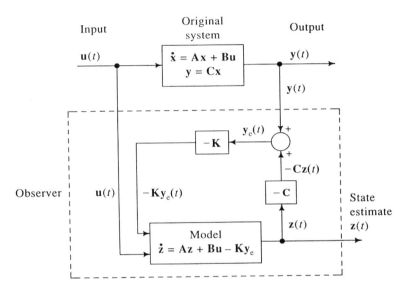

Figure 9.2 Closed-loop observer.

This observer has the mathematical representation

$$\dot{\mathbf{z}} = \mathbf{Az} + \mathbf{Bu} - \mathbf{K}(\mathbf{y} - \mathbf{Cz})$$

$$= (\mathbf{A} + \mathbf{KC})\mathbf{z} + \mathbf{Bu} - \mathbf{Ky} \ . \tag{9.11}$$

Simple calculation shows that the error $\mathbf{x}_e$ satisfies the relation

$$\dot{\mathbf{x}}_e = \dot{\mathbf{x}} - \dot{\mathbf{z}} = (\mathbf{Ax} + \mathbf{Bu}) - ((\mathbf{A} + \mathbf{KC})\mathbf{z} + \mathbf{Bu} - \mathbf{Ky})$$

$$= \mathbf{Ax} - \mathbf{Az} - \mathbf{KCz} + \mathbf{KCx} = (\mathbf{A} + \mathbf{KC})(\mathbf{x} - \mathbf{z}) \ ,$$

that is,

$$\dot{\mathbf{x}}_e = (\mathbf{A} + \mathbf{KC})\mathbf{x}_e \ . \tag{9.12}$$

Assume now that the original system is completely observable. Then Theorem 9.2 implies that there exists a matrix **K** such that **A** + **KC** has any desired characteristic polynomial. Select $p(\lambda)$ to be a polynomial having roots with negative real parts; then the homogeneous Equation (9.12) becomes asymptotically stable. Hence, $\mathbf{x}_e(t) \to 0$ as $t \to \infty$. A simplified scheme for this observer is presented in Figure 9.3.

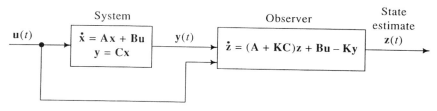

Figure 9.3   Simplified closed-loop observer.

Observer (9.11) is illustrated in the following example.

Example 9.3

Consider the system

$$\dot{\mathbf{x}} = \begin{pmatrix} 0 & \omega \\ -\omega & 0 \end{pmatrix} \mathbf{x} + \begin{pmatrix} 0 \\ 1 \end{pmatrix} u,$$

$$y = (1, 1)\mathbf{x} ,$$

which was the subject of our earlier Examples 9.1 and 9.2. Since **C** is a two-dimensional row vector, **K** must be a two-dimensional column vector. If k_1 and k_2 denote the components of **K**, then

$$\mathbf{A} + \mathbf{KC} = \begin{pmatrix} 0 & \omega \\ -\omega & 0 \end{pmatrix} + \begin{pmatrix} k_1 \\ k_2 \end{pmatrix}(1, 1) = \begin{pmatrix} k_1 & k_1 + \omega \\ k_2 - \omega & k_2 \end{pmatrix}$$

with characteristic polynomial

$$\varphi(\lambda) = \det \begin{pmatrix} k_1 - \lambda & k_1 + \omega \\ k_2 - \omega & k_2 - \lambda \end{pmatrix} = \lambda^2 - \lambda(k_1 + k_2) + (k_1\omega - k_2\omega + \omega^2) .$$

The roots of this polynomial have negative real parts if and only if

$$k_1 + k_2 < 0$$

$$k_1 - k_2 + \omega > 0 .$$

Select, for example, $k_1 = k_2 = -\omega$, then $\varphi(\lambda) = \lambda^2 + 2\lambda\omega + \omega^2$ with the single eigenvalue $\lambda = -\omega$. Hence system (9.11) has the form

$$\dot{\mathbf{z}} = \begin{pmatrix} -\omega & 0 \\ -2\omega & -\omega \end{pmatrix} \mathbf{z} + \begin{pmatrix} 0 \\ 1 \end{pmatrix} u + \begin{pmatrix} \omega \\ \omega \end{pmatrix} y \; .$$

9.3   *Reduced-Order Observers*

The observer discussed in the previous section gives the estimates of all state variables. This results in some redundancy, since certain linear combinations of these state variables (specified by the rows of matrix $\mathbf{C}$) are already known. In this section a new method will be introduced to eliminate this redundancy. It uses a lower dimensional observer, which gives only the information required to recover the entire state. For the original development of reduced-order observers, see [29]. Our discussion will focus on continuous systems, but discrete systems can be treated in an analogous manner. The details are left to the reader as an exercise.

Consider the time-invariant linear continuous system

$$\dot{\mathbf{x}} = \mathbf{A}\mathbf{x} + \mathbf{B}\mathbf{u}$$

$$\mathbf{y} = \mathbf{C}\mathbf{x} \; , \tag{9.13}$$

where we assume that matrix $\mathbf{C}$ has linearly independent rows. That is, if $\mathbf{y}$ is p-dimensional and $p \leq n$, then it is assumed that $rank(\mathbf{C}) = p$. Let matrix $\mathbf{D}$ be selected such that matrix

$$\mathbf{T} = \begin{pmatrix} \mathbf{D} \\ \mathbf{C} \end{pmatrix} \tag{9.14}$$

is nonsingular. Since the rows of $\mathbf{C}$ are linearly independent, such $\mathbf{D}$ exists. Introduce the new variable $\tilde{\mathbf{x}} = \mathbf{T}\mathbf{x}$, which can be partitioned as

$$\tilde{\mathbf{x}} = \begin{pmatrix} \mathbf{z} \\ \mathbf{y} \end{pmatrix} \; ,$$

where $\mathbf{z}$ is $(n-p)$-dimensional, and $\mathbf{y}$ is the output of the original system. Hence, we may assume without loss of generality that p components of the state of the original system can be measured directly. When matrices $\mathbf{A}$ and $\mathbf{B}$ are partitioned accordingly, the system can be rewritten as

$$\begin{pmatrix} \dot{\mathbf{z}} \\ \dot{\mathbf{y}} \end{pmatrix} = \begin{pmatrix} \mathbf{A}_{11} & \mathbf{A}_{12} \\ \mathbf{A}_{21} & \mathbf{A}_{22} \end{pmatrix} \begin{pmatrix} \mathbf{z} \\ \mathbf{y} \end{pmatrix} + \begin{pmatrix} \mathbf{B}_1 \\ \mathbf{B}_2 \end{pmatrix} \mathbf{u} \; . \tag{9.15}$$

Multiplying the lower part by a matrix $\mathbf{K}$ and adding the resulting equation to the upper part yields the relation

$$\dot{\mathbf{z}} + \mathbf{K}\dot{\mathbf{y}} = (\mathbf{A}_{11} + \mathbf{K}\mathbf{A}_{21})\mathbf{z} + (\mathbf{A}_{12} + \mathbf{K}\mathbf{A}_{22})\mathbf{y} + (\mathbf{B}_1 + \mathbf{K}\mathbf{B}_2)\mathbf{u}$$

$$= (\mathbf{A}_{11} + \mathbf{K}\mathbf{A}_{21})(\mathbf{z} + \mathbf{K}\mathbf{y})$$

$$+ (-\mathbf{A}_{11}\mathbf{K} - \mathbf{K}\mathbf{A}_{21}\mathbf{K} + \mathbf{A}_{12} + \mathbf{K}\mathbf{A}_{22})\mathbf{y} + (\mathbf{B}_1 + \mathbf{K}\mathbf{B}_2)\mathbf{u} \ .$$

Introduce now the new variable

$$\bar{\mathbf{x}} = \mathbf{z} + \mathbf{K}\mathbf{y} \ ,$$

then the above equation shows that it satisfies relation

$$\dot{\bar{\mathbf{x}}} = (\mathbf{A}_{11} + \mathbf{K}\mathbf{A}_{21})\bar{\mathbf{x}} + (-\mathbf{A}_{11}\mathbf{K} - \mathbf{K}\mathbf{A}_{21}\mathbf{K} + \mathbf{A}_{12} + \mathbf{K}\mathbf{A}_{22})\mathbf{y}$$

$$+ (\mathbf{B}_1 + \mathbf{K}\mathbf{B}_2)\mathbf{u} \ . \tag{9.16}$$

Note that $\mathbf{u}$ and $\mathbf{y}$ are measurable; only $\bar{\mathbf{x}}$ is unknown. This unknown state can be observed by merely modeling this system as

$$\dot{\bar{\mathbf{z}}} = (\mathbf{A}_{11} + \mathbf{K}\mathbf{A}_{21})\bar{\mathbf{z}} + (-\mathbf{A}_{11}\mathbf{K} - \mathbf{K}\mathbf{A}_{21}\mathbf{K} + \mathbf{A}_{12} + \mathbf{K}\mathbf{A}_{22})\mathbf{y}$$

$$+ (\mathbf{B}_1 + \mathbf{K}\mathbf{B}_2)\mathbf{u} \tag{9.17}$$

and measuring the state $\bar{\mathbf{z}}$ of this system. Let $\bar{\mathbf{x}}_e$ denote the error $\bar{\mathbf{x}} - \bar{\mathbf{z}}$, then subtract Equations (9.16) and (9.17) to get

$$\dot{\bar{\mathbf{x}}}_e = (\mathbf{A}_{11} + \mathbf{K}\mathbf{A}_{21})\bar{\mathbf{x}}_e \ . \tag{9.18}$$

The reader can verify that if the original system (9.13) (or equivalently (9.15)) is completely observable, then the reduced system (9.17) is also completely observable. Therefore, Theorem 9.2 implies that there exists a matrix $\mathbf{K}$ such that system (9.18) is asymptotically stable, from which we conclude that $\bar{\mathbf{x}}_e(t) \to 0$ as $t \to \infty$. Hence, state $\bar{\mathbf{x}}$ is observable, and then, the unmeasurable part $\mathbf{z}$ of the state of the original system is obtained as

$$\mathbf{z} = \bar{\mathbf{x}} - \mathbf{K}\mathbf{y} \ ,$$

which follows from the definition of vector $\bar{\mathbf{x}}$. An example for reduced observers will be presented in the next section.

9.4   The Eigenvalue Separation Theorem

Consider the time-invariant linear continuous system

$$\dot{\mathbf{x}} = \mathbf{Ax} + \mathbf{Bu}$$

$$\mathbf{y} = \mathbf{Cx} \; , \qquad\qquad\qquad (9.19)$$

and assume that it is completely controllable. Assume furthermore that the entire state is available for feedback. This feedback structure is shown in Figure 9.4. The input of the feedback system is

$$\bar{\mathbf{u}} = \mathbf{u} + \mathbf{K}_C\mathbf{x} \; ;$$

therefore,

$$\dot{\mathbf{x}} = \mathbf{Ax} + \mathbf{B}(\mathbf{u} + \mathbf{K}_C\mathbf{x}) = (\mathbf{A} + \mathbf{BK}_C)\mathbf{x} + \mathbf{Bu} \; . \qquad (9.20)$$

From Theorem 9.2 we know that the eigenvalues of this new coefficient matrix $\mathbf{A} + \mathbf{BK}_C$ can be placed in any desired positions.

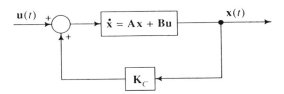

Figure 9.4   Feedback controller.

When the state is not available, we propose making use of the observer introduced in Section 9.2:

$$\dot{\mathbf{z}} = (\mathbf{A} + \mathbf{K}_O\mathbf{C})\mathbf{z} + \mathbf{Bu} - \mathbf{K}_O\mathbf{y} \; . \qquad\qquad (9.21)$$

Then, the observed state $\mathbf{z}$ is fed back to the system, as shown in Figure 9.5. Therefore, in this case,

$$\dot{\mathbf{x}} = \mathbf{Ax} + \mathbf{B}(\mathbf{u} + \mathbf{K}_C\mathbf{z}) = \mathbf{Ax} + \mathbf{BK}_C\mathbf{z} + \mathbf{Bu}$$

and

$$\dot{\mathbf{z}} = (\mathbf{A} + \mathbf{K}_O\mathbf{C})\mathbf{z} + \mathbf{B}(\mathbf{u} + \mathbf{K}_C\mathbf{z}) - \mathbf{K}_O(\mathbf{Cx}) =$$

$$- \mathbf{K}_O\mathbf{Cx} + (\mathbf{A} + \mathbf{K}_O\mathbf{C} + \mathbf{BK}_C)\mathbf{z} + \mathbf{Bu} \; .$$

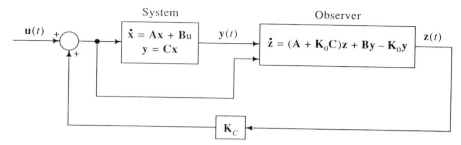

Figure 9.5   Feedback controller with observer.

Thus, we obtain the following system:

$$\begin{pmatrix} \dot{\mathbf{x}} \\ \dot{\mathbf{z}} \end{pmatrix} = \begin{pmatrix} \mathbf{A} & \mathbf{BK}_C \\ -\mathbf{K}_O\mathbf{C} & \mathbf{A} + \mathbf{K}_O\mathbf{C} + \mathbf{BK}_C \end{pmatrix} \begin{pmatrix} \mathbf{x} \\ \mathbf{z} \end{pmatrix} + \begin{pmatrix} \mathbf{B} \\ \mathbf{B} \end{pmatrix} \mathbf{u} \ .$$

Next we replace $\mathbf{z}$ by the new variable $\mathbf{x}_e = \mathbf{x} - \mathbf{z}$ using the transformation

$$\mathbf{T} = \begin{pmatrix} \mathbf{I} & \mathbf{O} \\ \mathbf{I} & -\mathbf{I} \end{pmatrix} \ .$$

Then easy calculation shows that

$$\dot{\mathbf{x}} = \mathbf{Ax} + \mathbf{BK}_C(\mathbf{x} - \mathbf{x}_e) + \mathbf{Bu} = (\mathbf{A} + \mathbf{BK}_C)\mathbf{x} - \mathbf{BK}_C\mathbf{x}_e + \mathbf{Bu}$$

and

$$\dot{\mathbf{x}}_e = \dot{\mathbf{x}} - \dot{\mathbf{z}} = [\mathbf{Ax} + \mathbf{BK}_C(\mathbf{x} - \mathbf{x}_e) + \mathbf{Bu}]$$

$$-[-\mathbf{K}_O\mathbf{Cx} + (\mathbf{A} + \mathbf{K}_O\mathbf{C} + \mathbf{BK}_C)(\mathbf{x} - \mathbf{x}_e) + \mathbf{Bu}]$$

$$= (\mathbf{A} + \mathbf{K}_O\mathbf{C})\mathbf{x}_e \ .$$

These equations can be summarized as

$$\begin{pmatrix} \dot{\mathbf{x}} \\ \dot{\mathbf{x}}_e \end{pmatrix} = \begin{pmatrix} \mathbf{A} + \mathbf{BK}_C & -\mathbf{BK}_C \\ \mathbf{O} & \mathbf{A} + \mathbf{K}_O\mathbf{C} \end{pmatrix} \begin{pmatrix} \mathbf{x} \\ \mathbf{x}_e \end{pmatrix} + \begin{pmatrix} \mathbf{B} \\ \mathbf{O} \end{pmatrix} \mathbf{u} \qquad (9.22)$$

$$\mathbf{y} = (\mathbf{C}, \mathbf{O}) \begin{pmatrix} \mathbf{x} \\ \mathbf{x}_e \end{pmatrix} \ . \qquad (9.23)$$

This derivation and Equations (9.22) and (9.23) have the following consequences:

1. The observation error $\mathbf{x}_e$ is uncontrollable from the input $\mathbf{u}$. This is as expected, since the error tends to zero as $t \to \infty$ regardless of what input is selected.

2. The characteristic polynomial of the composite system is the product of the characteristic polynomials of matrices $\mathbf{A} + \mathbf{BK}_C$ and $\mathbf{A} + \mathbf{K}_O\mathbf{C}$.

The second property is known as the *eigenvalue separation theorem*, and it follows from the simple fact that the characteristic polynomial of the coefficient matrix of system (9.22) can be factored as

$$\det \begin{pmatrix} \mathbf{A} + \mathbf{BK}_C - \lambda\mathbf{I} & -\mathbf{BK}_C \\ \mathbf{O} & \mathbf{A} + \mathbf{K}_O\mathbf{C} - \lambda\mathbf{I} \end{pmatrix} =$$

$$\det (\mathbf{A} + \mathbf{BK}_C - \lambda\mathbf{I}) \cdot \det(\mathbf{A} + \mathbf{K}_O\mathbf{C} - \lambda\mathbf{I}) \ .$$

This means that the insertion of an observer into a feedback system does not affect the eigenvalues of the original system. That is, an observer does not change stability or dynamic response. It does, however, add additional modes.

Example 9.4

Consider the system

$$\dot{\mathbf{x}} = \begin{pmatrix} 0 & \omega \\ -\omega & 0 \end{pmatrix} \mathbf{x} + \begin{pmatrix} 0 \\ 1 \end{pmatrix} u,$$

$$y = (1,1)\mathbf{x} \ ,$$

which is known from our Examples 5.4 and 6.2 to be completely controllable and completely observable.

Since $y = x_1 + x_2$ is the measurable output, introduce the new state variables x_1 and y. Since

$$\dot{x}_1 = \omega x_2 = \omega(y - x_1) = -\omega x_1 + \omega y$$

and

$$\dot{y} = \dot{x}_1 + \dot{x}_2 = \omega x_2 - \omega x_1 + u = \omega(y - x_1) - \omega x_1 + u$$

$$= -2\omega x_1 + \omega y + u \ ,$$

we have the transformed system

$$\begin{pmatrix} \dot{x}_1 \\ \dot{y} \end{pmatrix} = \begin{pmatrix} -\omega & \omega \\ -2\omega & \omega \end{pmatrix} \begin{pmatrix} x_1 \\ y \end{pmatrix} + \begin{pmatrix} 0 \\ 1 \end{pmatrix} u$$

$$y = (0,1) \begin{pmatrix} x_1 \\ y \end{pmatrix} . \tag{9.24}$$

This system is not asymptotically stable, since the roots of its characteristic polynomial

$$(-\omega - \lambda)(\omega - \lambda) - \omega(-2\omega) = \lambda^2 + \omega^2$$

are $\pm j\omega$.

First we apply the feedback (9.20) for system (9.24) with $\mathbf{K}_C = (k_1, k_2)$. Then

$$\mathbf{A} + \mathbf{B}\mathbf{K}_C = \begin{pmatrix} -\omega & \omega \\ -2\omega & \omega \end{pmatrix} + \begin{pmatrix} 0 \\ 1 \end{pmatrix} (k_1, k_2)$$

$$= \begin{pmatrix} -\omega & \omega \\ -2\omega + k_1 & \omega + k_2 \end{pmatrix}$$

with characteristic polynomial

$$(-\omega - \lambda)(\omega + k_2 - \lambda) - \omega(-2\omega + k_1) = \lambda^2 - \lambda k_2 + \omega^2 - \omega(k_1 + k_2) .$$

Note that by selecting $k_2 = -\omega$ and $k_1 = 3\omega/2$, the eigenvalues become $\frac{1}{2}(-\omega \pm j\omega)$. Hence the system becomes asymptotically stable.

The reduced observer for the unmeasurable state component x_1 will be constructed next. Use relations (9.15) to see that in this case,

$\mathbf{A}_{11} = -\omega, \mathbf{A}_{12} = \omega, \mathbf{A}_{21} = -2\omega, \mathbf{A}_{22} = \omega, \mathbf{B}_1 = 0$, and $\mathbf{B}_2 = 1$.

If $\mathbf{K}_C = (k_C)$, then

$$\mathbf{A}_{11} + \mathbf{K}_C \mathbf{A}_{21} = -\omega + k_C(-2\omega) .$$

Since this is a 1×1 matrix, its only eigenvalue is $-\omega - 2\omega k_C$, which is negative, for example, by selecting $k_C = 0$. Use equation (9.17) to find that the reduced observer has the form

$$\dot{z} = -\omega \bar{z} + \omega y , \tag{9.25}$$

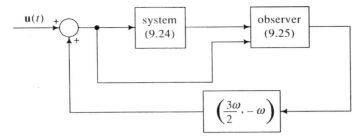

Figure 9.6   Feedback controller with observer for Example 9.4.

which is the first equation of the systems model (9.24). The combination of the resulting feedback and reduced observer is shown in Figure 9.6.

In many applications, the original system may change due to such effects as warming up, wearing out, fatigue, and so on. In such cases the observer will no longer be an exact copy, which can be source of many problems: instability, low speed, poor performance, and so on. The usual way to overcome this difficulty is to design an adaptive observer, which follows the changes of the original system. The mathematical details are not presented here; they can be found in adaptive control literature. However, some basic ideas of adaptive control systems will be discussed in Chapter 10.

9.5   *Applications*

This section is devoted to the discussion of some applications of feedback systems in engineering and the social sciences.

9.5.1   *Dynamic Systems in Engineering*

1. The reduced-order observer for our *harmonic motion model* was derived in Example 9.4. We will not do anything else with it.

2. For the *second-order mechanical system*, assume that the original model parameters are $K = M = 1$, $B = 2$. This would make $\zeta = 1 = w_n$, with $s_{1,2} = -1$, and would produce, according to Equation (3.74), a step response of

$$\theta(t) = 1 - (1 + t)e^{-t} .$$

Now suppose your boss says this is not fast enough: he wants poles at $-2 \pm j2$, i.e., $\zeta = \sqrt{2}/2$ and $w_n = 2\sqrt{2}$. How can you move the poles without changing the physical elements M, B, and K?

Measure the output position and measure or compute the output velocity, and use these with feedback to place the poles where they are desired. The original system is described with equation

$$\dot{\mathbf{x}} = \begin{pmatrix} 0 & 1 \\ -\frac{K}{M} & -\frac{B}{M} \end{pmatrix} \mathbf{x} + \begin{pmatrix} 0 \\ \frac{1}{M} \end{pmatrix} u \ .$$

If we feed back the states through vector $\mathbf{k}^T$ and add this to the input u, we will find that the new input to the system is

$$\bar{u} = u + \mathbf{k}^T \mathbf{x} \ ,$$

so the feedback system has the form

$$\dot{\mathbf{x}} = \left\{ \begin{pmatrix} 0 & 1 \\ -\frac{K}{M} & -\frac{B}{M} \end{pmatrix} + \begin{pmatrix} 0 \\ \frac{1}{M} \end{pmatrix} (k_1, k_2) \right\} \bar{\mathbf{x}} + \begin{pmatrix} 0 \\ \frac{1}{M} \end{pmatrix} u \ .$$

The coefficient matrix is

$$\tilde{\mathbf{A}} = \begin{pmatrix} 0 & 1 \\ -\frac{K}{M} + \frac{k_1}{M} & -\frac{B}{M} + \frac{k_2}{M} \end{pmatrix}$$

with characteristic polynomial

$$\tilde{\varphi}(\lambda) = -\lambda \left(-\frac{B}{M} + \frac{k_2}{M} - \lambda \right) - \left(-\frac{K}{M} + \frac{k_1}{M} \right)$$

$$= \lambda^2 + \lambda \left(\frac{B}{M} - \frac{k_2}{M} \right) + \left(\frac{K}{M} - \frac{k_1}{M} \right) \ .$$

Or, in the notation of Section 9.1,

$$\tilde{\varphi}(\lambda) = \lambda^2 + p_1 \lambda + p_0 \ .$$

Therefore, we obtain equations

$$\frac{B}{M} - \frac{k_2}{M} = p_1$$

and

$$\frac{K}{M} - \frac{k_1}{M} = p_0 \ ,$$

so

$$k_2 = B - p_1 M$$

and
$$k_1 = K - p_0 M .$$

Now, at the beginning of this problem we said that $K = M = 1$, $B = 2$, and that we wanted

$$\lambda_{1,2} = -2 \pm j2 .$$

Therefore, the desired characteristic polynomial is

$$\tilde{\varphi}(\lambda) = (\lambda + 2 - j2)(\lambda + 2 + j2)$$

$$= \lambda^2 + 4\lambda + 8 ,$$

so
$$p_0 = 8, \qquad p_1 = 4 .$$

Therefore, our desired feedback gains are

$$k_2 = -2 \qquad \text{and} \qquad k_1 = -7 .$$

Therefore, we have shown that feedback can alter the dynamics of a system. In this example we have moved the poles from one place in the s-plane to another in order to establish certain desired system properties.

3. For our *second-order L-R-C circuit*, assume that we would like to use feedback to move the poles or adjust the sensitivity of the system, but we cannot gain access to measure the state variables. (Perhaps the circuit is modeling something sealed inside a container, such as the human skull.) What can we do? Well, we can build an observer and use its state variables for control. The original system is

$$\mathbf{A} = \begin{pmatrix} a_{11} & a_{12} \\ a_{21} & a_{22} \end{pmatrix} \qquad \mathbf{b} = \begin{pmatrix} \frac{1}{L} \\ 0 \end{pmatrix}$$

$$\mathbf{c}^T = (0, 1) .$$

After applying feedback, we have the new coefficient matrix

$$\mathbf{A} + \mathbf{k}\mathbf{c}^T = \begin{pmatrix} a_{11} & a_{12} + k_1 \\ a_{21} & a_{22} + k_2 \end{pmatrix} .$$

Next we want to make sure the observer is stable. The characteristic polynomial is

$$\tilde{\varphi}(\lambda) = (a_{11} - \lambda)(a_{22} + k_2 - \lambda) - a_{21}(a_{12} + k_1)$$

$$= \lambda^2 - \lambda(a_{22} + k_2 + a_{11}) + (a_{11}a_{22} + a_{11}k_2 - a_{21}a_{12} - a_{21}k_1) \ .$$

Therefore, the feedback system will be stable if

$$k_2 < -(a_{11} + a_{22})$$

and, because $a_{11} < 0$ (we are not allowing negative resistance or inductance values),

$$k_2 < \frac{a_{21}}{a_{11}}k_1 + \frac{a_{21}a_{12} - a_{11}a_{22}}{a_{11}} \ .$$

The feasible region is shown is Figure 9.7.

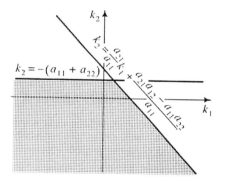

Figure 9.7   Region of stability for Application 9.5.1-3.

4. If we construct an observer for the *transistor circuit*, we have

$$\mathbf{A} + \mathbf{k}\mathbf{c}_2^T = \begin{pmatrix} -\alpha & 0 \\ \beta & 0 \end{pmatrix} + \begin{pmatrix} k_1 \\ k_2 \end{pmatrix}(1,0)$$

$$= \begin{pmatrix} k_1 - \alpha & 0 \\ k_2 + \beta & 0 \end{pmatrix} \ .$$

Since the matrix is lower triangular, the eigenvalues are $\lambda_1 = 0$ and $\lambda_2 = k_1 - \alpha$. If $k_1 < \alpha$, then the system is stable but the stability is not asymptotic. If $k_1 > \alpha$, then the system is unstable. The zero eigenvalue shows in general that no asymptotical stability occurs. Physically, it means that there is no feedback that makes the error of the initial state converge to zero as $t \to \infty$.

5. The behavior of the *two-tank hydraulic system* of Application 3.5.1-5 is very sensitive to the resistance R_1. Suppose that this resistance is increasing due to rust, or deposits of cholesterol or algae. How can you make the output less sensitive to changes in R_1? By using feedback. Our original system was described with

$$\mathbf{x}(t) = \begin{pmatrix} h_1(t) \\ h_2(t) \end{pmatrix}, \mathbf{A} = \begin{pmatrix} -\frac{1}{R_1 A_1} & \frac{1}{R_1 A_1} \\ \frac{1}{R_1 A_2} & -\left(\frac{1}{R_1 A_2} + \frac{1}{R_2 A_2}\right) \end{pmatrix}, \mathbf{b} = \begin{pmatrix} \frac{1}{A_1} \\ 0 \end{pmatrix},$$

and assume now that $\mathbf{c}^T = (0, 1/R_1)$.

Let us now measure the height of the fluid in each tank as shown in Figure 9.8, and use this in a feedback loop:

$$\tilde{u} = u + \mathbf{k}^T \mathbf{x} ,$$

where $\tilde{u}$ is the input with feedback and u is input without feedback. The feedback model is the following:

$$\dot{\mathbf{x}} = (\mathbf{A} + \mathbf{b}\mathbf{k}^T)\tilde{\mathbf{x}} + \mathbf{b}u .$$

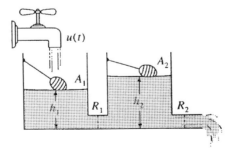

Figure 9.8 The two-tank system of Application 3.5.1-5 with the addition of water-level sensors to be used for feedback.

From

$$\tilde{H}(s) = \mathbf{c}^T (s\mathbf{I} - \mathbf{A} - \mathbf{b}\mathbf{k}^T)^{-1}\mathbf{b}$$

we have

$$\tilde{H}(s) = \begin{pmatrix} 0, \dfrac{1}{R_1} \end{pmatrix} \begin{pmatrix} s + \frac{1}{R_1 A_1} - \frac{k_1}{A_1} & -\frac{1}{R_1 A_1} - \frac{k_2}{A_1} \\ -\frac{1}{R_1 A_2} & s + \frac{1}{R_1 A_2} + \frac{1}{R_2 A_2} \end{pmatrix}^{-1} \begin{pmatrix} \frac{1}{A_1} \\ 0 \end{pmatrix} .$$

From these equations we can find that the transfer function is

$$\tilde{H}(s) =$$

$$\frac{R_2}{s^2(R_1^2 R_2 A_1 A_2) + s(R_1 R_2 A_1 + R_1^2 A_1 + R_1 R_2 A_2 - k_1 R_1^2 R_2 A_2) + C_1} ,$$

where C_1 is equal to $R_1 - k_1 R_1 R_2 - k_1 R_1^2 - k_2 R_1 R_2$. Now, if we want to talk about the sensitivity of a system, we must specify the sensitivity of what, to what, evaluated at what frequency or time. A good choice for this system is to look at the sensitivity of the steady-state value of the step-response with respect to R_1. To find this, let us first take the derivative of the step-response, $\tilde{H}(s)/s$ (which we will call $SR(s)$), with respect to R_1:

$$\frac{\partial SR(s)}{\partial R_1} =$$

$$\frac{-R_2[2s^2(R_1 R_2 A_1 A_2) + s(R_2 A_1 + 2R_1 A_1 + R_2 A_2 - 2k_1 R_1 R_2 A_2) + C_2]}{s[s^2(R_1^2 R_2 A_1 A_2) + s(R_1 R_2 A_1 + R_1^2 A_1 + R_1 R_2 A_2 - k_1 R_1^2 R_2 A_2) + C_1]^2} ,$$

where C_2 is equal to $1 - k_1 R_2 - 2k_1 R_1 - k_2 R_2$. To make this problem tractable, let us now make some simple numerical substitutions: let $R_1 = R_2 = A_1 = A_2 = 1$.

$$\frac{\partial SR(s)}{\partial R_1} = \frac{-[2s^2 + s(4 - 2k_1) + 1 - 3k_1 - k_2]}{s[s^2 + s(3 - k_1) + 1 - 2k_1 - k_2]^2} .$$

We are only interested in the steady-state, or low-frequency, characteristics of the step-response, so we can use the final value theorem derived in Application 3.5.1-9 as

$$\lim_{t \to \infty} f(t) = \lim_{s \to 0} sF(s)$$

to find

$$\lim_{s \to 0} s \frac{\partial SR(s)}{\partial R_1} = \frac{3k_1 + k_2 - 1}{[1 - 2k_1 - k_2]^2} = S_{R_1}^{SR} .$$

We can get the transfer function, step-response, and sensitivities for the original system without feedback by merely setting $k_1 =$

$k_2 = 0$ in the above equations. We will find that the sensitivity S of the steady-state value of the step-response, SR, with respect to R_1 for the system without feedback is

$$S_{R_1}^{SR} = -1 \ .$$

However, if we let k_1 and k_2 assume reasonable values, such as 10, then the sensitivity of the feedback system becomes

$$S_{R_1}^{SR} = 0.046 \ ,$$

which is much smaller. This shows that feedback reduces the sensitivity of the steady-state value of the step-response with respect to R_1.

In general, adding feedback transfers sensitivity from the hard-to-change elements in the forward path to the easily changeable elements in the feedback path.

6.   We will now discuss our *multiple input electrical system*. We know that its coefficient matrix has the form

$$\mathbf{A} = \begin{pmatrix} -\frac{R}{L_1} & 0 & -\frac{1}{L_1} \\ 0 & 0 & -\frac{1}{L_2} \\ \frac{1}{C} & \frac{1}{C} & 0 \end{pmatrix} \ .$$

Instead of applying feedback, we wish now to select the values of parameters L_1, L_2 and R such that the eigenvalues of the system be $\lambda_1 = \lambda_2 = \lambda_3 = -1$.

In Application 4.4.1-6 we have computed the characteristic polynomial of $\mathbf{A}$:

$$\varphi(\lambda) = \lambda^3 + \lambda^2 \frac{R}{L_1} + \lambda \left(\frac{1}{CL_1} + \frac{1}{CL_2} \right) + \frac{R}{CL_1 L_2} \ .$$

If $\lambda_1 = \lambda_2 = \lambda_3 = -1$, then

$$\varphi(\lambda) = (\lambda + 1)^3 = \lambda^3 + 3\lambda^2 + 3\lambda + 1 \ ,$$

therefore, the parameters satisfy equations

$$\frac{R}{L_1} = 3, \qquad \frac{1}{C} \left(\frac{1}{L_1} + \frac{1}{L_2} \right) = 3, \qquad \text{and} \qquad \frac{R}{CL_1 L_2} = 1 \ .$$

For any fixed value of C, the solutions of these equations are

$$L_1 = \frac{3}{8C}, \qquad L_2 = \frac{3}{C}, \qquad \text{and} \qquad R = \frac{9}{8C} \ ,$$

and with these numerical values all eigenvalues become -1.

7. Again, for the *stick-balancing problem*, assume we would like to use feedback to move the poles or adjust the sensitivity, but we have no way of measuring the position or angle of the stick on the person's hand, so we build an observer.
The original system is described with

$$\mathbf{A} = \begin{pmatrix} 0 & 1 \\ g & 0 \end{pmatrix}, \qquad \mathbf{b} = \begin{pmatrix} 0 \\ -g \end{pmatrix}, \qquad \text{and} \qquad \mathbf{c}^T = (1,0) \ .$$

The following observer can be obtained:

$$\mathbf{A} + \mathbf{kc}^T = \begin{pmatrix} 0 & 1 \\ g & 0 \end{pmatrix} + \begin{pmatrix} k_1 \\ k_2 \end{pmatrix} (1,0)$$

$$= \begin{pmatrix} k_1 & 1 \\ g + k_2 & 0 \end{pmatrix} \ .$$

What are the requirements to make the observer stable? Since

$$\varphi(\lambda) = (k_1 - \lambda)(-\lambda) - (g + k_2) = \lambda^2 - \lambda k_1 - (g + k_2) \ ,$$

the asymptotical stability conditions are, therefore,

$$k_1 < 0 \qquad \text{and} \qquad g + k_2 < 0 \ ,$$

which implies

$$k_2 < -g \ .$$

8. For *two sticks on a cart* problem, assume we have the same problem, namely that we want to use feedback to alter the dynamics or sensitivity but we cannot measure the state variables. So we build an observer.
The new coefficient matrix is

$$\mathbf{A} + \mathbf{kc}^T = \begin{pmatrix} 0 & 0 & 1 & 0 \\ 0 & 0 & 0 & 1 \\ a_1 & a_2 & 0 & 0 \\ a_3 & a_4 & 0 & 0 \end{pmatrix} + \begin{pmatrix} k_1 \\ k_2 \\ k_3 \\ k_4 \end{pmatrix} (1,0,0,0)$$

$$= \begin{pmatrix} k_1 & 0 & 1 & 0 \\ k_2 & 0 & 0 & 1 \\ a_1 + k_3 & a_2 & 0 & 0 \\ a_3 + k_4 & a_4 & 0 & 0 \end{pmatrix} \ .$$

Now let us prove that the observer can be made stable by finding the roots of the characteristic polynomial:

$$\varphi(\lambda) = \det \begin{pmatrix} k_1 - \lambda & 0 & 1 & 0 \\ k_2 & -\lambda & 0 & 1 \\ a_1 + k_3 & a_2 & -\lambda & 0 \\ a_3 + k_4 & a_4 & 0 & -\lambda \end{pmatrix}$$

$$= \det \begin{pmatrix} k_1 - \lambda & 0 & 1 \\ a_1 + k_3 & a_2 & -\lambda \\ a_3 + k_4 & a_4 & 0 \end{pmatrix} - \lambda \det \begin{pmatrix} k_1 - \lambda & 0 & 1 \\ k_2 & -\lambda & 0 \\ a_1 + k_3 & a_2 & -\lambda \end{pmatrix}$$

$$= \lambda^4 + \lambda^3(-k_1) + \lambda^2(-a_4 - a_1 - k_3)$$

$$+ \lambda(k_1 a_4 - k_2 a_2) + (a_1 a_4 + k_3 a_4 - a_2 a_3 - a_2 k_4) .$$

We want all roots to be negative, for example, $\lambda_1 = \lambda_2 = \lambda_3 = \lambda_4 = -1$, then

$$\varphi(\lambda) = (\lambda + 1)^4 = \lambda^4 + 4\lambda^3 + 6\lambda^2 + 4\lambda + 1 .$$

Comparing the like coefficients

$$-k_1 = 4$$

$$-a_4 - a_1 - k_3 = 6$$

$$k_1 a_4 - k_2 a_2 = 4$$

$$a_1 a_4 + k_3 a_4 - a_2 a_3 - a_2 k_4 = 1 ,$$

which can be easily solved for the unknowns k_1, k_2, k_3, and k_4,

$$k_1 = -4$$

$$k_3 = -a_4 - a_1 - 6$$

$$k_2 = \frac{-4a_4 - 4}{a_2}$$

$$k_4 = \frac{1}{a_2}(-a_4^2 - 6a_4 - a_2 a_3 - 1) .$$

Hence, selecting the above values for the feedback, the feedback system becomes asymptotically stable.

9. The *electronic heating system* mathematically is the special case of our earlier Application 3 with $L = 1$, therefore, the results of that case automatically can be applied.

10. Mathematically, the *nuclear reactor* problem is the same as the *electric L-R-C circuit* of Application 9.5.1-3, so we will do nothing else with it.

9.5.2 *Applications in the Social Sciences and Economics*

1. Consider first the linearized *predator–prey* model (6.16), which is repeated here for the sake of convenience:

$$\dot{G}_\delta = -\frac{bc}{d}W_\delta + \frac{c}{d}u$$

$$\dot{W}_\delta = \frac{da}{b}G_\delta$$

$$y = W_\delta \ . \tag{9.26}$$

Assume that the goat population cannot be measured; only the input u and the number of wolves y can be observed.

As an example, an observer for this system will be constructed. The general form is given by Equation (9.11). Note first that in this case

$$\mathbf{A} = \begin{pmatrix} 0 & -\frac{bc}{d} \\ \frac{da}{b} & 0 \end{pmatrix}, \qquad \mathbf{b} = \begin{pmatrix} \frac{c}{d} \\ 0 \end{pmatrix}, \qquad \text{and} \qquad \mathbf{c}^T = (0,1) \ .$$

In order to find the suitable feedback, we have to construct matrix $\mathbf{k}$. In our case, $\mathbf{k}$ is a two-dimensional column vector. If k_1 and k_2 denote the coefficients of $\mathbf{k}$, then

$$\mathbf{A} + \mathbf{k}\mathbf{c}^T = \begin{pmatrix} 0 & -\frac{bc}{d} \\ \frac{da}{b} & 0 \end{pmatrix} + \begin{pmatrix} k_1 \\ k_2 \end{pmatrix}(0,1) = \begin{pmatrix} 0 & -\frac{bc}{d} + k_1 \\ \frac{da}{b} & k_2 \end{pmatrix} \ .$$

Note that the characteristic polynomial of this matrix is

$$\varphi(\lambda) = -\lambda(k_2 - \lambda) - \frac{da}{b}\left(-\frac{bc}{d} + k_1\right) = \lambda^2 - k_2\lambda + \left(ac - k_1 \cdot \frac{da}{b}\right) \ .$$

From Application 4.4.1-3 we know that the roots of $\varphi(\lambda)$ have negative real parts if and only if

$$k_2 < 0 \qquad \text{and} \qquad ac - k_1 \frac{da}{b} > 0 \ .$$

For example, select $k_1 = 0$ and $k_2 = -1$. Then

$$\mathbf{A} + \mathbf{kc}^T = \begin{pmatrix} 0 & -\frac{bc}{d} \\ \frac{da}{b} & -1 \end{pmatrix} \ ,$$

and, hence, an observer (9.11) is given as

$$\dot{\mathbf{z}} = \begin{pmatrix} 0 & -\frac{bc}{d} \\ \frac{da}{b} & -1 \end{pmatrix} \mathbf{z} + \begin{pmatrix} \frac{c}{d} \\ 0 \end{pmatrix} u - \begin{pmatrix} 0 \\ -1 \end{pmatrix} y \ . \tag{9.27}$$

2. Assume that the population of at least one age group in the *cohort population* model (3.115) cannot be measured. Then an observer is supposed to be constructed analogously to the previous case. The details are left to the reader as an exercise.

3. In the case of the *arms races* model (6.18), assume again that nation 1 is unable to observe the armament level of the other nation, and, therefore, it is willing to use an appropriate observer of the form (9.11). In this case,

$$\mathbf{A} = \begin{pmatrix} -b & a \\ c & -d \end{pmatrix}, \qquad \mathbf{b} = \begin{pmatrix} \alpha \\ \beta \end{pmatrix}, \qquad \text{and} \qquad \mathbf{c}^T = (1,0) \ .$$

Matrix $\mathbf{K}$ is a column vector again with unknown coefficients k_1 and k_2, and, therefore,

$$\mathbf{A} + \mathbf{KC} = \begin{pmatrix} -b & a \\ c & -d \end{pmatrix} + \begin{pmatrix} k_1 \\ k_2 \end{pmatrix} (1,0) = \begin{pmatrix} -b + k_1 & a \\ c + k_2 & -d \end{pmatrix} \ .$$

The characteristic polynomial of this matrix is as follows:

$$\varphi(\lambda) = (-b + k_1 - \lambda)(-d - \lambda) - a(c + k_2)$$

$$= \lambda^2 + \lambda(d + b - k_1) + (bd - k_1 d - ac - ak_2) \ .$$

The resulting observer is asymptotically stable if and only if

$$d + b - k_1 > 0 \qquad \text{and} \qquad bd - k_1 d - ac - ak_2 > 0 \ .$$

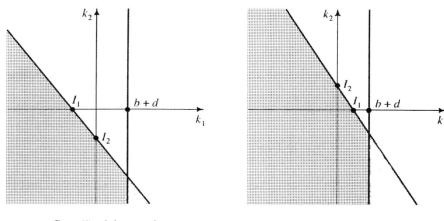

Case (*i*): $bd - ac \le 0$ Case (*ii*): $bd - ac > 0$

Figure 9.9   Feasible set for (k_1, k_2).

The feasible set (k_1, k_2) is shown in Figure 9.9.
The k_1 and k_2 intercepts are

$$I_1 = b - \frac{ac}{d} \quad \text{and} \quad I_2 = \frac{bd}{a} - c .$$

Note that in the case when $bd - ac > 0$, $k_1 = k_2 = 0$ belongs to
the feasible set. If $\mathbf{K} = \mathbf{O}$, no feedback is needed; the model itself
is asymptotically stable. If $bd - ac \le 0$, then, for example,

$$k_1 = b \quad \text{and} \quad k_2 = -2c$$

is an appropriate selection, and the resulting observer has the form

$$\dot{\mathbf{z}} = \begin{pmatrix} 0 & a \\ -c & -d \end{pmatrix} \mathbf{z} + \begin{pmatrix} \alpha \\ \beta \end{pmatrix} u - \begin{pmatrix} b \\ -2c \end{pmatrix} y . \qquad (9.28)$$

4.   Consider next *the warfare* model (6.19), where we assume that
     each nation can monitor only its own casualties. An observer is,
     therefore, needed to observe the casualties of the other nation. We
     can proceed similar to the previous model. The details are omitted.

5.   The linear *epidemics* model of Application 6.4.2-5 can be written
     as

$$\dot{\mathbf{x}} = \begin{pmatrix} 0 & -\alpha\bar{x} \\ 0 & \alpha\bar{x} - \beta \end{pmatrix} \mathbf{x} + \begin{pmatrix} 0 \\ -1 \end{pmatrix} u$$

$$y = (0, 1)\mathbf{x} .$$

Since this system is not observable, an observer cannot be constructed. We illustrate this fact here by examining the possible eigenvalue locations of the matrix $\mathbf{A} + \mathbf{KC}$. In our case,

$$\mathbf{A} + \mathbf{KC} = \begin{pmatrix} 0 & -\alpha\bar{x} \\ 0 & \alpha\bar{x} - \beta \end{pmatrix} + \begin{pmatrix} k_1 \\ k_2 \end{pmatrix} (0, 1)$$

$$= \begin{pmatrix} 0 & k_1 - \alpha\bar{x} \\ 0 & k_2 + \alpha\bar{x} - \beta \end{pmatrix} .$$

The eigenvalues of this matrix are $\lambda_1 = 0$ and $\lambda_2 = k_2 + \alpha\bar{x} - \beta$. Since 0 is an eigenvalue for all values of k_1 and k_2, the system cannot be made asymptotically stable.

6.   In the case of the *Harrod-type national* economy model presented in Application 6.4.2-6, we have the matrices

$$\mathbf{A} = (1 + r - rm), \quad \mathbf{B} = (-r), \quad \text{and} \quad \mathbf{C} = (m) .$$

Since they are 1×1, matrix $\mathbf{K}$ of the observer feedback must be also 1×1. In this case,

$$\mathbf{A} + \mathbf{KC} = 1 + r - rm + km ,$$

and since the system is discrete, the feedback system is asymptotically stable if and only if

$$-1 < 1 + r - rm + km < 1 ,$$

that is,

$$\frac{-2 + rm - r}{m} < k < \frac{rm - r}{m} .$$

7.   We face a similar situation in the case of the *linear cobweb* model of Application 6.4.2-7. Since now

$$\mathbf{A} = \begin{pmatrix} b \\ a \end{pmatrix}, \quad \mathbf{B} = (1), \quad \text{and} \quad \mathbf{C} = (b) ,$$

a feedback $\mathbf{K} = (k)$ makes the system asymptotically stable if and only if

$$-1 < \frac{b}{a} + bk < 1 .$$

These relations are equivalent to the inequalities

$$-\frac{a+b}{ab} < k < \frac{a-b}{ab} \; .$$

8. The Application 6.4.2-8 of *interrelated markets* is not always observable, as was shown. Therefore, in such cases, no observer can be constructed.

Consider again the numerical example of Application 7.4.2-8, which had the form

$$\dot{\mathbf{p}} = \begin{pmatrix} -3 & 2 \\ 0 & -8 \end{pmatrix} \mathbf{p} + \begin{pmatrix} 1 & 0 \\ 0 & 1 \end{pmatrix} \mathbf{u}$$

$$y = \begin{pmatrix} \dfrac{1}{2}, \dfrac{1}{2} \end{pmatrix} \mathbf{p} \; .$$

If a feedback is constructed with matrix

$$\begin{pmatrix} k_1 \\ k_2 \end{pmatrix} \; ,$$

then the modified coefficient matrix is as follows:

$$\mathbf{A} + \mathbf{KC} = \begin{pmatrix} -3 & 2 \\ 0 & -8 \end{pmatrix} + \begin{pmatrix} k_1 \\ k_2 \end{pmatrix} \begin{pmatrix} \dfrac{1}{2}, \dfrac{1}{2} \end{pmatrix}$$

$$= \begin{pmatrix} -3 + \dfrac{k_1}{2} & 2 + \dfrac{k_1}{2} \\ \dfrac{k_2}{2} & -8 + \dfrac{k_2}{2} \end{pmatrix} \; .$$

The characteristic polynomial of this matrix has the form

$$\varphi(\lambda) = \left(-3 + \frac{k_1}{2} - \lambda \right)\left(-8 + \frac{k_2}{2} - \lambda \right) - \frac{k_2}{2}\left(2 + \frac{k_1}{2} \right)$$

$$= \lambda^2 - \lambda \left(-8 + \frac{k_2}{2} - 3 + \frac{k_1}{2} \right)$$

$$+ \left[\left(-3 + \frac{k_1}{2} \right)\left(-8 + \frac{k_2}{2} \right) - \frac{k_2}{2}\left(2 + \frac{k_1}{2} \right) \right]$$

$$= \lambda^2 - \lambda \left(\frac{k_1 + k_2}{2} - 11 \right) + \left(24 - 4k_1 - \frac{5}{2}k_2 \right) \; .$$

The eigenvalues have negative real parts if and only if

$$\frac{k_1 + k_2}{2} - 11 < 0 \qquad \text{and} \qquad 24 - 4k_1 - \frac{5}{2}k_2 > 0 \ .$$

Note that $k_1 = k_2 = 0$ is feasible, which means that the original system is asymptotically stable. That is, it can serve as an observer without any feedback. However, faster speed can be obtained by the appropriate selection of k_1 and k_2.

9. In the case of an *oligopoly*, assume that the time scale is discrete and the firms behave according to the Cournot assumptions. The resulting dynamic model was derived in Section 3.5.2 as

$$\mathbf{x}(t + 1) = \mathbf{A}_C\mathbf{x}(t) + \mathbf{b}_C u(t) \ , \tag{9.29}$$

where

$$\mathbf{A}_C = \begin{pmatrix} 0 & -\frac{1}{2} & -\frac{1}{2} & \cdots & -\frac{1}{2} \\ -\frac{1}{2} & 0 & -\frac{1}{2} & \cdots & -\frac{1}{2} \\ -\frac{1}{2} & -\frac{1}{2} & 0 & \cdots & -\frac{1}{2} \\ \vdots & \vdots & \vdots & \ddots & \vdots \\ -\frac{1}{2} & -\frac{1}{2} & -\frac{1}{2} & \cdots & 0 \end{pmatrix}, \qquad \mathbf{b}_C = \frac{1}{2a}\begin{pmatrix} b_1 - b \\ b_2 - b \\ \vdots \\ b_N - b \end{pmatrix}$$

and

$$u(t) \equiv 1 \ .$$

We have shown in Section 4.4.2 that this system is asymptotically stable if and only if $N = 2$, and it can be stabilized for arbitrary number N of firms by adaptive expectations. We conclude this section by investigating the possibility of a feedback-type stabilizator.

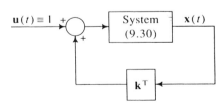

Figure 9.10   Feedback for the oligopoly model.

Figure 9.10 shows the feedback structure. Note that the feedback matrix is a row vector in order to have a scalar feedback that can be added to the original input of the system. The resulting system has the form

$$\mathbf{x}(t+1) = \mathbf{A}_C\mathbf{x}(t) + \mathbf{b}_C(u(t) + \mathbf{k}^T\mathbf{x}(t))$$

$$= (\mathbf{A}_C + \mathbf{b}_C\mathbf{k}^T)\mathbf{x}(t) + \mathbf{b}_Cu(t) , \qquad (9.30)$$

which is asymptotically stable if and only if all eigenvalues of matrix $\mathbf{A}_C + \mathbf{b}_C\mathbf{k}^T$ are inside the unit circle. Simple calculation shows that

$$\mathbf{A}_C + \mathbf{b}_C\mathbf{k}^T =$$

$$\begin{pmatrix}
\frac{1}{2a}(b_1-b)k_1 & \frac{1}{2a}(b_1-b)k_2 - \frac{1}{2} & \frac{1}{2a}(b_1-b)k_3 - \frac{1}{2} & \cdots & \frac{1}{2a}(b_1-b)k_N - \frac{1}{2} \\
\frac{1}{2a}(b_2-b)k_1 - \frac{1}{2} & \frac{1}{2a}(b_2-b)k_2 & \frac{1}{2a}(b_2-b)k_3 - \frac{1}{2} & \cdots & \frac{1}{2a}(b_2-b)k_N - \frac{1}{2} \\
\frac{1}{2a}(b_3-b)k_1 - \frac{1}{2} & \frac{1}{2a}(b_3-b)k_2 - \frac{1}{2} & \frac{1}{2a}(b_3-b)k_3 & \cdots & \frac{1}{2a}(b_3-b)k_N - \frac{1}{2} \\
\vdots & \vdots & \vdots & \ddots & \vdots \\
\frac{1}{2a}(b_N-b)k_1 - \frac{1}{2} & \frac{1}{2a}(b_N-b)k_2 - \frac{1}{2} & \frac{1}{2a}(b_N-b)k_3 - \frac{1}{2} & \cdots & \frac{1}{2a}(b_N-b)k_N
\end{pmatrix} .$$

Note that the eigenvalues of $\mathbf{A}_C + \mathbf{b}_C\mathbf{k}^T$ depend on the particular selection of the coefficients $k_1, k_2, \ldots, k_N$.

For the sake of simplicity, consider the special case when $b_1 = b_2 = \cdots = b_N$. Let b^* denote the common value and assume that $b^* \neq b$. Select the k_i values as

$$k_1 = k_2 = \cdots = k_N = \frac{a}{b^* - b} .$$

Then

$$\mathbf{A}_C + \mathbf{b}_C\mathbf{k}^T = \begin{pmatrix}
\frac{1}{2} & 0 & 0 & \cdots & 0 \\
0 & \frac{1}{2} & 0 & \cdots & 0 \\
0 & 0 & \frac{1}{2} & \cdots & 0 \\
\vdots & \vdots & \vdots & \ddots & \vdots \\
0 & 0 & 0 & \cdots & \frac{1}{2}
\end{pmatrix} = \frac{1}{2}\mathbf{I}$$

with eigenvalues being equal to $1/2$. Hence the feedback system is asymptotically stable.

In the general case when the b_i values are different, we can proceed in the following way. Rewrite system (9.29) as

$$\mathbf{x}(t+1) = \mathbf{A}_C\mathbf{x}(t) + (\mathbf{b}_C - \mathbf{1}) + \mathbf{1}u(t) ,$$

where $\mathbf{1} = (1, 1, \ldots, 1)^T$. Then the above feedback rule results in the system

$$\mathbf{x}(t+1) = \mathbf{A}_C\mathbf{x}(t) + (\mathbf{b}_C - \mathbf{1}) + \mathbf{1}(u(t) + \mathbf{k}^T\mathbf{x}(t))$$

$$= (\mathbf{A}_C + \mathbf{1}\mathbf{k}^T)\mathbf{x}(t) + (\mathbf{b}_C - \mathbf{1}) + \mathbf{1}u(t) . \qquad (9.31)$$

By selecting $k_1 = k_2 = \cdots = k_N = 1/2$, $\mathbf{A}_C + \mathbf{1}\mathbf{k}^T$ becomes $(1/2)\mathbf{I}$ again. Hence, system (9.31) is asymptotically stable.

Problems

1. Apply the algorithm of Theorem 9.1 for system

$$\dot{\mathbf{x}} = \begin{pmatrix} 1\ 1 \\ 2\ 2 \end{pmatrix}\mathbf{x} + \begin{pmatrix} 1 \\ 0 \end{pmatrix}u$$

and polynomial $p(\lambda) = \lambda^2 + 2\lambda + 1$.

2. Apply the algorithm of Theorem 9.1 for system

$$\dot{\mathbf{x}} = \begin{pmatrix} 0\ 1\ 2 \\ 0\ 1\ 1 \\ 0\ 0\ 2 \end{pmatrix}\mathbf{x} + \begin{pmatrix} 1 \\ 2 \\ 1 \end{pmatrix}u$$

and polynomial $p(\lambda) = \lambda^3 + 1$.

3. Apply the algorithm of Theorem 9.1 for system

$$\dot{\mathbf{x}} = \begin{pmatrix} 0\ 0\ 0 \\ 0\ 1\ 0 \\ 0\ 0\ 2 \end{pmatrix}\mathbf{x} + \begin{pmatrix} 1 \\ 1 \\ 1 \end{pmatrix}u$$

and polynomial $p(\lambda) = \lambda^3$.

4. Apply the algorithm of Theorem 9.1 for system

$$\dot{\mathbf{x}} = \begin{pmatrix} 2\ 1 \\ 0\ 2 \end{pmatrix}\mathbf{x} + \begin{pmatrix} 1 \\ 1 \end{pmatrix}u$$

and polynomial $p(\lambda) = \lambda^2 + \lambda + 1$.

5. Apply the algorithm of Theorem 9.1 for system

$$\dot{\mathbf{x}} = \begin{pmatrix} 1 & 0 & 1 \\ 0 & 1 & 1 \\ 1 & 1 & 1 \end{pmatrix} \mathbf{x} + \begin{pmatrix} 0 \\ 1 \\ 0 \end{pmatrix} u$$

and polynomial $p(\lambda) = \lambda^3 + 2\lambda - 1$.

6. Apply the corollary of Theorem 9.1 for system

$$\dot{\mathbf{x}} = \begin{pmatrix} 1 & 1 \\ 2 & 2 \end{pmatrix} \mathbf{x} + \begin{pmatrix} 1 \\ 0 \end{pmatrix} u$$

and polynomial $p(\lambda) = \lambda^2 + 2\lambda + 1$. Select $\mathbf{c}^T = (1, 0)$.

7. Apply the corollary of Theorem 9.1 for system

$$\dot{\mathbf{x}} = \begin{pmatrix} 0 & 1 & 2 \\ 0 & 1 & 1 \\ 0 & 0 & 2 \end{pmatrix} \mathbf{x} + \begin{pmatrix} 1 \\ 2 \\ 1 \end{pmatrix} u$$

and polynomial $p(\lambda) = \lambda^3 + 1$. Select $\mathbf{c}^T = (1, 1, 1)$.

8. Apply the corollary of Theorem 9.1 for system

$$\dot{\mathbf{x}} = \begin{pmatrix} 0 & 0 & 0 \\ 0 & 1 & 0 \\ 0 & 0 & 2 \end{pmatrix} \mathbf{x} + \begin{pmatrix} 1 \\ 1 \\ 1 \end{pmatrix} u$$

and polynomial $p(\lambda) = \lambda^3$. Select $\mathbf{c}^T = (1, 1, 1)$.

9. Apply the corollary of Theorem 9.1 for system

$$\dot{\mathbf{x}} = \begin{pmatrix} 2 & 1 \\ 0 & 2 \end{pmatrix} \mathbf{x} + \begin{pmatrix} 1 \\ 1 \end{pmatrix} u$$

and polynomial $p(\lambda) = \lambda^2 + \lambda + 1$. Select $\mathbf{c}^T = (1, 1)$.

10. Apply the corollary of Theorem 9.1 for system

$$\dot{\mathbf{x}} = \begin{pmatrix} 1 & 0 & 1 \\ 0 & 1 & 1 \\ 1 & 1 & 1 \end{pmatrix} \mathbf{x} + \begin{pmatrix} 0 \\ 1 \\ 0 \end{pmatrix} u$$

and polynomial $p(\lambda) = \lambda^3 + 2\lambda - 1$. Select $\mathbf{c}^T = (0, 1, 0)$.

11. Show that the algorithm of Theorem 9.1 cannot be applied for matrix

$$A = \begin{pmatrix} 0 & 0 & 0 \\ 0 & 0 & 0 \\ 0 & 0 & 1 \end{pmatrix}.$$

12. Construct an observer (9.12) for system

$$\dot{x} = \begin{pmatrix} 1 & 1 \\ 2 & 2 \end{pmatrix} x + \begin{pmatrix} 1 \\ 0 \end{pmatrix} u$$

$$y = (1, 0)x .$$

13. Construct an observer (9.12) for system

$$\dot{x} = \begin{pmatrix} 0 & 1 & 2 \\ 0 & 1 & 1 \\ 0 & 0 & 2 \end{pmatrix} x + \begin{pmatrix} 1 \\ 2 \\ 1 \end{pmatrix} u$$

$$y = (1, 1, 1)x .$$

Place the roots of the observer at -1, -2, and -3.

14. Construct an observer (9.12) for system

$$\dot{x} = \begin{pmatrix} 0 & 0 & 0 \\ 0 & 1 & 0 \\ 0 & 0 & 2 \end{pmatrix} x + \begin{pmatrix} 1 \\ 1 \\ 1 \end{pmatrix} u$$

$$y = (1, 1, 1)x .$$

15. Construct an observer (9.12) for system

$$\dot{x} = \begin{pmatrix} 2 & 1 \\ 0 & 2 \end{pmatrix} x + \begin{pmatrix} 1 \\ 1 \end{pmatrix} u$$

$$y = (1, 1)x .$$

16. Construct an observer (9.17) for system

$$\dot{x} = \begin{pmatrix} 1 & 1 \\ 2 & 2 \end{pmatrix} x + \begin{pmatrix} 1 \\ 0 \end{pmatrix} u$$

$$y = (1, 0)x .$$

17. Construct an observer (9.17) for system

$$\dot{x} = \begin{pmatrix} 2 & 1 \\ 0 & 2 \end{pmatrix} x + \begin{pmatrix} 1 \\ 1 \end{pmatrix} u$$

$$y = (1, 1)x .$$

18. Repeat Example 9.4 for system

$$\dot{x} = \begin{pmatrix} 1 & 1 \\ 2 & 2 \end{pmatrix} x + \begin{pmatrix} 1 \\ 0 \end{pmatrix} u$$

$$y = (1, 0)x .$$

19. Repeat Example 9.4 for system

$$\dot{x} = \begin{pmatrix} 2 & 1 \\ 0 & 2 \end{pmatrix} x + \begin{pmatrix} 1 \\ 1 \end{pmatrix} u$$

$$y = (1, 1)x.$$

20. Construct observers and reduced-order observers for discrete systems.

21. Let $\phi_1(t, \tau)$ and $\phi_2(t, \tau)$ be the fundamental matrices of a system $\dot{x}(t) = A(t)x(t) + B(t)u(t)$ and the corresponding feedback system $\dot{z}(t) = (A(t) + B(t)K(t))z(t) + B(t)u(t)$, respectively. Show that

$$\phi_2(t, \tau) = \phi_1(t, \tau) + \int_\tau^t \phi_1(t, s)B(s)K(s)\phi_2(s, \tau)ds .$$

22. Let A, B, K be constant matrices. Let $R_1(s)$ and $R_2(s)$ denote the resolvent matrices of the system $\dot{x} = Ax + Bu$ and the corresponding feedback system $\dot{z} = (A + BK)z + Bu$. Show that

$$R_2(s) = [I - R_1(s)BK]^{-1}R_1(s) .$$

23. Prove that the time invariant system $\dot{x} = Ax + Bu$ is completely controllable if and only if the corresponding feedback system $\dot{z} = (A + BK)z + Bu$ is completely controllable. That is, state feedback does not destroy controllability.

24. Let $\dot{x} = Ax + Bu$ be a time invariant linear system, and assume that with some matrices A_1, R and invertible Q,

$$AQ - QA_1 = BR .$$

Find the state feedback such that the characteristic polynomials of matrices A_1 and $A + BK$ coincide.

25. Find the feedback matrix K which minimizes $\|A + BK\|_\infty$.

chapter ten

Advanced Topics

In this chapter, the fundamentals of four modern areas of systems theory are briefly outlined. In the first section we discuss conditions that guarantee that the state of a system is always nonnegative. This property is important because in many applications the state must not be negative. For example, production output, population, armament level, water flow, and so on must always be nonnegative. The second section contains the description of a special filter that minimizes the mean square error of the final state. The third section is devoted to adaptive control systems, where the control is based on continuous measurements of the state and/or output. In the last section, the basics of neural networks and neural computing are outlined.

10.1 Nonnegative Systems

Consider first the time-invariant discrete system

$$\mathbf{x}(t+1) = \mathbf{A}\mathbf{x}(t), \qquad \mathbf{x}(0) = \mathbf{x}_0 \tag{10.1}$$

with zero input, where $\mathbf{A}$ is an $n \times n$ constant matrix.

First we state a necessary and sufficient condition for the nonnegativity of this system.

DEFINITION 10.1   *A matrix $\mathbf{A}$ is nonnegative if all elements of $\mathbf{A}$ are nonnegative. This property is denoted as $\mathbf{A} \geq \mathbf{O}$.*

THEOREM 10.1
The state $\mathbf{x}(t)$ of system (10.1) is nonnegative for all $t \geq 0$ with arbitrary nonnegative initial state $\mathbf{x}_0$ if and only if $\mathbf{A} \geq \mathbf{O}$.

PROOF

(a) Assume first that $\mathbf{A} \geq \mathbf{O}$. We will use finite induction to verify that $\mathbf{x}(t) \geq \mathbf{0}$ for all $t \geq 1$, if $\mathbf{x}_0 \geq \mathbf{0}$. For $t = 1$,

$$\mathbf{x}(1) = \mathbf{A}\mathbf{x}(0) = \mathbf{A}\mathbf{x}_0 \geq \mathbf{0} ,$$

since both $\mathbf{A}$ and $\mathbf{x}_0$ are nonnegative. Assume next that for some t, $\mathbf{x}(t) \geq \mathbf{0}$. Then

$$\mathbf{x}(t + 1) = \mathbf{A}\mathbf{x}(t) \geq \mathbf{0} ,$$

which proves that $\mathbf{x}(t) \geq \mathbf{0}$ for all t.

(b) Assume now that $a_{kl} < 0$ with some k and l, where $\mathbf{A} = (a_{ij})$. Select $\mathbf{x}_0 = (0, \ldots, 0, 1, 0, \ldots, 0)^T$, where the lth component equals unity and all other components are equal to zero. Then

$$\mathbf{x}(1) = \mathbf{A}\mathbf{x}(0) = (a_{1l}, \ldots, a_{kl}, \ldots, a_{nl})^T ,$$

which is not nonnegative, since the lth element is negative. ∎

Assume that in system (10.1) the zero input is replaced by a constant, that is, the system is described with the difference equation

$$\mathbf{x}(t + 1) = \mathbf{A}\mathbf{x}(t) + \mathbf{b}, \qquad \mathbf{x}(0) = \mathbf{x}_0 , \tag{10.2}$$

where $\mathbf{b}$ is a constant vector. In this more general case, Theorem 10.1 can be extended as follows.

THEOREM 10.2

The state $\mathbf{x}(t)$ of system (10.2) is nonnegative for all $t \geq 0$ with arbitrary nonnegative initial state $\mathbf{x}_0$ if and only if $\mathbf{A} \geq \mathbf{O}$ and $\mathbf{b} \geq \mathbf{0}$.

REMARK 10.1 The nonlinear version of Equation (10.2) can be written as

$$\mathbf{x}(t + 1) = \mathbf{f}(\mathbf{x}(t)), \qquad \mathbf{x}(0) = \mathbf{x}_0 .$$

Starting from any nonnegative initial state $\mathbf{x}_0$, the state remains nonnegative for all future times if and only if $\mathbf{f}(\mathbf{x}) \geq \mathbf{0}$ for all $\mathbf{x} \geq \mathbf{0}$. ∎

Consider next the continuous system

$$\dot{\mathbf{x}} = \mathbf{A}\mathbf{x}, \qquad \mathbf{x}(0) = \mathbf{x}_0 \tag{10.3}$$

with zero input. Before deriving necessary and sufficient conditions for the nonnegativity of this system, a special class of matrices is introduced.

DEFINITION 10.2   *Matrix* $\mathbf{A} = (a_{ij})$ *is called a Metzler matrix if* $a_{ij} \geq 0$ *for all* $i \neq j$.

THEOREM 10.3

The state $\mathbf{x}(t)$ *of system (10.3) is nonnegative for all* $t \geq 0$ *with arbitrary nonnegative initial state* $\mathbf{x}_0$ *if and only if* $\mathbf{A}$ *is a Metzler matrix.*

PROOF

(a) Assume first that for all $j \neq i$, $a_{ij} > 0$. We will prove that starting from a nonnegative initial state, no component of the state becomes negative. The continuity of functions $\mathbf{x}_i(t)$ implies that if any component of $\mathbf{x}$ becomes negative, it has to be zero before. Assume that for some $t_0 \geq 0$, $\mathbf{x}(t_0) \geq 0$, but with some i, $x_i(t_0) = 0$.

If $x_j(t_0) = 0$ for all $j \neq i$, then for all $t \geq t_0$, $\mathbf{x}(t) = \mathbf{0}$. Otherwise, Equation (10.3) implies that

$$\dot{x}_i(t_0) = \sum_{j \neq i} a_{ij} x_j(t_0) > 0 ,$$

since each term is nonnegative and at least one term is positive. Hence x_i must not become negative.

Since the solution of Equation (10.3) depends continuously on the elements of $\mathbf{A}$, the weaker condition $a_{ij} \geq 0$ $(j \neq i)$ is also sufficient.

(b) Assume now that with some $k \neq l$, $a_{kl} < 0$. Select $\mathbf{x}_0 = (0, \ldots, 0, \alpha, 0, \ldots, 0)^T$, where the lth component equals $\alpha > 0$ and all other components equal zero. Then

$$\dot{x}_k(0) = \sum_{j=1}^{n} a_{kj} x_j(0) = a_{kl} \cdot x_l(0) = a_{kl}\alpha < 0 ,$$

which implies that $x_k(t) < 0$ for small positive values of t.   ∎

Consider next the slightly more general case, when the input is constant. Then Equation (10.3) is modified as

$$\dot{\mathbf{x}} = \mathbf{A}\mathbf{x} + \mathbf{b}, \qquad \mathbf{x}(0) = \mathbf{x}_0 , \tag{10.4}$$

where $\mathbf{b}$ is a constant vector. In this case, the previous theorem can be extended as follows.

THEOREM 10.4

The state $\mathbf{x}(t)$ of system (10.4) is nonnegative for all $t \geq 0$ with arbitrary nonnegative initial state $\mathbf{x}_0$ if and only if $\mathbf{b} \geq 0$ and $\mathbf{A}$ is a Metzler matrix.

REMARK 10.2   The nonlinear version of Equation (10.4) can be written as

$$\dot{\mathbf{x}} = \mathbf{f}(\mathbf{x}), \qquad \mathbf{x}(0) = \mathbf{x}_0 \ .$$

Similar to Theorem 10.3, one may easily prove that starting from arbitrary nonnegative initial state $\mathbf{x}_0$, the state remains nonnegative for all future times if and only if

$$f_i(x_1, \ldots, x_{i-1}, 0, x_{i+1}, \ldots, x_n) \geq 0 \qquad (i = 1, 2, \ldots, n)$$

for all $x_j \geq 0$ $(j = 1, 2, \ldots, n)$, where f_i denotes the ith component of function $\mathbf{f}$.   ∎

The practical consequence of the above theorems is that if you are modeling a system where you know the state variables must always be nonnegative, then you should check the above conditions to make sure the state really remains always nonnegative.

In the next part of this section, conditions will be presented for the existence of nonnegative equilibria of systems (10.2) and (10.4).

It is known from Section 3.1 that a vector $\bar{\mathbf{x}}$ is an equilibrium state of the discrete system (10.2) if and only if it satisfies equation

$$\bar{\mathbf{x}} = \mathbf{A}\bar{\mathbf{x}} + \mathbf{b} \ . \tag{10.5}$$

Assume that $\mathbf{B} = \mathbf{I} - \mathbf{A}$ is invertible, then

$$\bar{\mathbf{x}} = \mathbf{B}^{-1}\mathbf{b} \ .$$

Obviously, $\bar{\mathbf{x}} \geq 0$ for all nonnegative vectors $\mathbf{b}$ if and only if $\mathbf{B}^{-1} \geq \mathbf{O}$.

A vector $\bar{\mathbf{x}}$ is the equilibrium state of the continuous system (10.4) if and only if

$$\mathbf{A}\bar{\mathbf{x}} + \mathbf{b} = \mathbf{0} \ . \tag{10.6}$$

Assume that $\mathbf{A}$ is nonsingular, then

$$\bar{\mathbf{x}} = -\mathbf{A}^{-1}\mathbf{b} \ .$$

This vector is nonnegative for all nonnegative $\mathbf{b}$ if and only if

$$\mathbf{B}^{-1} \geq \mathbf{O} \qquad \text{with} \qquad \mathbf{B} = -\mathbf{A} \ .$$

Note that in the above two cases the nonnegativity of the inverse of a matrix, rather than the nonnegativity of the matrix itself, guarantees the nonnegativity of the equilibrium. In the following theorem, necessary and sufficient conditions will be presented for the nonnegativity of the inverse of a real square matrix.

THEOREM 10.5

Let $\mathbf{B}$ be an $n \times n$ real matrix. The inverse $\mathbf{B}^{-1}$ exists and is nonnegative if and only if there exists an $n \times n$ real matrix $\mathbf{D}$ such that

(i) $\mathbf{D} \geq \mathbf{O}$,

(ii) $\mathbf{I} - \mathbf{DB} \geq \mathbf{O}$, and

(iii)   all eigenvalues of $\mathbf{I} - \mathbf{DB}$ are inside the unit circle.

PROOF

(a) If $\mathbf{B}^{-1}$ exists and is nonnegative, then select $\mathbf{D} = \mathbf{B}^{-1}$. Then $\mathbf{I} - \mathbf{DB} = \mathbf{O}$, and, therefore, Conditions (i), (ii), and (iii) are obviously satisfied.

(b) Assume now the existence of matrix $\mathbf{D}$ satisfying Conditions (i), (ii), and (iii). Note first that Condition (iii) and Example 1.24 imply that $\mathbf{I} - (\mathbf{I} - \mathbf{DB}) = \mathbf{DB}$ is invertible. Therefore, $\mathbf{D}$ and $\mathbf{B}$ are both nonsingular. Example 1.24 also implies that

$$(\mathbf{DB})^{-1} = \mathbf{I} + (\mathbf{I} - \mathbf{DB}) + (\mathbf{I} - \mathbf{DB})^2 + \cdots \geq \mathbf{O}$$

since each term in the right-hand side is nonnegative as the consequence of Condition (ii). Hence, from (i) we conclude that

$$\mathbf{B}^{-1} = (\mathbf{D}^{-1}\mathbf{DB})^{-1} = (\mathbf{DB})^{-1}\mathbf{D} \geq \mathbf{O},$$

which completes the proof.   ∎

DEFINITION 10.3   An $n \times n$ real matrix $\mathbf{B} = (b_{ij})$ is called an M-matrix if $\mathbf{D} = diag(b_{11}, b_{22}, \ldots, b_{nn})^{-1}$ satisfies conditions (i), (ii), and (iii) of Theorem 10.5.

Note that for an M-matrix,

$$\mathbf{I} - \mathbf{DB} = \begin{pmatrix} 0 & -\frac{b_{12}}{b_{11}} & \cdots & -\frac{b_{1n}}{b_{11}} \\ -\frac{b_{21}}{b_{22}} & 0 & \cdots & -\frac{b_{2n}}{b_{22}} \\ \vdots & \vdots & \ddots & \vdots \\ -\frac{b_{n1}}{b_{nn}} & -\frac{b_{n2}}{b_{nn}} & \cdots & 0 \end{pmatrix} ; \tag{10.7}$$

therefore, $b_{ii} > 0$ for all i and $b_{ij} \leq 0$ for all $j \neq i$. Consequently, if $\mathbf{B}$ is an M-matrix, then $-\mathbf{B}$ is a Metzler matrix.

Metzler matrices are also closely related to nonnegative matrices, since if $\mathbf{A}$ is a Metzler matrix, then $\mathbf{A} + \alpha\mathbf{I} \geq \mathbf{O}$, where α is sufficiently large. Select, for example,

$$\alpha = \max_i \{ -a_{ii} \mid a_{ii} < 0 \} .$$

As the conclusion of this section, a sufficient condition is presented for an $n \times n$ real matrix to be an M-matrix.

THEOREM 10.6
Let $\mathbf{B}$ be an $n \times n$ real matrix such that

(i) $b_{ii} > 0$ for all i, and $b_{ij} \leq 0$ for all $j \neq i$;

(ii) $b_{ii} > \sum_{j \neq i} |b_{ij}|$ for all i.

Then $\mathbf{B}$ is an M-matrix.

PROOF   Matrix $\mathbf{D} = diag(b_{11}, b_{22}, \ldots, b_{nn})^{-1}$ is nonnegative, and relation (10.7) implies that $\mathbf{I} - \mathbf{DB} \geq \mathbf{O}$. That is, matrix $\mathbf{D}$ satisfies Conditions (i) and (ii) of Theorem 10.5. From (10.7) we have

$$\|\mathbf{I} - \mathbf{DB}\|_\infty = \max_i \sum_{j \neq i} \frac{|b_{ij}|}{|b_{ii}|} = \max_i \frac{1}{b_{ii}} \sum_{j \neq i} |b_{ij}| < 1 .$$

Therefore, Theorem 1.8 implies that for all eigenvalues λ of $\mathbf{I} - \mathbf{DB}$,

$$|\lambda| \leq \|\mathbf{I} - \mathbf{DB}\|_\infty < 1 .$$

Therefore, matrix $\mathbf{D}$ satisfies Condition (iii), which completes the proof.
∎

Finally, we note that some of the systems discussed in Section 3.5 satisfy the conditions of Theorems 10.2 and 10.4. In particular, Applications 3.5.2-2 and 3.5.2-6 have nonnegative coefficient matrices and Applications 3.5.1-4, 3.5.1-5, 3.5.1-7, 3.5.1-8, 3.5.1-10, 3.5.2-3, and 3.5.2-8 are based on Metzler matrices. In these applications, the states always remain nonnegative.

10.2   The Kalman–Bucy Filter

In the previous chapters, only *deterministic* systems were discussed. That is, we assumed that all inputs could be specified exactly and all outputs could be measured without measurement errors. In practice, these assumptions are rarely satisfied, since input and output components are usually corrupted by all manner of unpredictable fluctuations and disturbances.

The most common approach for analyzing the effect of such *noise* is based on probabilistic or statistical models, where random elements are added to the input and output components. The resulting *stochastic system* has the form

$$\dot{\mathbf{x}}(t) = \mathbf{A}\mathbf{x}(t) + \mathbf{B}\mathbf{u}(t) + \mathbf{B}_1\mathbf{w}(t)$$

$$\mathbf{y}(t) = \mathbf{C}\mathbf{x}(t) + \mathbf{v}(t) , \tag{10.8}$$

where $\mathbf{w}(t)$ is the *input noise* and $\mathbf{v}(t)$ is the *output noise*. The components of $\mathbf{w}(t)$ may represent fluctuation in the input signal, unknown disturbances to the system, or their combinations. Vector $\mathbf{v}(t)$ represents uncertainties or deviations in the output measurements. For the sake of simplicity, only time-invariant systems are considered.

Assume that the following conditions hold:

1.   System $(\mathbf{A}, \mathbf{B}, \mathbf{C})$ is completely controllable and completely observable.

2.   The processes $\mathbf{w}$ and $\mathbf{v}$ are assumed to have zero means and to be white, that is, for all t,

$$E[\mathbf{w}(t)] = \mathbf{0}$$

$$E[\mathbf{v}(t)] = \mathbf{0} , \tag{10.9}$$

and for all t and τ,

$$E[\mathbf{w}(t)\mathbf{w}^T(\tau)] = \mathbf{Q}\delta(t - \tau)$$

$$E[\mathbf{v}(t)\mathbf{v}^T(\tau)] = \mathbf{R}\delta(t - \tau) \qquad (10.10)$$

where $\mathbf{Q}$ and $\mathbf{R}$ are given (positive semidefinite) covariance matrices, and

$$\delta(t - \tau) = \begin{cases} 1 \text{ if } t = \tau \\ 0 \text{ otherwise .} \end{cases}$$

3. Processes $\mathbf{w}$ and $\mathbf{v}$ are uncorrelated with one another, that is, for all t and τ,

$$E[\mathbf{w}(t)\mathbf{v}^T(\tau)] = \mathbf{O} . \qquad (10.11)$$

4. The initial state $\mathbf{x}_0$ of the system at time t_0 is assumed to be a random variable with

$$E[\mathbf{x}_0] = \bar{\mathbf{x}}_0$$

and

$$E[(\mathbf{x}_0 - \bar{\mathbf{x}}_0)(\mathbf{x}_0 - \bar{\mathbf{x}}_0)^T] = \mathbf{S} , \qquad (10.12)$$

where $\bar{\mathbf{x}}_0$ is a given vector and $\mathbf{S}$ is a given (positive semidefinite) covariance matrix.

Assume that for all $t \in [t_0, T)$ the noisy output measurements $\mathbf{y}(t)$ are available where $T > t_0$ is given. The problem is now to estimate the final state $\mathbf{x}(T)$ as accurately as possible. By selecting the mean square principle for measuring the goodness of the estimator, we wish to find the estimator $\hat{\mathbf{x}}(T)$ of $\mathbf{x}(T)$ that minimizes the mean square error

$$E[\|\mathbf{x}(T) - \hat{\mathbf{x}}(T)\|^2] = \sum_{i=1}^{n} E[(x_i(T) - \hat{x}_i(T))^2] , \qquad (10.13)$$

where $\| \cdot \|$ is the l_2-norm, n is the size of $\mathbf{A}$, $\mathbf{x} = (x_i)$ and $\hat{\mathbf{x}} = (\hat{x}_i)$.

In Section 9.2, a solution for the deterministic version of the same problem was introduced by constructing an observer, which consisted of the model of the original system and a correction term being proportional to the output error. It can be proven (see, for example, [12]) that the best state estimator in the above stochastic case has a similar form, as given in the following theorem.

THEOREM 10.7

Under the above conditions, the best least square estimator of the state of system (10.8) is given by the state $z(t)$ of the Kalman–Bucy filter:

$$\dot{z}(t) = (\mathbf{A} + \mathbf{K}(t)\mathbf{C})z(t) + \mathbf{B}u(t) - \mathbf{K}(t)\mathbf{y}(t), \qquad z(t_0) = \bar{\mathbf{x}}_0 , \quad (10.14)$$

where

$$\mathbf{K}(t) = -\mathbf{P}(t)\mathbf{C}^T\mathbf{R}^{-1} \qquad (10.15)$$

with $\mathbf{P}(t)$ *being the solution of the matrix Riccati equation*

$$\dot{\mathbf{P}}(t) = \mathbf{A}\mathbf{P}(t) + \mathbf{P}(t)\mathbf{A}^T - \mathbf{P}(t)\mathbf{C}^T\mathbf{R}^{-1}\mathbf{C}\mathbf{P}(t) + \mathbf{B}_1\mathbf{Q}\mathbf{B}_1^T,$$

$$\mathbf{P}(t_0) = \mathbf{S} \qquad (10.16)$$

The scheme of the Kalman–Bucy filter is shown in Figure 10.1.

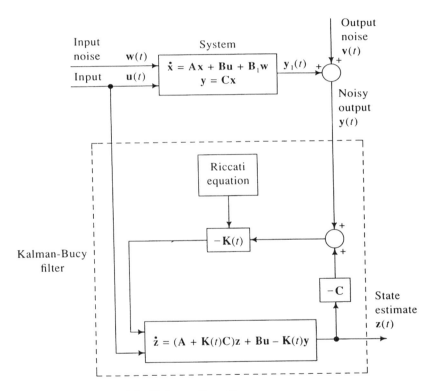

Figure 10.1 Scheme of the Kalman–Bucy filter.

The application of the Kalman–Bucy filter is limited because of the time dependence of matrix $\mathbf{K}(t)$, which makes the computation and, therefore, the whole construct very complicated in many cases. However, this difficulty can be eliminated in the following way. Note first that as $t \to \infty$, the solution $\mathbf{P}(t)$ of the Riccati equation converges to a steady state, which is the solution for $\dot{\mathbf{P}} = 0$. That is, this steady state $\mathbf{P}$ solves equation

$$\mathbf{O} = \mathbf{AP} + \mathbf{PA}^T - \mathbf{PC}^T\mathbf{R}^{-1}\mathbf{CP} + \mathbf{B}_1\mathbf{QB}_1^T \ . \qquad (10.17)$$

Then $\mathbf{K}(t)$ also converges as $t \to \infty$, and the limit matrix $\mathbf{K}$ is as follows:

$$\mathbf{K} = -\mathbf{PC}^T\mathbf{R}^{-1} \ . \qquad (10.18)$$

If we substitute $\mathbf{K}$ for matrix $\mathbf{K}(t)$ in (10.14), then the resulting *steady-state filter* has the form

$$\dot{\mathbf{z}} = (\mathbf{A} + \mathbf{KC})\mathbf{z} + \mathbf{Bu} - \mathbf{Ky} \ .$$

This concept is especially useful when the original system (10.8) is asymptotically stable (or can be stabilized), since for large values of t the effect of the initial state of the system dies out, and, therefore, $\mathbf{x}(t)$ and $\mathbf{y}(t)$ are stationary processes. In addition, the algebraic Riccati equation (10.17) is much easier to solve than the Riccati differential equation (10.16).

10.3   *Adaptive Control Systems*

In recent years, increasing attention has been given to systems that are capable of accommodating unpredictable changes, whether these changes arise within the system or externally. This property is called *adaptation* and is a fundamental characteristic of living organisms, since they attempt to maintain physiological equilibrium in order to survive under changing environmental conditions. In the system theory literature, there is no unified definition for adaptive control systems. Therefore, we will consider a system adaptive if it satisfies the following criteria:

1.   continuously and automatically measures the dynamic characteristics of the system;

2.   compares the measurements to the desired dynamic characteristics;

3.   modifies its own parameters in order to maintain desired performance regardless of the environmental changes.

An adaptive control system, therefore, consists of three blocks: performance index measurement, comparison–decision, and adaptation mechanism. It is always assumed that there is a closed-loop control on the performance index. A common configuration of an adaptive system is illustrated in Figure 10.2. An important class of adaptive systems, *model reference adaptive systems* is easy to implement. One particular scheme for these systems is given in Figure 10.3. Note that the set of given performance indices is replaced by a reference model. The output of this model and that of the adjustable system are continuously compared by a typical feedback comparator, and the difference is used by the adaptation mechanism either to modify the parameters of the adjustable system or to send an auxiliary input signal to minimize the difference between the performance indices of the two systems.

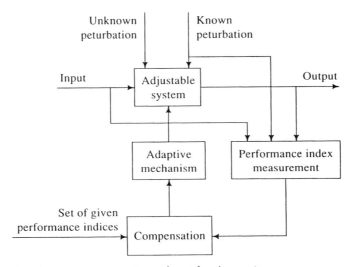

Figure 10.2   Common configuration of an adaptive system.

Another often-used class of adaptive systems is given by the *adaptive model-following control systems*. These control systems also use a model that specifies the design objectives, as illustrated in Figure 10.4. The mathematical model is formulated as follows. Assume that the reference model is given as

$$\dot{\mathbf{x}} = \mathbf{A}_M \mathbf{x} + \mathbf{B}_M \mathbf{u}_M \qquad (10.19)$$

and the plant to be controlled is

$$\dot{\mathbf{y}} = \mathbf{A}_P \mathbf{y} + \mathbf{B}_P \mathbf{u}_P . \qquad (10.20)$$

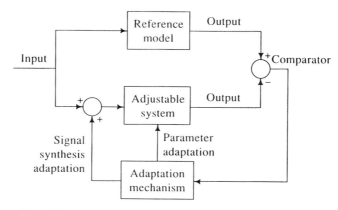

Figure 10.3   A model reference adaptive system.

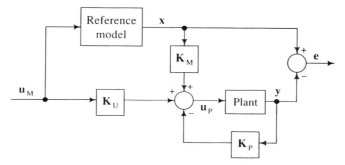

Figure 10.4   An adaptive model-following control system.

The plant control input is given by the relation

$$\mathbf{u}_P = -\mathbf{K}_P\mathbf{y} + \mathbf{K}_M\mathbf{x} + \mathbf{K}_U\mathbf{u}_M \ . \tag{10.21}$$

In this formulation, $\mathbf{A}_M$, $\mathbf{B}_M$, $\mathbf{A}_P$, $\mathbf{B}_P$ are given constant matrices; $\mathbf{x}$ and $\mathbf{y}$ are the states of the reference model and the plant; and $\mathbf{u}_M$ and $\mathbf{u}_P$ are their inputs. The coefficient matrices $\mathbf{K}_P$, $\mathbf{K}_M$, and $\mathbf{K}_U$ are unknowns; they are defined so that if the error vector $\mathbf{e} = \mathbf{x} - \mathbf{y}$ is initialized as $\mathbf{e}(0) = \mathbf{0}$, then it remains zero for all future time periods. We can subtract Equation (10.20) from (10.19) and substitute relation (10.21) to obtain the following inhomogeneous differential equation:

$$\dot{\mathbf{e}} = (\mathbf{A}_M - \mathbf{B}_P\mathbf{K}_M)\mathbf{e} + (\mathbf{A}_M - \mathbf{A}_P + \mathbf{B}_P(\mathbf{K}_P - \mathbf{K}_M))\mathbf{y}$$

$$+ (\mathbf{B}_M - \mathbf{B}_P\mathbf{K}_U)\mathbf{u}_M \ . \tag{10.22}$$

Perfect model following requires, therefore, that

$$\mathbf{A}_M - \mathbf{A}_P + \mathbf{B}_P(\mathbf{K}_P - \mathbf{K}_M) = \mathbf{O}$$

$$\mathbf{B}_M - \mathbf{B}_P\mathbf{K}_U = \mathbf{O} \ , \tag{10.23}$$

since these equations imply that for all real vectors $\mathbf{y}$ and $\mathbf{u}_M$ of appropriate dimensions, Equation (10.22) becomes homogeneous, and so the solution of the resulting homogeneous equation with zero initial condition is the zero vector for all $t \geq 0$. We can rewrite Equation (10.23) as

$$\mathbf{B}_P(\mathbf{K}_P - \mathbf{K}_M) = \mathbf{A}_P - \mathbf{A}_M$$

$$\mathbf{B}_P\mathbf{K}_U = \mathbf{B}_M \ . \tag{10.24}$$

The necessary and sufficient condition for the existence of matrices $\mathbf{K}_P$, $\mathbf{K}_M$, and $\mathbf{K}_U$ that satisfies Equation (10.24) is the following:

$$rank(\mathbf{B}_P) = rank(\mathbf{B}_P, \mathbf{A}_P - \mathbf{A}_M) = rank(\mathbf{B}_P, \mathbf{B}_M) \ . \tag{10.25}$$

These conditions mean that all columns of both matrices $\mathbf{A}_P - \mathbf{A}_M$ and $\mathbf{B}_M$ are in the subspace spanned by the columns of matrix $\mathbf{B}_P$. Note that Equation (10.24) can be solved by using Gauss elimination (see, for example, [42]).

Usually the initial condition of the error vector $\mathbf{e}$ differs from zero. In such cases we require that $\mathbf{e}(t) \to \mathbf{0}$ as $t \to \infty$, that is, Equation (10.22) is asymptotically stable. We know from Chapter 4 that this additional condition holds if and only if all eigenvalues of matrix $\mathbf{A}_M - \mathbf{B}_P\mathbf{K}_M$ have negative real parts.

Assume that the *rank* conditions (10.25) hold. Then from Equation (10.24) we can obtain at least one solution for $\mathbf{K}_P - \mathbf{K}_M$ and $\mathbf{K}_U$. Denote these solutions by $\mathbf{R}^*$ and $\mathbf{K}_U^*$. Assume that the *rank* of matrix $(\mathbf{B}_P, \mathbf{A}_M\mathbf{B}_P, \ldots, \mathbf{A}_M^{n-1}\mathbf{B}_P)$ is n, where $\mathbf{A}_M$ is assumed to be $n \times n$. Then Theorem 9.2 implies that there exists a matrix $\mathbf{K}_M^*$ such that all eigenvalues of $\mathbf{A}_M - \mathbf{B}_P\mathbf{K}_M^*$ have negative real parts, that is, system (10.22) is asymptotically stable. Then the selection

$$\mathbf{K}_P = \mathbf{K}_M^* + \mathbf{R}^*$$

$$\mathbf{K}_M = \mathbf{K}_M^*$$

$$\mathbf{K}_U = \mathbf{K}_U^*$$

gives an asymptotically stable, perfect model-following control system. Equation (10.24) are illustrated in the following example.

Example 10.1

Assume that

$$\mathbf{A}_P = \begin{pmatrix} 1 & 2 \\ 2 & 1 \end{pmatrix}, \quad \mathbf{B}_P = \begin{pmatrix} 1 & 2 \\ 1 & 2 \end{pmatrix},$$

$$\mathbf{A}_M = \begin{pmatrix} 0 & 1 \\ 1 & 0 \end{pmatrix}, \quad \mathbf{B}_M = \begin{pmatrix} 2 & 1 \\ 2 & 1 \end{pmatrix}.$$

Here Equation (10.24) has the form

$$\begin{pmatrix} 1 & 2 \\ 1 & 2 \end{pmatrix} \begin{pmatrix} r_{11} & r_{12} \\ r_{21} & r_{22} \end{pmatrix} = \begin{pmatrix} 1 & 1 \\ 1 & 1 \end{pmatrix},$$

$$\begin{pmatrix} 1 & 2 \\ 1 & 2 \end{pmatrix} \begin{pmatrix} k_{11} & k_{12} \\ k_{21} & k_{22} \end{pmatrix} = \begin{pmatrix} 2 & 1 \\ 2 & 1 \end{pmatrix},$$

where matrix $\mathbf{K}_P - \mathbf{K}_M$ is denoted by (r_{ij}) and $\mathbf{K}_U$ is denoted as (k_{ij}). Expanding the above operations, we get the following system of linear equations:

$$r_{11} + 2r_{21} = 1$$

$$r_{12} + 2r_{22} = 1$$

$$k_{11} + 2k_{21} = 2$$

$$k_{12} + 2k_{22} = 1,$$

where the repeated equations are omitted. It is easy to see that $r_{11} = r_{12} = 1$, $k_{11} = 2$, $k_{12} = 1$, $r_{21} = r_{22} = k_{21} = k_{22} = 0$ solve these equations. Hence, we may select

$$\mathbf{K}_P - \mathbf{K}_M = \begin{pmatrix} 1 & 1 \\ 0 & 0 \end{pmatrix} \quad \text{and} \quad \mathbf{K}_U = \begin{pmatrix} 2 & 1 \\ 0 & 0 \end{pmatrix}.$$

There are still infinitely many possibilities for selecting matrix $\mathbf{K}_M$, since only $\mathbf{K}_P - \mathbf{K}_M$ is specified. We wish to make this selection so that matrix $\mathbf{A}_M - \mathbf{B}_P\mathbf{K}_M$ has eigenvalues with only negative real parts. Note that in this case $n = 2$ and the $rank$ of matrix

$$(\mathbf{B}_P, \mathbf{A}_M \mathbf{B}_P) = \begin{pmatrix} 1 & 2 & 1 & 2 \\ 1 & 2 & 1 & 2 \end{pmatrix}$$

is unity; therefore, Theorem 9.2 cannot be applied to find an appropriate matrix $\mathbf{K}_M$. However in this special case an easy method can be used. Try to select $\mathbf{K}_M$ so that

$$\mathbf{A}_M - \mathbf{B}_P \mathbf{K}_M = -\mathbf{I},$$

that is,

$$\mathbf{B}_P \mathbf{K}_M = \mathbf{A}_M + \mathbf{I}.$$

If $\mathbf{K}_M = (\bar{k}_{ij})$, then this equation has the form

$$\begin{pmatrix} 1 & 2 \\ 1 & 2 \end{pmatrix} \begin{pmatrix} \bar{k}_{11} & \bar{k}_{12} \\ \bar{k}_{21} & \bar{k}_{22} \end{pmatrix} = \begin{pmatrix} 1 & 1 \\ 1 & 1 \end{pmatrix}.$$

It is easy to see that $\bar{k}_{11} = 1$, $\bar{k}_{12} = 1$, $\bar{k}_{21} = \bar{k}_{22} = 0$ are solutions. Therefore, the selection of

$$\mathbf{K}_U = \begin{pmatrix} 2 & 1 \\ 0 & 0 \end{pmatrix}, \qquad \mathbf{K}_M = \begin{pmatrix} 1 & 1 \\ 0 & 0 \end{pmatrix}, \qquad \mathbf{K}_P = \begin{pmatrix} 2 & 2 \\ 0 & 0 \end{pmatrix}$$

is satisfactory in order to construct an asymptotically stable adaptive system.

We can check our results by computing first matrix $\mathbf{A}_M - \mathbf{B}_P \mathbf{K}_M$. In this case,

$$\begin{pmatrix} 0 & 1 \\ 1 & 0 \end{pmatrix} - \begin{pmatrix} 1 & 2 \\ 1 & 2 \end{pmatrix} \begin{pmatrix} 1 & 1 \\ 0 & 0 \end{pmatrix} = \begin{pmatrix} -1 & 0 \\ 0 & -1 \end{pmatrix} = -\mathbf{I}.$$

Furthermore, check identities

$$\mathbf{B}_P(\mathbf{K}_P - \mathbf{K}_M) = \begin{pmatrix} 1 & 2 \\ 1 & 2 \end{pmatrix} \begin{pmatrix} 1 & 1 \\ 0 & 0 \end{pmatrix} = \begin{pmatrix} 1 & 1 \\ 1 & 1 \end{pmatrix} = \mathbf{A}_P - \mathbf{A}_M$$

and

$$\mathbf{B}_P \mathbf{K}_U = \begin{pmatrix} 1 & 2 \\ 1 & 2 \end{pmatrix} \begin{pmatrix} 2 & 1 \\ 0 & 0 \end{pmatrix} = \begin{pmatrix} 2 & 1 \\ 2 & 1 \end{pmatrix} = \mathbf{B}_M.$$

As the above example illustrates, Theorem 9.2 gives sufficient but not necessary conditions for the existence of the desired feedback. We note here that a new sufficient and necessary condition was introduced by [42].

In the conclusion of this section, a practical example based on [30] and [19] is presented which illustrates how adaptive systems really work. Figure 10.5 shows a typical state variable feedback control system with a time delay.

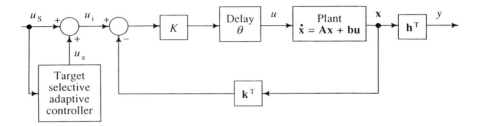

Figure 10.5 A typical state variable feedback control system with a time delay and a target-selective adaptive controller.

For this control scheme, the *target-selective adaptive controller* constructs an adaptive signal that depends on the frequency, amplitude, and waveform of the target movement, as well as on the time delay and dynamics of the plant. When this adaptive signal is applied to the time-delay system, it allows zero-latency tracking and improves dynamic performance. The system input $u_i(t)$ is composed of two parts: the reference source, $u_s(t)$, and the adaptive signal, $u_a(t)$. When $u_s(t)$ is not a known target waveform, $u_a(t)$ is turned off; $u_i(t)$ then equals $u_s(t)$ and the closed-loop transfer function becomes

$$\frac{Y(s)}{U_i(s)} = \frac{\mathbf{h}^T(s\mathbf{I} - \mathbf{A})^{-1}\mathbf{b}Ke^{-s\theta}}{1 + \mathbf{k}^T(s\mathbf{I} - \mathbf{A})^{-1}\mathbf{b}Ke^{-s\theta}} \ . \tag{10.26}$$

The $e^{-s\theta}$ term in the numerator is a pure time delay that remains in spite of the feedback. The similar term in the denominator produces phase lag that reduces the allowable gain. Of the other symbols, $Y(s)$ represents the scalar output, $U_i(s)$ the scalar system input, $\mathbf{I}$ the $n \times n$ identity matrix, $\mathbf{A}$ the $n \times n$ system matrix, K the scalar gain, $\mathbf{k}^T$ the $1 \times n$ feedback vector, $\mathbf{h}^T$ the $1 \times n$ output coefficient vector, and $\mathbf{b}$ the $n \times 1$ input coefficient vector. Superscript T indicates the transpose operation. The dimensions of the vectors and matrices are such that the numerator and the denominator of Equation (10.26) are scalars. The feedback vector $\mathbf{k}^T$ and the gain K must be selected to achieve stability.

Next we introduce four examples of compensation for plant time delays in systems with predictable inputs. The first compensates for the time delay and plant dynamics, the second compensates for the time delay and provides arbitrary pole placement, the third compensates for the time delay without requiring control gain changes, and the fourth compensates for the time delay while leaving the transient response unchanged.

In the first example, *compensation for time delay and system dynamics*, the system output is made identically equal to the reference input: $y(t) = u_s(t)$. The system input, $u_i(t)$, is the sum of $u_s(t)$, the reference source, and $u_a(t)$, the generated adaptive signal. When $u_s(t)$ is not a known predictable target waveform, $u_a(t)$ is turned off. When $u_s(t)$ is a known predictable target waveform, $u_a(t)$ augments $u_s(t)$ to achieve zero-latency tracking.

Applying the requirement $Y(s) = U_s(s)$ to Equation (10.26) produces

$$U_s(s) = \left[\frac{\mathbf{h}^T(s\mathbf{I} - \mathbf{A})^{-1}\mathbf{b}Ke^{-s\theta}}{1 + \mathbf{k}^T(s\mathbf{I} - \mathbf{A})^{-1}\mathbf{b}Ke^{-s\theta}} \right] (U_s(s) + U_a(s)) \ .$$

For notational simplicity, we omit the function's argument when it is complex frequency, s. Solving for U_a yields

$$U_a = \left[\frac{e^{s\theta}}{\mathbf{h}^T(s\mathbf{I} - \mathbf{A})^{-1}\mathbf{b}K} + \frac{\mathbf{k}^T(s\mathbf{I} - \mathbf{A})^{-1}\mathbf{b}}{\mathbf{h}^T(s\mathbf{I} - \mathbf{A})^{-1}\mathbf{b}} - 1 \right] U_s \ . \tag{10.27}$$

The time delay θ, the matrix $\mathbf{A}$, and the vectors $\mathbf{b}$, $\mathbf{k}^T$, and $\mathbf{h}^T$ must be known. If $u_s(t)$ can be estimated, then $u_a(t)$ can be computed in advance. For this example, the output was made equal to the input; we compensated for both the time delay and the plant dynamics. However, it may be unnecessary, or computationally efficient in real-time computer control, to compensate completely for the system dynamics.

This second example, *compensation for time delay with pole adjustment*, demonstrates that it is possible to cancel the effects of the time delay and also place the poles at any desired location. Let the desired new forward gain coefficient be K_a and the desired new feedback vector be $\mathbf{k}_a^T$. Substituting these requirements in Equation (10.26) yields

$$Y = \left[\frac{\mathbf{h}^T(s\mathbf{I} - \mathbf{A})^{-1}\mathbf{b}K_a}{1 + \mathbf{k}_a^T(s\mathbf{I} - \mathbf{A})^{-1}\mathbf{b}K_a} \right] U_s$$

$$= \left[\frac{\mathbf{h}^T(s\mathbf{I} - \mathbf{A})^{-1}\mathbf{b}Ke^{-s\theta}}{1 + \mathbf{k}^T(s\mathbf{I} - \mathbf{A})^{-1}\mathbf{b}Ke^{-s\theta}} \right] (U_s + U_a) \ .$$

Solving for U_a yields

$$U_a = \left[\frac{K_a}{K}e^{s\theta} + \frac{\mathbf{k}^T(s\mathbf{I} - \mathbf{A})^{-1}\mathbf{b}K_a}{1 + \mathbf{k}_a^T(s\mathbf{I} - \mathbf{A})^{-1}\mathbf{b}K_a} - 1 \right] U_s \ . \tag{10.28}$$

The first term of the right-hand side is the relationship of U_a to future values of the reference input. The remaining two terms represent the differential relationship between U_a and current value of U_s. For a known input reference U_s, one can readily compute U_a. Thus, the system response can be modified to have a desired characteristic response and no time delay.

It may be necessary to *cancel the effects of the time delay without inserting new gains*, that is, $\mathbf{k}_a^T = \mathbf{k}^T$ and $K_a = K$. For this requirement, we obtain a simplified case of Equation (10.28)

$$U_a = \left[e^{s\theta} - \frac{1}{1 + \mathbf{k}^T (s\mathbf{I} - \mathbf{A})^{-1}\mathbf{b}K} \right] U_s \ . \qquad (10.29)$$

This form has simple implementation requirements and lends itself easily to real-time computer control. It is used when the closed-loop system time delay is unacceptable but the system pole locations are not critical.

In the case of *compensation without changes in pole locations*, the auxiliary input from Equation (10.28) acts not only to cancel the effects of $e^{-s\theta}$ on the closed-loop system numerator, but also eliminates the effect of $e^{-s\theta}$ on the pole locations. To leave the closed-loop poles in the same location as in Equation (10.26), the system response to known targets may be specified as

$$Y = \left[\frac{\mathbf{h}^T (s\mathbf{I} - \mathbf{A})^{-1}\mathbf{b}K}{1 + \mathbf{k}^T (s\mathbf{I} - \mathbf{A})^{-1}\mathbf{b}K e^{-s\theta}} \right] U_s \ .$$

Substituting this requirement into Equation (10.26) yields

$$Y = \left[\frac{\mathbf{h}^T (s\mathbf{I} - \mathbf{A})^{-1}\mathbf{b}K}{1 + \mathbf{k}^T (s\mathbf{I} - \mathbf{A})^{-1}\mathbf{b}K e^{-s\theta}} \right] U_s$$

$$= \left[\frac{\mathbf{h}^T (s\mathbf{I} - \mathbf{A})^{-1}\mathbf{b}K e^{-s\theta}}{1 + \mathbf{k}^T (s\mathbf{I} - \mathbf{A})^{-1}\mathbf{b}K e^{-s\theta}} \right] (U_s + U_a) \ .$$

Solving for U_a produces

$$U_a = [e^{s\theta} - 1]U_s \ . \qquad (10.30)$$

Note that this is not the same result obtained by placing a predictor of $u_s(t + \theta)$ before the summing junction in Figure 10.5. Such a predictor would leave the effect of the time delay in the denominator.

Input signal waveforms may be predictable for human tracking of certain visual target waveforms and robotic tracking of objects on a moving platform. Both applications have large signal-processing time delays. Observations of human tracking indicate that input adaptation does occur and zero-latency tracking results. Although present robotic visual systems do not use such adaptive techniques, it may be advantageous.

The human eye movement control system performs in a manner suggesting target-selective adaptive control. When a target starts moving, there is a 150 msec delay before the eye starts moving, as shown in Figure 10.6 (upper). When the target stops, the eye continues to follow the predicted target for 150 msec; see Figure 10.6 (lower). However, when a human tracks a predictable target, the brain identifies the target within one half-cycle and generates an adaptive signal, $u_a(t)$, that makes the phase error approach zero as shown in Figure 10.7.

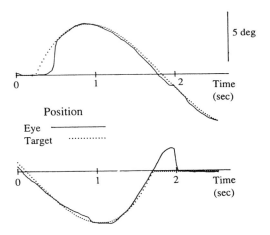

Figure 10.6   Performance of the human eye movement control system.

This change to zero-latency tracking is a result of control signal changes and not to changes in plant characteristics. The extraocular plant — consisting of the eyeball, the extraocular muscles, the nerve fibers, and the suspensory tissues — cannot change quickly. Neurophysiological studies suggest that changes in the plant or controller take hours or days to occur. Thus, the rapid performance change is being caused by the brain, presumably by changing the system input, $u_i(t)$.

The full model for the eye movement control system is shown in Figure 10.8. The smooth pursuit branch of this model acts as a velocity tracking system. The dynamics of the extraocular plant are very fast compared to the dynamics of the smooth pursuit branch, and the limiter does not affect the operation of the adaptive controller. For the

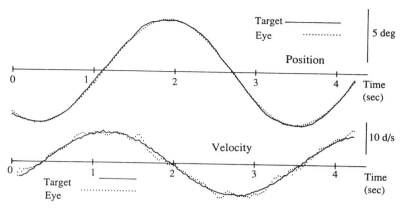

Figure 10.7   Human tracking of a moving target.

human eye movement system, the order of the system and the control and output vectors are 1 so that the following values are appropriate.

$$A = \frac{1}{\tau}$$

$$b = -\frac{1}{\tau}$$

$$h = 1$$

$$k = 1 \ .$$

The system's input, $u_i(t)$, is the sum of the target reference signal, $u_s(t)$, and the adaptive signal, $u_a(t)$, that must be computed. To obtain zero-latency tracking, $y(t)$ must equal $u_s(t)$. Putting all this information into Equation (10.26) gives

$$U_s = \frac{(s + \frac{1}{\tau})^T(\frac{1}{\tau})Ke^{-s\theta}}{1 + (s + \frac{1}{\tau})^T(\frac{1}{\tau})Ke^{-s\theta}}(U_s + U_a) \ .$$

Solving for U_a gives

$$U_a = \frac{e^{s\theta}}{K}(\tau s + 1)U_s \ . \tag{10.31}$$

The $e^{s\theta}$ term shows that predictions must be made. However, the smooth pursuit system is a velocity tracking system, not a position tracking system, so the controller must be able to predict future values

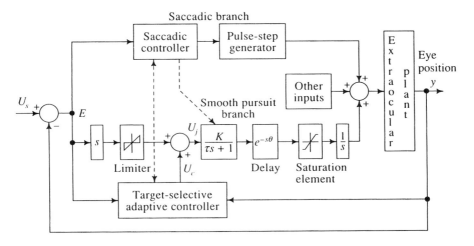

Figure 10.8   The target selective adaptive control model for human eye movement.

of target velocity. For example, if $\dot{u}_s(t)$ is the present target velocity, it must be able to produce $\dot{u}_s(t + \theta)$, where θ is the time delay of the smooth pursuit system. Also, the controller must modify this prediction to compensate for the dynamics of the system in accordance with Equation (10.32). Therefore, the compensation signal, U_c, of Figure 10.8 becomes

$$u_c(t) = \frac{1}{K} \left[\frac{d}{dt} \tau \dot{u}_s(t + \theta) + \dot{u}_s(t + \theta) \right] . \tag{10.32}$$

This compensation signal allows the smooth pursuit system to overcome the time delay. To synthesize this signal, the adaptive controller must be able to both predict future values of the target velocity and compute first derivatives. These are reasonable computations for the human brain. Therefore, Equation (10.32) is the algorithm that is in the box of Figure 10.8 labeled Target Selective Adaptive Controller.

We used six predictable waveforms and seven techniques for predicting, including a Kalman filter similar to that of Section 10.2. All yielded behavior comparable to human tracking. So we concluded that humans can predict certain waveforms and they do use mental models of their eye tracking systems. These mental models adapt for variations due to fatigue, age, and temperature. The behavior of warming up is just fine-tuning the mental model. When the mental model is for the combination of the human and the machine being controlled, the whole model must be adaptive. For example, when switching from a Lear jet to a Piper Cub, the pilot must change his or her mental model of the airplane.

10.4   *Neural Networks*

The field of artificial neural networks is arguably the fastest growing field in artificial intelligence. An artificial neural network is a massively parallel, adaptive computer system usually having multiple inputs and multiple outputs. During the past few years, neural networks have been used in a wide variety of applications, such as signature recognition in banks, loan underwriting in mortgage companies, planning and control of robot arm trajectories, process control, analyzing infrared images of asteroids, and nonlinear optimization [10].

Neural network technology has several advantages over conventional methods. Neural networks can deal with noisy and imprecise data, learn automatically from training data, adapt to a changing environment, degrade gracefully in the face of component failure, generalize to new situations, and (once trained) execute quickly. However, neural networks also suffer several weaknesses. The first is a lack of semantic interpretability. The information is stored as values of the interconnecting weights, and it is impossible to understand the behavior of a network by looking at the weight values. Second, input training sets can be faulty because of undesired or unwanted information, inappropriate training parameters, or bad initialization of connection weights. Unfortunately, it is difficult to detect such problems. Third, testing and validation are difficult with neural networks. The cost of testing a large hardware network may exceed the cost of manufacture.

Neural networks can be used in control systems. Traditional control systems have controllers, a controlled system, and a feedback loop. Typical controllers include proportional plus integral (PI) controllers, proportional plus integral plus derivative (PID) controllers, Smith Predictor controllers, and Model Reference Adaptive Controllers as shown in Section 10.3. As systems grew bigger, multiple controllers and multiple feedback loops have been used. However, the size and complexity of newer systems is pushing the limits of traditional techniques. Many advanced control systems are being built with knowledge-based systems such as expert systems and neural networks. There are many examples of neural networks in control systems [1]. One reason for their popularity is their adaptive nature. Examples of desired behavior are presented to a neural network and it learns to control the process. In a head-to-head test of a neural network, a self-tuning regulator, and a Lyapunov model reference controller [24], the neural network was the most robust in the face of model mismatches. It was second for control effort, tracking error, and noise rejection. It finished last only for convergence speed. Therefore, neural networks can now be used as adaptive controllers in many control processes.

There are dozens of different types of neural networks [16]. The main differences are in their method of training and weight adaptation. We will now explain the type called *backpropagation*. We choose it because it is the most common type used in control systems [1] and it can be used to illustrate most of the techniques used in other types of networks.

The extremely simple neural network of Figure 10.9 has two nodes in the input layer, two nodes in its solitary hidden layer, and two nodes in the output layer. The theoretical basis of this kind of neural networks is given by the famous Kolmogorov's theorem (see, for example, [20]) that states that any continuous function $f : [0,1]^n \rightarrow R^m$ can be implemented exactly by a three-layer neural network having n elements in the input layer, $(2n + 1)$ elements in the hidden layer, and m elements in the output layer. We mention that the elements in the hidden and output layers have special (usually irrational) nonlinear transfer functions. If one approximates these nonlinear functions by linear relations, the number of linear terms might be large. Therefore, neural networks may have thousands of nodes in each layer and perhaps many hidden layers. The weights between the layers are adjustable. This network is fully connected, that is, every node is connected to every node in the adjacent layers. Many networks are not fully connected. To explain how this network learns, assume that all the weights are initially 0.5. Apply a 0 to input-1 and a 1 to input-2 and specify that the desired outputs are 1 and 0, respectively, for output-1 and output-2. This network will have the values shown in Table 10.1.

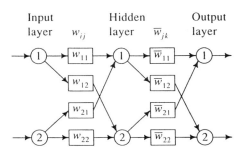

Figure 10.9 A simple neural network.

To explain our notation, w_{ij} represents the weight between the ith input node and the jth hidden node, and $\bar{w}_{jk}$ represents the weight between the jth hidden node and the kth output node.

As a first step in training this network, we will change the weights

Table 10.1 Initial Values for a Simple Neural Network

Input node values	Weights between input and hidden layers, w_{ij}	Hidden node values, h_j	Weights between hidden and output layers, $\bar{w}_{jk}$	Output node values, x_k	Desired outputs, d_k	Errors, E_k
0	0.5	0.5	0.5	0.5	1.0	0.5
	0.5		0.5			
	0.5		0.5			
1	0.5	0.5	0.5	0.5	0.0	-0.5

between the output layer and the hidden layer with the following equation:

$$\bar{w}_{jk}(t+1) = \bar{w}_{jk}(t) + \beta E_k(t)x_k(t) \ , \qquad (10.33)$$

where β is the learning rate, $x_k(t)$ is the actual output, and $E_k(t)$ is the error, which is defined as the difference between the desired and actual outputs, i.e., $E_k(t) = d_k(t) - x_k(t)$. This equation is an approximation to the Wiener–Hopf equation. For computation simplicity, it uses an estimate of the gradient of the error with respect to the weights instead of the actual gradient [19]. In the neural network literature, this equation is often called the Delta Rule. Using this equation and a learning rate β of 0.5, we can change the weights between the output layer and the hidden layer (the $\bar{w}_{jk}$'s), and the state of the network will change to the one shown in the second section of Table 10.2. One application of this equation is shown with the circles and arrows on Table 10.2.

That worked for the $\bar{w}_{jk}$'s but we cannot use Equation (10.33) for the weights between the input layer and the hidden layer (the w_{ij}s) because we do not know what the error is. The technique named backpropagation by Rumelhart et al. [39] assigns a weighted share of the blame to each of the w_{ij}'s. The backpropagation weight-changing formula is

$$w_{ij}(t+1) = w_{ij}(t) + \beta h_j(t) \sum_k \bar{w}_{jk}(t)E_k(t) \ . \qquad (10.34)$$

Using this equation we find that the w_{ij}'s do not change in this cycle, because the symmetry of our network causes the term

$$\sum_k \bar{w}_{jk}(t)E_k(t)$$

in Equation (10.34) to be zero.

But let's not give up yet. Let us present the input again and let the network learn some more. After the second presentation of the input we

Table 10.2 Changing the Weights between the Hidden and Output Layers

Time	Input node values	Weights, w_{ij}	Hidden node values, h_j	Weights, $\bar{w}_{ij}$	Output node values, x_k	Desired outputs d_k	Errors, E_k
Initial values	0	0.5	0.5	0.5	0.5	1.0	0.5
		0.5		0.5			
		0.5		0.5	β		
	1	0.5	0.5	0.5	0.5	0.0	−0.5
After using Eq. 10.33	0	0.5	0.5	0.625			
		0.5		0.375			
		0.5		0.625			
	1	0.5	0.5	0.375			
After using Eq. 10.34	0	0.5	0.5	0.625			
		0.5		0.375			
		0.5		0.625			
	1	0.5	0.5	0.375			

have

$$x_1 = 0.625 \quad \text{and} \quad E_1 = 0.375 \;,$$

$$x_2 = 0.375 \quad \text{and} \quad E_2 = -0.375$$

as shown in the first section of Table 10.3. So once again we use Equation (10.33) to change weights between the hidden and output layers and we get the results shown in the second section of Table 10.3. Next we apply Equation (10.34), backpropagation, and this time the w_{ij} do change as shown in the third section of Table 10.3. One application of this equation is shown with the circles and arrows on Table 10.3. This concludes our section on the simple weight-adjusting equations.

Next we show the network learning the desired pattern. Table 10.4 shows repeated application of the desired input–output pattern. We repeated it over and over again until the network finally learned. After 40 presentations of that input–output pattern (40 training cycles or 40 epochs), output-1 is close to 1 and output-2 is close to 0, as desired.

The example of Table 10.4 used a learning rate β of 0.5. With a learning rate of 0.25, the network converged slower. However, with a learning rate of 1.0, output-1 oscillated as shown in Table 10.5. In other

Table 10.3   The Second and Third Training Cycles

Time	Input node values	Weights, w_{ij}	Hidden node values, h_j	Weights, $\bar{w}_{ij}$	Output node values, x_k	Desired outputs d_k	Errors, E_k
2nd cycle. Calculate new h_j, x_k and E_k	0 1	0.5 0.5 0.5 0.5	0.5 β 0.5	0.625 0.375 0.625 0.375	0.625 0.375	1.0 0.0	0.375 −0.375
After using Eq. 10.33	0 1	0.5 0.5 0.5 0.5	0.5 0.5	0.742 0.304 0.742 0.304			
After using Eq. 10.34	0 1	0.523 0.523 0.523 0.523	0.5 0.5	0.742 0.304 0.742 0.304			
3rd cycle. Calculate new h_j, x_k and E_k	0 1	0.523 0.523 0.523 0.523	0.523 0.523	0.742 0.304 0.742 0.304	0.777 0.319	1.0 0.0	0.223 −0.319
After using Eq. 10.33	0 1	0.523 0.523 0.523 0.523	0.523 0.523	0.829 0.254 0.829 0.254			
After using Eq. 10.34	0 1	0.541 0.541 0.541 0.541	0.523 0.523	0.829 0.254 0.829 0.254			

words, the network did not learn. Problems like this are very common with simple neural networks.

Table 10.4   Forty Cycles of Training with $\beta = 0.5$

Time	Input node values	Weights, w_{ij}	Hidden node values, h_j	Weights, $\bar{w}_{ij}$	Output node values, x_k	Desired outputs d_k	Errors, E_k
0	0	0.5	0.5	0.5	0.5	1.0	0.5
		0.5		0.5			
		0.5		0.5			
	1	0.5	0.5	0.5	0.5	0.0	−0.5
1	0	0.5	0.5	0.625	0.625	1.0	0.375
		0.5		0.375			
		0.5		0.625			
	1	0.5	0.5	0.375	0.375	0.0	−0.375
5	0	0.545	0.545	0.875	0.955	1.0	0.045
		0.545		0.216			
		0.545		0.875			
	1	0.545	0.545	0.216	0.236	0.0	−0.236
20	0	0.518	0.518	0.963	0.997	1.0	0.003
		0.518		0.073			
		0.518		0.963			
	1	0.518	0.518	0.073	0.075	0.0	−0.075
40	0	0.510	0.510	0.979	0.999	1.0	0.001
		0.510		0.041			
		0.510		0.979			
	1	0.510	0.510	0.041	0.042	0.0	−0.042

The input–output pattern that we have been using is simple, but hardly worthwhile. Useful problems would surely have more complicated input–output patterns. For example, suppose we want to implement the Boolean functions AND and Exclusive OR as described in Table 10.6. If both inputs are 0, then we want both outputs to be 0, whereas if both inputs are 1 then we want only output-1 to be 1. If only one input is 1, then we want only output-2 to be 1. Now this example is starting to show the powerful pattern-recognition capabilities of a neural network. It can detect many different patterns in the input data. An analogy would be a neural network that accepted as inputs a person's height, weight, age, temperature, blood pressure, pulse, and cholesterol levels, and produced as outputs recommendations of normal, influenza, hypertension, etc. But back to our simple six-node network, first we apply pattern number 1, i.e., two 0s as the input and two 0's as the desired output, and allow the weights to adapt. Then we apply pattern num-

Table 10.5 Output Oscillations Caused by Increasing β to 1.0

Time	Input node values	Weights, w_{ij}	Hidden node values, h_j	Weights, $\bar{w}_{ij}$	Output node values, x_k	Desired outputs d_k	Errors, E_k
1	0	0.5 0.5 0.5	0.5	0.5 0.5 0.5	0.5	1.0	0.5
	1	0.5	0.5	0.5	0.5	0.0	-0.5
2	0	0.5 0.5 0.5	0.5	0.75 0.25 0.75	0.75	1.0	0.25
	1	0.5	0.5	0.25	0.25	0.0	-0.25
5	0	0.686 0.686 0.686	0.686	0.972 0.120 0.972	1.334	1.0	-0.334
	1	0.686	0.686	0.120	0.165	0.0	-0.165
20	0	0.629 0.629 0.629	0.629	0.501 0.027 0.501	0.629	1.0	0.370
	1	0.629	0.629	0.027	0.034	0.0	-0.034
40	0	0.966 0.966 0.966	0.966	0.619 0.010 0.619	1.197	1.0	-0.197
	1	0.966	0.966	0.010	0.020	0.0	-0.020

ber 2 and let the weights adapt. We continue with patterns 3 and 4. It might seem that with each new pattern the weights will change, destroying previous learning. But the hope is that with repeated application of these patterns, the network will learn to differentiate between the inputs. Unfortunately, this simple network does not learn and, as shown in Table 10.7, during the fifth cycle the outputs become ridiculously large.

To overcome these and other problems, neural network researchers have proposed dozens of additions to our basic Equations (10.33) and (10.34). The first addition that we present is the activation function, for which we will use a saturation, or limiting, element (many other functions are being used). This is one of the few properties of artificial neural networks that is analogous to a property of biological neural networks. Real neurons have a maximum firing rate. As the input becomes more intense they gradually approach their upper limit. Rumelhart et al. [39] proposed using a sigmoidal activation function for every node in the network. The following equation explains this property using a node

Table 10.6   Training File for AND and EXOR Functions

Pattern number	Input-1	Input-2	Desired output-1, logic AND	Desired output-2, logic EXOR
1	0	0	0	0
2	1	0	0	1
3	0	1	0	1
4	1	1	1	0

Table 10.7   Attempted Training with Multiple Examples

Cycle and pattern numbers	Input node values	Output node values	Desired outputs	Errors
1 – 1	0	0.0	0.0	0.0
	0	0.0	0.0	0.0
1 – 2	1	0.5	0.0	−0.5
	0	0.5	1.0	0.5
1 – 3	0	0.375	0.0	−0.375
	1	0.625	1.0	0.375
1 – 4	1	0.666	1.0	0.334
	1	1.623	0.0	−1.623
4 – 1	0	0.0	0.0	0.0
	0	0.0	0.0	0.0
4 – 2	1	0.004	0.0	−0.004
	0	0.102	1.0	0.898
4 – 3	0	0.105	0.0	−0.105
	1	3.092	1.0	−2.092
4 – 4	1	−0.276	1.0	1.276
	1	−20.79	0.0	−20.79
5 – 1	0	24.92	0.0	−24.92
	0	22730.5	0.0	−22730.5

in the output layer as an example. The limited output is

$$x_k = \frac{1}{1 + e^{-\sum_j \bar{w}_{jk}\mathbf{h}_j}} \; . \tag{10.35}$$

Thus, the output nodes constrained by sigmoidal activation functions will be restricted to the range of plus or minus 1. When sigmoidal activation functions (Equation (10.35)) were added to all of the nodes in our network, the problem of runaway outputs was solved, as shown in Table 10.8.

However, also as shown in Table 10.8, this network failed to learn. In particular, for pattern 1, output-2 is stuck at a high value (0.693) when it should be 0. The network is stuck in a local minima. One way to ameliorate this problem is to randomize the weights; either initially or after the network gets stuck. For this simple network, randomizing the initial weights works. For example, when we let the initial w_{ij}'s equal 0.5, -0.3, -0.5, and 0.3, and the initial $\bar{w}_{jk}$'s equal 0.3, -0.5, -0.5, and 0.3 (which was an arbitrary choice), we get the results shown in Table 10.9. The network learns the desired responses.

There is another less *ad hoc* technique for helping a neural network to escape from local minimum: adding momentum terms. Remember, our task was to vary the weights and search for values that reduced the error between the actual and desired outputs to the smallest possible value. However, we should not expect the error function to look like a bowl. It is just as likely that it has many hills and valleys, as shown in Figure 10.10.

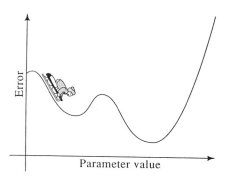

Figure 10.10   Error function of a neural network.

The network of Table 10.8 was stuck in a valley. Adding momentum might help it escape. To see how, imagine a boy on a shiny sled sliding down a snowy slope. When they get to a small valley, they get stuck.

Table 10.8   Network Training with $\beta = 0.5$, but with the Addition of Sigmoidal Activation Functions

Cycle number	w_{ij}	$\bar{w}_{jk}$	Pattern number	Output-1	Desired output-1	Output-2	Desired output-2
1	0.490	0.305	1	0.622	0.0	0.622	0.0
	0.493	0.305	2	0.622	0.0	0.622	1.0
	0.490	0.464	3	0.573	0.0	0.623	1.0
	0.493	0.464	4	0.530	1.0	0.666	0.0
10	0.470	0.132	1	0.266	0.0	0.492	0.0
	0.481	0.132	2	0.274	0.0	0.480	1.0
	0.470	0.312	3	0.266	0.0	0.489	1.0
	0.481	0.312	4	0.258	1.0	0.539	0.0
100	1.445	2.501	1	0.782	0.0	0.477	0.0
	1.492	2.501	2	0.257	0.0	0.475	1.0
	1.445	0.151	3	0.262	0.0	0.484	1.0
	1.492	0.151	4	0.624	1.0	0.540	0.0
1000	4.625	5.968	1	0.018	0.0	0.693	0.0
	4.627	5.968	2	0.022	0.0	0.661	1.0
	4.625	-2.739	3	0.022	0.0	0.661	1.0
	4.627	-2.739	4	0.968	1.0	0.070	0.0

Now imagine that the boy attacks the slope again, this time starting higher up the hill. This time they will be going faster when they reach the bottom; they will have more momentum. It is quite likely that they will continue across the small valley, up the hill, over the crest, and into the deeper valley on the other side. The principle is that if you are going in a good direction, then keep going in that direction. To accelerate training and help the network escape from local minimum, we will add momentum terms to our weight-adjustment equations. The momentum terms are proportional to the amount of the previous weight adjustment. With the addition of the momentum terms, Equations (10.33) and (10.34) become

$$\bar{w}_{jk}(t+1) = \bar{w}_{jk}(t) + \beta E_k(t)x_k(t) + \alpha(\beta E_k(t-1)x_k(t-1)) \quad (10.36)$$

and

$$w_{ij}(t+1) = w_{ij}(t) + \beta h_j(t)\sum_k \bar{w}_{jk}(t)E_k(t)$$

$$+ \ \alpha(\beta h_j(t-1)\sum_k \bar{w}_{jk}(t-1)E_k(t-1)) \ . \quad (10.37)$$

Typically, the momentum coefficient α is set between 0.5 and 0.9. Table 10.10 shows the neural network with the momentum terms ($\alpha = 0.5$)

Table 10.9 Network Training with $\beta = 0.5$ and Sigmoidal Activation
Functions, but with Arbitrarily Assigned Initial Weights

Cycle number	w_{ij}	$\bar{w}_{jk}$	Pattern number	Output-1	Desired output-1	Output-2	Desired output-2
1	0.482	0.349	1	0.525	0.0	0.475	0.0
	−0.308	−0.438	2	0.525	0.0	0.429	1.0
	−0.494	−0.468	3	0.446	0.0	0.481	1.0
	0.303	0.319	4	0.424	1.0	0.491	0.0
10	0.480	0.123	1	0.247	0.0	0.517	0.0
	−0.280	−0.756	2	0.275	0.0	0.463	1.0
	−0.677	−0.415	3	0.237	0.0	0.506	1.0
	0.100	0.316	4	0.252	1.0	0.511	0.0
100	1.077	1.721	1	0.048	0.0	0.502	0.0
	0.757	−3.961	2	0.238	0.0	0.480	1.0
	−1.852	−0.126	3	0.214	0.0	0.492	1.0
	−1.898	−0.206	4	0.629	1.0	0.536	0.0
1000	5.756	3.452	1	0.001	0.0	0.062	0.0
	5.752	−8.013	2	0.022	0.0	0.946	1.0
	−5.501	6.746	3	0.022	0.0	0.946	1.0
	−5.503	6.625	4	0.966	1.0	0.063	0.0

learning the input–output pattern. It learned faster and better than the network of Table 10.9.

We will now show one more technique to help the network stay out of local minimum, the addition of bias terms to the weight-adjustment equations. We add a bias term to each node in the hidden and output layers. We choose bias values randomly within the range 0.0 to 1.0. Table 10.11 shows the training behavior of our network with the addition of bias terms. It has learned in spite of setting the all the initial weights to 0.5.

To generalize all that we have learned about neural networks, let us now examine a sensitivity analysis of our network. We started with our best network, namely that with $\beta = 0.5$, $\alpha = 0.5$, two units in the hidden layer, bias values randomly selected between plus and minus 1.0, and initial values of 0.5 for all nodes. Then we varied each parameter throughout its feasible range and examined the residual error after 100 training cycles. Figure 10.11 shows the results.

Figure 10.11 shows the normalized error in the network,

$$E = \frac{0.5 \sum_k (d_k - x_k(t))^2}{\text{number of output units}},$$

after 100 training cycles. Figures showing these errors after 500 and 1000 training cycles had similar shapes but less variation. Which means that if you are willing to wait 1000 cycles for your network to learn, then

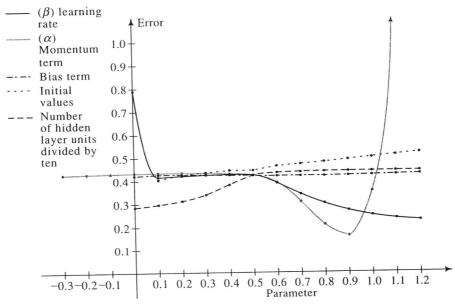

Figure 10.11 Sensitivity analysis of a neural network.

you can get away with nonoptimal parameters. Analogously, if you have all year to teach your dog to shake hands, then reward her with lettuce. But if you want her to learn in a day, reward her with beef jerky.

In this sensitivity analysis we varied the learning rate β from 0 to 1.2 and found that as β increased from 0 to 0.1, the error fell rapidly. Thereafter, increases in β produced small decreases in error.

Changing the momentum term α from -0.3 to $+0.6$ had little effect. But changes from 0.6 to 0.9 decreased the error. Finally, as α increased to 1.0 and beyond, the error jumped and produced overflow errors. The network is very sensitive to α in this region.

We changed the number of units in the hidden layer, n, from 1 to 12. The error was smallest for $n = 1$. It is a common finding in the neural network literature that the error as a function of number of units in the hidden layer is bowl-shaped [4]. With too few units, the hidden layer lacks sufficient richness; with too many units, the noise increases because the net is underconstrained.

In this neural network, the bias values were randomly selected between $\pm b$. For the sensitivity analysis we varied b from 0 to 1.2. This had almost no effect on the error. Changes in the initial values of the nodes also had small effect.

In this sensitivity analysis, we only varied one parameter at a time. However, there are interactions between the parameters. For example,

Table 10.10 Network Training with $\beta = 0.5$, Sigmoidal Activation Functions, and Arbitrarily Assigned Initial Weights, but with the Addition of Momentum Terms

Cycle number	w_{ij}	$\bar{w}_{jk}$	Pattern number	Output-1	Desired output-1	Output-2	Desired output-2
1	0.495	0.262	1	0.759	0.0	0.701	0.0
	−0.306	−0.530	2	0.746	0.0	0.652	1.0
	−0.494	−0.565	3	0.677	0.0	0.680	1.0
	0.301	0.232	4	0.634	1.0	0.681	0.0
10	0.475	−0.278	1	0.249	0.0	0.514	0.0
	−0.368	−1.120	2	0.276	0.0	0.458	1.0
	−0.662	−0.780	3	0.240	0.0	0.503	1.0
	0.073	−0.002	4	0.250	1.0	0.523	0.0
100	0.219	−0.241	1	0.042	0.0	0.378	0.0
	−0.957	−4.055	2	0.201	0.0	0.673	1.0
	−2.154	−0.714	3	0.197	0.0	0.656	1.0
	−2.031	−0.068	4	0.692	1.0	0.458	0.0
1000	−4.978	−0.336	1	0.001	0.0	0.046	0.0
	−4.994	−7.937	2	0.018	0.0	0.954	1.0
	−5.291	−7.970	3	0.018	0.0	0.954	1.0
	−5.302	7.181	4	0.972	1.0	0.053	0.0

as we have shown before, if we change both the bias terms and the initial weights to zero, then the network does not learn.

The results shown in Figure 10.11 are specific for the network and problem that we were studying. With a different desired input–output behavior, the sensitivity analysis would yield different results.

Many other variations of the basic weight-adjustment equations have been tried. Some had no effects and others had significant effects. For example, eliminating h_j from Equation (10.34) had almost no effect on performance. Also, dividing the second term of Equation (10.34) by the number of elements in the output layer, as is commonly done, had little effect, whereas multiplying the second term of Equation (10.34) by the derivative of the activation function, Equation (10.35), did enhance performance. Also, a technique called simulated annealing greatly improves the performance of neural networks. However, we will not consider any more enhancements of our basic equations.

In this section we used a neural network as a tool to perform a task: we trained a neural network to implement the Boolean functions AND and Exclusive OR. A lot of confusion exists about neural networks because two extremely diverse groups are using neural networks: (1) tool users, who use neural networks to accomplish tasks like pattern recognition and controlling systems, and (2) modelers, who use them to help understand biological systems. Most importantly, the two groups have

Table 10.11 Network Training with $\beta = 0.5$, Sigmoidal Activation Functions, Momentum Terms, and all Initial Weights Equal to 0.5, but with the Addition of Random Bias Terms

Cycle number	w_{ij}	$\bar{w}_{jk}$	Pattern number	Output-1	Desired output-1	Output-2	Desired output-2
1	0.492	0.243	1	0.770	0.0	0.896	0.0
	0.495	0.249	2	0.805	0.0	0.822	1.0
	0.492	0.446	3	0.734	0.0	0.836	1.0
	0.491	0.442	4	0.838	1.0	0.901	0.0
10	0.448	−0.467	1	0.194	0.0	0.636	0.0
	0.457	−0.492	2	0.264	0.0	0.379	1.0
	0.438	−0.078	3	0.296	0.0	0.399	1.0
	0.450	−0.129	4	0.146	1.0	0.561	0.0
100	−0.160	−0.233	1	0.317	0.0	0.504	0.0
	−0.069	−0.302	2	0.194	0.0	0.523	1.0
	−0.209	−0.168	3	0.283	0.0	0.586	1.0
	−0.128	0.010	4	0.158	1.0	0.524	0.0
1000	−5.392	−3.322	1	0.001	0.0	0.058	0.0
	−5.393	−7.627	2	0.029	0.0	0.946	1.0
	−5.165	−7.884	3	0.016	0.0	0.978	1.0
	−5.161	7.250	4	0.926	1.0	0.051	0.0

little in common. For example, the most popular algorithm among tool users is the backpropagation algorithm. But few modelers of biological systems would use this algorithm, because (1) it is not likely that any one neuron is going to be able to tell any other neuron that it is wrong (therefore, how can the error be determined?) and (2) in backpropagation, neural networks information flows in two directions: first input information flows forward through the network and then error information flows backward through the network. Biological neural networks do not have such bidirectional information flows.

This has been a very simple primer on neural networks. There are many techniques besides backpropagation for weight adjustment. Furthermore, many more enhancements could be added to our basic weight-adjustment equations. This example only used six nodes, and neural networks sometimes use millions of nodes. Our reason for including this simple example was merely to illustrate the idea in a manageable size problem. Modern controllers often use a combination of neural networks and rule-based expert systems. The neural networks learn, adapt, and control the process during normal operation. If a pipe or a value breaks, the neural network should fail, because that type of example should not have been included in the neural network training file. However, the engineers should be good at figuring out what to do in case of failures. For example, they could prescribe turning off an upstream valve if a pipe breaks. This type of knowledge would be best suited for a rule-based system that would take over from the neural network in case of abnormal events.

References

[1] Antsaklis, P.J. (1990) Special Issue on Neural Networks in Control Systems. *IEEE Control Systems Magazine,* Vol. 10, No. 3.

[2] Argyros, I.K. and F. Szidarovszky (1993) *The Theory and Applications of Iteration Methods.* CRC Press, Boca Raton, FL.

[3] Arrow, K.J. (1960) Price-Quantity Adjustments in Multiple Markets with Rising Demand. In: K.J. Arrow et al., eds., *Mathematical Methods in the Social Sciences.* Stanford University Press, Stanford, CA.

[4] Bahill, A.T. (1991) *Verifying and Validating Personal Computer-Based Expert Systems,* Prentice-Hall, Englewood Cliffs, NJ.

[5] Bahill, A.T., J.R. Latimer, and B.T. Troost (1980) Linear Homeomorphic Model for Human Movement. *IEEE Transactions on Biomedical Engineering,* Vol. BME-27, No. 11, pp. 631–639.

[6] Bailey, N.T.J. (1957) *The Mathematical Theory of Epidemics.* Hafner, New York.

[7] Chestnut, H. (1987) Cooperative Security System Description. *Proceedings of the 1987 IEEE International Conference on Systems, Man, and Cybernetics.* Oct. 20–23, 1987, Alexandria, Virginia, pp. 890–894.

[8] Collatz, L. (1966) *Functional Analysis and Numerical Mathematics.* Academic Press, New York.

[9] Courtiol, B. (1975) Applying Model Reference Adaptive Techniques for the Control of Electromechanical Systems. *Proceedings of the Sixth IFAC Congress,* Vol. ID, Boston, pp. 58.2-1–58.2-9.

[10] DARPA (1988) *DARPA Neural Network Study.* AFCEA International Press, Fairfax, Virginia.

[11] Deshpande, P.B. and R.H. Ash (1981) *Elements of Computer Process Control with Advanced Control Applications.* Instrument Society of America, Research Triangle Park, NC.

[12] Fleming, W.H. and R.W. Rishel (1975) *Deterministic and Stochastic Optimal Control.* Springer-Verlag, Berlin/Heidelberg/New York.

[13] Foster, K.R. (1991) Prepackaged Math. *IEEE Spectrum,* Vol. 28, No. 11, pp. 44–50.

[14] Gandolfo, G. (1971) *Economic Dynamics: Methods and Models.* North-Holland, Amsterdam.

[15] Gantmacher, F.R. (1959) *The Theory of Matrices.* Vols. I, II. Chelsee, New York.

[16] Grossberg, S. (1988) Nonlinear Neural Networks: Principles, Mechanisms, and Architectures. *Neural Networks,* Vol. 1, pp. 17–61.

[17] Haberman, R. (1977) *Mathematical Models.* Prentice-Hall, Englewood Cliffs, NJ.

[18] Hartman, P. (1982) *Ordinary Differential Equations.* Birkhauser-Verlag, Basel/Boston/Berlin.

[19] Harvey, D.R. and A.T. Bahill (1985) Development and Sensitivity Analysis of Adaptive Predictor for Human Eye Movement Model. *Transactions of the Society for Computer Simulation,* pp. 275–292.

[20] Hecht-Nielsen, R. (1990) *Neurocomputing.* Addison-Wesley, Reading, MA.

[21] Hetrick, D.L. (1971) *Dynamics of Nuclear Reactors.* University of Chicago Press, Chicago.

[22] Jamshidi, M. (1983) *Large-Scale Systems: Modeling and Control.* North-Holland, New York.

[23] Jamshidi, M. M. Tarokh, and B. Shafai (1992) *Computer-Aided Analysis and Design of Linear Control Systems.* Prentice-Hall, Englewood Cliffs, NJ.

[24] Kraft, L.G. and D.P. Campagna (1990) A Comparison between CMAC Neural Network Control and Two Traditional Adaptive Control Systems. *IEEE Control Systems Magazine,* Vol. 10, No. 3, pp. 36–43.

[25] Lanchester, F. (1916) *Aircraft in Warfare, the Dawn of the Fourth Arm.* Constable, London.

[26] Le Page, W.R. (1961) *Complex Variables and the Laplace Transform for Engineers.* McGraw-Hill, New York.

[27] Leslie, P.H. (1945) On the Use of Matrices in Certain Population Mathematics. *Biometrika,* Vol. 33, pp. 183–212.

[28] Lotka, A. (1956) *Elements of Mathematical Biology.* Dover, New York.

[29] Luenberger, D.G. (1971) An Introduction to Observers. *IEEE Transactions on Automatic Control,* Vol. AC-16, pp. 596–602.

[30] McDonald, J.D. and A.T. Bahill (1983) Zero-Latency Tracking of Predictable Targets by Time-delay Systems. *International Journal of Control,* 38: pp. 881–893.

[31] Melsa, J.L. and D.G. Schultz (1969) *Linear Control Systems.* McGraw-Hill, New York.

[32] Miller, R.K. and A.N. Michel (1982) *Ordinary Differential Equations.* Academic Press, New York.

[33] Nagai, M. (1983) Analysis of Rider and Single-Track-Vehicle System: Its Application to Computer-Controlled Bicycles. *Automatica,* Vol. 19, pp. 737–740.

[34] Nagai, M. (1986) Directional/Lateral Control of Two-wheeled Vehicle at Low Speeds. *Transactions of Society of Automotive Engineers of Japan,* No. 32, pp. 113–118, (in Japanese).

[35] Okuguchi, K. and F. Szidarovszky (1990) *The Theory of Oligopoly with Multi-Product Firms.* Springer-Verlag, New York.

[36] Ortega, J. and W. Poole (1981) *Numerical Methods of Differential Equations.* Pitman, Marshfield, Mass.

[37] Papandreou, A.G. (1952) *Fundamentals of Model Construction of Macro-Economics.* Serbinis Press, Athens.

[38] Richardson, L.F. (1960) *Arms and Insecurity.* Boxwood Press, Pittsburgh Quadrangle Book, Chicago.

[39] Rumelhart, D.E., G.E. Hinton, and R.J. Williams (1986) Learning Representations by Back-Propagation Errors. *Nature,* 323: pp. 533–536.

[40] Saaty, T.I. (1968) *Mathematical Methods in Arms Control and Disarmament.* Wiley, New York.

[41] Szidarovszky, F., A.T. Bahill, and S. Molnar (1990) On Stable Adaptive Control Systems. *Pure Mathematics and Applications,* Vol. 1, No. 2-3, pp 115–121.

[42] Szidarovszky, F. and S. Yakowitz (1978) *Principles and Procedures of Numerical Analysis.* Plenum, New York.

[43] Wilkinson, J. (1965) *The Algebraic Eigenvalues Problems.* Oxford University Press, London.

[44] Yakowitz, S. and F. Szidarovszky (1989) *Introduction to Numerical Computations* (2nd ed.). Macmillan, New York.

Index